Environmental Horticulture

Science and Management of Green Landscapes

This book is dedicated to the memory of Ian Cameron and Joan Hitchmough. Ian for inspiring Ross about the natural landscape and encouraging a passion for naturalistic gardening, even when this went too far and the asiatic rhododendrons, acers and clematis replaced the much loved lawn! Joan, for tolerating James' garden tyranny from an early age, thus unwittingly unleashing a new breed of horticultural 'plant community' to the wider world.

Environmental Horticulture

Science and Management of Green Landscapes

Ross W.F. Cameron

*Senior Lecturer in Landscape Management, Ecology and Design
Department of Landscape
University of Sheffield
UK
r.w.cameron@sheffield.ac.uk*

and

James D. Hitchmough

*Professor of Horticultural Ecology
Department of Landscape
University of Sheffield
UK
j.d.hitchmough@sheffield.ac.uk*

CABI is a trading name of CAB International

CABI	CABI
Nosworthy Way	745 Atlantic Avenue
Wallingford	8th Floor
Oxfordshire OX10 8DE	Boston, MA 02111
UK	USA
Tel: +44 (0)1491 832111	Tel: +1 (0)617 682 9015
Fax: +44 (0)1491 833508	E-mail: cabi-nao@cabi.org
E-mail: info@cabi.org	
Website: www.cabi.org	

© R.W.F. Cameron and J.D. Hitchmough, 2016. All rights reserved. No part of this publication may be reproduced in any form or by any means, electronically, mechanically, by photocopying, recording or otherwise, without the prior permission of the copyright owners.

A catalogue record for this book is available from the British Library, London, UK.

Library of Congress Cataloging-in-Publication Data

Names: Cameron, Ross W. F., author. | Hitchmough, James, author.
Title: Environmental horticulture : science and management of green landscapes / Ross W.F. Cameron and James D. Hitchmough, Department of Landscape, University of Sheffield, UK.
Description: Boston, MA : CAB International, [2016] | Includes bibliographical references and index.
Identifiers: LCCN 2015036042 | ISBN 9781780641386 (alk. paper)
Subjects: LCSH: Sustainable horticulture. | Urban gardening. | Ecological landscape design.
Classification: LCC SB319.95 .C34 2016 | DDC 635/.048--dc23 LC record available at http://lccn.loc.gov/2015036042

ISBN-13: 978 1 78064 138 6

Commissioning editor: Rachael Russell
Associate editor: Alexandra Lainsbury
Production editor: Tim Kapp

Typeset by SPi, Pondicherry, India
Printed and bound in the UK by CPI Group (UK) Ltd, Croydon, CR0 4YY

Contents

Acknowledgements	xi
1 Introduction to Environmental Horticulture: Issues and Future	**1**
1.1 Defining Environmental Horticulture	1
1.2 Horticulture Involves Human Agency	3
Environmental horticulture and relationships with pure (purist) ecology	4
What genotypes should be planted?	4
1.3 Future Directions	6
Horticulture's role in improving the functionality of vegetation	7
Conclusions	8
References	8
2 Environmental Horticulture: Benefits and Impacts	**9**
2.1 Introduction	9
2.2 Microclimate Modification	10
Urban heat islands	10
Cooling buildings and energy efficiency	12
Building insulation in winter	17
Wind amelioration	18
2.3 Noise Amelioration	19
2.4 Carbon Sequestration and Release	20
2.5 Water Management	24
Water management and sustainable urban drainage systems (SUDS)	24
Trees and rainfall capture	26
Turf and surface runoff	27
Green roofs/walls and impact on runoff	28
Urban water use	28
Grey water	29
2.6 Air Pollution	33
2.7 Pesticides and other Interactions with Chemicals	35
2.8 Non-Native 'Alien' Species	36
Conclusions	38
References	39
3 Green Space and Well-Being	**43**
3.1 Introduction	43
3.2 The Evidence for Green Space Affecting Health and Well-Being	44
Physiological health and physical fitness	44
Psychological health benefits	48
Health benefits and underlying mechanisms	51
Health benefits – not proven?	54
Health risks	55
3.3 Green Space and the Work Environment	56
3.4 Proximity, Scale and Type of Green Space	57

3.5	Horticulture as Therapy	61
3.6	Social Horticulture	61
	Green space and child development	62
	Green space – role in reducing crime and antisocial behaviour	63
	Green space and traffic calming	64
3.7	Environmental Horticulture and Healthy Diet	64
3.8	What Sort of Green Landscapes Should be Promoted?	65
3.9	Social Attitudes to Green Space and Values	68
	Conclusions	70
	References	71

4 Environmental Horticulture and Conservation of Biodiversity — 75

4.1	Introduction	75
4.2	A Definition for Biodiversity	76
4.3	Urban Ecology and Habitats	76
	Mosaics, networks and patches	77
	Promoting urban biodiversity	77
4.4	What Sort of Species Inhabit Urban Areas?	82
	Plants	83
	Mammals	84
	Birds	84
	Amphibians	85
	Reptiles	88
	Invertebrates	88
4.5	Management for Biodiversity	89
	Invasiveness	91
	Trend for planting more natives	92
4.6	Environmental Horticulture in Key Urban Wildlife Habitats	95
	Urban trees and woodland	98
	Parks and gardens	101
	Community gardens and allotments	107
	Wetlands and ponds	108
	Roads, railways and verges	109
	Brownfield sites, vacant plots and wastelands	109
	Conventional walls and roofs	111
	Green roofs and walls	111
4.7	Urban Biodiversity and Humans	116
	Conclusions	117
	References	118

5 Landscape Trees, Shrubs and Woody Climbing Plants — 121

5.1	Introduction	121
5.2	Woody Plant Production	123
	Sexual propagation	123
	Vegetative propagation	124
	Growing-on	129
5.3	Retail and Markets	130
	Sustainable production	131
5.4	Establishment	131
	Size/age of transplant	134
	Soil conditions	134
	Urban soils	136

		Mycorrhizae	137
		Planting depth and protection	138
	5.5	Maintenance	140
		Weed control	140
		Mulches	140
		Pruning	143
	5.6	Right Plant, Right Place	144
		Wind tolerance	145
		Wet soils and flooding tolerance	145
		Drought tolerance and xeriscaping	147
	5.7	Pests and Pathogens	148
	5.8	Urban Forests	152
		Conclusions	154
		References	155
6	**Herbaceous Plants and Geophytes**		**157**
	6.1	Introduction	157
	6.2	Distinguishing between Herbaceous Plants and Geophytes?	157
	6.3	Patterns of Growth in Herbaceous Plants and Geophytes	158
		Herbaceous plants	158
		Geophytes	159
	6.4	The Role of Herbaceous Plants and Geophytes in Designed Landscapes	159
	6.5	Contemporary Options for Herbaceous and Geophyte Planting in Public Landscapes	160
	6.6	Spatial Arrangements for Herbaceous Planting	161
		Block-based planting	161
		Repeating plantings of small blocks or individuals	162
		The number of plants forming each block	162
		The number of layer structures to be built in	162
		Incorporating geophytes into herbaceous and other plantings	163
	6.7	Plant Selection	164
		Robustness in herbaceous plants and geophytes	164
		Structural form in herbaceous/geophyte plant selection	167
	6.8	Phenology	167
	6.9	Attractiveness to Invertebrates	168
	6.10	Palatability to Molluscs	169
	6.11	Plant Establishment	169
		Specification, plant procurement, production systems	169
		Selecting the nursery product	170
		Timing of planting	170
		Planting protocols	170
		Mulching	171
	6.12	Longer-Term Maintenance	172
		Weed control	172
		Controlling long-term community development	173
	6.13	Manipulating Attractiveness	174
		Conclusions	174
		References	174
7	**Semi-Natural Grasslands and Meadows**		**175**
	7.1	Introduction	175
	7.2	The Role of Grass-Based Plant Communities in Urban Spaces	175
	7.3	Changing the Mowing Regime to Create a Spring Meadow	176

7.4	Changing the Mowing Regime to Occasional Flail Cutting	176
	Characteristics of occasional flail-cut grass	176
	Diversifying flail-cut grass	177
7.5	Changing from Gang Mowing to a Meadow Regime	179
7.6	Creating 'Meadow' Communities from Scratch	183
7.7	Choice of Plant Community	183
	Meadows	183
	Steppe	183
	Prairie	184
7.8	Designing a Seed Mix	185
7.9	Seed Management and Establishment	186
	Site preparation for sowing	186
	Sowing practice	187
	Post-sowing maintenance	188
	Conclusions	190
	References	190

8 Bedding and Annual Flowering Plants — 192

8.1	Introduction	192
8.2	Commercial Production of Bedding Plants	193
8.3	Propagation and Production Factors	196
	Seed	196
	Germination	198
	Media	198
	Nutrition	199
	Temperature	200
	Light	200
	Irrigation	202
	Growth regulation – chemical and management tools	203
8.4	Sustainable Production	205
8.5	Transport and Retail Stages	206
8.6	Establishment in the Landscape	211
	Irrigation	211
	Nutrition	216
8.7	Annual Flower Beds from 'Direct Sowing'	216
8.8	Cornfield Annuals and Annual 'Meadows'	217
8.9	Pests and Pathogens	217
	Conclusions	220
	References	221

9 Lawn and Sports Turf — 223

9.1	Introduction	223
9.2	Role of Turf in the Landscape	224
9.3	Grass Genotypes – Physiology and Traits	225
9.4	Grass Genotypes for More Sustainable Management Practices	229
9.5	Cultural Procedures	230
	Grass sward from seed	230
	Seed bed preparation	230
	Seeding and base fertilizer application	232
	Establishing a sward with turf	233

	9.6	Lawn Maintenance Practices	234
		Mowing	234
		Mowing of less-intensively used swards	236
		Grass mowing machinery	237
		Alternative and low energy mowers and other turf machinery	238
		Grass clippings	238
		Thatch and mat	239
		Aeration and drainage	239
		Turf reinforcement	240
	9.7	Shade	241
	9.8	Nutrient Management	241
		Nitrogen, phosphate and potassium	242
		Leaching and runoff	243
	9.9	Irrigation	244
		Irrigation control and scheduling	245
		Water quality and salinity stress	247
		Lawns and sustainable water use	248
		Attitudes to turf and water conservation	249
	9.10	Pesticide Use and Integrated Pest Management	250
	9.11	Less Intensive Management	252
		Genotype selection for 'low input' systems	252
		Growth regulation of the sward	253
		Artificial turf	253
		Areas where less intensive management is warranted	253
		Future directions	254
	Conclusions		256
	References		257
10	**New Green Space Interventions – Green Walls, Green Roofs and Rain Gardens**		**260**
	10.1	Introduction	260
	10.2	Green Walls	260
		Green façades	261
		Living walls	262
		Bio-walls	270
	10.3	Green Roofs	271
		Green roof typology	272
		Weights and load bearings	274
		Substrate technology	276
		Irrigation	278
		How 'green' are green roofs?	279
	10.4	Rain Gardens	279
		Water capture and infiltration	280
		Planting	281
		Pollutant control	282
	Conclusions		282
	References		283
11	**Interior Landscapes**		**284**
	11.1	Introduction	284
	11.2	Purpose and Function	285
		Health and well-being aspects of interior plant diplays	287
		Modifying the interior aerial environment	288

11.3	Interior Plant Requirements	290
	Temperature	291
	Irradiance	292
	Air quality	294
	Growing media	294
	Irrigation	295
11.4	Acclimatization to Interior Environments	295
11.5	Pests and Pathogens	296
11.6	Managing the Interior Landscape	296
11.7	Environmental Sustainability	296
	Conclusions	299
	References	299

Index   301

Acknowledgements

The authors would like to thank the Horticultural Trades Association, the Horticultural Development Council, the Royal Horticultural Society of the UK and the Department for Environment, Food and Rural Affairs for contributing information and supporting research that contributed to this book.

1 Introduction to Environmental Horticulture: Issues and Future

JAMES D. HITCHMOUGH

> **Key Questions**
> - What is environmental horticulture?
> - What is urban green infrastructure?
> - How does environmental horticulture differ from conventional ecological thinking and concepts?
> - What is meant by native and non-native flora – what are the potential problems associated with defining plant populations by national/political boundaries?
> - What might you consider as 'ethical' and 'non-ethical' planting?
> - What are some of the challenges that the environmental horticultural profession faces?

1.1 Defining Environmental Horticulture

This chapter attempts to put environmental horticulture into context. What are the core values of this discipline? Where does it come from? Where does its future lie? The expression 'environmental horticulture' first begins to appear in the literature and educational course terminology in the 1980s, particularly in North America. It is gradually adopted as a term for the subset of horticulture that is concerned with the use and management of plants in public and semi-public environments. In some parts of the world it replaces the term 'urban horticulture', although in many cases these descriptors co-exist for long periods of time, and in essence cover the same territory. In the UK, the words 'environmental' and 'urban' never really catch on as course descriptors, with the much older term 'amenity horticulture' tending to persist right up to the present day. These days in North America and elsewhere, 'landscape horticulture' is often the preferred term used to cover the planting and maintenance of landscape plants in public or private space.

At one level these terms are just about marketing and branding, trying to present a modern, culturally responsive face to compete in the educational marketplace for students. At another level these newer names signify changing ideas within this branch of horticulture. In the UK the term 'amenity horticulture' first emerges in the 1970s to denote the horticulture that is concerned with public and semi-public landscape spaces, as opposed to various forms of crop production. This particular term develops at the same time as recreation and leisure management and reflects a view of the world where this strand of horticulture exists to provide leisure or amenity benefits to citizens through the cultivation of plants in public landscapes (largely urban) visited by the public. The 1970s and 1980s witnessed substantial changes in how horticulture interacted with plants in public spaces. The emergence of landscape architecture as the dominant design discipline in public spaces greatly reduced the role of horticulturalists in plant use decisions, particularly on new build, or other capital intensive projects. There were also significant changes within, for example, local authority parks departments, which led to a diminution of horticultural ambition and capacity.

The first of these changes in the UK were a result of the Bains report (1972) which identified just how inefficient the delivery of many local authority services were, including parks. This led to experimentation with new forms of service delivery that has continued because of changing political philosophies, up to the present day (Byrne, 1994). The first of these was the shift from the standard day labour model, an approach modelled on the hierarchical organization of private estates of the landed gentry in the 19th century towards incentive bonus schemes. The former traditional organizational structures are based around chains of command, designed to highlight who is responsible for what, whilst developing horticultural skill. These systems did not necessarily maximize work rate, and hence the Bains report led to incentive bonus schemes. These were derived from 20th century industrial work study and placed much more emphasis on productivity; all tasks were allotted a time tariff for how long it should take to complete them. Staff received bonus payments where they demonstrated they had undertaken these tasks more quickly than the tariff. This attempt at modernization of service delivery also involved new forms of organization; horticultural service providers ceased to be fully in charge of their own future, in many cases becoming components of much larger departments, concerned with amenity and leisure in a much broader context. These changes can be seen in the marked shift in the content and tone of the articles published in the *Parks and Recreation Journal* in the UK during this time; from being dominated by detailed horticultural matters to much greater emphasis on the 'management of the recreational experience' in which horticulture becomes a relatively minor part.

With the benefit of hindsight, incentive bonus schemes proved to be unsuccessful in improving productivity in a way that was useful, with a tendency to be highly bureaucratic and to corrupt work priorities by favouring tasks that attracted the highest bonus payments.

In other parts of the world, such as North America, which have different horticultural traditions, based less on the gardenesque style of park of 19th century Britain, urban horticulture develops in the 1970s. From the personal perspective of the author, at this time, urban horticulture seemed more modern in its perspective. The name explicitly recognized that *urban* places were often more biologically challenging places in which to grow plants than were 'gardens' or 'green sward parks'. This was due to, for example, higher levels of atmospheric pollution, soils destroyed by engineering and construction activities, sealed surfaces, changed urban climatic regimes, and sometimes hostile social contexts such as vandalism and other forms of antisocial behaviour. Urban horticulture also explicitly made connections with the human social, cultural and psychological realm. Horticulture is engaged in and practised because it can make us feel better about our lives; it provides complex stimuli in both time and space that constructively moderates the fundamentally highly artificial experience of living in cities. In cities (and indeed in rural areas too), horticulture, whether practised by green-space professionals or home gardeners, is likely to provide the most immediate experience of 'nature', the patterns and processes associated with interactions between the physical world of our planet and the living organisms that have evolved in response to this.

Environmental horticulture is intrinsically linked too to the management of urban green space or green infrastructure, although these terms (see Box 1.1) may encompass woodland and other less intensively managed areas. As such they may not be the exclusive 'domain of the horticulturalist'.

What about 'environmental horticulture'? The exact origin of this term is uncertain, but it seems to represent an attempt to reposition horticulture to become more open to many of the ideas that developed from the 1990s onwards about biodiversity and ecological processes as well as the previously discussed human-centred ideas. In countries where the emergence of ecological consciousness had begun to lead to a wedge between ecological and horticultural thought, environmental horticulture provided a new paradigm whereby the practices involved in cultivation could be applied to systems that might, for example, consist of entirely native plants that would be seen as appropriate by ecologists. There is also an undercurrent that environmental horticulture should and does embrace more 'environmentally benign' actions and procedures that conform to an environmentally sustainable agenda. This includes reduced use of pesticides, use of biological control methods and seeking alternatives to peat and other 'non-sustainable' resources. Environmental horticulture has also been linked to ecosystem services, and the ability to deliver benefits to humans through the use of plants. This is particularly relevant to an urban environment, where 'horticultural' landscapes and plantings may be used to regulate water flow, improve water quality, alter microclimate or provide cultural services through opportunities for education, physical exercise, contemplation or creativity. Although environmental horticulture is not large-scale commercial field crop production, it does embrace food/human linkages through the likes of community gardens, allotments, edible walls and guerrilla gardening.

> **Box 1.1. What is meant by open space, green space and green infrastructure?**
>
> *Open space* is any open piece of land that is undeveloped (has no buildings or other built structures) and is generally accepted as being accessible to the public. Typical open spaces are school playgrounds, public squares and plazas, pathways, public seating areas, and vacant areas of even brownfield (ex-industrial) sites.
>
> *Green space* is a form of open space. This is land that is partly or completely covered with grass, trees, shrubs, or other vegetation. Green space includes parks, community gardens and cemeteries. Green space, however, also comprises land that may not be always open to the public, such as private gardens and certain sports facilities, e.g. golf courses or soccer pitches.
>
> *Green infrastructure* is the term used to combine different forms (typologies) of green space and is often used in conjunction with urban green spaces, i.e. as a juxtaposition to built (grey) infrastructure. It implies a matrix of green spaces and typologies, and increasingly one that is intentionally or strategically planned. It also alludes to the services such spaces should provide to local (human) populations. This includes its role in housing and economic growth, the regeneration of urban areas and the protection (or creation) of environmental assets and the underpinning of the sustainability of a town or city.
>
> Natural England, the Government body in England, UK, tasked with nature conservation and the management of National Nature Reserves defines green infrastructure as: 'a network of multi-functional green space, both new and existing, both rural and urban, which supports the natural and ecological processes and is integral to the health and quality of life of sustainable communities'.
>
> Included within green infrastructure typology are:
>
> - Natural and semi-natural urban green spaces
>   - woodland and shrubs, grassland (e.g. downland and meadow), heath or moor, wetlands, open and running water, wastelands and disturbed ground, bare rock habitats (e.g. cliffs and quarries)
> - Parks and gardens
>   - Urban parks, country and regional parks, formal gardens
> - Amenity green space
>   - Informal recreation spaces, housing green spaces, domestic gardens, village greens, urban commons, pocket parks, other incidental space, green roofs, green walls
> - Green (and blue) corridors
>   - Rivers and canals, including their banks, road and rail corridors, cycling routes, pedestrian paths, bridleways and other rights of way
> - Other areas
>   - Allotments, community gardens, urban farms, cemeteries and churchyards

1.2 Horticulture Involves Human Agency

Whilst environmental horticulture covers a wide range of theoretical and practical territory, at its core lie the same practices and understandings of cultivating plants to achieve clearly defined goals. Implicit in the idea of cultivation is that human decision making, 'agency', is consciously applied, both as thought and action. This agency may vary from being very occasional through to frequent, but potentially significantly impacting the life of a given plant and/or synthetic plant community that it is part of. This agency process often commences through making decisions on which plants can be used in which environment, indeed in a rational world environmental conditions of the planting site should be one of the main factors in plant selection, and this is discussed further subsequently in this chapter. Agency is applied in varying degrees to different circumstances, often involving the manipulation of water, nutrients, light and physical removal of plant tissues to control the rate and form of growth, the degree to which plants flower and fruit, and how plants are likely to be perceived by people. The use of these manipulation levers can be either intensive or extensive, sophisticated or crude, i.e. there are inherent gradients across which decision making can or must be made, depending on the needs of the location or context, and the resources that are available to decision makers.

At the low intensity end of the spectrum rather blunt forms of management, such as the non-selective cutting off of plant parts, may be used to nudge a plant or a plant community in a chosen direction. This issue of choice and decision making is critical to understanding the horticultural mindset. Within horticulture, choice, agency, or decision making, call it what you will, is seen as an intrinsic part of the process; essentially as a 'good', whilst recognizing that no one input or outcome is appropriate in all situations.

Environmental horticulture and relationships with pure (purist) ecology

The involvement of human agency within environmental horticulture is the prime distinction from parallel disciplines that are concerned with plants, for example conservation ecology or restoration ecology. Depending on who is practising these activities, and in what context, there is also a gradient in terms of how much human agency can be applied, but in general it is much less in total, and often restricted to forms of management which work to kick-start or direct an ecological process, for example, increasing light at ground level in a plant community to benefit particular species by canopy removal. These processes inevitably disadvantage some species in order to benefit others. There are winners and losers, at least temporally.

Horticulturalists tend to find this a little disturbing; in the horticultural paradigm there is a desire to achieve some form of equivalence of benefit, i.e. managing in order to avoid the creation of obvious losers. This notion of 'care' is much more remote in conservation ecology or restoration ecology than in environmental horticulture, even when the latter are practising these same activities. Indeed, because these former disciplines are philosophically deeply rooted in the idea that human agency typically corrupts or damages the organisms and processes that we call nature, they are fundamentally uncomfortable or even hostile to the application of human agency to vegetated systems. This is most strongly developed in large, recently settled (by Europeans) countries with a strong 'wilderness construct'. It is least developed (but still present) in countries with a long and obvious inter-relationship between the natural vegetation and people through various forms of low intensity agricultural exploitation and management.

Within ecological science dialogues, 'gardening', i.e. the intense and prolonged application of agency to plants, is often used as a pejorative, rather than a positive concept. These polar views explain why traditional ecological science is sometimes philosophically uncomfortable with the horticultural utilization of plants; it seems at best pointless and at worst almost decadent.

Horticulture, by contrast, is not at all embarrassed by human agency; it recognizes that human beings can obtain great pleasure and sometimes much deeper psychological states such as meaning, through application of their own agency and that of others. By adding another trophic level of interaction (human agency) to the ecological food web, it is possible to achieve endpoints that are impossible through the rather narrow and blinkered processes that underpin conventional ecosystem development. Within the limits of what is possible, horticulture can choose, it does not have to be insular and dogmatic!

But choose what? Within ecology and botany, the development of increasingly comparative taxonomy linked to effective field survey and vegetation sampling from the early 18th century onwards, allows the construction of reliable lists of the native plants that make up the plant communities of a given region. These lists or 'floras' do, within a strictly ecological view of the world, circumscribe our choices for us. Deviation from these lists, in terms of species selection, is increasingly seen in the contemporary world of biodiversity as inappropriate, unethical or even plain 'bad'. In the 'wilderness' countries recently settled by Europeans – such as the USA, Australia and New Zealand – this process has been in operation to some degree since at least the 1970s, but is increasingly prevalent even in long-settled countries as a result of biodiversity legislation effected through the planning process.

What genotypes should be planted?

For any given planting site the range of plants prescribed by the natural distribution of native species as a result of 'natural' ecological processes is often relatively small. In the UK for example, if the planting site was the whole of the British Isles, it would correspond to approximately 1100 native species. If a county was chosen, for example Northumberland, plant genotypes native to the area would decrease to about 1000, not much of a drop, because in small countries many species are found across the entire territory. If, however, a heavily shaded planting site in woodland in Northumberland was the chosen area, and hence required heavily shade-tolerant understorey plants, the number of native species would fall to <300. If the criterion was then applied that the plants chosen had to be particularly attractive to human beings, this number would fall to <30. Continuing the interrogation that any species identified needed to be in flower or otherwise attractive in autumn, the numbers would drop to almost zero.

Hence in the Northumberland context, if one wanted to use species that flower in autumn, it would have to be accepted that this is simply not possible or one would have to look to the floras of other locations, where in response to local evolutionary pressures, plants exist that do what is desirable. In climates where summer rainfall is very high, a niche is created for some woodland plants to flower in autumn because the soil is moist enough to support this and there is less competition for pollinating insects, hence a genotype such as the eastern North American *Aster divaricatus* could be utilized. There will nearly always be a series of locations elsewhere in the world with a climate analogous to that of the site that needs to be planted, each with its own distinctive flora that is sufficiently fit to grow well on the designated site. For Northumberland, one looks to the montane species of western China and Japan, western and eastern North America. This larger pool of species throws up choices such as *Actaea racemosa, Aster divaricatus, Heuchera villosa, Rudbeckia fulgida* (all North America), *Begonia grandis, Impatiens omeiana, Saruma henryi, Nipponanthemum nipponicum, Saxifraga fortunei* (China and Japan) and so on. The quid pro quo in this process of meeting aesthetic aspirations is that these species may not be as well-fitted to their planting environment, and hence as robust, as the species native to Northumberland. Hence a balance must be struck between aesthetic and functional fitness, and where resources are too few to 'care' for the non-native species of lower fitness, then native or other more fitted species that do not meet the original aesthetic specification must be used.

Context therefore becomes very important in making these judgements; non-native species are more likely to be used in urban areas – with relatively high expectations of what plantings should look like and relatively abundant resources for management – than in rural locations. In large countries, for example the USA, an interesting situation has arisen that has fed the nativism concept in plant selection. Because the USA is both very large and biologically at the richer end of the plant diversity spectrum, it is easier to meet aesthetic planting specifications from the politically native flora of particular (very large) nation states. This leads to a certain assumption that working purely within the native flora is entirely possible and why do other countries not do the same? This argument plays less well at an intellectual and practical level in small nation states such as the UK or the Netherlands and where there are relatively low levels of botanical diversity. The flip side to the USA position, which promotes the exclusive use of native plants, is that just because a plant is politically native does not mean it is necessarily well-fitted to use anywhere within this politically defined envelope. *Rudbeckia fulgida* is native to the USA, but is highly unfit for those parts of the country that have severe winter or low rainfall.

Whilst horticulture has a strong relationship with the use of plants from other parts of the world for the reasons given above, in large floras, for example South Africa, it is often possible to work purely within the native flora, although again subject to the inevitable issues of poor fitness for distant species, unless altitude and other factors reduce the expected loss of fitness. In this latter sense, the issues facing those who use native species drawn from beyond the local region and those using non-politically native species from further afield are the same. A fascination with the floras of other places, whether within the nation state or not, not simply as part of intellectual curiosity but as something that might be used and useful, is fundamental to the horticultural paradigm.

This extends to the manufacture of new genotypes through collection in the wild, breeding or selection; horticulture has an endless appetite for the new, whether the plants are originally native or non-native. This is in stark contrast to ecological thought, which is often horrified by the manufacture of the new and would like to believe that species represented in lists (floras etc.) are intrinsically right for that location and always have been. When one considers these issues in the context of geological time, it is clear that floras represent only ephemeral moments in time, not immortal certainties.

Can one apply moral or ethical notions to these contrasting paradigms? Debates in the media appear to assume that one can do this. Over the past 20 years there has been a tendency to see species drawn from regional lists as somehow right and appropriate, perhaps even good, and species from outside of these regions as inappropriate, perhaps even unethical. Such positions are entirely human constructs; one cannot find evidence for the righteousness of this within ecological science per se. One can measure the negative consequences of these choices in terms of species that are insufficiently or too well-fitted and have naturalized amongst extant native species – although even here it is often difficult to measure harm per se (Thompson, 2014). The basis of measuring harm is that a native plant is better for the native animals that depend on that vegetation. Whilst in highly specialized ancient floras, such as that of South Africa, this may well be the case. It is looking increasingly difficult, however, to argue this as a generalization in Northern Europe. A recent paper (Hanley *et al.*, 2014) on the 'goodness' of native plants for generalist and highly specialized pollinators (bee species) found that the idea of localness at the scale of the nation state is often ecologically meaningless because many plants and pollinators have very wide overlap over their ecological history (Fig 1.1); a UK native bee or butterfly might be just as comfortable with the plants of distant portions of the Palearctic (the temperate band of vegetation running from Western Europe to Japan north of 33°).

1.3 Future Directions

So what of the future, the world in which environmental horticulture has to practise? Many of the core attitudes and values of horticulture were fashioned in a time when resources were much more abundant than they are now. This includes energy, water and the affordability of human labour. Diminishing resource availability has not made the end points traditionally valued by horticulture irrelevant; however, they have made it much more difficult to deliver on these at the required level. In addition to the resources issue, there is also the question of horticulture having to deal with and respond to new policy drivers, for example biodiversity and sustainability legislation, and the ongoing reduction in various biocides used when staffing levels are very low to manage weed competition with horticultural plants. To this must be added changing attitudes towards education and human relationships with the environment, as reflected for example in a crisis within horticultural recruitment and training at all levels but particularly at graduate and postgraduate. The challenge in the 21st century is for horticulture to successfully deliver planted landscapes in this context.

Approaches to delivery vary, from various forms of rationing through to more radical approaches in which completely new types of vegetation are employed. Rationing is the simplest and most obvious means to reduce resource input, where the area of most intensive establishment and management of vegetation is reduced to meet the budget available. Traditional 'staples' such as annual bedding plants are shrunk down to small roundels in traffic islands sponsored by local businesses. This process tends ultimately to result in landscapes composed primarily of trees and mown grass, since everything else horticultural has relatively higher establishment and management costs. This not only reduces the potential for seasonal change, drama and engagement for people but also reduces the amount of pollen and nectar, and foraging space and volume available to pollination and herbivorous invertebrates that ultimately build urban food chains. Rationing, no matter how thoughtfully undertaken, ultimately leads to impoverishment over much of the urban estate.

The next step in a rationing process is to look critically at how to manage existing horticultural vegetation to achieve similar outputs with fewer inputs. The problem may be viewed through a social or cultural lens, determining the vegetation aesthetic threshold that is sufficiently appealing to people. For this, technical understanding may be employed, for example reduction in watering or fertilizing frequency.

The only other option is to look at substituting alternative forms of planting for more expensive energy-consumptive horticulture to maintain seasonal change and richness. In purely economic terms few if any of these substitutions other than simple forms of native woodland can be cheaper to manage

Fig. 1.1. A native butterfly *Aglais urticae* (small tortoiseshell) feeding off the nectar of a non-native plant *Aubrieta deltoidea* in a UK garden. Although some schools of thought suggest that the planting of such non-native plants should not be encouraged, the species in question are unaware of the political boundaries. Indeed, these species may overlap in their distributions in other parts of Europe, and have some degree of 'ecological fitness' due to this.

than gang-mown grass. The cheapest forms of managed herbaceous vegetation, such as rough grass flail-cut once a year, cost approximately the same as 30 cuts per year gang-mown grass. The argument to spend more than is necessary to maintain mown grass has generally therefore to be made on enhanced experiences for people or habitats for wildlife.

These substitution vegetation types are generally based upon or reflect natural or semi-natural plant communities, for example woodland, heathland-scrub, wetlands, and meadow-like vegetation. In all cases these substitution vegetation types can be created either with entirely native species: the restoration/conservation ecology approach; or with a horticultural approach using non-native species, such as in the Landscape Laboratories at the Swedish Agricultural University at Alnarp, near Malmo; or indeed through a combination of native and non-native species. The work of the author of this chapter is strongly grounded in the horticultural end of this gradient; taking non-native species, derived from prairies and other exotic plant communities and constructing essentially horticultural plant communities that look wild and are managed as if they were native plant communities.

The critical thing in most cases is that horticultural conceptualization of even an entirely native woodland system is likely to lead to different forms of management and a different physical end point than that a conservation ecologist would produce. For example, there is likely to be more interest in a hands-on than a hands-off approach and more emphasis on aesthetic impact at seasonal points in time rather than species diversity per se. The Heem Parks of Amstelveen in Amsterdam (Netherlands) are an excellent example of how native nature can be managed to fit more comfortably within horticultural value sets in which 'beauty' is a key element. Utilizing these substitution vegetation types requires a different skill set for horticulturalists. In particular, it requires a greater understanding of ecological processes, and how to manipulate these through reduced soil productivity, plant density, and other ecological 'levers' to achieve desired outcomes. This capacity to conceptualize and read plantings through an ecological lens is a challenging new skill.

Horticulture's role in improving the functionality of vegetation

One factor that will further encourage a drift to these more ecological types of horticultural vegetation is the need to increase the functional performance of vegetation in urban places. Historically, horticulture has mainly seen vegetation in terms of cosmetic beauty. This is a critical factor to maintain and indeed obtain public support, but there is also a need to gain function, particularly when vegetation is to be used on a large scale (as is the idea with almost all of the substitution vegetation types mentioned above). Hence urban designed woodlands have to fix carbon, reduce local heat island effects, support as much animal biodiversity as possible and be harvestable for wood to meet local needs. Woodlands based on riparian species that tolerate cycles of flooding can be used to clothe and provide these functions at the same time as acting as infiltration swales for urban runoff. The same parallels can be drawn with herbaceous vegetation, such as the species used for sustainable drainage swales, which are temporally very wet then gradually dry out until they are replenished by a new stormwater event. These plants must, irrespective of where the species come from, look attractive for as long as possible, and in particular must not wilt and collapse when subject to reduced moisture availability in the height of summer. This mixture of ecological and functional characteristics force a different approach to plant selection and plant use.

Because of climate change and associated sustainability and biodiversity agendas, vegetation design, establishment and management skills are likely to become increasingly important in urban areas offering the potential for a horticultural urban renaissance. However, this will only be the case where horticultural skills are honed to match the current political and policy agenda. Environmental horticulture will maximize the employability of its graduates by looking to forge more relationships with other disciplines who are also key actors in the development and management of urban landscapes and have shared interests, such as in landscape architecture, architecture and engineering. Because of the changes from the 1970s onwards, discussed earlier in this chapter, horticultural input into urban planting design has much diminished, often to its detriment, for example, by a lack of critical thinking in selecting suitable species for a given project. This situation is not going to change; however, there would appear to be a huge opportunity for horticulture to work with landscape architecture as a partner in developing new landscapes that are as sustainable as possible rather than cosmetic.

Communication and interaction with less conventional partners will also become paramount.

Environmental horticulturalists need to be able to liaise with architects and engineers, for example to further break down the barriers between grey and green infrastructure, so in future it may be possible literally to entirely cover a building with a green mantel of vegetation (not just provide a rather tokenistic panel or two on the front façade). Similarly, better dialogue with sociologists and health experts is required for horticultural activities to be more effectively implemented to address issues around mental health, well-being, physical activity, crime avoidance and enhanced social integration.

To make these collaborations work, however, horticulture will need to be able to speak at least some of the language of these other disciplines, and in particular it will need to demonstrate that, in addition to having experience of plants in practice, it also has a research-based understanding of vegetation and establishment in the city. Due to the decline of the research-focused undergraduate and postgraduate horticulture sector in the UK over the past 20 years, these sorts of understandings are now thinly spread. This text attempts to assist with this process, with each chapter grounded in the latest research.

Finally, although much of the research pertains to the management of the public green realm, reference is made frequently to the private domestic garden, as there is obvious overlap in terms of agendas and practices, but also because private gardens contribute a significant component of urban green infrastructure. These private gardens may not come under the jurisdiction of the professional environmental horticulturalist, but much of the content of this book is also relevant to the keen amateur gardener who wishes to understand more about the scientific principles underpinning garden management.

Conclusions

- The term 'environmental horticulture' was coined in the 1980s and broadly correlates with the management of landscape plants in public and semi-public arenas. Over the subsequent years it has also become linked with a philosophy for more environmentally sustainable practices within urban horticulture.
- Environmental horticulture has a number of similarities with other forms of green space or environmental management, but differs in that it actively promotes the cultivation of plants, and does not necessarily restrict itself to the use of native genotypes. It tends to bring humans and their perceptions and activities more actively into the framework of green space design and management.
- Environmental horticulture recognizes that human beings can obtain great pleasure and sometimes much deeper psychological states such as meaning, through application of their own interventions in the green space, and indeed from that of others.
- There is an acknowledgement within the discipline that the urban environment is unique in terms of its conditions and pressures, and that plant choice in this environment needs to reflect this, not only in terms of fitness, but also in the wider rationale of public appreciation or functionality.
- Although environmental horticulture will frequently use native plant species, it recognizes that certain non-native plants in particular contexts provide value too. This can include offering some services to native fauna species, such as pollen and nectar resources. Although alien invasive plants should be avoided in plantings, the concept that only plant genotypes located within political boundaries (countries) are appropriate for use should also be challenged.
- Those wishing to develop a career in environmental horticulture need to be adept at plant husbandry skills, but also are increasingly required to embrace other disciplines and widen their skills base. Horticulturalists of the 21st century need to work in partnership with landscape architects, ecologists, land or civil engineers, health and social work professionals and indeed many other disciplines.

References

Bains Committee (1972) *The New Local Authorities: Management and Structure*. Her Majesty's Stationery Office, London.

Byrne, T. (1994) *Local Government in Britain*, 6th edn. Penguin Politics, London.

Hanley, M.E., Awbi, A.J. and Franco, M. (2014) Going native? Flower use by bumblebees in English urban gardens. *Annals of Botany*, 113, 799–806, doi:10.1093/aob/mcu006.

Thompson, K. (2014) *Where Do Camels Belong? Why Invasive Species Aren't All Bad*. Greystone Books, Vancouver.

2 Environmental Horticulture: Benefits and Impacts

Ross W.F. Cameron

> **Key Questions**
> - What are ecosystem services and disservices?
> - How do plants impact on the microclimate around them?
> - What role do plants play in relation to buildings?
> - How is air movement and noise affected by vegetated screens and shelter belts?
> - How do urban plants influence the 'carbon debate'?
> - Why is vegetation important in flood avoidance?
> - Air quality – what are the 'pros and cons' of urban trees?
> - What are the disservices associated with urban vegetation and its management?

2.1 Introduction

Environmental horticulture has a key role to play in effective ecosystem service provision, especially within an urban context. The term ecosystem service has been developed to help understand and quantify the benefits derived from natural resources and systems. The Millennium Ecosystem Assessment (Anon., 2005) divided these services into four groups and classified them as:

- Provisioning services: the products obtained from ecosystems, including food, fibre, fuel, genetic resources, biochemicals, natural medicines, pharmaceuticals, ornamental resources and fresh water.
- Supporting services: the services that are necessary for the production of all other ecosystem services, including water and nutrient cycling, photosynthesis and soil formation.
- Regulating services: the benefits obtained from the regulation of ecosystem processes, including the regulation of air quality, climate, water, erosion, pests and disease and natural hazards.
- Cultural services: the non-material benefits people obtain from ecosystems through spiritual enrichment, education, reflection, recreation and aesthetic experiences – thereby taking account of the values people place on landscapes.

Horticultural activities, however, can contribute to ecosystem disservices too, and a number of their positives roles are prone to over-emphasis or exaggeration ('greenwashing'). This chapter aims to highlight the genuine current, as well as the potential future, benefits derived from environmental horticultural activities, whilst highlighting some of the associated drawbacks.

Overall, the principal benefits of vegetation within an urban context include:

- reducing urban air temperatures, thus helping to mitigate urban heat island effects;
- lowering the surface temperature of buildings during summer thereby reducing the reliance on mechanized air conditioning;
- screening out aerial particulate matter and improving air quality;
- de-activating chemical compounds in the soil, and recycling nutrients;
- intercepting precipitation, reducing rates of water runoff, mitigating flash-flooding and removing pollutants from water courses;
- regulating water flow and providing potential storage of water in wetlands and watersheds;
- acting as a physical barrier to wind, noise and visual intrusions;

- providing habitat and food sources for wildlife (i.e. promoting urban biodiversity) (see Chapter 4);
- enhancing human health and well-being (see Chapter 3);
- providing food, fibre and medicinal resource for humans;
- providing aesthetic enhancement and opportunities for recreation; and
- promoting educational opportunities and enhancing cultural experiences.

A number of these environmental services and disservices are covered in this chapter.

2.2 Microclimate Modification

The composition of plant communities and rate of plant development is dictated strongly by climate, yet plants themselves affect climate. In urban areas the provision of vegetation influences the local microclimate, by altering temperature, humidity and wind speed or direction. The precise role that green infrastructure plays in influencing a city's cooling, or reducing energy loads on buildings and improving human thermal comfort, has warranted attention over the last two decades. This is being driven largely by concerns over climate change, urban expansion and increasing density of housing/offices, and the desire for more environmentally-sustainable buildings.

Urban heat islands

Urban areas are typically 2°C warmer than neighbouring rural areas (Fig. 2.1) a phenomenon known as the urban heat island effect. Differentials of up to 11°C, however, between city centre districts and their outlying rural areas have been recorded at specific times and locations. The intensity of this 'heat island effect' is influenced by a range of factors peculiar to each city, including population size, ratio of inner city vegetation to buildings/hard surfaces, traffic volumes, albedo of construction materials and the density of high-rise buildings. The primary use of buildings also affects the heat generated (e.g. rate of occupancy, number of computers and other electrical heat generating equipment, manufacturing processes, as well as the sources and duration of central heating and ventilation systems within individual buildings). Microclimates across the urban locale vary widely; for example, railway lines are amongst the hottest areas, contributing to local heat island effects over the day, but cooling rapidly in the evening and remaining relatively cool overnight. Conversely, high-rise buildings trap short-wave solar radiation, which is absorbed and radiated out as warm, long wave radiation thus creating day-time heat pockets which develop into nocturnal heat islands. Heat islands form around consistent sources of heat, such

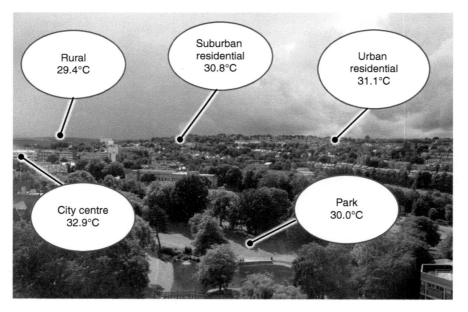

Fig. 2.1. The 'urban heat island effect' – the effects of land typology on air temperatures.

as major roads with heavy volumes of traffic, or where there is a predominance of low albedo building materials (i.e. those that reflect a low proportion of incoming light (irradiance)). For example, tarmac roads and dark bitumen roofs absorb high levels of short-wave radiation and conduct long-wave radiation back, adding to the localized heat generation. As buildings themselves act as wind shields, urban areas have typically lower wind speeds than more open countryside – the lack of air movement causing 'canyon effects' where warm air gets trapped at street level. Urban buildings, particularly shops and offices, are also dependent on mechanized air conditioning to provide internal cooling, but which as a consequence produce surplus heat externally. These effects create localized hot air pockets within the urban matrix.

Overall, the compound effect of these factors plays a major role in urban temperature flux, which in turn results in significant increases in interior temperatures and discomfort to humans. The urban heat island effect will become more prominent with climate change, and with the further expansion and increased density of urban infrastructure predicted for many cities. These effects have direct implications for human health as mortality rates rise during 'heat waves'. This is especially so in towns and cities where there is little respite from the heat at night. Elderly and very young citizens or those with respiratory problems or heart disease are most vulnerable to excessive heat. Effects can be dramatic in countries that are unaccustomed to heat waves. During the 2003 heat wave event in central Europe, weekly mortality rates rose by 17% in England and Wales and by 60% in France (Wilby, 2007).

Increasing the component of green infrastructure within cities mitigates against heat island effects and is advocated to help moderate the more extreme temperature profiles predicted by climate change and increased urbanization. Plants provide cooling to their immediate environment but the mechanisms to do so vary with the type of vegetation. Cooling is provided by direct shading of the ground and other surfaces, evaporation of leaf surface moisture and transpiration of internal water, including that being transferred through the plant's water columns, i.e. ultimately from the soil. Plant groupings also modify air flow and promote insulating layers of still ('dead') air within a building envelope, and their foliage alters the albedo of the land surface (i.e. the amount of irradiance that is reflected back into space). Plants that are transpiring are also photosynthesizing, i.e. converting light energy into sugars and hence creating biomass. Photosynthetic inefficiencies are such, however, that a proportion of irradiance captured by the leaf is lost as heat, e.g. 40–60% depending on plant species and prevailing environmental conditions.

The cooling influence of different typologies within the green infrastructure has been studied, including for urban forests, street trees, parks, lawns and meadows, green roofs, domestic and public gardens and green walls. Modelling approaches suggest that for heavily urbanized cities in temperate climates, increasing the proportion of green infrastructure by 10% could reduce mean urban air temperatures by 2.5°C, thereby reducing the frequency and magnitude of the urban heat island events encountered (Gill *et al.*, 2007).

Within any given form of green infrastructure, the predominant plant type and interactions with other factors such as soil moisture content are likely to strongly influence the cooling potential. Even relative contributions of these cooling mechanisms will depend on plant form, species, canopy cover, moisture availability, seasonality and plant vigour. Shading is often quoted as the most significant aspect of plant cooling, suggesting that the greater the cover/volume of foliage the more effective the cooling. Studies on urban trees indicated that direct shading accounted for most of the cooling capacity under the tree canopy (80%) with evapotranspiration and other tree characteristics having significant, but less influence. Evapotranspiration though may play a part in reducing air temperatures across a wider area, with cooling effects being recorded up to 100 m from street trees. Individual street trees can provide over 950 MJ (almost 270 kWh) cooling per day due to evapotranspiration alone (Huang *et al.*, 1990). Indeed, city cooling may be one of the key justifications for keeping large-stature trees within urban conurbations, despite increasing pressures for space and perceived conflicts with infrastructure. Cooling influences are not restricted to urban trees alone and other forms of vegetation aid localized cooling, through the planting of shrubs, climbers, turf etc. next to, or on, a building. Trees, however, have the advantage in that the shade they cast may cover an entire building if the specimens are large enough and well-positioned.

The process of evapotranspiration releases moisture vapour into the surrounding air, and under warm conditions increases the relative humidity of the air. Excessive humidity, however, combined with high temperatures tends to reduce human comfort levels. This is especially so where there is little air movement to encourage heat loss (and moisture evaporation) from the human skin. So in theory, dense, enclosed plant canopies that reduce air movement and increase local humidity may actually increase human discomfort levels. This aspect, however, has received little research. It would be prudent, however, in the meantime to design landscapes that optimize shading and evapotranspirational cooling, whilst also encouraging some air movement through the vegetation.

Comparisons between different types of green infrastructure have been made with respect to their cooling potential. Using hard concrete surfaces as a reference point, grass swards reduced ground surface temperatures by up to 24°C, whereas shade from trees only cooled surfaces by up to 19°C (Armson et al., 2012). Trees, however, were shown to have a more pronounced effect (5–7°C) on air temperatures recorded above the surface of the ground. Similarly, in studies simulating green roof scenarios, vegetation influenced the surface temperature of the substrate, but had no perceptible impact on air temperatures measured above the surface (Blanusa et al., 2013). In this study, the choice of plant species influenced both plant canopy temperature and substrate surface temperatures, with the lowest temperatures recorded with *Stachys byzantina* – a low-growing perennial species with hairy (pubescent), silver-coloured leaves and a relatively high transpiration rate (Fig. 2.2). The presence of the silver leaves was thought to affect the heat balance by reflecting more of the incoming solar irradiance. Other xerophytic (dry-adapted) Mediterranean species with grey or pubescent foliage may have similar attributes, but further research is required to substantiate this.

In those regions where it is difficult to grow full-canopy trees, due to arid soils and high temperatures, Yilmaz et al. (2008) argue that the presence of (irrigated) grass is beneficial as an alternative form of green infrastructure. In these more extreme environments, their work demonstrated that grass swards were able to reduce both ground surface temperatures (11.8°C cooler compared to asphalt/concrete) and, importantly, also air temperatures 2 m above the surface of the grass (7.5°C cooler).

The cooling influence of green infrastructure has not only been demonstrated for climatic properties, but on human thermal comfort directly (Shashua-Bar et al., 2011). Studies using semi-enclosed courtyards evaluated the effects of vegetation and synthetic shading mesh on human perception of heat. Different combinations of shading and vegetation types influenced the responses reported by the participants of the study. Optimizing the design resulted in human discomfort being reduced both in the duration of the discomfort felt and in its severity. For example, when combined with grass, overhead shading by trees or mesh resulted in comfortable conditions over all hours of the day. Overall, trees were more effective than the synthetic mesh, but only marginally. With grass alone and no shading, 'hot' conditions in the courtyard were restricted to a short period in mid-afternoon, a considerable improvement over unshaded paving. This was attributed to the grass reducing the radiant energy reflected back from the ground surface. Many cultures, of course, have long understood the value of providing shade trees in hot climates, whether this is within the market squares of Mediterranean towns or at the meeting place of African villages.

Cooling buildings and energy efficiency

The potential for urban planting to reduce energy consumed through artificial air conditioning has been explored in warm climates, including the southern and western states of the USA. Studies in Sacramento, California (USA), indicated that just a 25% increase in urban tree cover could reduce energy consumption associated with cooling domestic properties by 40% per year (Huang et al., 1987). The existing urban forest of Sacramento is thought to provide 12% of the city's cooling requirement

Fig. 2.2. *Stachys byzantina* (lambs' ears). Investigations indicated that *S. byzantina* was more effective at surface cooling than *Sedum*, *Hedera* or *Bergenia*.

already with large-stature, mature shade trees being most beneficial. More optimistically, Akbari et al. (1997) revealed energy savings of around 30% from existing urban trees in California. Others have gone further, with estimations of 50% reductions in cooling energy requirements at certain times, through selective placement of tree and shrub planting (Meerow and Black, 1993). Indeed, scaling up these datasets across the USA would result in overall a 20% reduction in cooling energy demand, amounting to over US$10 billion savings per annum (Akbari et al., 2001). Similarly, the provision of four carefully-positioned 'shade' trees per house would result in annual savings of carbon emissions of up to 41,000 tonnes per city (Akbari, 2002). The results are not restricted solely to the hottest regions. Even in more temperate climates such as the UK, there is evidence that office blocks located within well-vegetated locations do not need mechanized air conditioning to maintain internal temperatures <24°C, whereas those buildings with limited green infrastructure in their immediate neighbourhood require continuous artificial cooling during warm periods to maintain the same temperature (Kolokotroni et al., 2006)

Although the value of trees in providing localized cooling is now recognized, there are conflicts with enhanced tree planting in urban areas. Urbanization is encouraging more compact housing (high densification) with less room for large tree species. Compact developments often feature medium- to high-density building footprints, smaller open spaces and buildings located closer to roads and pathways. Where trees are still encouraged, smaller growing species (e.g. *Malus*, *Sorbus*, and *Cercis*) are often advocated over the taller, larger canopy species, resulting in trees growing in parallel to the building rather than providing shade to the roof.

To avoid urban sprawl, planners are encouraging the use of more compact developments, with mixed use functions, e.g. housing alongside retail or business structures. These trends are likely to continue; urban areas in the conterminous USA having doubled in size between 1969 and 1994, and are set to increase by a further 50% by 2050. More pressure on urban space reduces opportunities for new tree plantings and increased density of build infrastructure constrains both tree roots and canopies, providing insufficient room for healthy canopy development and root growth and creating a greater likelihood of conflicts with the built infrastructure. The close proximity of roots to cables, pipes, building foundations and pathways can lead to structural damage. Similarly, the encouragement of large canopies does not sit comfortably with the desire to maximize irradiance levels for energy generation through roof-mounted solar panels. Allowing the development of solar technology within residential areas, whilst also retaining the cooling effect that trees provide for buildings, will require better designed green infrastructure (i.e. taking greater account of aspect, wind direction and diurnal shading patterns). As solar panels themselves cause localized microclimatic warming, other forms of green infrastructure that do not shade the panels directly, but provide nearby cooling, are worth investigating (e.g. green roofs and/or green walls).

Green walls are often cited as providing effective cooling to buildings, although the type of green wall may influence the extent of the cooling (and the amount of resources required to do so). Green walls tend to be divided into two main categories: (i) 'green facades' which are composed of climbing plants (vines) grown in soil at ground level, and (ii) 'living wall systems' where the plants are rooted into cellular modules or placed in hydroponic systems (Fig. 2.3) (see also Chapter 10). Green walls shield buildings from high temperatures by reducing short-wave heat gain, and depending on the system can provide a layer of 'dead' air as insulation between the wall and the vegetation. Living wall systems in China have been shown to reduce exterior wall temperatures by a maximum of 20.8°C, and the corresponding interior side of the wall by 7.7°C. Air temperatures between the wall and vegetation were on average 3.1°C cooler than ambient air. Most studies on green walls focus on wall surface temperatures, with maximum differences between vegetated and non-vegetated walls cited as 2.7°C (Netherlands), 8.3°C (Greece), 11.6°C (Singapore), 15.2°C (Spain), 18°C (Japan) and 20°C (Italy) (Cameron et al., 2014). As with the influence of nearby trees, effective wall cover on buildings is likely to reduce the reliance on mechanized air-conditioning units. For example, computer models using *Hedera helix* (English ivy) have suggested that solar gain (heat due to sunlight) is reduced on a south-facing wall by up to 37%.

Further studies in Japan have shown that maximum temperature differences between vegetated and non-vegetated walls varied due to the plant species being utilized, although some of the differences were explained by variation in the percentage of

Fig. 2.3. Both green façades (left) and living walls (right) are grown for their ability to cool a building as well as for aesthetic appeal.

canopy cover over the wall, rather than any other plant trait. Cameron *et al.* (2014) demonstrated that plants could cool walls through both shading and evapotranspiration, but the proportional contribution of these two factors altered between the species being used (Fig. 2.4). *Hedera*, *Jasminum* and *Lonicera* provided greater cooling through the shade they cast on the wall. In contrast, *Fuchsia* promoted cooling mainly through evapotranspiration, with this accounting for 3°C cooling compared to only 1.5°C associated with shading. Cooling effects with *Stachys* and *Prunus* were equally attributable to shading and evapotranspiration. With the silver pubescent-leaved *Stachys*, greater reflection of irradiance may also have contributed to the cooling influence (i.e. the leaves providing a greater albedo effect than conventional green leaves).

The precise location of plants with respect to an individual building, or the directional aspect of a wall that a green façade is placed upon, strongly determines the extent of any benefit. Planting can restrict the flow of cool air to a wall, particularly in the early and later parts of the day, but alternatively other data has shown that reducing air flow and trapping shaded (hence cooler) air at later times during the day is desirable (Yoshimi and Altan, 2011). In the northern hemisphere, facades that are south-facing benefit from being planted with deciduous climbers where leaf cover blocks excessive solar irradiance in summer, yet the absence of foliage in winter enables irradiance to penetrate through to the building, thus contributing to heating the building during the coldest months. Conversely, evergreen foliage may be used on a north aspect to provide insulation all year round.

Green roofs too have a role in reducing interior temperatures, frequently through the enhanced insulation effect provided by the substrate placed on the roof, in addition to any evapotranspirational cooling from the vegetation. Deeper substrate depths improve the insulation; however, the depth of green roof substrate is dictated in practice by the weight load placed on the roof (i.e. thinner substrates are preferred from an engineering perspective). Although substrate depth (and moisture content) are often quoted as the main factors providing thermal insulation to the building, others have argued that the volume of vegetation (e.g. larger, more dense canopies) may be as, or more, important (Theodosiou, 2003). Although green roof vegetation harbours the potential for urban temperature reduction, the extent to which they contribute to urban cooling compared to other vegetation types or landforms is unclear. Extensive green roofs, with shallow substrate and intermittent or limited plant canopy cover, and which host mostly drought-adapted species are likely to provide less cooling influence compared to deeper substrates that support large plants

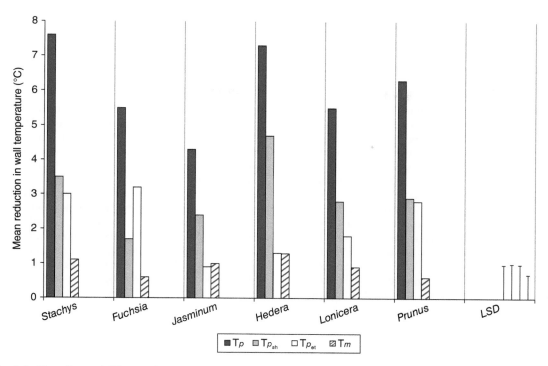

Fig. 2.4. The effects of different plant species in reducing wall temperatures in a controlled environment with a point heat source. Reduction in wall temperature (°C) attributed to whole planted troughs Tp with derived values for shade (Tp_{sh}), evapotranspiration (Tp_{et}) and evaporation from medium (Tm). LSD bars also given for each value respectively, from left to right. (Source: Cameron et al., 2014.)

(as are commonly found on intensive green roofs or roof gardens). Larger plants not only afford more shade to the roof directly, but are likely to provide greater cooling via evapotranspiration, due to their more extensive leaf surface areas; this is as long as soils retain their moisture (via deeper substrates or artificial irrigation).

Most evaluations of green roofs to provide city-level cooling have been carried out through simulated models. Work based on Chicago, USA, suggested green roofs could reduce urban temperatures by 2–3°C (in the evening) compared to temperatures simulated in the absence of green roofs. Other simulations suggest more modest or no benefits. Extensive roofs using turf grass were considered in models for New York, USA. These reported that the peak ambient temperature at 2 m height was decreased by an average of 0.3–0.55°C. In Tokyo, Japan, on the other hand, the models generated suggested that extensive green roofs planted with grass on medium- and high-rise buildings would have a negligible effect at street level. Similarly, in Hong Kong simulations of both intensive and extensive green roofs on 60 m-high buildings also showed no tangible effects on the ground (Ng et al., 2012). These results inferred that when the building height is greater than the width of the street canyon below it, the ability to provide ground-level cooling is low.

The role for green roofs appears more positive when evaluating an individual building's energy use. Even here though, white reflective roofs (roofs composed of artificial materials with a high albedo) can outperform green roofs in terms of energy savings within the building, at least under some circumstances. Simulations showed that buildings equipped with reflective white roofs presented lower net energy consumption than those with green roofs (Sailor et al., 2012), at least in those comparisons depicting a warm climate scenario and where the maximum substrate profile of the green roof was limited to a depth of 300 mm (i.e. representing most 'extensive' (shallow substrate) green roof systems). For simulations depicting cooler climates, however, vegetated green roofs proved more valuable than

Environmental Horticulture: Benefits and Impacts

reflective surfaced roofs in terms of energy savings over the entire year. This is largely because they not only afford some cooling effects in summer, but also a thermal insulating effect in winter, thereby reduced heat loss during the colder months of the year (Fig. 2.5). Buildings equipped with green roofs where the vegetation is characterized by a high leaf area index (LAI – leaf area m^{-2} of roof surface) present a much lower energy consumption value for cooling than buildings with a low LAI value. Even in the warmer climates where it is assumed white roofs are generally more effective than green, those green roofs with good vegetation cover and high LAI (e.g. a value ≥5) seem to have an advantage over their white roof counterparts (Sailor *et al.*, 2012). To this effect, it must be assumed that 'intensive' green roofs with deeper substrates, which support the growth of shrubs, large herbaceous plants and even small trees, will provide markedly more insulation and cooling potential than extensive systems. As such, maximizing the amount and size of vegetation would enhance the environmental performance of a building, but further research is required to determine what the environmental costs of providing more intensive green roof systems would be in terms of the construction of the building. Deeper substrates with greater water holding capacity results in heavier roofs, and this has implications for the weight loading of the building and the design/costs of the construction.

Although vegetation can cool via shading, evapotranspiration cooling is compromised by limited water availability to the plant. Additional irrigation, therefore, undoubtedly has a significant role in maintaining evapotranspiration effects (McPherson *et al.*, 1989) and plant growth (Cameron *et al.*, 2008); the latter ultimately affecting shading potential too. During dry periods, evapotranspiration cooling may need to be maintained by continuing to supply irrigation from potable or other water sources (although the use of potable water for this may not be deemed a priority when domestic water supplies themselves are put under pressure during heat waves and drought periods). Lower quality water and grey water (see below) sources are likely to become more important in the future to ensure city plantings retain their cooling influence during such critically warm periods. Certainly, maintaining some form of irrigation during dry periods has additional advantages

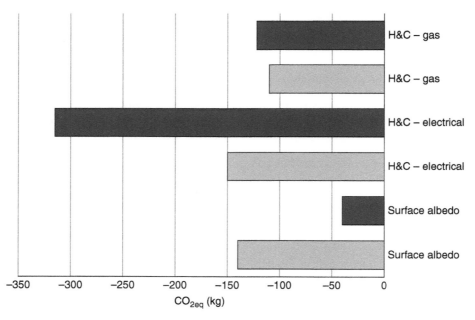

Fig. 2.5. The climate change mitigation potential of white (pale bars) and green roof (dark bars) systems. Data includes mitigation of energy from heating and cooling (H&C) from either gas or electrical fuel sources, or due to the surface albedo effect. Green roofs provide the greatest advantage when electricity is used to both heat the building in winter and cool it in summer. (Modified from Susca *et al.*, 2011.)

as when rain subsequently arrives, moist soil allows more effective infiltration of water compared to that of dry soil; the latter tending to promote surface runoff rather than penetration into the soil. The volume of water required to maintain evapotranspiration cooling relates to factors such as vegetation form, species, soil type, hydrogeology, topography, local microclimate and regional weather patterns. For example, turf grass surrounding a building can significantly reduce the requirement for air conditioning but only when well-watered. Conversely, shrubs (a mix of shade and evapotranspiration) require less water to maintain the same cooling potential (McPherson et al., 1989). Even within shrubs though, species respond differently to a drying soil, with some genotypes, e.g. *Cornus alba* 'Elegantissima', closing their stomata rapidly and keeping them shut, and hence ceasing any cooling effect via evapotranspiration for a prolonged time period (Cameron et al., 2006, 2008). Others, on the other hand, retain some photosynthetic capacity by keeping their stomata at least partially open, even under quite notable soil moisture deficits; *Forsythia* × *intermedia* 'Lynwood' aligns itself to this strategy. Yet others, such as *Cotinus coggygria* 'Royal Purple', close their stomata quickly, but also open again fairly rapidly should the moisture status subsequently improve, even if only marginally. This species therefore, is a promising candidate for 'city-cooling' scenarios. So species selection affects stomatal performance, and hence evapotranspirational cooling, but the development of comprehensive databases on species choice in this regard remains insufficient.

Building insulation in winter

Green infrastructure has the advantage of influencing a building's energy dynamics in winter as well as summer. Again, appropriate choice of vegetation and careful positioning is paramount to optimize insulation and reduce energy loss during cold weather. This is accomplished through a number of mechanisms. Vegetation acts as a windbreak helping to maintain a positive microclimate surrounding the building through the trapping of warm air. It reduces the speed of air masses moving over a building and thus the rate at which heat is lost. Finally, vegetation slows the velocity of air directly and reduces the effect of cold draughts entering the building and warm air leaking out. Dense planting close to a building, such as that which occurs with climbing perennials and wall shrubs, creates dead-air space, reducing air movement in the immediate vicinity of walls (Hutchison and Taylor, 1983). The physical presence of the vegetation and the associated dead air space provides a barrier which reduces cold air infiltration through cracks and apertures by up to 40%. Climbing plants useful for such scenarios include *Clematis montana*, *Parthenocissus quinquefolia* (Virginia creeper), *Solanum crispum*, *Lonicera* spp. (honeysuckle) and for slightly warmer climates, *Passiflora caerulea* (passionflower), *Solanum jasminoides* and *Bougainvillea* cultivars.

In addition, about 8% of convective heat loss is preventable with shelter belts composed of trees and hedges located in a manner to directly intercept any particularly strong or cold winds (Fig. 2.6). Strategically placed vegetation, which takes account of prevailing wind directions and solar irradiance, has been shown to reduce overall winter energy costs by 17–25% (Liu and Harris, 2008). The benefits of shelter belts and other woodland features are most apparent in the winter (Fig. 2.7). Even woodland shelter belts located some distance from the building can have a marked effect, for example Wang (2006) showed that trees 50 m away reduced energy costs by 4.5% when they were in position to intercept the prevailing winds.

The benefits of green walls may be even more marked. Due to the challenges of attaining identical buildings for replicated controlled trials, Jane Taylor, a PhD student at the University of Reading, UK, developed a 'mini-city' of 20 brick cuboids over a uniform 500 m² site. These cuboids each enclosed a heated water tank that was maintained at 16°C throughout the winter period and was monitored for energy use. Half the cuboids were covered with vegetation (*Hedera helix*) and comparisons made on

Fig. 2.6. A dense shelter belt of × *Cupressocyparis leylandii* (Leyland cypress) is planted to the north-west of a building to help reduce heat loss from the prevailing winds in winter. The flat topography results in an exposed landscape where trees are one of the few natural features to offer protection from the wind.

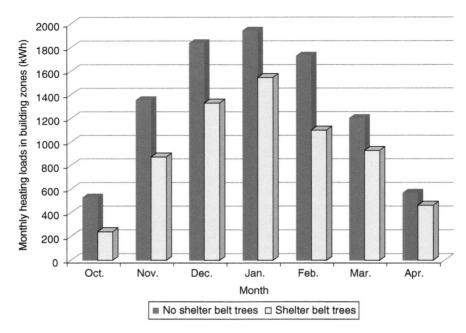

Fig. 2.7. Comparison of monthly heating loads for the zones of a building either with or without a shelter belt of trees. (Source: Lui and Harris, 2008.)

the energy consumed between the vegetated and non-vegetated 'buildings'. A mean energy saving of 37% was achieved over one winter due simply to the presence of the insulating plant material (3.7 kWh per week for vegetated cuboids compared to 5.9 kWh per week for non-vegetated). Interestingly, over discrete periods of severe weather – such as sub-zero temperatures, strong wind or heavy rain – the green walls further increased the energy efficiency by 40–50% and enhanced the wall surface temperatures by up to 3°C (Cameron *et al.*, 2015). Further comparisons are required on real 'to-scale' buildings, but the principle around the insulating potential of plants is made clear by this and allied studies.

Wind amelioration

In urban open spaces, trees and hedges can increase the microclimatic temperature on the leeward side of the wind by 1°C, with potential for further warming on those aspects that face the sun (i.e. south facing in the northern hemisphere) where solar irradiance heats pockets of the retained air. Hedges and other forms of screening minimize evaporative losses from landscape plants and allied soils/substrates, and reduce abrasion caused by the wind to the foliage and flowers. Similarly, they help reduce water loss from ponds, fountains and other water features, and hence the requirement to continually top these up with supplementary water.

Trees, of course, can be damaged by the wind and hence pose a potential risk to human health and property. During severe storm events falling trees and broken boughs are linked to human deaths and serious injury. Winter storms that crossed central Europe in 1999 were estimated to have killed 137 people, a proportion of which were killed by either dropped branches or entire trees falling onto houses and roads (Lopes *et al.*, 2009).

Urban trees often have a greater disposition to wind damage compared to those in forest situations due to a lack of protection from neighbouring trees, and the fact that wind forces may be made even stronger by the channelling of air along roads and urban canyons. The integral strength of street trees may also be compromised by a range of urban stress factors, including reduced or restricted root space ('root-run'), drought stress, urban heat, compacted soil, air pollution, road salt damage and root damage through trenching activities for cables and pipes. Greatest storm damage in higher latitudes is often associated with large deciduous trees, which have retained a full canopy of leaves at the onset of autumnal storms. The increased resistance to the wind, due to the presence of a full canopy results in the whole tree being thrown over. It has been estimated that the majority of tree falls are linked to wind speeds in excess of 7 m s^{-1} for 6 h or more (Lopes *et al.*, 2009). Other aspects, such as heavy precipitation that loosens the soil around tree roots or ice storms that add weight to the tree branches, also increase the likelihood of trees being thrown over.

2.3 Noise Amelioration

Hedges and shelter belts provide a service through their ability to absorb, diffuse and deflect noise (noise attenuation). Shelter belts of vegetation protect domestic properties from nearby industrial facilities, roads or railways, where there are consistent or intermittent noise sources. Fang and Ling (2003) suggested that dense plantings of shrubs, where the plants were a few metres higher than the receiver of the noise were optimal for reducing noise. They studied tree belts composed of 35 different species, and those genotypes characterized by dense foliage and branches correlated with greatest noise reduction (Fig. 2.8). Particularly valuable are shrubs and trees which possess a low forking habit as these have the greatest effect on noise reduction. In contrast, shelter belts where the forking and branching was higher up in the canopy had few obstructions to absorb noise at ground level, and acoustic waves passed along below the canopy. Other plant traits, however, interact with sound waves. Greater density, height, length and width of shelter belts help diffuse noise more effectively, whereas increasing the leaf size and branching characteristics aid absorption of sound waves. Overall, endorsing those plants that minimize visual sight of the noise source are likely to reduce the noise effect, and in practice a belt of trees under-planted with a dense shrub layer may be considered the best design to intercept sound waves as they move away from the source of the noise.

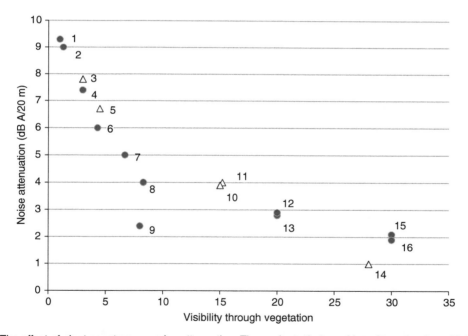

Fig. 2.8. The effect of plant genotype on noise attenuation. Those plants that provide a dense barrier, with little visibility through them, often also reduce noise most effectively. Note with *Ficus macrocarpa* (Δ), however, how factors such as planting design, canopy height, planting density etc. affect the noise attenuation. (Modified from Fang and Ling, 2003.)

Plant species: 1, *Bambusa dolichoclada*; 2, *Garcinia subelliptica*; 3, 5, 10, 11, 14, *Ficus macrocarpa*; 4, *Ilex aquifolium*; 6, *Nageia nagi*; 7, *Ravenala madagascariensis*; 8, *Nerium indicum*; 9, *Araucaria heterophylla*; 12, *Pongamia pinnata*; 13, *Gardenia jasminoides*; 15, *Camellia japonica*; 16, *Podocarpus macrophyllus*.

Even relatively small-scale green infrastructure interventions may aid noise control. Concrete embankments and metal baffles along urban roadways specifically designed to retain noise within the confines of the roadway are often augmented with climbing or trailing plants. These 'soften' the effects of the vehicle noise by absorbing some of the sound waves, and avoiding noise reverberation off the walls. Using an urban courtyard as a scenario for modelling noise, Van Renterghem *et al.* (2013) studied the effect of surrounding green roofs, green facade walls and low, vegetated screens at the edges of flat roofs to mitigate the effects of adjacent road noise. The study suggested green roofs have the greatest potential for attenuating noise, and on certain roof designs could reduce noise by 7.5 decibels. As is the case with thermal insulation of green roofs, a number of advantages associated with noise attenuation were linked with the substrate rather than the plants. The benefits and positioning of green walls was more dependent on the surface materials of the surrounding buildings. The models used predicted that green facades worked best when they were positioned high up on the walls surrounding the courtyard. This did depend on the particular materials used for the buildings though. When the construction materials located near the street level were softer (with more noise absorbance capacity), however, then it was better to position the green facades around the courtyard itself. The vegetated screens on roof edges were only effective when the supporting structures were made from absorbent materials as opposed to rigid materials. The latter could actually increase noise levels directed into the courtyard. In overall design terms, soft roof edge screens combined with either green roofs or walls were the most effective means to reduce noise. This research, however, highlights the need to 'blend' the right physical materials with the green infrastructure to optimize any benefits.

Noise is both a physical and psychological factor, with perceptions of noise being a source of stress as well as actual volume. As such, the presence of natural elements seems to have a moderating influence on people's noise responses. High quality neighbourhoods with a good proportion of attractive parks and other green spaces have been shown to lower dissatisfaction with traffic noise to a significant extent (Kastka and Noack, 1987). Having green spaces available close to one's home and where they can be easily accessed has reduced noise-related stress symptoms compared to those living in areas with little accessible green space (Gidlöf-Gunnarsson and Öhrström, 2007). These authors concluded that availability of nearby green areas may provide a protective factor that partially moderates the adverse impact of road traffic noise. Others too found that adding vegetation had a positive effect on people's perceptions of 'noise-contaminated' environments (Fig 2.9). Yang *et al.* (2011) exposed participants to the same level of road traffic noise whilst either viewing a busy road scene, or conversely a park scene. Using both questionnaires and emotional tests using an electroencephalogram (EEG) they found that 90% of the subjects believed that landscape plants contributed to noise reduction with 55% over-rating the plants' actual ability to attenuate noise.

2.4 Carbon Sequestration and Release

Since the 1990s the concept of planting trees to sequester atmospheric carbon dioxide (CO_2) has been in vogue. Tree planting schemes have been advocated in an attempt to mitigate against carbon release through a range of anthropogenic activities, with 'trade-off' schemes, including the sponsorship of tree planting to offset air travel, for example. These 'off-setting' approaches have had their critics, but the notion of reafforestation and careful soil management to aid carbon sequestration has remained. The Intergovernmental Panel on Climate Change (IPCC) has estimated that tree planting on a global scale could sequester as much as $1.1–1.6 \times 10^{15}$ g CO_2 per year, but this is dwarfed by current global annual emissions (3.45×10^{16} g CO_2) (Anon., 2010). Within an urban context, horticultural activities are

Fig. 2.9. Increasing green infrastructure around roads may do little to directly affect noise in the immediate vicinity of the road, but it may have positive benefits on people's perception of noise in such situations.

now scrutinized more closely to help promote sustainable approaches to design and management, and the advent of a greater component of urban green space is generally seen as a virtue. The advantages in terms of sequestering atmospheric carbon, however, are strongly determined by the specifics of the activities involved. Indeed, the advantages and disadvantages of certain activities are still difficult to assess as they often depend on incomplete calculations or certain assumptions being made within life cycle analyses (LCAs). The point too that the global climate is changing also infers that activities now that are considered positive, need not necessarily remain so. For example, the desire to increase soil carbon pools, which are currently being encouraged through management activities (adding organic manures, not removing dead wood or other vegetation from sites and reducing soil tillage), may only enhance soil carbon levels for these to be vulnerable to oxidation later (i.e. released back into the atmosphere). This may be a scenario if mean soil temperatures increase, or microorganism dynamics alter with climate change.

Green space provides two principal locations for carbon pools to be stored: in the living vegetation and in the soil. The soil is the larger of these two carbon sinks, for example within residential landscapes 83% of carbon is found in the top 600 mm of soil with 16.5% in trees and shrubs and only 0.5% in grass and herbaceous plants. Studies in Leicester, UK, suggested that trees held 97.3% of the total 'above ground' carbon in the urban green space, with relatively little in shrubs and herbaceous plants (Davies *et al.*, 2011). Trees in both public civic space and private gardens contribute to the greatest carbon pool (Fig. 2.10). Across urban landscapes, green infrastructure is thought to contribute approximately 3.2×10^7 g C ha^{-1} (Davies *et al.*, 2011) but most of this is concentrated in areas of urban forest or where there are street trees (2.9×10^8 g C ha^{-1}), with typical domestic gardens storing 0.75–2.5×10^7 g C ha^{-1} depending on the amount and size of plantings in the garden. Total carbon sequestered in urban trees in the USA is estimated to be 7.0×10^{14} g (Nowak and Crane, 2002). In woody plants the carbon in dry biomass is fairly consistent across species at 49–50% (Whitford

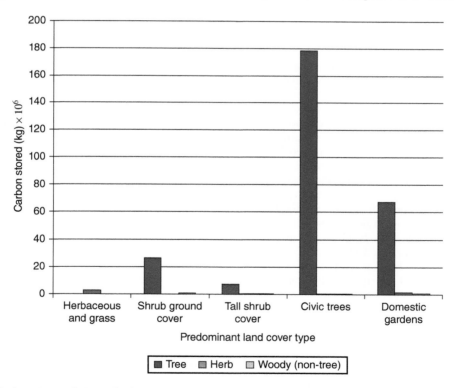

Fig. 2.10. Carbon storage in trees, herbaceous plants (including grass) and other woody (non-tree) vegetation as determined by predominant land cover type in the city of Leicester, UK. (Modified from Davies *et al.*, 2011.)

et al., 2001), although bulbs for example can be lower (e.g. 35%), depending on their growth stage. Comparisons using five shrub species indicated carbon storage in biomass is $4.7–7.2 \times 10^6$ g C ha^{-1}, and atmospheric carbon uptake is $0.56–0.80 \times 10^6$ g C ha^{-1} per year (Jo, 2002).

Woody plants provide significant carbon sequestration improvement over transient planting schemes due to soil carbon being reduced significantly by soil disturbance and tillage during replanting. Sequestration therefore would seem to be optimized by long-term, stress-tolerant species, which require minimum soil disturbance and management requirements, typical of perennial plantings that reflect permanent woodland or long-term meadow systems. Trees/shrubs sequester and store carbon at different rates, depending on age, lifespan, growth rate, species type, health and local conditions, including microclimate, hydrology, topography, soil type etc. Plants contribute to soil carbon directly through root exudates, senescing roots and organic molecules directed to mycorrhizal fungal hyphae and other microorganisms, dead leaves and fruit and wood from decaying/dying plants. As indicated above, carbon accumulation rate as well as its distribution within the soil profile, differs according to plant genotype, even varying between different tree species of similar stature. This may relate to aspects such as volume of leaf litter and its rate of decomposition, volume of carbon-based solutes leaked from roots (mucilage), the location of roots (surface versus deep-rooting species) and the 'turn-over' of roots, i.e. the proportion that die off every year and are replaced by new roots. Soil texture, structure, temperature and overall biological activity will influence these parameters. Such factors may not only determine the amount of carbon stored, but its longer-term stability too. The action of macro-and microorganisms further influences the forms of carbon held within the soil carbon pool (Heath *et al.*, 2005).

Overall carbon balance (life cycle analysis) within green space is principally influenced by vegetation management, aspects such as fossil fuel consumption with respect to lawn mowers, hedge-trimmers and chainsaws but also the use of vehicles to visit sites and remove debris, dispose of biomass, manage diseases and pests, and other effects due to local geographical/sociological factors.

Although few comprehensive life cycle analyses have been carried out for urban green space it is assumed that practices that align with permaculture principles and sustainable forms of irrigation are likely to have the least impact in terms of CO_2 release. This includes: embedding management strategies that minimize soil cultivation and foster passive incorporation of organic matter into the soil via mulching, and that move away from the use of potable water, using recycled water where possible; strategies, such as integrated pest management, that lower rates of artificial chemical use; designing the landscape to encourage more heterogeneous vegetation cover. Even practices designed to aid carbon sequestration, such as tree planting, may have surprisingly long pay-back times in terms of carbon balance. Due to the embedded energy associated with tree production and the volume of soil CO_2 released during planting, it has been estimated that newly planted trees become 'carbon neutral' only after 3–10 years (Schlosser *et al.*, 2003).

Pesticides, fertilizers and potable water used in the landscape provide their own carbon footprint. The production/use of artificial fertilizers contribute significantly to greenhouse gas emissions (Howarth *et al.*, 2002), particularly through the Haber–Bosch process that is used to convert atmospheric nitrogen into ammonia and then used in inorganic nitrogen fertilizers. In contrast, manufacture of phosphate and potassium fertilizers is 10–20-fold less energy intensive. The use of composted organic matter in the garden offers a lower carbon cost alternative for supplementing nitrogen than using an energy-intensive inorganic nitrogen fertilizer such as ammonium nitrate (NH_4NO_3). The incorporation of green and other appropriate organic wastes into the soil is also likely to reduce energy consumption embedded in moving such materials off-site; they frequently have the additional advantage of improving soil structure in the process.

The use of fertilizers and pesticides, combined with petrol-driven mowers, has been estimated to negate the carbon sequestration potential of lawn turf. Domestic lawns in the USA are thought to hold approximately 4.96×10^{14} g C, but fertilizer application and mowing may release $2.5–7.6 \times 10^{12}$ g C each year. In effect, the carbon footprint of lawn maintenance would mean that any benefits of carbon storage within the lawn would be lost after 66–199 years of maintenance (shorter if the lawns are irrigated with potable water too). Lawn fertilizer use alone has a significant effect on greenhouse gas emissions (Livesley *et al.*, 2010) with lawns emitting up to ten times more nitrous oxide (N_2O) than neighbouring agricultural grassland, the higher emissions being due to more frequent irrigation and higher soil temperatures encountered in urban lawns. Naturalized grassland (meadows) though have a low environmental footprint, in contrast to heavily-maintained lawns. The former requires a single annual cut, compared to the high frequency mowing associated with formal lawns (e.g. weekly during optimum growth). Although accurate estimations of carbon use through mowing are difficult to assess, largely due to differences in type and energy supply to the mower, the frequency of cutting, climatic factors and what is the ultimate fate of the grass clippings (Reid et al., 2010), it is thought that petrol-powered mowing may release 50% more carbon than the lawn itself can sequester. Mowers with two-stroke engines particularly are considered detrimental with a 7-fold higher by-product emission compared to four-stroke engines (Volckens et al., 2007).

Despite intensively managed lawns being associated with high fertilizer and herbicide inputs (and in warmer, affluent countries, high water-use through artificial irrigation) there are situations where the laying down of a lawn stimulates a positive carbon balance – at least in terms of *soil* carbon. Where urbanization results in housing development on land previously degraded or heavily cultivated, the creation of a new lawn may act as a catalyst to replace soil organic carbon that was originally lost via oxidation during the former activities (Zirkle et al., 2011). Pouyat et al. (2009) claim that the relatively high organic carbon component within residential soils in the USA is due to lawn management. They relate this to management activities that typically include supplements of water and nutrients which maximizes the productivity of the grass. Indeed, they state that soils of residential lawns appear to have the highest density of carbon in urban landscapes – higher than many forest soils in the USA. Soil carbon pools, though, vary widely with location and climate. For example, urbanization (including the promotion of gardens and other green space) in the north-east USA is thought to reduce overall soil carbon pools. This is largely due to the parent soil here being either brown podzols (woodland) or dark mollisols (prairie grasslands), as such soils have a relatively large natural organic carbon content. In contrast, on more arid soils in the southern states (where carbon storage is naturally lower), urbanization can increase soil carbon content through more prevalent lawn culture. In such situations the age of the lawn often determines the levels of stored carbon (Fig. 2.11).

Outwith turf management per se other high carbon costs are associated with construction materials. The constituents of paving, footpaths, walls etc. may have high embodied energy associated with their transport (many being bulky heavy materials), or, as in the case of cement, energy used in their manufacture.

What happens to these materials over time though is also intriguing. Ex-industrial brownfield sites have a high proportion of broken rubble and concrete, and this may itself absorb CO_2 from the air as it weathers. Calcium and magnesium from the rubble react with the CO_2 and precipitate out as insoluble carbonates and thus help lock up atmospheric carbon. So it is not only organic processes that are useful for sequestering carbon in urban soils.

Actively promoting the planting of new street and other urban trees for energy conservation and carbon sequestration reasons comes with an economic cost. Street trees require training, pruning and pest/pathogen control as well as encountering costs with initial planting and final removal. The costs of vegetation maintenance have been compared directly to the benefits of carbon sequestration and energy saving. Depending on the tree species planted, the cost of reducing carbon averaged across planting locations is in the range of US$3133–8888 per tonne of carbon (tC) (Kovacs et al., 2013). Long-lived, trouble-free tree species such as *Platanus* × *acerifolia* (London plane) are deemed most cost-effective due to their long lifespan and large canopy; the marginal cost of carbon reduction for the species was in the range of US$1553–7396 per tC across planting locations. Benefits improve significantly when the adjacent buildings are lower in stature and more residential in nature. This is due to trees providing greater energy efficiencies in residential two-storey buildings because they shade a greater proportion of these buildings in comparison to taller office blocks, where perhaps only the first 2–3 storeys out of a 20 storey-plus building would be shaded.

Carbon sequestration is optimized when the tree genotypes selected possess traits such as fast growth rates, high density wood, large trunk diameters and large indices of total biomass (e.g. height × breadth × wood density). Longevity and general resistance to pathogens and pests means that the wood resists decay, and there is little requirement

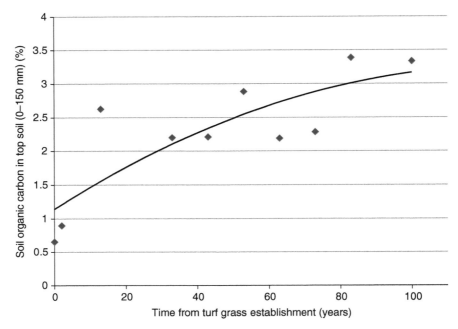

Fig. 2.11. Soil organic carbon in the top 150 mm of lawns in Houston, Texas (USA) as affected by the age of the lawn. (Modified from Selhorst and Lal, 2013.)

for intensive maintenance. An additional bonus relates to the ability to 'fix' the carbon in the wood for a further period by conversion into timber or furniture, again limiting the amount of carbon released at/near the end of the tree's natural lifespan. In addition to *Platanus × acerifolia*, other species that are valued for their carbon sequestration/low maintenance characteristics include: *Juglans nigra* (black walnut), *Liquidambar styraciflua* (American sweetgum), *Pinus ponderosa* (western yellow pine), *P. resinosa* (red pine), *P. strobus* (white pine), *Pseudotsuga menziesii* (Douglas fir), *Quercus coccinea* (scarlet oak), *Q. robur* (English oak), *Q. rubra* (red oak), *Q. virginiana* (Virginia live oak) and *Taxodium distichum* (bald cypress).

2.5 Water Management

Water management and sustainable urban drainage systems (SUDS)

Urban areas are particularly prone to flooding due to the occurrence of impermeable surfaces that do not allow rainwater to soak into the soil, but rather encourage it to run off the surface (surface flow). As a result, reductions in urban green space, including the paving over of gardens to increase 'off-road' car parking space have increased the risks of urban flooding. In the UK it has been estimated that the proportion of total garden area paved over has expanded from 28% to 48% over a 10-year period from 2000. In the USA impermeable land surfaces equate to 110,000 km^2; an area of land equivalent to the entire state of Ohio (Elvidge *et al.*, 2004). Urbanization along with land-use changes (e.g. deforestation in upland areas and river catchment basins) and climate change are predisposing greater numbers of cities to more frequent and severe flooding events. In China, 62% of all cities experienced some flood event between 2008 and 2010, with 137 individual cities experiencing flooding three times or more.

Traditional methods of dealing with high precipitation rates and flood risk in urban areas have been to increase the number of drains, widen drain pipes, divert river flows, deepen channels and develop a complex network of storm channels and culverts that function to drain stormwater away. Not only are such interventions expensive, they may now not have the capacity to deal with the enhanced flows associated with further urbanization, or indeed from future climatic changes that will induce more severe stormwater events. In addition, adapting cities simply through 'hard' engineered approaches has adverse effects on the ecological function of water courses, including negative factors such as increased erosion, the concentration of pollutants in waterways and loss of habitat to wildlife. Culverting, which is the diversion of water through a tunnel or pipe, has been commonly used in urban design since the mid-20th century. This has been used to contain water flows and facilitate building development over a river course. Indeed, many cities that expanded during the industrial revolution due to their riverside location (water being used as a source of energy, within manufacturing industrial processes and as a transport corridor) have now covered over part, or all, of their waterways. Culverting, however, eliminates light from the waterway, excluding any plant growth and inhibiting animals from migrating through the 'barren zone'. Canalization (the straightening and deepening of river courses) also provides little suitable habitat for native stream flora and fauna. Canalization removes natural features such as gravel beds and permeable soil banks, thus reducing opportunities for plants to establish and leading to the loss of feeding and refuge sites for animal life. Deeper, straight water channels alter water temperature and flow rate, again interfering with natural colonization of the river. Such factors affect terrestrial animal behaviour as well as aquatic life forms, as many species use water courses as highways and networks within their territories (blue/green corridors and networks).

Urban stormwater runoff is a significant component of non point-source water pollution, being the third most important factor contributing to overall lake and river pollution. As it flows across the land surface, runoff water accumulates rubbish, detritus and chemical contaminants from roads and other hard surfaces. This may include toxic hydrocarbons from engine oil, diesel, garden pesticides and other 'domestic' chemical compounds. Sewers and drains too may overflow and contaminate water courses with organic pollutants. Rapid, high energy flows erode river banks leading to increased sedimentation of the water course, which in turn reduces light levels (irradiance) in the water column and alters the chemistry of the water, with adverse effects on aquatic plants and invertebrates. Canalizing and physically channelling waterways may help divert water at one location but rarely slows the rate of passage, and potentially can exacerbate problems for sites downstream.

Problems with water flow tend to occur during either of two scenarios: (i) intense storms with high precipitation rates but which occur over a relatively short period, or (ii) prolonged periods of moderate rainfall, which saturate the soil profile and natural aquifers, but then cause rivers and other watercourses to overflow. Either scenario may induce surface flows that exceed the drainage capacity of the urban network. Green spaces, though, provide storm attenuation 'services' to the urban matrix in a variety of ways:

- Vegetation intercepts intense precipitation, holds water temporarily within the plant leaf canopy thus reducing peak flow and eases demand on the city drains.
- In addition, vegetation mitigates flood risk by increasing infiltration into the soil and diminishing the potential for surface flow.
- Vegetation reduces the physical impact of precipitation on soil particles, helping to retain the integrity of the soil structure.
- Lessening surface flow stops rivulets from forming, minimizing soil particle migration (erosion).

Bare soil is very prone to erosion, especially if there is a slope. Plants are frequently used to stabilize the soil, and turf or woody plants that establish a strong root system quickly are indispensable in this regard. *Salix* (willow), *Cornus* (dogwood) *Ribes* and *Forsythia* are all woody plant genera that readily root from hardwood cuttings, and stem sections can be inserted into slopes directly during the dormant phase to encourage a rapid coverage of 'protective' greenery. Reeds, bulrushes and aquatic iris are used to similar effect on river banks in an attempt to stop bankside erosion.

In surface runoff, green spaces reduce the dissolved pollution load in the surface water-flow, by aiding the capture of soil particulates and any embedded contaminants; the larger the single area of green cover the greater the reduction of pollutants. The ability for vegetation to act effectively against flooding events is influenced by factors such as rainfall intensity and duration, previous soil moisture content (saturated soils allowing little capacity for further storage, whilst excessively dry soil inhibits infiltration and encourages surface runoff), canopy morphology, plant type and species, and the presence or absence of leaves on the branches. Deciduous woodlands, for example, can have very different hydrological flow patterns between winter and summer.

The increasing use of vegetation as part of sustainable urban drainage systems (SUDS) demonstrates the significant role vegetation plays in providing solutions to runoff-related problems. Increasingly, more sustainable and integrated approaches are adopted to reduce surface flow rates, temporarily retain floodwater and increase infiltration rates to the parent soil. This not only includes using vegetation to trap and slow rainwater runoff, but in addition encourages the evaporation and transpiration of water back into the atmosphere directly. In essence this enables the storage capacity of the soil to become 'recharged' quickly. SUDS also have the advantage of incorporating swales and raingardens into their design, and these divert water away from the main drainage channels by facilitating a period of temporary flooding on land that specifically accommodates this.

Rates of runoff in housing with larger gardens has been found to be one-third less than that from dense urban development (Pauleit and Duhme, 2000). Using equivalent 9 m^2 plots, Armson *et al.* (2013) showed that turf almost completely eliminated runoff, whereas a tree planted within asphalt reduced it by 62% compared to control plots composed entirely of asphalt. Infiltration of water into the tree pit itself was thought to improve the retention rates associated with the tree plots, as the reduction in runoff was more than interception from the tree alone could have produced. Reductions relative to the canopy area were also greater than estimated by many previous studies, suggesting the planting pit was playing a pivotal role.

Trees and rainfall capture

Trees are often cited as one of the best forms of vegetation to reduce runoff, but various factors influence runoff between different plant species and morphological forms. Precipitation falling on a tree is characterized in three ways: (i) through-fall, which includes water passing through the canopy or water that is caught then drips off; (ii) stem-flow, which is water caught in the canopy that flows down the branches and trunk; and (iii) interception, which is water that is held in the canopy and subsequently lost via evaporation or absorbed and utilized in cell hydrology and/or enters the transpiration stream.

In addition to variations in the precipitation itself (raindrop size, frequency, duration wind direction etc.) the morphology of the tree, including size/ spread of crown, trunk surface area (rough and smooth bark types), branch structure and number of branch/leaf nodes, foliar characteristics including leaf size, shape and texture, leaf distribution and density all affect moisture retention (Fig. 2.12). Trees also dictate other important indirect factors such as the composition of the herb layer, depth of leaf litter and soil structure. Leaf litter, for example, helps act as a 'sponge' for moisture on the ground. Soil and plant water status determine the proportion of water that runs off the soil surface and how much is absorbed by soil and roots. The presence of a leaf canopy in winter is important; broadleaved evergreen trees intercept 27% of annual gross rainfall compared to a deciduous species (15%), largely due to difference in winter precipitation capture (Xiao and McPherson, 2011) (Fig. 2.13). Similarly in practical terms during winter, an evergreen conifer such as *Pinus wallichiana*, with a large total surface area due to its needle-like leaves, may have considerably greater capacity to store rainwater than, say, a pollarded *Populus* or *Salix* with only a few, bare vertical branches present.

Trees seem to be particularly important in urban areas as the soil's capacity to absorb water when acting alone may be compromised by both compaction and soil sealing. Overall though, when modelling runoff, infiltration has been deemed more important than evapotranspiration, although such processes are interrelated. Changes in soil infiltration capacity give large impacts on the total water balance and ponding. Water losses through evapotranspiration can cause small relative impacts on

Fig. 2.12. Variation in leaf size, branch arrangement and canopy density between tree species affects the amount of precipitation that is intercepted or falls through.

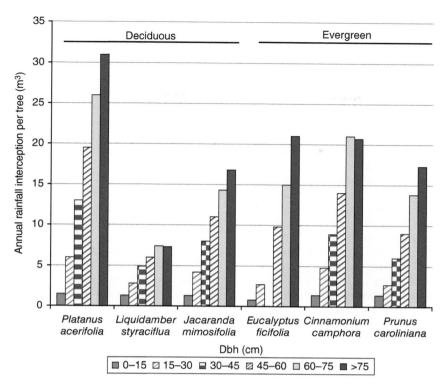

Fig. 2.13. Tree size and character (leaf/branch habit and duration of leaf retention) affect rainfall interception. Larger trees (increasing diameter at breast height, dbh) capture more rainfall, but also note differences between deciduous species). (Modified from Xiao and McPherson, 2002.)

the total volume and water balance, but the difference is deemed insignificant for the capacity of the stormwater system (Berggren et al., 2013). In an attempt to capture and retain rainwater in street situations the use of 'structured soil' in tree planting pits has become prevalent. These structured soils, mix gravel aggregates with soil to improve the drainage and storage characteristics of the medium within the pit. They tend to improve porosity and moisture-holding capacity over often poorly structured, urban soils.

Turf and surface runoff

In developed countries, particularly those with temperate climates (although also some with warmer climates such as USA and Australia), increased urbanization through extensive, but low-density, development has correlated with a net increase in the land surface area covered by lawn. It is estimated that the total area of turf grass in the USA, including residential, commercial and institutional lawns, golf courses, and parks, occupies 163,800 km^2 (±35,850 km^2) (Milesi et al., 2005). This area of turf grass helps alleviate some of the problems associated with the comparable increases in impermeable soil coverage (asphalt, concrete, stone paving etc.) brought on by urbanization. As with other vegetation, grass foliage intercepts and absorbs raindrop impact and provides resistance to water runoff. The grass sward protects the soil structure, reduces chemical movement and filters out pollutants from runoff. Associated microbial activity in the rhizosphere has the potential to degrade organic chemicals relatively rapidly, thus improving the runoff water quality and reducing groundwater contamination. Lawn management, however, can impact on the benefits, for example many lawns are over-fertilized with resultant potential for leaching of nitrate and phosphate ions, especially if fertilizer application precedes heavy irrigation or precipitation.

It is due to such factors that nitrogen pollution of water courses can be higher in residential areas than equivalent rural forested areas, or even agricultural land (Kaushal et al., 2008). Urban development results in the stripping-off of topsoil, leaving a compacted subsoil characterized by a low proportion of organic matter present and a poor pore structure – factors that reduce water retention and infiltration. Rainfall simulations have shown that runoff and pollution levels are influenced markedly by whether a lawn is established on topsoil or subsoil, and how lawns are fertilized. In comparison, it took 25 min for runoff to become apparent on lawns laid on topsoil, compared to only 13 min for those laid above subsoil (Cheng et al., 2014). The total volume of runoff water and sediment loss were significantly larger in subsoil lawns too (11,900 ml and 3700 mg, respectively) compared to that in top-soil lawns (2200 ml and 900 mg, respectively). Although nutrient losses were relatively low in this study, leachate values were still significantly higher from subsoil than topsoil lawns, and from lawns treated with inorganic fertilizer compared to those applied with organic forms. Clearly, amelioration of poorly-structured urban soils before the establishment of vegetation pays dividends in terms of off-setting problems with nutrient and water applications.

Green roofs/walls and impact on runoff

Green roofs are frequently advocated due to their ability to detain and retain stormwater flows. Some studies show retention rates of up to 70% during light to medium precipitation events (i.e. when soil medium is unsaturated) (Hutchinson et al., 2003). The ability to detain/retain water depends again on factors such as water-holding capacity of the substrate (affected by substrate type, structure and depth), precipitation intensity and the amount of rainfall in the preceding period. Deeper substrates provide an advantage in water-holding capacity (although there are implications for the weight load on the roof with deeper substrates). Comparisons of green roof modules (the trays in which moisture reservoirs, substrate and plants are normally held within) indicated that depth of substrate was important. Studies in the absence of vegetation demonstrated that the amount of precipitation intercepted and accumulated increased with deeper substrate, for example 200 mm depth captured 83% of precipitation, whereas 120 mm depth only stored 63% (Nardini et al., 2012). Developing green roof substrates that are lightweight, yet provide good characteristics in terms of water-holding capacity and water supply (to the plants), remains a key objective for this sector.

The contribution of vegetation per se on green roofs remains unclear. VanWoert et al. (2005) reported that vegetation contributed relatively little to the amount of runoff reduction by green roofs. In contrast, Schroll et al. (2011) considered that the vegetation had a significant effect on runoff when compared with substrate-only roofs. Dunnett et al. (2008) indicated that short or prostrate plants appeared to shed the greatest amount of water, whereas the broader-leaved grasses intercepted and retained the largest volumes. Lowest runoff rates were associated with species such as *Silene uniflora*, *Anthoxanthum odoratum* and *Trisetum flavescens*, with a mixture of four grass species (*Anthoxanthum odoratum*, *Festuca ovina*, *Koeleria macrantha* and *Trisetum flavescens*) also showing promise. Studies comparing modules planted with shrubs and modules planted with herbaceous perennials indicated that both types intercepted and stored more than 90% of rainfall during intense precipitation events, with no significant difference between the two vegetation types despite different substrate depths (Nardini et al., 2012). Larger plant canopies may hold more water on their surfaces, but active root systems allow for effective uptake into the transpiration stream. Indeed after a heavy rainfall event, the rapid depletion of substrate water through transpiration in itself re-establishes the substrate's capacity to cope with any follow-on rain events (Schroll et al., 2011). Defining plant species that have a large elasticity in stomatal control and that rapidly alter the volume of moisture transpired depending on the water available in the substrate would be useful in this respect for future roof designs.

Urban water use

Green spaces, and domestic gardens in particular, are associated with high volumes of water use, especially during dry weather. The proportion of potable water used for the domestic garden escalates with increased aridity, with water consumption being greatest in warm regions such as southern USA, Australia and southern Europe. In Western Australia, 56% of domestic water is within domestic gardens (Syme et al., 2004). In Spain 30% of total household water use is consumed within the garden, and this rises to 50% during the summer (Domene and Sauri, 2006). Estimates for the USA are even higher; landscape irrigation can vary from 40% to 70% of household water use (St Hilaire *et al*., 2008).

Recent plantings are particularly dependent on irrigation water to help new roots develop and allow the plant to establish effectively. During transplantation from a nursery field-site trees have been known to suffer up to 95% root loss, due to the vast majority of the root volume being associated with the numerous fine tertiary roots at the periphery of the root system. These are the roots that are lost through the excavation of the tree during transplanting. Once removed from nursery soil the remaining roots need to access large volumes of water to meet the demands of the canopy, once the plant is *in situ*. With container-grown specimens, the open, porous growing media frequently used will only provide moisture for a short period of time post-planting. Some of these growing media too, such as peat, become hydrophobic if they dry out, thus accentuating the problems of absorbing water from the surrounding soil. Shrinkage of the pot medium within the soil may mean that roots need to cross an air gap to access the parent soil, a process that is unlikely unless the gap is filled with water, or at least air of a very high humidity. Where evapotranspiration rates are high, permanent irrigation infrastructure (pipes, drip nozzles, pumps and weather or evaporation based controllers) will be required to ensure the new landscape plantings have longevity.

Climate change will increase demand for garden water in temperate regions. There will also be enhanced competition for this resource from industrial processes, food production and other domestic requirements, under most future scenarios. On a more positive note, it is likely that many landscapes are currently over-watered compared to the need to maintain ecosystem services such as carbon regulation, climate control, and preservation of aesthetic appearance (St Hilaire *et al*., 2008). Garden landscapes may be over-watered primarily due to the fear of losing plants to drought, but unintentionally this practice may pre-dispose specimens to greater susceptibility to stress long term, due to limited opportunities for the plants to express adaptive traits, such as deep rooting, effective stomatal control and cellular osmotic adjustment. Changes in management practice that facilitate such strategies lead to savings in water use. In lawns, organic versus intensive style of turf management can reduce water consumption up to tenfold. Mulching flower and shrub borders too minimizes moisture evaporation and has the added benefit of limiting gaseous emissions from the soil (notably nitrous oxides).

Water use has been correlated with affluence. Larger residential properties in Barcelona, Spain, where the owners could afford to water gardens extensively, developed a 'lush', 'Atlantic' style of gardening, with high water-demanding species being prominent (Domene *et al*., 2005). In contrast, smaller, less affluent properties were associated with vegetation more typical of the arid Mediterranean region. This results in neighbourhoods within the one city, where not only the landscape character is radically different, but so too is the range of wildlife that is supported. This ecological artefact is being driven, simply, by those house-owners who can afford the cost of water and those who cannot.

Climate change is likely to increase the demand for water use in green space. Water consumption for use in UK domestic gardens was estimated at 6.3 l per person per day in 1991; this is likely to rise to 15.9 l in 2021 even without taking account of climate change effects (Bisgrove and Hadley, 2002). In many parts of the world the pressure on water resource will increase with climate change, as rainfall patterns become more unpredictable. For example, the UK Climate Impacts Programme forecasts that summer rainfall in the southern UK will fall by 30–40% by 2090 while winter rainfall may rise by 30% (Hulme *et al*., 2002). Consequently, if droughts become more prevalent water storage will be necessary to maintain the benefits of green infrastructure. Water availability will become a key issue in heat island and climate change mitigation, and is an important consideration in the efficacy of urban planting to keep the locale cool. Stored supplies could aid urban cooling through evaporation from ponds, rills, and/or SUDS (Box 2.1). Use of stored runoff water in preference to mains-supplies saves energy too, as there is no energy-intensive process required to clean the water to the standard required for drinking water. Increasingly, water storage may be incorporated into new homes or at a local community level but for most existing homes storage is currently limited to water butts.

Grey water

There are opportunities to irrigate urban green spaces with grey water. Grey water is waste derived from domestic water use, but excludes sewage

Box 2.1. How garden design in the UK may change with climate change, and a desire for more sustainable use of resources. Scenario based on an East Anglian garden, east England, with: Representative Concentration Pathway (RCP) 8.5, radiative forcing rising to 8.5 W m^{-2} in 2100; mean temperature 5°C warmer than current; moderate winter precipitation with limited summer precipitation, and soil moisture deficits common in summer; frost uncommon and rarely below −4°C; growing season longer than current. RCPs are a standard set of scenarios depicting the impacts of climate change and use consistent starting conditions, historical data and projections across the various branches of climate science.

Key to Garden Features

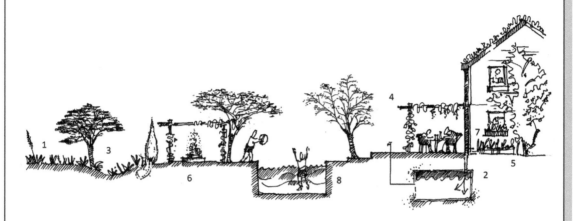

- Some lawns converted to dry steppe meadows, with bulb species used to extend the flowering period. Dry grasses and other stems and seeds heads are used to provide form in mid- to late summer after the peak flowering period. Spot plants include *Agave americana*, *Eremurus* and *Echium* spp. (1).
- Grass lawns replaced in some gardens by artificial synthetic lawns in an attempt to replicate the verdant appearance of the traditional lawn, and provide a low maintenance playing area for children. Provides poor value for local wildlife, however.
- In other gardens the lawn has also disappeared, to be replaced by gravel beds and hardy 'cornfield' annuals; the latter providing peak flowering May–June.
- Down-pipes from the house roof connected to an underground tank to store rainwater runoff that can be used in the garden during summer (2).
- Trees planted with a tubular watering pipe penetrating into the rootball to provide deep watering during hot periods. Although this encourages localized rooting around the pipe, the benefits outweigh the disadvantages.

- Plants bought from the garden centre and nursery will be preconditioned to drought stress during the production phase (but brought back up to full water status before shipping to the sales bench). Drying-down the nursery plants in a controlled way preconditions plants to subsequent drought phases and aids survival after planting in the garden; the establishment phases being the most vulnerable time for many ornamental plants.
- Gardens are screened to the south and west to provide midday shade. Drought-tolerant species such as *Citrus*, *Cercis siliquastrum* (Judas tree), *Prunus dulcis* (almond), *P. persica* (peach) and *Olea europaea* (olive) are commonly used. In some gardens these are located in shallow scrapes of which recycled grey water from the Local Authority communal tank is added when required (3).
- Patios are used more frequently, and have become important hubs for the social activities of the household. Shade is provided by species such as *Catalpa*, *Koelreuteria*, *Paulownia* and *Pinus pinea* where moisture availability allows (4).

Continued

Box 2.1. Continued.

- Wall climbers are planted around the air conditioning units of houses to improve their cooling efficiency further. A mixed collection of *Bougainvillea*, *Ipomea learii*, *Jasminum* and *Wisteria* provide both shade and evapotranspirational cooling to the unit. These are again linked to irrigation systems using recycled water to ensure plants remain adequately watered during heatwaves and continue to provide a cooling service to the house (5).
- Shallow swales and depressions are used to recreate 'dry' riverbed landscapes, with prominent use of gravel, stones and 'drift' wood. Planting is infrequent, but used to good effect to promote form (e.g. grasses such as *Miscanthus sinensis*) and brief interludes of colour via flowers (e.g. *Eschscholzia caespitosa*).
- Although water will be a challenging feature to manage due to high evaporation rates, rills, small pools, bubble fountains etc. become increasing important to provide relief from the heat and dryness experienced in summer. Windbreaks and sheltering walls help ensure these areas minimize wind movement and hence reduce the amount of moisture lost as spray, or evaporated from the water feature (6).
- Colourful patio plantings and window boxes are located close to the house to help ease of watering. Plants currently semi-tender become mainstream. *Bidens* for example, flowers from June to December (7).
- Borders and beds are more frequently mulched to reduce soil temperatures, and inhibit evaporation of moisture from the soil. Mulch will include organic materials, e.g. pine bark, but also inorganic materials such as pebbles, crushed glass etc.
- Shade becomes a more important element in the garden to help mitigate effects due to dry soil (although there will be less cloud cover – solar irradiance itself won't increase). Nevertheless, drought-sensitive species will be given some respite by providing shade/semi-shade especially during the middle of the day.
- More gardens accommodate a swimming pool to capitalize on the warmer summers now prevalent (8).

Plantings

- Warmer temperatures increase the range of plants that can be grown, although this is somewhat confounded by greater incidence of soil moisture deficit in summer. Evergreen sclerophyllous species (e.g. *Arbutus* spp., *Eucalyptus* spp. *Garrya elliptica*, *Laurus nobilis* and *Quercus ilex*) and smaller or pinnate leaved-species (*Acacia dealbata*, *Albizia julibrissin*, *Cytisus battandieri* and *Tamarix* spp.) as well as drought-tolerant pines (e.g. *Pinus aleppo*, *P. halepensis*, *P. pinaster* and *P. pinea*) and palms (e.g. *Phoenix canariensis* and *Trachycarpus fortunei*) become more common.
- Summer soil moisture deficits mean that conventional garden shrubs such as *Cotinus*, *Cotoneaster*, *Photinia* and *Syringa* have shorter internodes, and tend to be smaller and more compact plants than is currently the case. Even climbers such as *Clematis*, *Wisteria* and grape vines (*Vitis*) are less vigorous and have reduced shoot extension. This compaction, however, results in a more intense flower display, as the flowering nodes are not spaced so far apart.
- The drier climate not only reduces growth of plants, but increases the longevity of species adapted to such climates, such as Mediterranean shrub/sub-shrubs (*Ceanothus*, *Fremontodendron*, *Lavandula*, *Rosmarinus* and *Salvia* spp.) as growth is more in line with quiescent/dormancy phases, and there is less pathogen pressure brought on by wet weather.
- On light soils, mixed borders have lost their hybrid tea roses in favour of species roses and drought-tolerant climbers such as *Rosa* 'Veilchenblau', 'Alberic Barbier' and 'American Pillar'. Border dahlias have been replaced by *Osteospermum*. Grey and evergreen foliage becomes more prominent in mixed borders as greater reliance falls on Mediterranean shrubs and sub-shrubs.
- Herbaceous perennial borders still occur, but the range of species tends to omit those that require good levels of moisture such as *Aconitum, Astilbe, Dicentra, Dodecatheon, Galega, Heuchera, Hosta, Filipendula, Primula, Pulmonaria* and *Trollius*.
- Species such as *Ceratostigma plumbaginoides*, *Cotoneaster horizontalis*, *Hypericum olympicum*, *Rosmarinus officinalis* 'Corsican Blue', *Stachys byzantina* and *Vinca major* remain reliable ground coverers, but others such as *Ajuga*, *Bergenia*, *Polygonatum* and dwarf bamboos become less common.
- Rock gardens have lost many of their European and Asian alpine species, but now host xerophytes from Australia, North America and South Africa.

water. Normally it is collected from hand-basins, kitchen sinks, baths, showers, washing machines and dishwashers, but excludes waste from toilets. Grey water constitutes 50–80% of the total household wastewater. Water properties vary with the source from within the house. Kitchen grey water and the laundry grey water are higher in both organics and physical pollutants compared to bathroom grey water. Family lifestyle also strongly influences water volume and composition, providing variation between one household and the next. Grey water tends to show good biodegradability in terms of the chemical oxygen demand to biological oxygen demand (COD:BOD) ratios, but can have low nitrogen and phosphorous proportions (e.g. in the latter case if phosphate-free detergents are used).

Reclaimed grey water should fulfil four criteria – hygienic safety, aesthetics, environmental tolerance and economic feasibility – before it can be justified for re-use. A lack of appropriate water quality standards or guidelines, however, has hampered appropriate grey water re-use (Lazarova et al., 2003). Despite omitting waste stream from toilets, grey water frequently still contains faecal coliform microorganisms. Pathogenic organisms of concern associated with grey water re-use include enterotoxigenic bacteria *Escherichia coli*, *Salmonella*, *Shigella*, *Legionella*, and enteric viruses (Ottosson and Stenström, 2003). Faecal *Streptococci* and faecal coliforms are frequently assessed as indicator organisms to signal the presence of mammalian pathogens. Indeed, a number of studies have found that grey water is only mildly less contaminated (Casanova et al., 2001; Jefferson et al., 2004), and in some cases, more contaminated (Brandes, 1978) than waste that includes sewage. As it can contain hazardous agents, including pathogens and heavy metals, grey water should be treated before application to land where there is a risk of human contact.

The level of treatment and techniques employed to improve the quality of grey water vary with its potential re-use. For landscape irrigation, systems such as sand/gravel filtration or settlement and flotation are operated, but there is little data to determine how effective these processes are. To remove bacteria, biological systems tend to be used. These promote degradation of pathogens by the action of predatory or competitive microorganisms, and reduce the volume of organic content in the water as well as removing other pollutants too. Biological processes are often preceded by a physical pre-treatment step such as sedimentation, passage through septic tanks or filtering/screening. Biological action is encouraged by allowing grey water to flow over surfaces covered with beneficial microbial populations (biofilms), such as a membrane bioreactor or through pond/sludge systems. Holistic techniques that mimic natural processes, such as the use of constructed wetlands, are considered the most environmentally-friendly and cost-effective technologies for grey water treatment (Table 2.1), although even here coliform bacteria are reduced but not eliminated. Wetland plant species that aid in the remediating of grey water include *Phragmites australis* (common reed), *Juncus* spp. (rush), *Typha* spp. (bulrush, reedmace, cattail), *Scirpus* spp. (bulrush, club-rush), *Carex* spp. (sedge) *Salix* spp. (willow), *Glyceria maxima* (reed grass), *Iris pseudacorus* (yellow flag) and *Acorus calamus* (sweet flag).

How grey water is applied to the landscape will influence the risks involved. Overhead sprinklers should be avoided and even surface application may pose problems, whereas risks are likely to be reduced by application via sub-irrigation pipes or buried seep hoses. Investigations on turf grass irrigated with water containing bacteriophages (Enriquez et al., 2003) showed that there were

Table 2.1. The effect of a wetland treatment in reducing pollutant loads in grey water. (Source: Gross et al., 2007.)

Pollutant	Grey water	Effluent
Total suspended solids	158 mg l^{-1}	3 mg l^{-1}
Biological oxygen demand (BOD)	466 mg l^{-1}	0.7 mg l^{-1}
Chemical oxygen demand (COD)	839 mg l^{-1}	157 mg l^{-1}
Total nitrogen	34.3 mg l^{-1}	10.8 mg l^{-1}
Total phosphate	22.8 mg l^{-1}	6.6 mg l^{-1}
Anionic surfactants	7.9 mg l^{-1}	0.6 mg l^{-1}
Boron	1.6 mg l^{-1}	0.6 mg l^{-1}
Faecal coliform	5×10^7 / 100 ml	2×10^5 / 100 ml

lower transmission rates in sub-surface compared to surface irrigation. Even so, pathogens can accumulate in soil, for example *E. coli* is thought to be able to survive 230 days in warm soil, although these environmental conditions may not be as conducive for all potential pathogens.

The effects of grey water on plants and soil biology will depend on the constitution of the water, but common concerns relate to changes in soil pH (usually increasing pH), build-up of ionic levels (e.g. aluminium, sodium, zinc) and alterations in relative ratios of sodium to calcium and magnesium (the SAR ratio). Organic components may also disrupt/alter soil microbial populations. The phytotoxicity of grey water is mainly due to the anionic surfactant content that alters the microbial communities associated with the plant rhizosphere. Phytotoxic effects may vary between species. No negative effects of grey water have been noted for studies in plants, including *Lycopersicum esculentum* (tomato), *Lolium perenne* (ryegrass), *Beta vulgaris* var. *cicla* (Swiss chard) or *Daucus carota* (carrot), but kitchen and laundry waste waters have been shown to be toxic to both algae and *Salix* trees, while bathroom water was toxic to algae only (Eriksson *et al.*, 2006).

Increasing soil pH to above pH 9 resulted in a substantial reduction in transpiration rate in *Salix*. Non-replicated field observations of garden plants supplied with grey water have suggested that species such as those in *Chrysanthemum*, *Euonymus*, *Hibiscus* and *Juniperus* (juniper) have high tolerance levels, whereas *Prunus* spp., *Valeriana californica*, *Iris germanica* (bearded iris) and *Pinus mugo* (dwarf mountain pine) are intermediate in tolerance, with *Pinus sylvestris* (Scots pine), *Persea americana* (avocado) and *Citrus limonum* (lemon) showing higher sensitivity (Sharvelle *et al.*, 2012). In the case with *Lycopersicum*, plants irrigated with grey water showed no negative symptoms, but did increase uptake of key ions such as phosphate, sodium and iron compared to tap-water controls. Indeed, the application of grey water has been linked with a moderate fertilizer effect, with better growth compared to tap-water recorded in a number of plant species (e.g. Salukazana *et al.*, 2006).

2.6 Air Pollution

Although numerous reports cite the benefits of green infrastructure in mitigating aerial pollutants, it is one of the more controversial 'ecosystem service claims', in that research results are often not consistent. There is also a difference between scientifically significant results, and the ability to make a critical difference to urban pollutant loads in practice. The situation is made more complex in that there are a number of damaging air pollutants and the ability for vegetation to mitigate the effects varies with the particular particles or gases in question. Key urban air pollutants include: carbon monoxide (CO), volatile organic compounds (VOCs), nitrous oxides (NOx), sulfur dioxide (SO_2), ozone (O_3), benzene (C_6H_6), various heavy metals including lead (Pb), and particulate matter (PM). Particulate matter is a complex mixture of extremely small particles and liquid droplets, and constitutes a number of components, including acids (derived from nitrates and sulfates), organic chemicals, metals and soil or dust particles. Particles smaller than 10 μm (PM_{10}) are considered particularly damaging to humans as they generally pass through the throat and nose and enter the lungs. Many of the pollutants are generated by vehicle traffic as well as industrial processes. Plants are more effective at dealing with some pollutants than others, but even for a single element, trapping/absorption may be influenced by factors such as temperature, air flow, leaf presence and previous pollutant loading, as well as the plant species' morphology and physiology.

Numerous reports record the ability for green infrastructure to aid aerial pollutant capture (e.g. Jim and Chen, 2008), for example street trees absorb heavy metals, nitrogen dioxide, sulfur dioxide, and vehicle particulates. Climbing vegetation on walls has been found to trap particulate matter, reducing pollutants in the interior living space with the associated implications for human health (Ottelé *et al.*, 2010). More recent studies advocate that vegetation influences local air quality, if not necessarily at the city level. Although Pugh *et al.* (2012) acknowledge that green infrastructure might only reduce total pollutant levels by <5%, increased planting at a street level (street canyons where still air increases the concentration of air pollutants) can reduce pollutant levels in those canyons by as much as 40% for nitrogen dioxide and 60% for particulate matter. They argue that substantial street-level air quality improvements may be gained through action at the scale of a single street canyon or across city-sized areas of canyons. Models of urban air flows too suggest that cool air generated by green roofs flows down to street level and enters the street canyon, encouraging air movement at road level and dispersing pollutants within the canyon (Baik *et al.*, 2012).

Where there is insufficient space for trees, then green roofs may provide some role in mitigating local pollution levels (Speak *et al.*, 2012). These researchers focused on particulate matter of ≤10 μm (PM_{10}) and showed that effectiveness of capture was influenced by the distance of the green roof from the major pollutant sources. Responses between species varied due to differences in macro- and micro-morphology and the grasses *Agrostis stolonifera* and *Festuca rubra* were more effective than broadleaves *Plantago lanceolata* and *Sedum album* at PM_{10} capture. Modelling the performance of *Sedum* (the most common green roof species) in a scenario where it was used extensively on city roofs and where the city-centre area covered a total of 325 ha, it was shown that the annual potential PM_{10} removal was 0.21 tonnes, which is 2.3% of the 9.18 tonnes produced annually. In a similar manner, Yang *et al.* (2008) estimated that 1675 kg of air pollutants, such as NO_2, SO_2 and PM_{10}, were removed by 19.8 ha of green roofs in one year. A model for Toronto indicated that 58,000 kg of air pollutants could be removed if all the roofs in the city were converted to green roofs, with intensive green roofs having a higher impact than extensive green roofs (Currie and Bass, 2008). Intensive roofs have a deeper soil substrate than extensive roofs, which allows for a larger above-ground plant biomass and a wider variety of plants to be grown. While these studies offer promising results, they are based on modelling alone and, to date, relatively little empirical data on green roof removal of air pollution has been published.

Although it is clear that plants trap/absorb aerial pollutants, the key question becomes one of volume and how much green space is required to solicit a large enough impact to help reduce human health risks. Tallis *et al.* (2011), modelling the role of urban forests in London, indicate that the tree canopy removes 0.7–1.4% of PM_{10} (852–2121 tonnes) annually from the urban boundary layer. Increasing the urban tree component from 20 to 30% of total land area would only increase removal to 1.1–2.6% of the total by 2050 (1109–2379 tonnes). They suggest tree planting should therefore focus on the more polluted areas (e.g. beside roads) and use species that maximize particulate deposition all year round (e.g. coniferous species). Some research studies suggest that relatively wide belts of woodlands are required to elicit measurable benefits (Pataki *et al.*, 2011), at least for some of the pollutants encountered. Work on a forested, peri-urban national park near Mexico City showed that total annual air quality improvement due to park vegetation was detectable, but only accounted for approximately 0.02% of carbon monoxide, 1% of ozone and 2% of PM_{10} of annual concentrations for these pollutants (Baumgardner *et al.*, 2012). Setälä *et al.* (2013) claim the 'empirical evidence on the potential of urban trees to mitigate air pollution is meagre, particularly in northern climates with a short growing season'.

Their study monitored levels of nitrogen dioxide, anthropogenic VOCs and particle deposition using passive sampling systems in two Finnish cities. Concentrations of each pollutant in August (summer; leaf period) and March (winter; leaf-free period) were slightly but not significantly lower under tree canopies than in adjacent open areas. Furthermore, vegetation-related environmental variables (extent of canopy closure, number and size of trees, density of understorey vegetation) did not explain variation in pollution concentrations. They concluded that the ability of urban park/forest vegetation to remove air pollutants was insignificant, at least in northern climates. Yet, the claims that plants have a central role in mitigating urban air pollution is still one of the more common themes presented about their value to urban ecosystems, at least in the lay press.

Certain plant species may contribute to atmospheric pollution through being strong emitters of biogenic volatile organic compounds (BVOCs), such as isoprene and monoterpenes (Table 2.2). They have been implicated in contributing to the formation of photochemical smog and increasing levels of ozone in the urban environment (Peñuelas and Staudt, 2010). There is a complex relationship between BVOC emissions from trees and ozone formation and then ozone absorption by the trees themselves. Thus, the impact of trees which release BVOCs on urban ozone concentrations is difficult to estimate accurately. Nevertheless, trees that emit high BVOC concentrations may pose a risk and choice of tree species could be critical in determining the future quality of urban air.

Trees and grass of course, are emitters of one of the world's most common allergens – pollen. Pollen is the cause of hay fever (allergic rhinitis) which not only causes discomfort to millions of sufferers around the globe, but is also occasionally a precursor for more serious respiratory diseases such as asthma

Table 2.2. Variation in the form and mean concentration of biogenic volatile organic compounds (BVOCs) emitted from tropical and Mediterranean plant species. (Source: Bracho-Nunez et al., 2013.)

	BVOC ($\mu g\ g^{-1}\ h^{-1}$)				
	Isoprene	Mono-terpenes	Sesqui-terpenes	Methanol	Acetone
Tropical plant species					
Garcinia macrophylla	16.83	–	–	1.86	–
Hevea brasiliensis	–	21.33	–	2.20	–
Hevea guianensis	–	9.45	–	–	–
Hevea spruceana	–	52.50	–	–	0.66
Hura crepitans	–	–	–	20.16	–
Ocotea cymbarum	–	–	–	–	–
Pachira insignis	12.13	–	–	4.00	–
Pouteria glomerata	–	–	–	–	–
Pseudobombax munguba	–	–	–	13.53	–
Scleronema micranthum	–	0.35	–	1.53	–
Vatairea guianensis	63.20	–	–	6.05	–
Zygia juruana	13.73	–	–	–	–
Mediterranean plant species					
Brachypodium retusum	56.18	1.1	–	5.48	1.77
Buxus sempervirens	21.46	0.13	0.07	1.04	–
Ceratonia siliqua	0.52	7.94	0.03	0.79	–
Chamaerops humilis	18.93	0.14	–	–	–
Cistus albidus	–	0.3	0.63	8.2	–
Cistus monspeliensis	–	1.03	0.6	7.46	–
Coronilla valentina	0.43	0.75	–	13.48	0.61
Ficus carica	60.75	–	–	4.22	4.17
Olea europaea	0.12	1.21	0.09	1.03	0.11
Pinus halepensis	0.1	2.75	–	2.95	0.09
Prunus persica	0.04	0.36	–	4.01	0.32
Quercus afares	0.88	16.18	–	1.27	–
Quercus coccifera	0.64	5.77	0.24	4.11	0.25
Quercus suber	0.26	35.58	–	1.83	–
Rosmarinus officinalis	–	1.97	–	2.03	–
Spartium junceum	28.28	–	–	3.74	–

and allergic bronchitis. In some areas, one in four people are thought to suffer from hayfever. In these terms, trees and meadows must be seen to provide a considerable ecosystem disservice.

On balance, there is a high level of uncertainty about the role of urban vegetation in improving air quality (Pataki et al., 2011).

2.7 Pesticides and other Interactions with Chemicals

In many countries, maintenance of green space in the public realm is less reliant on chemical inputs than in the past. Increasingly, integrated control measures and more targeted applications of fertilizers and pesticides are aimed at improving environmental performance as well as reducing costs. This is perhaps in contrast to applications in the domestic context. Garden chemicals have widespread use, e.g. 50% of UK and 74% of USA homeowners use them (Grey et al., 2006). Their application, particularly on lawns, is apparently linked to home property values, with more affluent residents using them more frequently (Robbins et al., 2001). In the USA, local by-laws encouraging tidy well-maintained gardens increase the use of garden chemicals and actively promote intensive management (e.g. lawns regularly mown and weed free; Clayton, 2007). Urbanization has also increased the area of land devoted to lawns in the USA. It is estimated that nearly a quarter of urban cover in the USA is dedicated to lawn turf, which amounts to 164,000 km², three times the area of the most extensive irrigated crops

(Milesi *et al.*, 2005). This increase in the total area dedicated to lawns has coincided with greater emphasis on lawn maintenance with a subsequent expansion of fertilizer and pesticide use (Fissore *et al.*, 2011).

Garden and landscape pesticides are still significant contributors to non-point source water pollution in the USA. Nutrient and pesticide runoff contributes to eutrophication of rivers, lakes, and coastal zones. Different formulations vary widely in levels of toxicity, health effects and environmental impact. Risks involved are also dependent on correct or misappropriate use by home owners. Common garden pesticides with potential for mammalian toxicity include 2,4-D, dicamba, glyphosate and chlorpyrifos. Pesticides that are only mildly toxic to human health may be detrimental to the integrity of water resources, including the biological health of streams, fish, and macro-invertebrates. Direct application of pesticides may also be problematic. Neonicotinoids – a range of systemic insecticides (e.g. imidacloprid, clothianidin and thiomethoxam) – have been linked to dramatic falls in bee populations. Use in domestic settings may be higher (perhaps 120 times higher) than that used in commercial agriculture and horticulture. Direct contact with foliar neonicotinoid sprays is hazardous to pollinating insects, and residues can remain on leaf tissues for a number of days after spraying. As the chemicals are systemic they redistribute themselves around an individual plant, and chemical runoff is also thought to persist in the soil for a number of months. Despite this, clear evidence that neonicotinoids are directly responsible for bee declines is not evident, to date, and further research is warranted.

Most public green space has low fertilizer and pesticide use, but not all. The most intensively managed grass swards, such as bowling greens, golf courses (particularly golf greens and tees) and premier soccer pitches have higher than average chemical inputs (see Chapter 9). Kearns and Prior (2013) carried out a questionnaire on pesticide use on golf courses in Northern Ireland. They showed that almost 48% of the total area of golf courses assessed received a pesticide treatment in the year preceding the survey. Overall, treated areas of the course received an average of 2.2 kg ha^{-1} per annum, with most intensive application being on the greens with a mean of 7.5 kg ha^{-1} per annum. Herbicides comprised the largest component of pesticides (52%). Kearns and Prior (2013) commented that the rates of pesticide use were relatively high, for example higher than used in pasture management in agricultural systems (Fig. 2.14).

They highlighted three areas of risk/concern with these levels of application:

1. Although many of the pesticides used are estimated as low risk to humans, they noted that some golfers could access the sites quite shortly after application of the pesticides (21% of courses had removed 'sprayed areas' signs within 4 h of application) with potential for direct contact of the pesticide as golfers placed/lifted their ball from the turf.
2. Many courses had waterbodies present with the potential for applied chemicals to enter aquatic systems.
3. Almost 33% of all rough areas on the courses also received a pesticide treatment (mean 1.6 kg ha^{-1}); as these rough areas are potentially important as wildlife habitat in their own right, this was highlighted as a point of concern. Despite these datasets, many golf courses have tried to reduce their environmental impact over the last two decades due to original criticisms associated with high chemical use and non-sustainable use of irrigation water.

The surface runoff of pesticides, fertilizers and particulate matter from urban areas contributes to the pollution of water courses. Parts of the gardening community are, therefore, encouraging lower chemical use and organic approaches, largely due to concerns over these wider environmental issues. For example, the use of organic composts (green waste and spent mushroom) that release nitrogen and phosphates more slowly than liquid feeding with a commercially manufactured fertilizer, can reduce the risks of nutrients being washed out of the soil or potting media. Predatory species (e.g. spiders and hoverfly larvae) that predate on pests such as aphids are encouraged, not only by reducing the quantity of insecticides applied but also by using more imaginative design that promotes heterogeneous plant form and variable vegetation structure. Even the use of 'home-made' compost can increase overall invertebrate diversity, and provides opportunities for predators such as ground beetles, and these and similar 'holistic' measures have shown some success (Bell *et al.*, 2008). Furthermore, green spaces themselves reduce pollution through capture of particulates and dissolved pollutants.

2.8 Non-Native 'Alien' Species

One environmental disservice that has intense impact is that of introducing non-native plant species and plant pests/pathogens to new parts of the world. A large proportion of these have entered

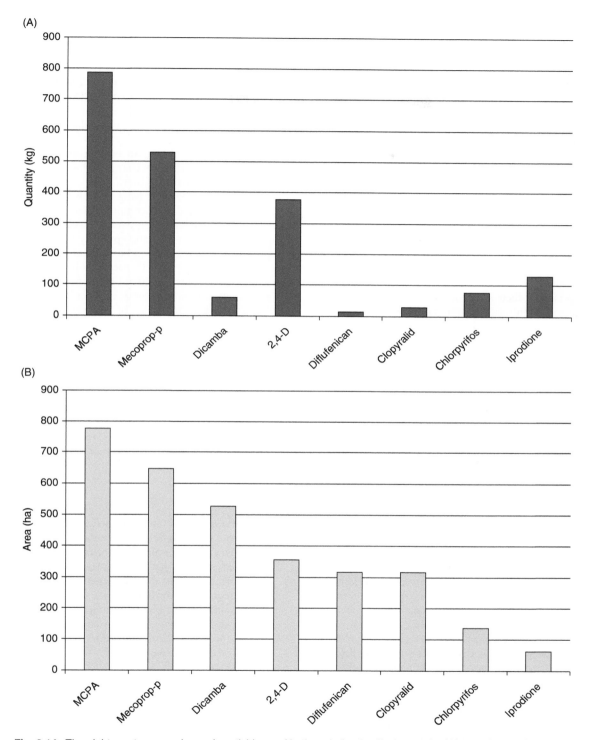

Fig. 2.14. The eight most commonly used pesticides on Northern Ireland golf courses, by (A) quantity and (B) area applied. The first six compounds are herbicides. (Modified from Kearns and Prior, 2013.)

through the garden trade. Many plant species stay restrained within the garden environment and rely on careful cultivation to maintain a presence in their new country. Others, however, can 'cross the garden fence' and become invasive alien weeds. Notable examples include *Acacia* spp. in South Africa, *Clematis vitalba* in Australia and New Zealand, *Solidago gigantea* in Japan, *Ailanthus altissima*, *Eichhornia crassipes*, *Lantana camara*, *Lonicera japonica* in the USA, and *Rhododendron* × *superponticum*, *Fallopia japonica* and *Impatiens glandulifera* in the UK, resulting in huge costs associated with eradication, e.g. over US$34 billion in the USA (Olson, 2006). Context is important here; some native species could also be described as invasive (e.g. *Pteridium aquilinum* (bracken) or *Cirsium arvense* (creeping thistle) in western Europe) and the arguments for and against the use of non-native flora vary based on the logic of defining species by their host country, a country's previous history of introducing non-native species and where such genotypes might be used (e.g. urban or rural areas). These arguments are outlined in more detail in Chapter 1.

Increasing international trade in plants also introduces non-native pests and pathogens. The ornamental trade is thought to contribute 90% of the human-assisted introductions of plant pests in the UK (Smith *et al.*, 2007). Most new plant pathogens tend to be found on ornamental plants (53%) with lower incidences on native species (15%) (Jones and Baker, 2007). The relatively high plant diversity found in urban environments, most notably gardens, provides opportunities for pest and pathogen species to move across to new host plant genotypes. Cultivated forms of wild plants commonly grown in gardens may have more invertebrate herbivores associated with them than has been recorded in the less-widespread, native wild form. With *Potentilla fruticosa*, for example, nine species of moth have been recorded feeding off the cultivated ornamental genotypes, but none on the native species itself (Owen, 2010). Increasing movement of people and goods, including plants and derived products across the globe, are contributing to greater incidences of pests and pathogens. This has corresponded to some countries tightening regulations about the movements of plants, including the requirement for sanitary certificates for imported plant materials, more inspections of incoming material by plant health officials and greater use of quarantine facilities to hold plants until they are shown to be free of key pests and pathogens.

In the UK alone, significant landscape tree species such as *Fraxinus excelsior* (ash), *Larix kaempferi* (Japanese larch) and *Pinus sylvestris* (Scots pine) are under threat from pathogens such as *Chalara fraxinea* (ash die-back), *Phytophthora ramorum* (sudden oak death) and *Dothistroma septosporum* (Dothistroma needle blight) respectively; the likely sources of initial infection coming in from overseas. Similar problems exist elsewhere.

The other driver for enhanced pest and pathogen incidences is climate change. Alterations in climate may be providing new niches for pathogenic organisms, or increasing the likelihood of pest species being able to complete their life cycle or colonize new territory. In parallel, such climatic shifts may induce greater susceptibility in the host plants, due to stress from abiotic factors (Cregg and Dix, 2001) – weakened specimens being more prone to pathogens. Greater occurrence of heat island effects and allied abiotic stresses have been linked to a higher incidence of pest species, such as *Podesia syringae* (clearwing moth) on *Fraxinus pennsylvanica* in the UK. Warmer winter temperatures and a lack of frost in urban centres means that certain tropical/semi-tropical pests, that were once restricted to protected environments such as glasshouses in temperate regions, now have the capacity to survive all year round outdoors. This includes the notable common pest species *Trialeurodes vaporariorum* (glasshouse whitefly), *Icerya purchasi* (cottony cushion scale) and *Tetranychus urticae* (red spider mite).

Conclusions

- Urban vegetation modifies temperature extremes; it provides cooling services in summer through shading, evapotranspiration, converting heat to chemical energy (photosynthesis), altering air movement, and through albedo effects. In winter, it reduces air movement over buildings and insulates the walls of buildings, thereby reducing heat loss, and improving energy conservation.
- Trees and other forms of vegetation can help block or diffuse noise as well as influence people's perception of noise. Further information is required on how urban landscapes can be designed more effectively to enhance this benefit.
- Urban vegetation sequesters atmospheric carbon, but the majority of carbon ends up fixed/stored in the soil. Hence, the ability for vegetation to help build up soil carbon levels is

an important factor in capturing atmospheric carbon. These advantages to carbon mitigation are offset by activities that use fossil fuels in the maintenance of green infrastructure, or encourage organic soils to oxidize and re-release CO_2 into the atmosphere.

- Vegetation is a significant component of sustainable urban drainage systems (SUDS), helping to reduce precipitation impacts on the soil, retaining and evapotranspiring rainwater, and reducing surface runoff flow rates. These aspects enable urban stormwater systems to cope better with high precipitation and flash-flood scenarios.
- The management of irrigation water will become increasingly important as urban climates warm, and pressure increases on potable water supplies. Alternative sources of water, such as grey water, are likely to be used more extensively, but with implications for plant and soil quality.
- Horticultural practices themselves need to be more sustainable in the future, with less reliance on non-sustainable raw materials and energy, and more considered selection of suitable genotypes for urban landscapes.

References

Akbari, H. (2002) Shade trees reduce building energy use and CO_2 emissions from power plants. *Environmental Pollution* 116, S119–S126.

Akbari, H., Kurn, D.M., Bretz, S.E. and Hanford, J.W. (1997) Peak power and cooling energy savings of shade trees. *Energy and Buildings* 25, 139–148.

Akbari, H., Pomerantz, M. and Taha, H. (2001) Cool surfaces and shade trees to reduce energy use and improve air quality in urban areas. *Solar Energy* 70, 295–310.

Anon. (2005) *Ecosystems and Human Well-Being*, Vol. 5. Island Press, Washington, DC.

Anon. (2010) *Global Forest Resources Assessment*. Food and Agriculture Organization of the United Nations (FAO) Forestry Paper 163.

Armson, D., Stringer, P. and Ennos, A.R. (2012) The effect of tree shade and grass on surface and globe temperatures in an urban area. *Urban Forestry and Urban Greening* 11, 245–255.

Armson, D., Stringer, P. and Ennos, A.R. (2013) The effect of street trees and amenity grass on urban surface water run-off in Manchester, UK. *Urban Forestry and Urban Greening* 12, 282–286.

Baik, J.J., Kwak, K.H., Park, S.B. and Ryu, Y.H. (2012) Effects of building roof greening on air quality in street canyons. *Atmospheric Environment* 61, 48–55.

Baumgardner, D., Varela, S., Escobedo, F.J., Chacalo, A. and Ochoa, C. (2012) The role of a peri-urban forest on air quality improvement in the Mexico City megalopolis. *Environmental Pollution* 163, 174–183.

Bell, J.R., Traugott, M., Sunderland, K.D., Skirvin, D.J., Mead, A., Kravar-Garde, L. and Symondson, W.O. (2008) Beneficial links for the control of aphids: the effects of compost applications on predators and prey. *Journal of Applied Ecology* 45, 1266–1273.

Berggren, K., Moghadas, S., Gustafsson, A.M., Ashley, R. and Viklander, M. (2013) *Sensitivity of urban stormwater systems to run-off from green/pervious areas in a changing climate.* NOVATECH 2013.

Bisgrove, R. and Hadley, P. (2002) *Gardening in the Global Greenhouse: the Impacts of Climate Change on Gardens in the UK*. The UK Climate Impacts Programme.

Blanusa, T., Vaz Monteiro, M.M., Fantozzi, F., Vysini, E., Li, Y. and Cameron, R.W.F. (2013) Alternatives to *Sedum* on green roofs: can broad leaf perennial plants offer better 'cooling service'? *Building and Environment* 59, 99–106.

Bracho-Nunez, A., Knothe, N., Welter, S., Staudt, M., Costa, W.R., Liberato, M.A.R. and Kesselmeier, J. (2013) Leaf level emissions of volatile organic compounds (VOC) from some Amazonian and Mediterranean plants. *Biogeosciences* 10, 5855–5873.

Brandes, M. (1978) Characteristics of effluents from gray and black water septic tanks. *Journal (Water Pollution Control Federation)* 1978, 2547–2559.

Cameron, R., Harrison-Murray, R., Fordham, M., Wilkinson, S., Davies, W., Atkinson, C. and Else, M. (2008) Regulated irrigation of woody ornamentals to improve plant quality and precondition against drought stress. *Annals of Applied Biology* 153, 49–61.

Cameron, R.W.F., Harrison-Murray, R.S., Atkinson, C.J. and Judd, H.L. (2006) Regulated deficit irrigation: a means to control growth in woody ornamentals. *Journal of Horticultural Science and Biotechnology* 81, 435–443.

Cameron, R.W.F., Taylor, J.E. and Emmett, M.R. (2014) What's 'cool' in the world of green façades? How plant choice influences the cooling properties of green walls. *Building and Environment* 73, 198–207.

Cameron, R.W.F., Taylor, J. and Emmett, M. (2015) A *Hedera* green façade – energy performance and saving under different maritime-temperate, winter weather conditions. *Building and Environment* 92, 111–121.

Casanova, L.M., Gerba, C.P. and Karpiscak, M. (2001) Chemical and microbial characterization of household graywater. *Journal of Environmental Science and Health, Part A*, 36, 395–401.

Cheng, Z., McCoy, E.L. and Grewal, P.S. (2014) Water, sediment and nutrient run-off from urban lawns established on disturbed subsoil or topsoil and managed with inorganic or organic fertilisers. *Urban Ecosystems* 17, 277–289.

Clayton, S. (2007) Domesticated nature: motivations for gardening and perceptions of environmental impact. *Journal of Environmental Psychology* 27, 215–224.

Cregg, B.M. and Dix, M.E. (2001) Tree moisture stress and insect damage in urban areas in relation to heat island effects. *Journal of Arboriculture* 27, 8–17.

Currie, B.A. and Bass, B. (2008) Estimates of air pollution mitigation with green plants and green roofs using the UFORE model. *Urban Ecosystems*, 11, 409–422.

Davies, Z.G., Edmondson, J.L., Heinemeyer, A., Leake, J.R. and Gaston, K.J. (2011) Mapping an urban ecosystem service: quantifying above-ground carbon storage at a city-wide scale. *Journal of Applied Ecology* 48, 1125–1134.

Domene, E. and Saurí, D. (2006) Urbanisation and water consumption: influencing factors in the metropolitan region of Barcelona. *Urban Studies* 43, 1605–1623.

Domene, E., Saurí, D. and Parés, M. (2005) Urbanization and sustainable resource use: the case of garden watering in the metropolitan region of Barcelona. *Urban Geography* 26, 520–535.

Dunnett, N., Nagase, A., Booth, R. and Grime, P. (2008) Influence of vegetation composition on run-off in two simulated green roof experiments. *Urban Ecosystems* 11, 385–398.

Elvidge, C.D., Milesi, C., Dietz, J.B., Tuttle, B.T., Sutton, P.C. Nemani, R. and Vogelmann, J.E. (2004) U.S. constructed area approaches the size of Ohio. *EOS Archives* 85, 233–236.

Enriquez, C., Alum, A., Suarez-Rey, E.M., Choi, C.Y., Oron, G. and Gerba, C.P. (2003) Bacteriophages MS2 and PRD1 in turfgrass by subsurface drip irrigation. *Journal of Environmental Engineering* 129, 852–857.

Eriksson, E., Baun, A., Henze, M. and Ledin, A. (2006) Phytotoxicity of grey wastewater evaluated by toxicity tests. *Urban Water Journal* 3, 13–20.

Fang, C.F. and Ling, D.L. (2003) Investigation of the noise reduction provided by tree belts. *Landscape and Urban Planning* 63, 187–195.

Fissore, C., Baker, L.A., Hobbie, S.E., King, J.Y., McFadden, J.P., Nelson, K.C. and Jakobsdottir, I. (2011) Carbon, nitrogen and phosphorus fluxes in household ecosystems in the Minneapolis-Saint Paul, Minnesota, urban region. *Ecological Applications* 21, 619–639.

Gidlöf-Gunnarsson, A. and Öhrström, E. (2007) Noise and well-being in urban residential environments: the potential role of perceived availability to nearby green areas. *Landscape and Urban Planning* 83, 115–126.

Gill, S.E., Handley, J.F., Ennos, A.R. and Pauleit, S. (2007) Adapting cities for climate change: the role of the green infrastructure. *Built Environment* 33, 115–133.

Grey, C.N.B., Nieuwenhuijsen, M.J., Golding, J. and ALSPAC Team (2006) Use and storage of domestic pesticides in the UK. *Science of the Total Environment* 368, 465–470.

Gross, A., Shmueli, O., Ronen, Z. and Raveh, E. (2007) Recycled vertical flow constructed wetland (RVFCW) – a novel method of recycling greywater for irrigation in small communities and households. *Chemosphere* 66, 916–923.

Heath, J., Ayres, E., Possell, M., Bardgett, R.D., Black, H.I., Grant, H. and Kerstiens, G. (2005) Rising atmospheric CO_2 reduces sequestration of root-derived soil carbon. *Science* 309, 1711–1713.

Howarth, R.W., Boyer, E.W., Pabich, W.J. and Galloway, J.N. (2002) Nitrogen use in the United States from 1961–2000 and potential future trends. *AMBIO* 31, 88–96.

Huang, Y.J., Akbari, H., Taha, H. and Rosenfeld, A.H. (1987) The potential of vegetation in reducing summer cooling loads in residential buildings. *Journal of Climate and Applied Meteorology* 26, 1103–1116.

Huang Y.J., Akbari H. and Taha, A.A. (1990) The wind shielding and shading effects of trees on residential heating and cooling requirements. In: *Proceedings of the Winter meeting of the American Society of Heating, Refrigerating and Air-Conditioning Engineers, Inc. Atlanta, Georgia*, p. 22.

Hulme, M., Jenkins, G.J., Lu, X., Turnpenny, J.R., Mitchell, T.D., Jones, R.G., Lowe, J. and Hill, S. (2002) *Climate Change Scenarios for the United Kingdom: the UKCIP02 Scientific Report*. Tyndall Centre for Climate Change Research, University of East Anglia, Norwich, UK, pp. 1–120.

Hutchison, B.A. and Taylor, F.G. (1983) Energy conservation mechanisms and potentials of landscape design to ameliorate building microclimates. *Landscape Journal* 2, 19–39.

Hutchinson, D., Abrams, P., Retzlaff, R. and Liptan, T. (2003) *Stormwater monitoring two ecoroofs in Portland, Oregon, USA*. City of Portland Bureau of Environmental Services Report.

Jefferson, B., Palmer, A., Jeffrey, P., Stuetz, R. and Judd, S. (2004) Grey water characterisation and its impact on the selection and operation of technologies for urban reuse. *Water Science and Technology* 50, 157–164.

Jim, C.Y. and Chen, W.Y. (2008) Assessing the ecosystem service of air pollutant removal by urban trees in Guangzhou (China). *Journal of Environmental Management* 88, 665–676.

Jo, H.K. (2002) Impacts of urban greenspace on offsetting carbon emissions for middle Korea. *Journal of Environmental Management* 64, 115–126.

Jones, D.R. and Baker, R.H.A. (2007) Introductions of non-native plant pathogens into Great Britain, 1970–2004. *Plant Pathology* 56, 891–910.

Kastka, J. and Noack, R. (1987) On the interaction of sensory experience, causal attributive cognitions and visual context parameters in noise annoyance. *Developments in Toxicology and Environmental Science* 15, 345–362.

Kaushal, S.S., Groffman, P.M., Band, L.E., Shields, C.A., Morgan, R.P., Palmer, M.A. and Fisher, G.T. (2008) Interaction between urbanization and climate variability amplifies watershed nitrate export in Maryland. *Environmental Science and Technology* 42, 5872–5878.

Kearns, C.A. and Prior, L. (2013) Toxic greens: a preliminary study on pesticide usage on golf courses in Northern Ireland and potential risks to golfers and the environment. In: Garzia, F., Brebbia, C.A. and Guarascio, M. (eds) *Safety and Security Engineering V*. WIT Press, Southampton, UK, pp. 173–182.

Kolokotroni, M., Giannitsaris, I. and Watkins, R. (2006) The effect of the London urban heat island on building summer cooling demand and night ventilation strategies. *Solar Energy* 80, 383–392.

Kovacs, K.F., Haight, R.G., Jung, S., Locke, D.H. and O'Neil-Dunne, J. (2013) The marginal cost of carbon abatement from planting street trees in New York City. *Ecological Economics* 95, 1–10.

Lazarova, V., Hills, S. and Birks, R. (2003) Using recycled water for non-potable, urban uses: a review with particular reference to toilet flushing. *Water Supply* 3, 69–77.

Liu, Y. and Harris, D.J. (2008) Effects of shelterbelt trees on reducing heating-energy consumption of office buildings in Scotland. *Applied Energy* 85, 115–127.

Livesley, S.J., Dougherty, B.J., Smith, A.J., Navaud, D., Wylie, L.J. and Arndt, S.K. (2010) Soil-atmosphere exchange of carbon dioxide, methane and nitrous oxide in urban garden systems: impact of irrigation, fertiliser and mulch. *Urban Ecosystems* 13, 273–293.

Lopes, A., Oliveira, S., Fragoso, M., Andrade, J. A. and Pedro, P. (2009) Wind risk assessment in urban environments: the case of falling trees during windstorm events in Lisbon. In: *Bioclimatology and Natural Hazards*. Springer, Netherlands, pp. 55–74.

McPherson, E.G., Simpson, J.R. and Livingston, M. (1989) Effects of three landscape treatments on residential energy and water use in Tucson, Arizona. *Energy and Buildings* 13, 127–138.

Meerow, A.W. and Black, R.J. (1993) *Enviroscaping to Conserve Energy: Guide to Microclimate Modification*. University of Florida Cooperative Extension Service, Institute of Food and Agriculture Sciences, EDIS, USA.

Milesi, C., Running, S.W., Elvidge, C.D., Dietz, J.B., Tuttle, B.T. and Nemani, R.R. (2005) Mapping and modeling the biogeochemical cycling of turf grasses in the United States. *Environmental Management* 36, 426–438.

Nardini, A., Andri, S. and Crasso, M. (2012) Influence of substrate depth and vegetation type on temperature and water run-off mitigation by extensive green roofs: shrubs versus herbaceous plants. *Urban Ecosystems* 15, 697–708.

Ng, E., Chen, L., Wang, Y. and Yuan, C. (2012) A study on the cooling effects of greening in a high-density city: an experience from Hong Kong. *Building and Environment* 47, 256–271.

Nowak, D.J. and Crane, D.E. (2002) Carbon storage and sequestration by urban trees in the USA. *Environmental Pollution* 116, 381–389.

Olson, L.J. (2006) The economics of terrestrial invasive species: a review of the literature. *Agricultural and Resource Economics Review* 35, 178.

Ottelé, M., van Bohemen, H.D. and Fraaij, A.L. (2010) Quantifying the deposition of particulate matter on climber vegetation on living walls. *Ecological Engineering* 36, 154–162.

Ottoson, J. and Stenström, T.A. (2003) Faecal contamination of greywater and associated microbial risks. *Water Research* 37, 645–655.

Owen J. (2010) *Wildlife of a Garden: a Thirty-Year Study*. Royal Horticultural Society, London.

Pataki, D.E., Carreiro, M.M., Cherrier, J., Grulke, N.E., Jennings, V., Pincetl, S. and Zipperer, W.C. (2011) Coupling biogeochemical cycles in urban environments: ecosystem services, green solutions and misconceptions. *Frontiers in Ecology and the Environment* 9, 27–36.

Pauleit, S. and Duhme, F. (2000) Assessing the environmental performance of land cover types for urban planning. *Landscape and Urban Planning* 52, 1–20.

Peñuelas, J. and Staudt, M. (2010) BVOCs and global change. *Trends in Plant Science* 15, 133–144.

Pouyat, R.V., Yesilonis, I.D. and Golubiewski, N.E. (2009) A comparison of soil organic carbon stocks between residential turf grass and native soil. *Urban Ecosystems* 12, 45–62.

Pugh, T.A., MacKenzie, A.R., Whyatt, J.D. and Hewitt, C.N. (2012) Effectiveness of green infrastructure for improvement of air quality in urban street canyons. *Environmental Science and Technology* 46, 7692–7699.

Reid, S.B., Pollard, E.K., Sullivan, D.C. and Shaw, S.L. (2010) Improvements to lawn and garden equipment emissions estimates for Baltimore, Maryland. *Journal of the Air and Waste Management Association* 60, 1452–1462.

Robbins, P., Polderman, A. and Birkenholtz, T. (2001) Lawns and toxins: an ecology of the city. *Cities* 18, 369–380.

Sailor, D.J., Elley, T.B. and Gibson, M. (2012) Exploring the building energy impacts of green roof design decisions – a modeling study of buildings in four distinct climates. *Journal of Building Physics* 35, 372–391.

Salukazana, L., Jackson, S., Rodda, N., Smith, M., Gounden, T., McLeod, N. and Buckley, C. (2006) Re-use of greywater for agricultural irrigation. In: *Proceedings of the Third International Conference on Ecological Sanitation*. Durban, South Africa.

Schlosser, W.E., Bassman, J.H., Wandschneider, P.R. and Everett, R.L. (2003) A carbon balance assessment for containerized *Larix gmelinii* seedlings in the Russian Far East. *Forest Ecology and Management* 173, 335–351.

Schroll, E., Lambrinos, J., Righetti, T. and Sandrock, D. (2011) The role of vegetation in regulating stormwater run-off from green roofs in a winter rainfall climate. *Ecological Engineering* 37, 595–600.

Selhorst, A. and Lal, R. (2013) Net carbon sequestration potential and emissions in home lawn turfgrasses of the United States. *Environmental Management* 51, 198–208.

Setälä, H., Viippola, V., Rantalainen, A.L., Pennanen, A. and Yli-Pelkonen, V. (2013) Does urban vegetation mitigate air pollution in northern conditions? *Environmental Pollution* 183, 104–112.

Sharvelle, S., Roesner, L.A., Qian, Y., Stromberger, M. and Azar, M.N. (2012) *Long-Term Study on Landscape Irrigation Using Household Graywater – Experimental Study*. The Urban Water Center, Colorado State University, USA.

Shashua-Bar, L., Pearlmutter, D. and Erell, E. (2011) The influence of trees and grass on outdoor thermal comfort in a hot-arid environment. *International Journal of Climatology* 31, 1498–1506.

Smith, R.M., Baker, R.H., Malumphy, C.P., Hockland, S., Hammon, R.P., Ostojá-Starzewski, J.C. and Collins, D.W. (2007) Recent non-native invertebrate plant pest establishments in Great Britain: origins, pathways and trends. *Agricultural and Forest Entomology* 9, 307–326.

Speak, A.F., Rothwell, J.J., Lindley, S.J. and Smith, C.L. (2012) Urban particulate pollution reduction by four species of green roof vegetation in a UK city. *Atmospheric Environment* 61, 283–293.

St Hilaire, R., Arnold, M.A., Wilkerson, D.C., Devitt, D.A., Hurd, B.H., Lesikar, B.J. and Zoldoske, D.F. (2008) Efficient water use in residential urban landscapes. *HortScience* 43, 2081–2092.

Susca, T., Gaffin, S.R. and Dell'Osso, G.R. (2011) Positive effects of vegetation: urban heat island and green roofs. *Environmental Pollution* 159, 2119–2126.

Syme, G.J., Shao, Q., Po, M. and Campbell, E. (2004) Predicting and understanding home garden water use. *Landscape and Urban Planning* 68, 121–128.

Tallis, M., Taylor, G., Sinnett, D. and Freer-Smith, P. (2011) Estimating the removal of atmospheric particulate pollution by the urban tree canopy of London, under current and future environments. *Landscape and Urban Planning* 103, 129–138.

Theodosiou, T.G. (2003) Summer period analysis of the performance of a planted roof as a passive cooling technique. *Energy and Buildings* 35, 909–917.

Van Renterghem, T., Hornikx, M., Forssen, J. and Botteldooren, D. (2013) The potential of building envelope greening to achieve quietness. *Building and Environment* 61, 34–44.

VanWoert, N.D., Rowe, D.B., Andresen, J.A., Rugh, C.L., Fernandez, R.T. and Xiao, L. (2005) Green roof stormwater retention. *Journal of Environmental Quality* 34, 1036–1044.

Volckens, J., Braddock, J., Snow, R. and Crews, W. (2007) Emissions profile from new and in-use handheld, two-stroke engines. *Atmospheric Environment* 41, 640–649.

Wang, F. (2006) Modelling sheltering effects of trees on reducing space heating in office buildings in a windy city. *Energy and Buildings* 38, 1443–1454.

Whitford, V., Ennos, A.R. and Handley, J.F. (2001) 'City form and natural process' – indicators for the ecological performance of urban areas and their application to Merseyside, UK. *Landscape and Urban Planning* 57, 91–103.

Wilby, R.L. (2007) A review of climate change impacts on the built environment. *Built Environment* 33, 31–45.

Xiao, Q. and McPherson, E.G. (2002) Rainfall interception by Santa Monica's municipal urban forest. *Urban Ecosystems* 6, 291–302.

Xiao, Q. and McPherson, E.G. (2011) Performance of engineered soil and trees in a parking lot bioswale. *Urban Water Journal* 8, 241–253.

Yang, F., Bao, Z.Y. and Zhu, Z.J. (2011) An assessment of psychological noise reduction by landscape plants. *International Journal of Environmental Research and Public Health* 8, 1032–1048.

Yang, J., Yu, Q. and Gong, P. (2008) Quantifying air pollution removal by green roofs in Chicago. *Atmospheric Environment* 42, 7266–7273.

Yilmaz, H., Toy, S., Irmak, M.A., Yilmaz, S. and Bulut, Y. (2008) Determination of temperature differences between asphalt concrete, soil and grass surfaces of the City of Erzurum, Turkey. *Atmósfera* 21, 135–146.

Yoshimi, J. and Altan, H. (2011) Thermal simulations on the effects of vegetated walls on indoor building environments. In: *Proceedings of Building Simulation 2011: 12th Conference of International Building Performance Simulation Association, Sydney, 14–16 November*. International Building Performance Simulation Association.

Zirkle, G., Lal, R. and Augustin, B. (2011) Modelling carbon sequestration in home lawns. *HortScience* 46, 808–814.

3 Green Space and Well-Being

Ross W.F. Cameron

> **Key Questions**
> - What is the biophilia hypothesis?
> - Does green space affect physiological health and opportunities for physical fitness?
> - What are the impacts of green space on psychological health and 'attention restoration'?
> - What health risks are associated with green space and its management?
> - How does proximity to green spaces, their size and quality affect any perceived health and well-being benefits?
> - What role does horticultural therapy have in the 21st century?
> - What is meant by 'social horticulture'?
> - How is a healthy diet influenced by environmental horticulture?
> - What are the features of restorative landscapes?

3.1 Introduction

Environmental horticulture is frequently associated with designing or managing a green space to improve opportunities for wildlife, or encouraging more sustainable approaches to the use of materials and management activities. Its *raison d'être*, however, is fundamentally about improving 'the habitat' for humans. This chapter explores the relationships between humans and green space and how the form of green space, and the activities that occur within it, can influence human well-being, including improvements to health (physical and psychological), social factors and self-fulfilment.

Humans rely on natural processes and systems to provide food, water, clothing, timber products, oxygen to breathe and sunlight to synthesize vitamin D. We also actively engage with nature to provide recreational opportunities or act as a medium for artistic or cultural expression. More of a controversial point, however, is do we *need* nature for our psychological and spiritual well-being? Is it theoretically possible for a human to live in an entirely artificial environment as long as their physical needs are met? Perhaps it is, but there is increasing evidence to suggest that individuals would be more prone to a range of physical and psychological disorders compared to someone who had greater exposure to natural processes and landscapes. This is the basis of the 'biophilia hypothesis', which indicates that exposure to nature is an intrinsic requirement for effective human development, and conversely disengagement with nature can lead to psychological problems. Wilson (1984) stated that biophilia is determined by biological needs and is an emotional or spiritual relationship between humans and nature. It has been cited as not only helping human development, including aspects such as cognitive skills, but facilitating a stronger appreciation of the natural world and an empathy for other living organisms (Fig. 3.1) – factors important to the long-term sustainability of natural resources, and hence, in the longer term, human survival.

It has been theorized that biophilia relates to our evolution, with human development accelerating once our ancestors moved out of forests and dwelt in the 'parkland-like' African savannahs. As much of our fundamental development as *Homo sapiens* relates to the savannah habitat, it is argued that we still have a close affinity with such landscapes, or components of them. Studies have shown that humans have preferences, even today, for open park-like landscapes and other iconic features typical of the

Fig. 3.1. Green space facilitates opportunities for direct engagement with nature, aspects considered important for both human psychological development, and for a wider understanding of ecological systems.

savannah. For example, these landscape features include flat top-trees (reminiscent of the thorn acacias – *Vachellia tortilis* – that epitomize the savannahs of East Africa), vantage points with vistas (mimicking the rocky outcrops or 'kopjes' – where game animals and/or potential threats could be viewed) and the element of water as typified by ponds and rivers (evolutionarily important as a source of drinking water and effective locations to trap/ambush game animals). Despite certain landscape features and environments being crucial in our development, this does not necessarily imply that these landscapes have a key role in determining human well-being now. There is increasing scientific evidence, however, to suggest they, or landscapes that mimic them in some form or another, do impact on how people feel and that they can influence health status and ultimately a sense of well-being (Fig. 3.2).

3.2 The Evidence for Green Space Affecting Health and Well-Being

Access to nature and engagement with natural/semi-natural and urban green landscapes are thought to directly influence physiological health, provide a range of positive psychological responses as well as enhance capacity for improved social relations. A number of cited positive benefits are summarized in Fig. 3.3.

Physiological health and physical fitness

Green space acts as a catalyst for physical activities, with a wide range of pursuits associated with natural and semi-natural landscapes (Fig. 3.4). These include formalized sports such as soccer, golf, baseball, rugby etc. but also more informal pastimes that include cycling, running, walking, gardening and conservation work. Conservation work, for example, involves physical exercise within a range of activities, notably scrub clearance, pond building, hay cutting, hedge laying and ditch clearing amongst others. These activities are linked to an enhanced health status and reduce the risks of significant 21st century diseases, including cardiovascular disease, stroke, type 2 diabetes, osteoporosis and certain forms of cancer (e.g. Mitchell and Popham, 2008; Coombes *et al.*, 2010). Moreover, exercising in green space per se appears to have additional benefits compared to comparable activities indoors, potentially through a range of psychological factors associated with the natural environment (Thompson-Coon *et al.*, 2011; Gladwell *et al.*, 2012). There is evidence that green spaces provide greater incentives for physical activity and those that participate in outdoor pursuits tend to maintain these activities for longer. In essence, they are seen less as a chore compared to the conventional forms of keeping fit, such as joining an indoor gymnasium or partaking in exercise routines at home. Indeed the term 'green gym' has been coined for those activities where people volunteer their labour for nature conservation work but gain a range of personal benefits in return – most notably letting individuals get back to physical and mental fitness through regular physical activity within an outdoor natural environment, such as a woodland or flower meadow. The social interactions, comradery, development of new skills and feelings of contributing to a worthwhile cause, typical of such

Fig. 3.2. Garden form often mimics nature. Humans have a preference for landscapes where two-thirds of the area is open to allow space for movement and good visibility, with one-third being composed of higher structural features for shelter and promoting a degree of enclosure. Winding paths and water features provide additional points of fascination.

activities, are thought to enhance further the positive psychological responses.

Gardening may have similar positive benefits in terms of promoting regular physical activity within a 'stimulating' environment. Gardening and similar activities may be popular (e.g. with the elderly) because they provide exercise in a non-demanding way, with little stigma attached to factors such as age, gender or level of physical fitness.

Exercise in the green space may motivate participants to undertake physical activities by increasing enjoyment and providing a better means to escape from 'everyday' problems and anxieties. As well as exercise per se they often provide social and entertainment value (Gladwell *et al.*, 2013). Exercise may also seem easier! A study by Focht (2009) showed that when participants were allowed to self-select their own walking speed, they walked faster outdoors than indoors, yet reported lower levels of perceived exertion. Such results are reinforced by other studies on walking and cycling (Akers *et al.*, 2012). Amazingly, in competitive sport too, the degree of green space may influence performance. DeWolfe *et al.* (2011) studied the impact of green landscaping on an athlete's performance and levels of cognitive and somatic anxiety using a population of 128 track and field athletes. Greener or more highly 'landscaped' arenas correlated with better individual performances. A greater proportion of athletes recorded their best performances at those stadia that scored highly for vegetation content, and conversely a higher proportion recorded their poorest performances at those sites deemed to have limited plant content. The authors attributed the results to the fact that the greener sites may have reduced the anxiety levels in the athletes to a greater extent before their performance, or indeed that it contributed to better levels of concentration and focus. Such results remain to be substantiated, but it does pose some interesting opportunities for

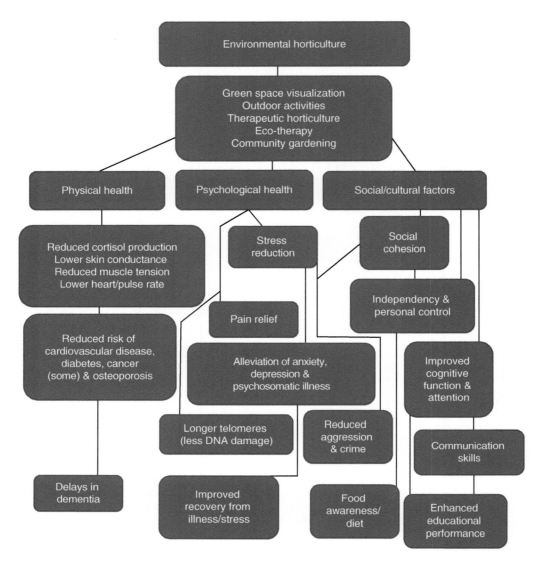

Fig. 3.3. Health, well-being and social benefits associated with environmental horticulture and green space.

the landscape sector, with respect to how soccer, athletic and even indoor sports arenas may be designed and planted-up in future.

The green landscape is becoming increasingly 'political', not least in terms of the politics, policies and social issues surrounding health. The opportunities afforded by green space, particularly with respect to accessible, user-friendly, safe green environments are seen as increasingly important factors in tackling sedentary lifestyles, obesity and associated diseases (particularly heart disease, stroke and cancer). In global terms, obesity is the fifth leading cause of death, and the number of people classified as clinically obese has doubled since 1980. In the UK, 20–25% of adults and 15–16 % of children are now considered clinically obese. This relates to issues around work and social trends where a greater proportion of time is dedicated to sedentary activities, particularly computer-based work and leisure activities, or time spent viewing television. For children the sedentary lifestyle increases with age, for example in England, data suggests that on average children of 4–7 years of age spent 6–7 h per day in sedentary activities, but for those between

Fig. 3.4. Both formalized and informal activity within green space can provide physical and psychological health benefits. (Lower image courtesy of Andy Clayden.)

12 and 15 years of age the sedentary time increases to 8–9 h per day. For leisure time with computer games, 41% of boys and 13% of girls reported they spend 2 or more hours per day on this activity alone. An increasingly sedentary lifestyle allied with other factors, such as a diet with a high fat content, are leading to problems with obesity and related illnesses. The cost of treating diseases linked to obesity exceeds £6 billion a year in the UK. As such, exploiting strategies that help prevent obesity and associated illness are deemed paramount and for some developed countries, policies specifically relating to green space are beginning to reflect this.

Increasing the levels of physical activity will enhance physical and cardiovascular-respiratory fitness, but even those activities that result in less physical exertion can have positive outcomes if maintained over prolonged periods. Gardening and other similar forms of active green pastimes provide both physiological and psychological health benefits, with prolonged engagement being particularly advantageous. Taken in their entirety, physiological health benefits associated with green space include:

- a more regular heart rate;
- reductions in blood pressure (Ulrich et al., 1991);
- reduced muscle tension (Tzoulas et al., 2007);
- lower skin conductance (Verlarde et al., 2007);
- lower cortisol (stress hormone) levels (van den Berg and Custers, 2010);
- lower urine adrenaline levels (Yamaguchi et al., 2006);
- reduced instances of digestive illness (Moore, 1981);
- enhanced physiological motor performance/ maintenance of mobility (Scholz and Krombholz, 2007); and
- improved cognitive function (Park et al., 2007).

As this body of evidence increases, health authorities are increasingly keen to adopt interventions that promote physical fitness. This may be as simple as encouraging citizens to partake in regular walking and cycling. Even here, however, the degree or extent of 'greenness' within the environment seems to have a role to play. Lachowycz and Jones (2011) for example, indicated that tree-lined walkways particularly, are associated with lower incidences of obesity, and people living in localities with a high proportion of tree-lined streets had lower body-mass indices than those in areas without street trees. (Street trees tend to be more common in affluent areas though and so there may also be a socio-economic factor to this finding.) For elderly people in Japan, access to green areas where they could walk has been shown to promote longevity of life (Takano et al., 2002). Other studies in Japan suggest that exposure to woodlands and forests (so-called 'forest-bathing') alter physiological indicators and improve immune responses. Groups of men were asked to visit natural forest settings or city environments for short durations (3 days) and conduct a number of walks in each. Exposure to the forest environments increased the activity and number of natural killer cells as well as perforin, granulysin, and granzyme A/B-expressing cells present in the subjects' bodies. Such cells are thought to enhance the body's immunological function and/or provide anti-cancer properties. The concentration of adrenaline in urine also decreased during the period in the forest.

These positive physiological responses were detected up to 7 days after the visit to the natural environments. In contrast, for those who visited the city there was no increase in natural killer cell activity or numbers, nor in the expression of selected intracellular anti-cancer proteins. Likewise in this 'city population', there was no decrease in adrenaline detected in the participant's urine. Rather than the observed responses being purely based on relaxation and reduced stress hormone production as most other studies imply, the researchers (Li et al., 2007) suggest that phytoncide chemicals (released from trees), such as alpha-pinene and beta-pinene, were detected in forest air, but not in the city air, and that these compounds may too partially explain the increase in activity of anti-cancer cells.

Conversely, arguments are made that ecological degradation and loss of green space is empirically associated with degradation in terms of both physical health (e.g. respiratory illnesses) and psychological health (e.g. depression).

Psychological health benefits

So a strong case can be made for the fact that green spaces could be used more effectively to encourage people into physical activity. But what about just viewing green or natural landscapes? The reality is that many of the benefits of green space are also stimulated by visual and other sensory responses associated with natural/semi-natural landscapes, even without the requirement for any physical activity. Moreover, providing artificial representations of natural scenes or green space, namely videos, photographs or paintings also elicit some positive psychological or psycho-physiological responses. Kahn *et al.* (2008) compared responses to real and artificial views of nature in an office setting. Using a population of 90 individuals they monitored reactions to vegetation viewed from a window to that of the same image projected via a plasma screen. A blank brick wall was used as a control setting. After a series of mental tasks the participants were seated within sight of the views and heart rate was recorded. Viewing the real scene through the window led to a more rapid recovery of heart rate compared to those individuals presented with the plasma screen, but in turn the plasma screen viewers recovered more rapidly than those viewing the brick wall. Both the window and plasma screen drew people's attention as frequently as one another, but the glass-window view held their attention for longer than the plasma screen.

Views of green spaces, gardens or natural landscapes such as rivers and forests affect our psyche. A study in Uruguay indicated that people living in tree-lined streets considered themselves happier and to have an improved social life, in comparison to those living in tree-less neighbourhoods (Gandelman *et al.*, 2012). For those suffering from moderate forms of depression, experience of nature can be a boon. Frequently, women, who have undergone surgery for breast cancer are prone to post-operative mental fatigue and depression, but research results illustrated that when these women undertook a restorative activity such as walking around a garden, patients showed higher rates of recovery from depression than control groups. Similarly, studies with other groups suffering mental health problems have demonstrated that walking can be beneficial, but the benefits can be enhanced when the surrounding environment is a green one (Roe and Aspinall, 2011). Evaluations of the 'stress hormone' cortisol in humans back up the notion that activities in green space reduce stress and anxiety. Cortisol levels dropped by 1.5 nmol l^{-1} for those participants who engaged with gardening after an imposed stressful event, compared to decreases of only 0.6 nmol l^{-1} for those who relaxed via reading after the event (van den Berg and Custers, 2010).

Theory is beginning to be translated into action. Natural sounds such as bird song and images of nature have been used to calm patients during surgical operations or to relieve pain (e.g. Hansmann *et al.*, 2007). Ulrich (2001) cites that viewing natural settings can produce significant restoration from physiological stress and anxiety within less than five minutes, as indicated by positive changes in blood pressure, heart activity, muscle tension, and brain electrical activity. As such, others argue that the presence of greenery around domestic dwellings, streetscapes and working environments are particularly important in reducing everyday stress loads (Kuo and Sullivan, 2001a, b; Grahn and Stigsdotter, 2003). Green spaces may have positive influences on physical development and agility and be a key environment for children's play. Children who engaged in forest 'kindergartens' (nurseries) have been deemed to have enhanced motor performance compared to those attending more conventional nursery settings, although factors such as duration of stay may confound some of these direct responses.

Others argue that urban green spaces are important components in engaging people with elements of the natural world, including those arenas that have particularly positive psychological impacts. Current issues include concerns about young people and their disengagement with nature due to changing lifestyles, perceptions about personal security in certain green spaces and loss of 'accessible' wild or semi-natural spaces. Yet, with respect to the first of these issues, evaluation of younger-aged city-dwellers showed that they responded positively to natural sounds such as birdsong. Young residents of urbanized areas in Sweden were exposed to various combinations of bird song, including:

1. Single species – *Passer domesticus* (house sparrow).
2. Single species – *Phylloscopus trochilus* (willow warbler).
3. Combined song from seven species – *Phylloscopus trochilus* (willow warbler), *Fringilla coelebs* (chaffinch), *Cyanistes caeruleus* (blue tit), *Parus major* (great tit), *Erithacus rubecula* (European robin), *Turdus merula* (blackbird) and *Dendrocopos major* (great spotted woodpecker).

Birdsong was played to the participants in different urban settings (three separate residential areas with varying amount of greenery). Results showed that urban settings combined with bird song were more highly appreciated than the settings alone. Increasing the number of bird species singing within the 'chorus' further enhanced the feelings of appreciation. The authors of the report (Hedblom *et al.*, 2014) claim that bird song contributes significantly to the positive values associated with urban green space. They conclude their study by urging planners to incorporate those habitats that maximize bird activity/diversity, so as to improve the recreational experience and positive emotions for citizens living within city zones.

The provision, location and type of urban green infrastructure can have surprising effects on the range of benefits encountered. For example, expanding the amount of roadside vegetation has a positive influence on car drivers, by allowing them to recover from stress more effectively, or to cope better with driving-related frustrations, anxieties and aggression (so-called 'road rage'). People also seem to be influenced directly by the loss of green space or its assets. The decline of 100 million trees across 15 states of the USA as a result of invasion by the emerald ash borer pest (*Agrilus planipennis*) has been linked to increased human mortality. The loss of trees is correlated with greater incidences of both cardiovascular disease (an additional 15,080 deaths) and lower-respiratory-tract illness (an additional 6113 deaths) (Donovan *et al.*, 2013).

As demonstrated above, much of the literature promotes the notion of green space being good for health and well-being, but does the natural/semi-natural landscape actually need to be green in colour? Although working with a relatively small population sample, Li *et al.* (2012) indicated that respondents exposed to a range of natural images (images of flowering plant communities of various monochromatic colours, e.g. red poppies versus purple/blue lavender) exhibited physiological improvements, namely decreases in systolic and diastolic blood pressure, heart rate, electrocardiogram readings and fingertip pulse rate and increased galvanic skin response, irrespective of the colour they were exposed to. Participants, however, who viewed green and purple/blue plantscapes had more positive psychological responses than those exposed to red, yellow, or white flowers. This was evidenced by lower ratings of irritability, fatigue and anxiety, and higher scores of vigour compared with the other groups. These results would appear to confirm that there is something unique to the softer, neutral hues of blue and green per se when eliciting restorative responses than perhaps other hues sometimes found in the landscape (see also Elsadek and Fujii, 2014). As such, the increasingly commonly used terms of 'green therapy' and 'blue therapy' (water/waterside activities) to describe health and well-being interventions are especially apt; no one yet has advocated 'red with a polka-dot of white therapy'! The sample sizes in these studies were small though, and this aspect of colour warrants further attention. Not least, to determine if there are any cultural or geographical differences; do people from countries where the flora are naturally more vibrant have the same preference for the more pastel blues and greens?

Li *et al.* (2012) suggested that landscape designers and horticulturists should give greater consideration to the choice of plant/plantscape colours used in different settings. They state:

> Green and purple plants can be used to create a relaxing environment for patients and medical workers and to improve energy, efficiency and self-confidence of employees in the office. In residential areas, public parks and university campuses, red and yellow plantscapes can be incorporated in children's activity areas, while green, purple and white plantscapes are appropriate in quiet recreational areas.

Psychological benefits of green space largely relate to the restorative effect from mental fatigue and stress (Kaplan and Kaplan, 1989). Restorative experiences figure in human emotional and self-regulatory processes through which individuals develop an identity of place. Studies by Korpela and Hartig (1996) indicated that individuals often went to their own personal favourite places to help them relax, calm down and clear their mind of any negative thought or events. These favourite places frequently incorporate elements of vegetation or water, or are landscapes of high scenic quality. Such 'preferred' places have coined the term 'therapeutic landscapes'. These restorative therapeutic landscapes do not need to be remote pristine wilderness per se, and 'everyday' natural environments such as local parks and open spaces (Bedimo-Rung *et al.*, 2005), street trees (van Dillen *et al.*, 2012), gardens (Cameron *et al.*, 2012) and fields and forests (Nilsson *et al.*, 2011; Park *et al.*, 2011) can provide respite from stress. The motivations to visit green space may not always be expressed simply in terms of seeking respite from stress factors, but this is often a consequence. Irvine *et al.* (2013) suggested that visitors to urban green space were motivated by a wide range of different factors, including opportunities for walking, 'getting fresh air', providing play opportunities for children, purposeful mental pursuits and social gatherings as well as providing informal, spontaneous leisure time. However, responses from the participants did not always align directly with their motivations. Responses included feeling relaxed, being revitalized, experiencing positive emotions, developing attachments towards a particular place, spiritual tranquillity and connection, cognitive satisfaction, an overall sense of health, and social connection. Interestingly, feeling relaxed was the most prevalent response, yet was not a priority for the visit in the first instance. In contrast, opportunities for social engagement were frequently quoted as a reason to visit the green space, but were then relatively rarely quoted as a key factor in the actual visit.

Comparisons between the restorative effects of outdoor green space and highly synthetic environments (urban, hard landscaped exterior areas and locations within buildings) also suggest there are relative benefits with respect to emotions such as anger, fatigue and sadness, as well as levels of attention and personal energy after there has been exposure to natural environments (Bowler et al., 2010). Although this review indicated positive effects on anxiety and tranquillity, there was more variance in these factors across different case studies, indicating that such responses can be promoted in certain types of man-made environments too.

The relationship between green space and relaxation/stress alleviation on societal health may, in reality, be as significant as that of promoting physical activity. Many forms of mental illness are activated by physiological stress. Short periods of stress are a natural response to dangerous or frightening situations, activating the hormone adrenaline to motivate the human body to deal with immediate threats. In contrast, frequent or prolonged bouts of stress are detrimental. These are linked to cardiovascular disease, hypertension, neuro-hormonal imbalance, type 2 diabetes, increased susceptibility to pathogen infection, anxiety and depression (Park et al., 2009).

Poor mental health is a growing problem in many developed countries. In the UK 18% of the population suffer anxiety disorders, depression, exhaustion or fatigue syndromes. One percent now suffer from a severe mental illness, such as bipolar disorder or schizophrenia. These mental health problems are estimated to cost £105.2 billion per annum in healthcare, welfare benefits and lost productivity. The reasons for these mental health problems are complex and varied, but general changes that may be contributing to poor mental health in society include:

- more time spent on sedentary activities;
- a more highly organized, pressurized and demanding urban lifestyle;
- less exposure to natural landscapes; and
- less opportunity for 'low-key' recreational activities.

A work environment that has in general terms moved away from active manual labour (often outdoors) to desk-based, mental tasks may be an additional factor (see section 3.3). Interestingly, work-related stress can also be diminished by access or viewing nature or natural features. A view of natural elements, such as plants and water features, off-sets some of the stress factors associated with the work environment. The inclusion of green/natural elements marginally improves overall general well-being and reduces the likelihood of employees feeling the need 'to quit their job'. In line with this, workers stated they had greater job satisfaction, more patience with colleagues and perceptions of better health simply due to the presence of plants.

Another perceived benefit of green space is its role in aiding social cohesion. Well-designed spaces contribute to feelings of safety and familiarity, and foster a sense of 'community'. For individuals living in inner-city apartment blocks, local green space provides an opportunity to develop stronger ties with neighbours and promote a greater sense of safety in their local environment. In some, but not all, residential areas, a greater proportion of green space has been linked to a greater sense of 'social safety'. The presence of local green space encourages people to leave the confines of their home and encourages social contact in a relaxed and informal way. These activities and improved social relations have further 'knock-on' positive effects particularly through increased informal surveillance of the neighbourhood, which in turn tends to reduce crime rates (Sullivan et al., 2004). Locally-managed, familiar and welcoming green areas are valuable too in that they help reduce the incidences of loneliness (and perceptions around a lack of social support) in elderly citizens and other vulnerable members of society.

Health benefits and underlying mechanisms

Many of the benefits of green space are linked with restoration from stress. Kaplan and Kaplan were the first to try and define the relationship between green space and physiological stress through their attention restoration theory (ART) (Kaplan, 1995). The ART states that nature or green space provides a restorative opportunity by acting at four levels: (i) allowing patients to 'be away' (removed from stress-inducing factors either physically or psychologically); (ii) promoting 'fascination' (effortless, interest-driven engagement with natural objects); (iii) providing 'extent' (scope for exploration and curiosity-driven discovery); and (iv) being 'compatible' (how closely the environment matches an individual's needs or inclinations at that time) (Table 3.1). In essence, nature tends to promote 'involuntary attention', where attention is captured by inherently intriguing or spontaneous stimuli, whereas 'directed attention' relies on intense, focused cognitive-driven processes. Prolonged or over-use of directed attention, though, leads to physiological stress, and involuntary (indirect) attention is important in providing respite from this. Grahn and Stigsdotter (2010) argue that how people collate and store information is important too. The human brain relies on three different processes. These are:

1. Subsymbolic processes (sensory, motor and somatic modes, as perceived by muscles, inner organs and skin).
2. Symbolic-imagery (visual pictures in a person's mind).
3. Symbolic-verbal (concepts discussed verbally and their interpretations).

Natural areas and green spaces have the advantage that they promote simple relationships that do not overload these sensory processes, particularly the symbolic-imagery and symbolic-verbal. In contrast, complex interactions (for example, such as people may experience during an intense conversation or argument) can overtax these processes. In effect, those environments (natural landscapes, wilderness,

Table 3.1. The key components to attention restoration theory. (Source: Kaplan, 1995.)

Component	
Being away	At first impression this seems to intimate the need to remove oneself away from the 'everyday' environments to those destinations that represent wilderness or the expansive qualities of the natural world – wildscapes, including mountains, coastlines, forests, meadows, lakes, streams etc. In reality, it is more closely aligned with the ability to 'escape' mentally through engagement with nature at a range of different levels and scales. In practical terms, it can mean involvement with green spaces that are more familiar or easily accessible (e.g. town park, garden or even a picture or photograph of a natural feature that becomes absorbing). In essence, any visual cues that offer an opportunity for respite from intensive, single-minded concentration, i.e. 'directed attention'.
Fascination	The ability of nature to provide absorbing processes and objects that help distract the mind from problems and anxious thoughts. Many of the fascinations afforded by the natural setting qualify as 'soft' fascinations: processions of caterpillars moving along the ground, shoals of swimming fish, cloud patterns, leaves blowing in the breeze, snow flake patterns, birdsong, sunsets etc. Attending to these patterns is effortless, and they do not overly tax the brain.
Extent	The impression of the vastness of nature – the ability to extend onwards. Again, in a similar manner to 'being away' although there are advantages with larger scales and extended wilderness, this aspect is about how the mind can extend its thinking. Even a relatively small area can provide a sense of extent. Trails and paths can be designed so that small areas seem much larger. Diversity of form or miniaturization are devices that can help create illusions of different worlds or timescales. Japanese gardens, for example, can mimic landscapes at a much grander scale, or remind viewers of cultural, symbolic or historical components. Landscapes include historical artefacts or geographical features that promote a sense of being connected to past eras and other environments and thus to a larger world.
Compatibility	For many individuals the natural environment is almost the 'default' environment. Functioning in the natural setting seems to require less effort than functioning in more 'civilized' urban settings, even though they have much greater familiarity with the latter. Kaplan and Kaplan (1989) relate this to the many 'patterns' humans use when relating to natural environments. These include a range of roles or activities, e.g. 'predatory' (hunting/fishing), 'locomotion' (boating/hiking/cycling), 'domestication/dominance of the wild' (gardening/caring for pets), 'observing' (bird watching, visiting zoos), 'survival' (fire building, constructing shelter), social interaction at meals (picnics, barbeques). People often approach natural areas with the purposes that these patterns fulfil already in mind, thus increasing compatibility.

parks, gardens) that optimize the simple processes are deemed to have greater restoration potential than those that require more complex sensory procedures (shopping malls, busy roadways, railway stations, airports, business and work environments) (Hartig et al., 2003).

Taking some of the arguments further, a subsequent question arises as to whether the type of green landscape itself has a strong influence over the ability to restore directed attention and reduce stress; indeed, are some landscapes more restorative than others? Despite Kaplan's claims that these restorative environments do not need to be remote wilderness, there is evidence to suggest that scale, form and quality of landscape may impact on the restoration potential (see sections 3.4 and 3.8).

Although there may be some underlying universal relationships between humans and their environment, inevitably there are individual and personal aspects of an individual's psyche that affects their relationship with the natural world. These too, may not be consistent. How people respond to a landscape varies with their own state of mental health and individual interests/perspectives. Although the benefits of green space are most often expressed with respect to individuals who have suffered some form of bereavement, mental breakdown or trauma, the benefits need not be exclusive to these circumstances, nor even work in the same way for people with similar experiences. Even individuals who have suffered trauma or prolonged stress may require different elements or components from the landscape as they proceed through progressive stages of recuperation (Stigsdotter and Grahn, 2002; Grahn and Stigsdotter, 2010). Similarly, people looking for restoration may have preferences that differ from those not suffering any mental health problems at all. Such studies suggest that individuals suffering from physiological stress had a preference for landscapes that promoted concepts of 'refuge' and 'nature'. Indeed, landscapes that provided 'refuge', 'nature' and were deemed 'rich in species', but had few 'social' components, were considered optimal from a restorative perspective for those suffering stress. Non-stressed individuals, on the other hand, had an affinity for landscapes that promoted feelings of 'serenity', and offered 'space' and 'nature'. Attributes associated with 'rich in species' and 'refuge' were considered less important. In populations of people either suffering or not suffering mental health problems, aspects relating to 'culture', 'prospect' and 'social' were least preferred.

Similar results were observed in follow-on studies investigating nine different small urban public green spaces in Copenhagen (Peschardt and Stigsdotter, 2013), with the exception that 'social' ranked quite highly. 'Social' was particularly high for one city site that was characterized by cafés and restaurants but limited green infrastructure, suggesting that the restorative components may not be exclusive to green elements alone.

Despite the large volume of data centred around ART and like-minded hypotheses, the notion that we as a species are attracted to green spaces, primarily out of some desire for mental restoration is challenged in some quarters. In their research, de Groot and van den Born (2003) indicated that significant sections of the population (≥50%) express preferences for landscapes where one experiences the 'full force of nature', with another third preferring 'wild landscapes' that they can directly interact with. These authors concluded that 'attractive' landscapes should be categorized by either visual or behavioural preference, i.e. some people are drawn to certain landscapes for personal interests, desires or to partake in their hobbies, e.g. active outdoor sports. These desires are not restricted to the truly adventurous, challenging and wild landscapes either. Even within relatively 'tame' urban parks the attraction may not always be for tranquillity and restoration, but perhaps rather for a sense of fun or intrigue. Visitors to urban parks were expected to experience greater pleasure in those landscapes that had a higher component of mature trees and less shrubby undergrowth present, as parks with these features were considered to be optimal for encouraging feelings of calmness and relaxation (Hull and Harvey, 1989). Preference within the city's parks, however, increased in line with feelings of pleasure and arousal rather than relaxation. Indeed, significant numbers of visitors preferred walking along paths with thick undergrowth which induced emotions of exhilaration and arousal as well as an element of fear. These arousal-inducing characteristics of the parks were very much opposite to the calming influence expected by the researchers. This indicates that although feelings of calm and relaxation are major components of people's emotional reactions to nature, more 'animated' responses of stimulation or awe are also important (Fig. 3.5). Ulrich (1990) sums up by stating: 'Enjoyment of green areas may help people to relax or may actually give them fresh energy.' This aspect of green space, however, has received a lot less attention

Fig. 3.5. Varying the typology of landscapes affects people's responses. Some landscapes may be seen as having a high restorative value (top, left), others social value (bottom, left) or even designed for exhilaration (right).

from researchers than the therapeutic components and warrants much more consideration.

Overall, the positive effects of green space on human health are now being advocated by health authorities and other policy makers. Increasing numbers of publications demonstrating strong evidence for the benefits of green space across a wide variety of health issues is resulting in a call for greater provision and access to green space, despite pressures on public budgets. In the UK, sections within both the National Health Service and private medical care industry back calls for more urban trees, parks, green roofs, community gardens and other forms of green infrastructure, based on the benefits to human health and well-being. Similar scenarios occur in other parts of Western Europe, the USA, Australia and Southeast Asia. Such voices are now 'mainstream' if not quite universal, and have moved on from being at the periphery of the health care system.

Health benefits – not proven?

So green is good for health? Is it that simple? Despite some uptake by mainstream medical authorities in promoting green space to maintain health ('prevention better than cure' approaches) and to help recovery from illness, the relationship between nature and human health/well-being remains controversial and not universally accepted (Lee and Maheswaran, 2011). Partly this is due to the evidence base being incomplete, and the difficulties in developing methodologies that categorically explain cause and effect. In terms of green space, it is not clear that all sectors of society respond in the same way. Cultural or educational background, type or severity of illness, age, occupation and personal interests may influence the extent of any benefits (Ottosson and Grahn, 2005).

In a Canadian study on teenagers (aged 11–17) the relationships between natural space and positive emotional well-being were weak and lacked consistency (Huynh *et al.*, 2013). Slightly more positive responses were observed for teenagers living in small cities, implying the environmental context (not least the predominant type of green space) may have some effect on the responses. Overall though, the data contrasts with that for adults and for other countries. Possibly, teenagers vary in their perceptions, use and interactions with green space compared to adults (not least because they may be seen as places to socialize, rather than relax; green space tranquillity may be viewed as 'boring'). The Canadian context itself may be a factor – many winter activities are indoor due to the climatic conditions, and local green spaces may not be deemed 'valuable' in the same way that more remote extensive 'wilderness' settings are. Other climatic factors such as prevalent temperatures, rainfall, light and day length have been suggested as moderating factors in other situations, although not always in an obvious, intuitive way. For example, local green space may be used more in wet weather, as alternative, more distant 'target' destinations are more popular on fair weather days. These generalizations also tend to overlook the fact that the specific type of green space may be important (see sections 3.4 and 3.8).

Poor or inconsistent correlations between green space and health have also been noted in studies in the Netherlands (Maas *et al.*, 2009a), the UK (Mitchell and Popham, 2007), Japan (Takano *et al.*, 2002) and New Zealand (Richardson *et al.*, 2010). In the case of New Zealand, it was considered that a lack of correlation between green space and mortality was explained by the fact that green space is generally abundant everywhere, with less social and spatial variation in its availability than found in other countries. Indeed, in common with the Canadian study, rural inhabitants may be indifferent to the benefits of green space due to it being 'commonplace'. Other contextual factors may be critical. The relationship between health/well-being and local green space seems to break down in more deprived areas, possibly due to reductions in quality of that green space. Geographical information systems (GIS) and other mapping systems used in a number of these studies are able to identify and locate extensive areas of inner city green space, but may omit more subtle details. These 'green oases' in the urban form may be represented in reality by flat expanses of grass, little variation in landscape typology, few trees, an abundance of litter and strong evidence of vandalism and other antisocial behaviour – aspects that are unlikely to make the sites attractive whatever the propensity of 'green'.

Indeed, optimizing the benefits of green space therefore may be contingent on improving the form, extent or quality of the green infrastructure (Mitchell and Popham, 2007). Similarly, they may depend strongly on the nature of the 'green' activity undertaken (Barton and Pretty, 2010) or ancillary factors such as ease of access, degree of motivation or perceptions of safety. Korpela *et al.* (2010) also suggest that although natural/semi-natural landscapes promote greater health benefits, there was no difference within a city context between green space (parks) and favourite places within the built infrastructure. Therefore, familiar places selected on the basis of preference or emotional attachment (whether they are green or not) may have some restorative value. In terms of memory recall, Hartig *et al.* (1991) found no difference in the ability to recall positive, negative or neutral memories between semi-natural (garden) environments and urban environments, despite improvement in mood in the former. Similarly, in their review of activities such as walking and running in natural and man-made/dominated environments, Bowler *et al.* (2010) found little difference in the environmental settings with respect to improvements in systolic and diastolic blood pressure and cortisol concentrations.

A key question that has arisen in recent years is: does the reason for your engagement with the green space influence your response to it? In stark contrast to the benefits casual users of green space may attain, Bingley (2013) claims that those working in such environments may not experience the 'restorative influences of the green idyll' or any positive effects may wear off over time. This can relate to risks posed by the vocation (machinery, working with or exposure to animals, inclement weather), or concerns/stresses generally experienced in line with other jobs. Concerns over low pay, lack of promotion opportunities, and job insecurity may undermine many of the benefits of working outdoors. Some green infrastructure jobs are also associated with long periods of isolation, a problem that is thought to contribute to high suicide rates in farmers.

Health risks

Despite the benefits being attributed to green space, the concept of some sort of universal Eden-like idyll should be challenged of course. Engagement with green space is not risk-free. One percent of US citizens suffer some sort of injury in their garden every month. Horticultural activities, including home-gardening, can result in injury due to certain pathogens, dermatitis, allergies and not least the misuse of tools. Powered machinery, particularly lawn mowers and hedge trimmers, contribute to injuries, commonly to the hand, foot or eye. Data from Dutch studies (van Duijne et al., 2008) indicate that 60% of lawn mower accidents relate to people cutting themselves on the blades or other sharp parts of the mower, which includes cases of users who touched the blades when the machine was operating, as well as during servicing of the machinery. Studies from the USA suggest that stones and other debris flying out from the blades are a significant source of injury, especially when hitting vulnerable parts of the body such as the eyes. Not surprisingly, children ≤15 years old had a proportionally higher risk of injury from mowers than adults. Men were also more likely to be injured than woman, although this data was on a total, not a proportionate basis (i.e. fewer women may operate lawnmowers).

Green activities such as gardening and conservation work are recommended as a means to keep fit, but they can also be a source of muscle or skeletal injuries, most notably in the elderly or those who undertake excessively long, repetitive or heavy lifting tasks without being physically fit first. Excessive digging, lifting, twisting, stooping, kneeling and squatting are the reasons for injury, with lower back pain being the most commonly cited source of pain. If precautions are not taken, gardening can be associated with increased risk of arthritic pain, heat stress, skin cancer and carpal tunnel syndrome (numbness in the hand and wrist). Exposure to pollen induces allergenic reactions; both tree species and grasses can elicit hay fever (pollen allergy), and in some cases increased risk of asthma and rhinitis. Increased incidences of hay fever and asthma in urban areas have been blamed on enhanced levels of air pollutants (e.g. aerial particulate matter and ozone), but also on the increased use of male trees that produce pollen (male trees being preferred over females in those species that readily drop fruit and result in unsightly fruit-littered pavements).

Green spaces – even highly managed spaces – can house flora (and some fauna) that pose health risks, albeit often marginal ones. Numerous garden plant genotypes are poisonous if ingested, including some commonly occurring species (e.g. *Aconitum, Convallaria, Delphinium, Digitalis, Laburnum, Pieris* spp.) while others cause skin irritation and dermatitis on contact (e.g. *Chrysanthemum, Euphorbia, Ginkgo, Narcissus, Primula, Ruta, Tanacetum* spp.). In the less formal semi-natural landscapes, species such as *Urtica dioica* (stinging nettle: Europe) and *Toxicodendron radicans* (poison ivy: North America) can be commonplace – both species causing skin irritation and rashes. Fatalities due to plant poisoning are low, however, with <5 per annum in the USA. Children are most susceptible to poisoning, but even here an Irish survey suggested household medications contributed to 65% of ingested poisons, household or gardening products 34% and plants only 1%, with no fatalities reported (Rfidah et al., 1991). These relatively low levels relate to the fact that many poisonous plants are not visually enticing to eat, have a bitter taste or possess emetic properties (i.e. induce vomiting before toxins are absorbed through the stomach lining).

In public green space, effective management is required to ensure trees remain safe, with regular inspections of large specimen trees in publicly accessible locations. Increasing severity of weather events – strong winds, intense rainstorms, flooding, but also drought – can all weaken the structural integrity of trees or their component parts. More than ever, trees need to be carefully planted and maintained through pruning and other activities to ensure, for example, roots are distributed effectively to optimize tree stability, and that boughs remain strong, secure and pathogen free. Inspections and activities need to be recorded and documented to ensure management plans are being implemented, and that there is evidence that these are appropriate and proportional to the potential risks. Litigation has been brought against a number of land management bodies, where they have been seen to be negligent in the maintenance of tree stocks or even in their ability to logically document the maintenance schedules. Despite the range of health risks potentially posed by green space, health benefits are usually considered to outweigh the drawbacks.

3.3 Green Space and the Work Environment

Exposure to greenery and green spaces may be advantageous in the workplace. The health benefits of interior plant displays in offices and other workplaces has been frequently cited (e.g. Park and Mattson, 2008). Some companies have provided gardens for their employees to sit out in at break times/lunch to encourage a sense of freedom and become a 'coping mechanism' to seek diversion during the working day. Green views from workplace windows also reduce stress with employees feeling less uptight compared to views of urban scenes. Lottrup et al. (2013) indicated physical and visual access to workplace greenery can have a significant positive effect on employees, with physical access to outdoor patios, gardened areas and lawns etc. particularly promoting positive attitudes. There were differences, however, between how male and female employees responded. For males, access and visual sight of green space improved the workplace attitude and decreased perceived levels of stress, whereas for females, attitudes improved but there was no significant effect on stress levels. These gender variations may partially relate to the different pressures/stressors men and woman commonly experience at work. It was found that more men than women went outdoors during the working day, and that women often reported 'being too busy' as a reason not to go outdoors. Compared to men, women tend to report more interpersonal stressors, more stress due to multiple roles, lack of career progress, and discrimination and stereotyping. Coping mechanisms for such stresses may also vary – women relying on social support from friends/colleagues rather than environmental influences.

3.4 Proximity, Scale and Type of Green Space

Engagement with green space and hence health benefits attributed to it may depend on proximity to the space, the size, form and quality of the green space. Relative closeness to green space has been shown to have both a positive (e.g. Richardson et al., 2013; Mass et al., 2009b) (Fig. 3.6) or no (e.g. Potwarka et al., 2008) influence on physical activity/health. Results may vary with the type of health issue, with increasing green space within neighbourhoods improving mental health, but having little impact on poor general health and obesity. Larger areas of green space may enhance health effects but in reality these are often scarce in the more deprived areas where health problems are greatest. Such inconsistent findings may reflect variations in user groups (e.g. children versus adults), dietary behaviour, degree, type or frequency of physical activity, or type and accessibility of green space.

Links between living in green neighbourhoods and mental health status are increasingly being made. Richardson et al. (2013) found that residents of the greenest urban neighbourhoods had significantly lower risks of mental health problems compared to those of areas with limited green space. Indeed, the relationship was linear; more green space equating to better mental health. A study of over 10,000 individuals in the UK indicated that people were happier when living in urban areas with greater amounts of green space. Compared with when they live in areas with less green space, they demonstrated higher levels of well-being and lower levels of mental distress (White et al., 2013). Data were controlled for other factors that would affect these parameters: income, employment status, marital status, health, housing type, and local-area-level variables such as crime rates. Although at an individual level the effects appear small, the cumulative effect across the population was deemed significant. Somewhat in contrast to this, Francis et al. (2012), investigating the quantity and quality of public open space in Australian neighbourhoods, concluded that quantity of open space was not a factor influencing psychological stress, but quality was. It should be noted that this study restricted itself to public open space, not the greenness of entire neighbourhoods.

Recent studies have capitalized on the monitoring of human cortisol levels to determine how populations and individuals respond to green space. Reductions in cortisol levels are associated with lower levels of physiological stress. Cortisol indicators are useful as concentrations can be determined readily from human saliva samples, i.e. relatively easy to attain and require limited intervention with the sample population. Studies that investigated levels of green space within deprived urban neighbourhoods in Scotland measured diurnal patterns of cortisol secretion (Roe et al., 2013). The sample group incorporated individuals not in work and aged between 35 and 55 (a demographic group prone to stress and mental health problems). Results from this study showed a significant reduction in cortisol levels over the day for those residents living in areas with a high proportion of green space, in contrast to those with limited green infrastructure;

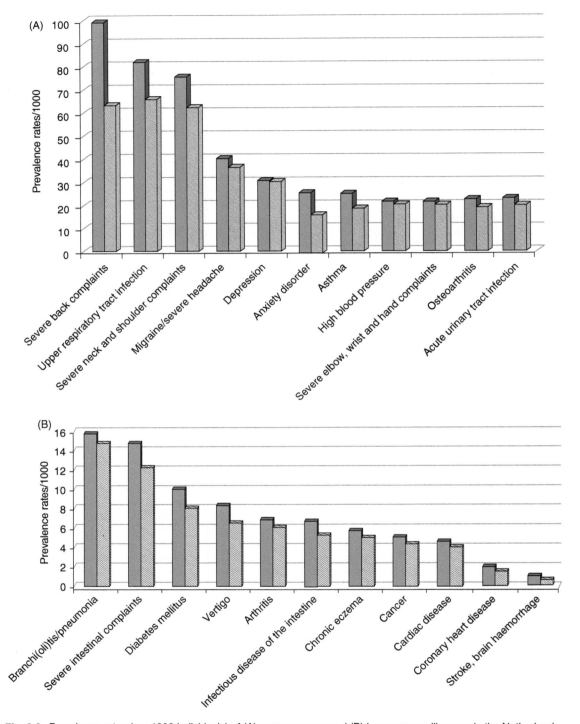

Fig. 3.6. Prevalence rates (per 1000 individuals) of (A) more common and (B) less common illnesses in the Netherlands, when living in neighbourhoods with 10% green space (solid) or 90% green space (hatched) within 1 km from home. (Modified from Maas *et al.*, 2009b).

thus indicating that living in areas with a higher percentage of green space is associated with lower stress. There were differences between the two genders, however, in terms of mean cortisol levels. High levels of neighbourhood green space seemed to provide women in this demographic with greater stress avoidance benefits than men. A point of concern from the research study was the indication that some women living in deprived areas with little green space were regularly experiencing substantial levels of stress, depression and other negative life events.

Proximity to green space has been correlated with other health benefits: relationships between green space and pregnancy/foetal development being a prime example. Separate studies in Barcelona and Oregon suggest that enhancing the proportion of greenery around family homes correlates with increases in the birth weights of newborn babies. In the US study, a 10% increase in tree-canopy within 50 m of a house reduced the number of babies born underweight by 1.42 per 1000 births (Donovan et al., 2011).

Sleep too, may be affected by the extent of vegetation cover in the neighbourhood. A survey using a database of 260,000 Australian citizens aged ≥45 years old demonstrated that increasing the proportion of green space in people's residential area correlated with improved sleep patterns, particularly duration of sleep achieved each night (Astell-Burt et al., 2013). Sleep deprivation adversely affects health criteria, including reduced longevity of life, cardiovascular disease, obesity, diabetes and poor self-rated health scores. Favourable mental health and active lifestyles are also considered to be drivers of a healthier duration of sleep (usually around 8 h per night). Increasing the proportion of green space correlated with reductions in the number of people suffering from medium-short sleep (6–7 h) (Fig. 3.7A) and short sleep (<6 h) (Fig. 3.7B). The cause and effect of this relationship is not known. Living in areas with large amounts of green space may encourage greater physical activity during daylight hours, thus improving sleep during the night period. The green space may be promoting relaxation and inducing less anxiety in residents, with positive consequences on sleep at night. Alternatively, it may be a more subtle relationship, as areas with large proportions of green spaces are likely to be quieter or have less air pollution than more urban ones, and thus could also be impacting on the quality of sleep.

Coombes et al. (2010) demonstrated that proximity to a green location was important in their study; people living further from a park or similar green space were less likely to use it, were less likely to meet the minimum guidelines for physical activity and were more likely to be overweight. Frequency of use was increased by closer proximity. In light of this, one of the UK Government's conservation bodies, Natural England, claimed that no residence should be located more than 300 m from an area of green space, i.e. encouraging planners to embed green areas more frequently within the urban framework. The proximity question is open to debate, however, as Schipperijin et al. (2010) found that only 56% of respondents used their nearest green space on a weekly basis.

The importance of local green space may vary with different user groups, for example it may have a higher priority for elderly people, the disabled or those with young children, where visiting green spaces located further afield may require greater organization. The convenience factor is also important for dog walkers, a high proportion being reliant on local spaces due to the requirement for regular use. Interestingly, a number of factors did *not* appear to affect the use of local green spaces, including:

- the range of facilities available or the different activities within the space;
- the shape and diversity of the area;
- maintenance levels;
- the dominant vegetation;
- an individual's preference for different green space elements;
- marital status or profession of the visitors;
- the levels of stress people were experiencing; and
- people's view on nature (Maas et al., 2008, 2009a, b).

Remarkably, those citizens without access to a domestic garden did not use their local green spaces more frequently – people with gardens utilizing local green space as, or more, frequently, than those without. This may imply though, that those who frequently use their garden are also more interested in spending time outside in general terms, and perhaps enjoy engaging in a wide variety of green environments – the home garden being utilized for some activities but more extensive green spaces providing opportunities for alternative experiences (Maas et al., 2008, 2009a, b).

In contrast to those groups who selected local green space, others actively preferred and sought

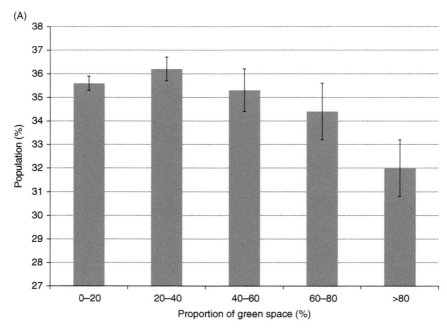

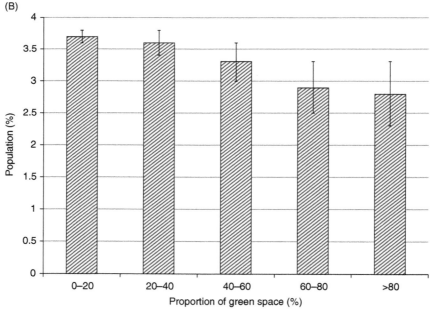

Fig. 3.7. The percentage of population with (A) medium-short sleep (6–7 h) or (B) short sleep (<6 h) based on area of residential green space. Bars, 95% confidence intervals. (Modified from Astell-Burt et al., 2013.)

larger green areas (larger areas >5 ha in general being more popular) even if travel distance/time was greater. This is an important point for city planners, as they need to find a balance between the provision of smaller, intimate, highly accessible, local green spaces and the larger, more extensive parks and woodlands within a reasonable distance of residents. Further analysis is also merited for determining the actual components of green space and assessing their key characteristics. Indeed, many studies fail to properly define the types of green spaces being discussed, with little attempt to document details or assess environmental or quality factors. More sophisticated approaches are needed, especially as some studies imply that size and attractiveness are associated with frequency of use.

3.5 Horticulture as Therapy

Horticultural activities, including domestic gardening, the growing of food on allotments and community gardening, provide an opportunity for restorative processes. Local government organizations and charities may specifically use the existing green estate to provide facilities for clients with health or social problems – closely supervised groups of individuals helping to maintain flower beds, propagate plants for sale or be involved in nature conservation work. Inherently, many individuals will engage with gardening and other similar activities to relax and 'be close to nature'. More formally, however, such activities are intentionally used as interventions in either Social and Therapeutic Horticultural (S&TH) or Horticultural Therapy (HT) programmes. Social & Therapeutic Horticulture has a more general focus on well-being improvements and is not necessarily set against clinical objectives. These courses are well-used with 21,000 clients engaging with S&TH every week in the UK.

Horticultural Therapy, on the other hand, has predefined clinical goals and helps clients learn new skills or regain lost skills. It aims to help clients improve their memory, initiate tasks, take on greater responsibility, enhance problem-solving skills, pay greater attention to detail or regain physical abilities such as coordination, balance and strength. Various client groups respond to HT: people with physical or mental disabilities or in rehabilitation from illness, injury, addiction or abuse (Haubenhofer et al., 2010). Both forms of green care may utilize specially designed therapeutic gardens, but their activities are also closely linked with hospitals and care homes where much of the interactions are relatively passive (e.g. viewing and hearing nature, smelling and touching plants). Nevertheless, even passive engagement with gardens and other forms of green space has been associated with less agitation and aggressive behaviour with, for example, people suffering from Alzheimer's disease and similar disorders that cause frustration (Detweiler et al., 2008).

Despite the popularity of HT and S&TH programmes, the scientific evidence for the benefits remains incomplete, largely due to the difficulties in conducting large, randomly-controlled experiments with small individual client groups at any one location, or attempting to account for clients with different mental health or social problems and different levels/forms of intervention (Sempik, 2007). Despite this, there have been a number of discrete studies that highlight the benefits of horticultural programmes. Gonzalez et al. (2010, 2011), working with individuals who suffered from clinical depression, demonstrated that mental health scores (meaned across five different assessment criteria) increased significantly after HT, compared to the same assessments conducted before the programme. In this study, large reductions in the severity of depression were observed over the first 4 weeks, and these were still evident 3 months after the HT finished. The value of HT in relation to clinical depression is that it is thought to activate the 'being away' and 'fascination' components of Kaplan's attention restoration theory, as well as provide social interactions and cohesion.

Likewise, positive responses have been cited for patients suffering from dementia. Gigliotti and Jarrott (2005) noted greater levels of social engagement in patients when HT approaches were adopted (78%) compared to more conventional occupational therapy activities (28%). Spring et al. (2013) reported that 'Gardening was a constructive, outdoor activity that promoted social interaction, physical activity and provided stimulating cognitive challenges for patients with dementia.' Interestingly, the authors note that it also provided positive benefits to the staff who worked in the hospital and indeed the people visiting the patients. This point is reflected in earlier research involving a children's hospital garden. In this case, 54% of visitors reported feeling more relaxed and less stressed, 12% felt refreshed and rejuvenated with 18% feeling more positive and able to cope after visiting the garden (Whitehouse et al., 2001). The garden was valued as a 'tonic' for the staff, almost as much as for the patients and their families. Even short visits were beneficial, as half of visitors spent less than 5 minutes in the garden at any one time.

3.6 Social Horticulture

Horticultural activities and restorative landscapes are linked to social and educational benefits, as well as those purely relating to health. Horticultural educational programmes are used to reduce truancy from school for disaffected pupils. Qualitative data suggests that such programmes can not only reduce the incidence of truancy, but aid academic performance (improved skills in English and Mathematics), promote self-discipline, self-esteem and team skills. Horticulture acts as a social 'leveller' in this context, with programmes considered useful in breaking down age-related barriers and bestowing students with positive role models from the older gardeners or demonstrators. Similarly in this light, horticulture is seen as a 'diversionary activity' – reducing the chances of young people becoming engaged in anti-social behaviour, including drug abuse and petty crime. Likewise with prison inmates, horticultural programmes have been linked with greater opportunity to take up vocational work after release and tend to reduce the chances of reoffending. Even within the prison environment the role plants play is an intriguing one. During disruptive periods, inmates frequently damage buildings or other infrastructure, but rarely destroy the plants they or their fellow inmates have grown. Perhaps, the hands-on involvement with plants per se may help reduce tension and alleviate bad temper.

Horticultural programmes allow individuals to engage with members of the wider community who share an interest in horticulture. They help to develop closer relationships between vulnerable (socially excluded) and non-vulnerable (socially included) citizens. Moreover, their value extends to placing emphasis on the abilities of vulnerable adults rather than their disabilities. Benefits are not centred around plant cultivation solely either, with clients contributing to the entire spectrum of horticultural activities. Those partaking in these programmes are likely to be involved with team building, training, promotional and marketing activities, as well as selling produce to the wider public. These processes elicit self-confidence and independence as well as contribute to physical and mental well-being (Sempik *et al.*, 2005). Parr (2007) suggested that over and above the benefits attributed to crop cultivation, the social interactions, and particularly the positive attitudes and experiences relayed by garden staff to clients, significantly contributed to healing and rehabilitation processes.

Social engagement seems to be a key component of allotment gardens too, especially for older people. Allotment gardens present a supportive environment that combats social isolation and contributes to the development of an individual's social network (Milligan *et al.*, 2004). Family relations improve when members of a family come together within community garden projects (Carney, 2012), with evidence of additional health benefits associated with greater intake of fresh fruit and vegetables. Eating fruit and vegetables in a 'several times a day category' increased from 18% to 85% for adults during the duration of the project (and from 24% to 64% for children). Furthermore, gardening and allied community activities centred round green space are utilized to integrate different ethnic communities. A gardening programme at an immigrant centre in Germany was successful at bringing individuals from different cultural backgrounds together, through their ambition to achieve common goals and to provide healthy, inexpensive food for their families.

Green space and child development

As with other groups, green space is perceived to aid self-esteem and a sense of empowerment in children, and helps foster their engagement with school. Gardens and other types of green space within the vicinity of the home strongly influence a child's development, particularly in terms of boosting capacity to focus and maintain attention. When children's cognitive functioning was compared before and after they moved from poor-quality (low volume green space) to better-quality housing (greater green space) differences emerged in attention capacity, even when the effects of the improved housing were taken into account. Similarly, more recent research from Barcelona indicated better cognitive development in school children when their schools were surrounded by green space (Dadvand *et al.*, 2015).

These points link to the philosophy behind the Forest School movement. This aims to encourage school pupils to engage with education through activities that take place in natural settings, such as woodlands and meadows, rather than conventional 'uninspiring' classrooms. The benefits in educational terms are not solely restricted to improving understanding of natural science but also to improving performance across the entire curriculum (Mathematics, Arts, English, etc.) – a point re-enforced by studies on teachers' use of school gardens. These were not only utilized to teach horticulture, plant and environmental sciences in the curriculum, but also languages, arts and ethics, with school gardens being seen as a resource across the academic, recreational, social and therapeutic spectrum. The impact of plants and greenery on attention, engagement and interest may also explain why children with attention deficit disorder (ADD) appear to respond better when playing in a green or natural environments. ADD children focused their attention and performed better after activities in green settings, and increasing the proportion of vegetation resulted in less severe attention deficit symptoms. For children in general, the provision of more natural play-settings is currently being advocated (Fig. 3.8). Increasing the green component through greater planting and introducing natural elements such as stones, wood and other fibres into play-parks stimulates more creative and imaginative play, with positive effects on language and collaborative skills (Fjørtoft and Sageie, 2000).

Green space – role in reducing crime and antisocial behaviour

Health benefits associated with green spaces lead to far-reaching secondary effects. Restoration from stress and reductions in anger or frustration are thought to be at least partially responsible for lower incidences of crime. Possibly the most striking example of the impact of green views is that of housing estates in Chicago, which demonstrated that domestic violence (aggression against a partner) increased by 25–35% when large landscape trees were removed from the view of some housing blocks, but no increase was found in comparable blocks where trees remained. (Data was normalized for housing stock type, and the resident's socio-economic background.) The higher levels of aggression encountered were attributed to enhanced stress and anxiety in those dwellings where there was no view of greenery (Kuo and Sullivan, 2001a, b).

Similarly, correlations have been made between the re-landscaping of derelict city plots and reductions in gun crime and vandalism in Philadelphia (Branas, 2011). Indeed, community landscape and garden projects are often now encouraged in inner city zones due to the fact that they have positive knock-on effects on the surrounding neighbourhoods. This includes residents taking more time and effort to maintain their streets and houses, lower incidences of vandalism and greater social cohesion within the local community. Such community responses, however, may be influenced by the wider socio-economic context. In low income, ethnic communities in New York, the development of community gardens resulted in a fourfold increase in other community-based activities, including crime prevention and improved neighbourhood services, whereas no such response was noted with those community gardens associated with more affluent residents. A key benefit of 're-greening' neighbourhoods and allied community engagement is it can break negative cycles. Locations that are poorly maintained, vandalized and purely utilitarian in function, send out a signal that those in charge do not place a high value on the area for the residents who live there, or indeed that the residents themselves have little respect for their environment and its potential. In contrast, well-designed, 'cared for' green spaces demonstrate endeavour and strong civic responsibility – a point reinforced by evidence that such green spaces themselves tend to suffer less from graffiti and litter.

Do all types of plants have equal merit when considering the 'crime and antisocial' dimension? The limited data available to date would suggest that trees in general prevent crime, whereas the role of shrubs is more ambivalent. Comparisons between neighbourhood crime rates and tree cover show that for the most part, increasing tree cover correlates with lower crime rates. Neighbourhoods, with large canopy trees were associated with reduced incidences of robbery, burglary, theft and shooting. This was the case for both trees grown on private and public land, but the magnitude of the response was 40% greater for public than for private land. Spatially adjusted models suggested that

Fig. 3.8. A number of initiatives encourage children's play within natural or naturalized settings, with increased avocation of equipment constructed from natural materials and fibres.

a 10% increase in tree canopy was associated with a 12% decrease in crime. The type and location of the urban greenery may affect crime rates, however, as the positive associations broke down where there was an extensive interface between industrial and residential properties. Smaller trees and shrub areas that concealed properties from roadways, or those associated with abandoned land, were more likely to be linked with higher rates of crime.

Green space and traffic calming

Traffic calming schemes include aspects such as where roads have been realigned, speed reduction measures have been introduced (bumps and chicanes) and elements of green space have been incorporated such as grass-covered verges or street trees. The introduction of traffic calming schemes has been linked with a reduction in traffic-related problems. For example, the introduction of 20 mile per hour speed restriction zones have seen reductions in road casualty rates of between 38% and 42%, depending on location. Similarly, city-wide traffic calming in Seattle has been associated with a 90% reduction in road traffic accidents. Road calming tends to alter driver behaviour, with subsequent reductions in driving speed and the number of collisions incurred. Although certain nuisance factors may increase with road calming schemes, additional positive outcomes noted have been increased pedestrian activity in these areas and improvements to physical health. There was no evidence, however, of enhanced psychological health.

Roadside trees have been linked with improving driver alertness and reduced numbers of collisions, but one downside of the presence of trees is higher fatality rates in those roads where trees are present (at least along faster dual carriageways and rural roads). Unlike other aspects of roadside 'furniture' such as hedges and metal crash-barriers, which to some extent absorb the impact of a vehicle collision, the tree trunks of mature trees can cause considerable damage when directly impacted by a car, thus increasing the chances of fatalities. Effective road design now still includes trees, but often these are set back some distance from the carriageways themselves. Urban traffic calming schemes in residential areas may also use the occasional tree, but often the verges are planted with low-level shrubs and herbaceous plants that are low maintenance, drought tolerant and beneficial to wildlife. This might include long-duration, summer-flowering species such as *Lavandula*, *Cistus*, *Penstemon* and *Ceanothus thyrsiflorus* var. *repens* which are prostrate enough to avoid impairing drivers'/pedestrians' views of the roadway.

Although urban traffic calming schemes can result in some disadvantages (e.g. increased traffic noise and vibration), greater public use and hence surveillance by pedestrians and cyclists of the roadways and immediate surrounding neighbourhoods has been instrumental in reducing the incidence of drug crime and prostitution in some locations (by 60% and 80% respectively, according to one study by Lockwood and Stillings (1998)).

3.7 Environmental Horticulture and Healthy Diet

Environmental horticultural programmes that encourage community gardens, allotments, home gardening or other forms of urban food production regularly improve diet and encourage healthier eating habits in those who participate. Fruit and vegetables provide essential vitamins, fibre, minerals and phytochemicals. Some fruit/vegetable genotypes are particularly beneficial due to their specific 'health-related' compounds, such as antioxidants, e.g. broccoli (*Brassica oleracea*), containing sulforaphane; blackberries (*Rubus fruticosus*), containing anthocyanins and phenolics; and tomatoes (*Lycopersicon esculentum*), containing lycopene. Evidence would suggest that the provision of more balanced diets, with less reliance on fat- or starch-based foods has considerable health enhancing attributes. Globally, inadequate consumption of fruit and vegetables is estimated to contribute to 2.6 million premature deaths a year. Encouraging people to consume up to 600 g per day of fresh fruit and vegetables has been linked to reductions in:

- cardiovascular heart disease by 31%;
- ischaemic stroke by 19%;
- stomach cancer by 19%;
- oesophageal cancer by 20%;
- lung cancer by 12%; and
- colorectal cancer by 2%.

Encouraging citizens to understand food and nutrition better, for example through being involved in its production at home or in a community garden, correlates closely with the adoption of healthier eating habits. School gardening projects may affect children's vegetable consumption, including improved recognition of, attitudes toward, preferences for, and willingness to taste vegetables. Effectively, the resistance to the 'eat your greens' message can be dissipated! Gardening also increases the variety of vegetables eaten. Due to concerns of obesity and an increasingly sedentary lifestyle in children, this group has been specifically targeted for intervention schemes, with positive results. Most interventions have been school-focused and success has been increased when a number of factors align themselves. This includes when:

- the initiatives have lasted ≥12 months;
- the whole school community has been engaged;
- the pupils are formally educated on the growing, cooking and nutritional aspects of food; and
- the teachers/catering staff have been provided with additional training.

Indeed, where gardening was effectively embedded in the curriculum, there is some evidence that the most successful projects were associated with those scenarios where teachers and pupils had ownership of the concepts and directed the development of the garden. In terms of healthy eating, backing up the initiatives in the home environment is also seen as a key component.

The value of urban food production varies significantly with context. In Cuba, the collapse of trade and support from the Soviet Union in the 1990s resulted in citizens being forced to produce their own food, through the cultivation of any spare space in the urban environment. An interest in horticulture arose almost overnight, largely driven by necessity. Similarly, for many developing countries, people are still dependent on their own production of fresh produce to meet nutritional needs. In contrast, in the developed world, home food production may be desirable due to cost savings and the benefits of providing regular, fresh, nutritional produce at a reasonable price. For more affluent people, the motivations relate more to perceptions of better quality food, or political ideology relating to the sustainable use of resources, or perhaps out of simple enjoyment of growing their own food and watching it come to fruition. In 1996 the United Nations acknowledged the value of urban food production to the health and well-being agenda. They defined the advantages as:

- increased quantity of food being made available;
- increased quality (freshness) of produce; and
- the potential to gain/supplement financial income by supplying local markets.

In the first of these categories, urban food is unlikely to rival larger-scale production from rural areas, but can provide essential supplies during times of crisis, when wider food insecurity issues are present or when there is a scarcity of produce from other traditional markets. Worldwide there are about 800 million people involved in urban food production either to meet their own requirements or to develop business opportunities through the sale of food. In less developed, poorer societies, 25–66% of the entire population may be involved in urban/peri-urban food production. It should be noted, however, that urban gardeners are not necessarily the poorest individuals in any society, but are more likely to be those who have had access/ownership of land, can attain a reliable water supply and who understand and engage with the local market dynamics. Urban food production systems are closely aligned to healthy diet, but as discussed elsewhere, well-being and societal benefits are manifest too. These include urban poverty alleviation, community engagement and cohesion, opportunities to engage with nature, feelings of self-fulfilment, waste management and recycling, and urban greening.

3.8 What Sort of Green Landscapes Should be Promoted?

Access to pristine natural wildness is not always feasible, and most people engage with green space on a day-to-day basis in an urban context. Yet if access, or even just a view of green space, can help provide a range of health and social benefits, what sort of urban green space needs to be provided to maximize the benefits? Tenngart Ivarsson and Hagerhall (2008) suggested that a number of urban green landscapes, including well-designed gardens, provided restorative value (Fig. 3.9). Evidence exists for beneficial effects associated with urban forests (Tsunetsugu et al., 2007, 2013), parks (Irvine et al., 2013), green roofs (Banting et al., 2005), green walls (Hop and Hiemstra, 2012), urban nature reserves (Keniger et al., 2013), allotments (van den Berg et al., 2010) and gardens (Cameron et al., 2012). Coombes et al. (2010) claim that formal green spaces (parks) make them particularly suitable for physical activities due to the network of pathways encouraging cycling, walking and jogging. Smaller green areas or components of larger green infrastructures may be used for allotment or community gardening, with physical activity being encouraged through digging, weeding and cutting/harvesting of crops. Exercise in waterside locations and extensively managed natural settings (mainly urban woodlands) have been

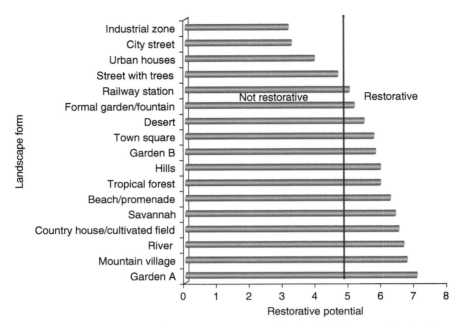

Fig. 3.9. The relative restorative value of different landscape types. A value of ≥5 denotes when landscapes were considered to be particularly positive in terms of restoration potential. Note the difference in score between the two gardens (gardens A and B) with different styles of landscape. (Modified from Tenngart Ivarsson and Hagerhall (2008), who compared results from three different studies.)

linked to providing greater restoration potential (Fig. 3.10) compared to green spaces in urban settings (mostly parks) or favourite places around buildings. Although the inference may be that large, heterogeneous green spaces optimize benefits, this does not suggest that smaller local areas have limited value. Indeed tree-lined streetscapes may provide greater benefits than more extensive areas of urban green space when quality aspects are improved, due to the fact that these are more immediately accessible to a greater proportion of the population. Irrespective of the scale of landscape space being considered, improvements in quality criteria tend to increase the restoration potential (van Dillen *et al.*, 2012). Quality was defined here as:

- increased accessibility;
- high levels of maintenance;
- greater variation in landscape type and form;
- more natural and more colourful landscape forms;
- clear arrangement and harmonious design;
- provision of shelter;
- absence of litter;
- perceptions of safety; and
- overall general impressions.

Similarly, improving quality aspects in local public open spaces (scored by assessments on pathways, lighting, sports and play grounds, the provision of shade and well-maintained lawns, presence of water features or a close locality to water, and the relative abundance of birdlife) were shown to have a positive effect on mental health ratings (Francis *et al.*, 2012). In this case, such factors were deemed more important than subjective aspects such as perceived friendliness, comfort or safety.

Close proximity to green space has been seen as a factor in promoting citizen health and well-being (see section 3.4), but studies in southern Sweden also tried to define quality of such spaces (Björk *et al.*, 2008). A large population-based study was used to investigate associations between recreational values of the local natural environment and aspects such as neighbourhood satisfaction, physical activity, obesity and well-being (Fig. 3.11). Residential postcode data and GIS techniques were used to assess objectively five 'recreational values' linked to the local green space (inner city zones were excluded), these being categorized as: 'serene', 'wild', 'lush', 'spacious' and 'culture'

Fig. 3.10. Water, woodland and a largely natural style are considered to provide high restorative components to the landscape.

(see Table 3.2 for definitions). These values were scored by each residence in the survey. Despite Sweden having a proportionally high amount of green space, the average citizen had access to green space within 300 m of their home that was generally defined as 'low-medium' in its recreational value. More than 70% of the population stated they lived in areas where they considered there was no recreational value, or only some form of culture within 300 m. For those where recreational values were higher, however, there were significant positive correlations to moderate physical activity per week, neighbourhood satisfaction, body mass index (for tenants, but not home-owners), but less strong correlations with self-rated physical and psychological health and self-rated vitality (Fig. 3.11).

Even within gardens, design aspects and landscape features may influence the restoration potential. This correlates with some authors' findings that different natural features seem to vary in their potential to promote restorative responses. Flowers and water correlated better with relaxation and related health indicators than factors such as the presence of animals, trees, hills, natural aromas or sounds (Ogunseitan, 2005). The relationship to water is intriguing; Barton and Pretty (2010) indicated that conducting green exercise activities within sight of water may enhance the health benefits compared to the same activities elsewhere (Fig. 3.12). Water may be one of those quality attributes that becomes subjective; older adults considering water features as calming and reflective, whereas parents of young children may view them with suspicion. Even the form of vegetation has been cited as influencing human responses. Pine trees (*Pinus thunbergii*) clipped in the Japanese Sukashi technique (an important component of inferring a natural style in Japanese garden flora) promoted a greater relaxing effect as measured by decreased cerebral blood flow compared to unpruned trees (Elsadek *et al.*, 2013). The positive effect due to the pruned trees was recorded in both genders of Japanese citizens. The fact that stylized trees (albeit to represent an elderly, wind-blown or weathered tree) was favoured in preference to untrained forms indicates

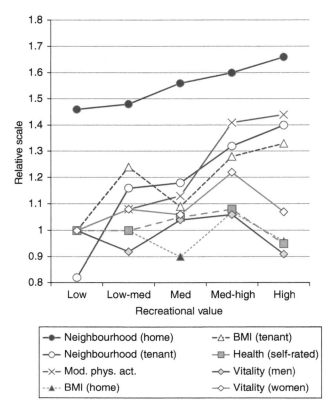

Fig. 3.11. Relationship between the recreational value of local green space and aspects such as overall neighbourhood satisfaction (for home-owners and tenants), duration of moderate physical activity, body mass index (BMI), and self-reported scores for health and vitality. (Modified from Björk et al., 2008.)

that a cultural or aesthetic dimension may also be influencing the psycho-physiological responses. This positive response and desire for nature or wilderness within garden style may account for the popularity of bonsai growing or alpine gardening which infer wilderness, but actually involve highly contrived design and construction processes.

3.9 Social Attitudes to Green Space and Values

If green space is vital to our well-being as much of this chapter alludes to, do we proportionally value it and cherish it? In the UK, 46% of the population are thought to use a park or other public green space at least once a week. Yet, with growing global population and rapidly urbanized societies, pressure to develop inner city green and brownfield sites and to build on previously 'green belt' land at the urban fringe is increasing. Integrating existing, or embedding new green infrastructure, however, becomes vital to ensuring quality of life attributes are maintained. This is the challenge facing city planners. How should more housing and infrastructure be provided, whilst guaranteeing enough green space to ensure recreational and health benefits? What form, scale and quality of green space is required?

Despite some of the universal concepts described above (e.g. the need for restorative space), designing effective urban green space remains challenging and needs to account for different user groups, cultural factors, as well as local circumstances and conditions (climatic, social, economic, different or even conflicting ecosystem services). There have been mis-matches as to what sort of urban green space is desired between teenagers and adults (Gearin and Kahle, 2006). In China (Guangzhou City), preference is given to green space that helps

Table 3.2. The recreational value of natural green space based on description of sites and typical habitats/characteristics in southern Sweden. (Modified from Björk et al., 2008.)

Recreational value	Description	Habitats and Characteristics
Serene	A place of peace, silence and care. Sounds of wind, water, birds and insects. No rubbish, no weeds, no disturbing people.	Includes: broadleaved forest, mixed forest, pastures, inland marshes, wet mires, other mires, water courses, lakes and ponds. Excludes: noise >30 dB and artillery ranges.
Wild	A place of fascination with wild nature. Plants seem self-sown. Lichen and moss-grown rocks, old paths.	Includes: forest, thickets, bare rock, inland marshes, wet mires, other mires, water courses, lakes, ponds and slopes >10°. Excludes: noise >40 dB and at least 800 m distance from wind power machinery.
Lush	A place rich in animal and plant species.	Includes: mixed forest, marshes, mires, beaches, dunes, sand plains, bare rock, all registered 'key biotopes', pasture land of regional interest, biodiversity areas, bird biotopes, Natura 2000 sites and National Parks.
Spacious	A place offering a restful feeling of 'entering another world', a coherent whole, like a beech forest.	Includes: beaches, dunes, sand plains, bare rock, sparsely vegetated areas, burnt areas, natural grassland, moors and heathland, forest >25 ha, slopes >10°, farmland pointed out in a National Plan and preserved coastal zones. Excludes: noise >40dB.
Culture	The essence of human culture. A historical place offering fascination with the course of time.	Includes: non-urban parks, farmland pointed out in a National Plan, areas of national interest for cultural preservation, nature reserves.

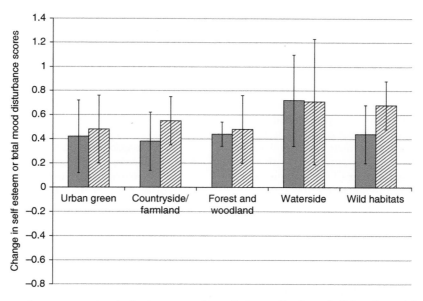

Fig. 3.12. A range of green space typologies have a positive effect on self-esteem (solid) and mood (hatched), although the presence of water seems to provide additional benefits. Bars, 95% confidence intervals. Data, a summary of ten UK studies. Reprinted (adapted) with permission from Barton and Pretty, 2010. Copyright (2010) American Chemical Society.

resolve environmental quality factors such as air pollution, noise abatement and thermal comfort, with concerns over species conservation and habitat provision ranked much lower (Jim and Chen, 2006). In contrast, studies in Finland indicated that citizens there preferred urban green space that provided naturalness, closeness with nature, tranquillity and feelings of being in a forest (Tyrväinen et al., 2007). In both these studies though, preference tended to be in favour of naturalistic styles of landscape. So despite the limited concern for species conservation/habitat provision, Chinese citizens still had a strong affinity to styles that reflect the natural world closely. In New Orleans, USA, aesthetic value and the provision of shade from green space were considered the most important aspects, with wildlife habitat, urban cooling and the potential to increase property values also considered significant positive features. Less of a priority was mitigation of wind, dust and stormwater runoff.

In contrast to the 'universal desire' for landscapes that mimic nature, van den Berg and van Winsum-Westra (2010) demonstrated that when it came to preference for garden styles, a certain component of the population desired gardens with formal structure, rather than more informal naturalistic styles. This may be due to respondents relating gardens as distinct from other forms of green space (stronger cultural influences for example) or that any biophilic drives are being weakened by urban living (disengagement with nature or 'nature deficit syndrome'; Bixler and Floyd (1997)). In the UK, it might be argued that significant proportions of the population see their garden as some form of reflection of nature and a place to engage with/appreciate other species, whereas to others it is simply an 'outside' room, reflecting their tastes and values, with the leaf blower, the power-jet washer and patio heater taking the place of the vacuum cleaner, mop and domestic fire/radiator. Others of course, just see it as a place to park their car. So preferences and motivations vary. These studies illustrate that what individuals seek from their public green space also varies to some degree, but that the health benefits are likely to be significant (if not necessarily consciously realized by the participants). Planners need to take this on board and should ensure that both frequent and heterogeneous forms of green space are provided within the urban matrix, so as to maximize the health and well-being benefits for the human population.

Conclusions

- Environmental horticulture has a key role to play in optimizing the health and well-being benefits associated with green infrastructure.
- There is growing evidence that green space promotes physical activity and encourages individuals to commit to more regular or prolonged exercise. This not only has implications for improved physical fitness, but also is thought to be able to help offset key health risks associated with a sedentary lifestyle (e.g. heart disease, type 2 diabetes, high blood pressure and stroke).
- Similarly, views and activities within green space reduce psycho-physiological stress and improve aspects of mental health. This includes aiding resilience to anxiety and depression, and promoting positive thoughts, as well as some evidence of delaying diseases such as dementia.
- Positive effects on mental well-being have been closely linked with the attention restoration theory (ART), where key elements of 'being away', 'fascination', 'extent' and 'compatibility' are thought to influence recovery/recuperation from directed attention and reduce the likelihood of psycho-physiological stress occurring.
- Although many of the benefits around green space are associated with stress reduction and restorative effects, other emotional motivators, such as stimulation from wild or challenging natural/semi-natural environments may also aid the human psyche, although these aspects are much less well researched.
- More research and evaluation is also required on what types of green space provide the greatest benefits, and how does scale, proximity and quality of the space influence the potential advantages.
- Environmental horticultural activities that promote home or community food-growing schemes provide the potential for both nutritional and social benefits, as well as improving awareness about the natural world.
- Social benefits of green space include opportunities for community cohesion, social integration, crime reduction, educational attention and engagement.
- Horticultural therapy (HT) is one of a number of eco-therapy programmes available to people with special needs. These programmes are popular, with good anecdotal evidence of their success, but 'clinical' evidence is still unsubstantiated.

- Increments in the proportion and quality of vegetation enhance the work environment and can foster employee satisfaction.
- Green space does not necessarily provide universal benefits; there are 'down-sides', including risks associated with allergens, toxins, physical injury and repetitive strains. Not all demographic groups necessarily respond to green space in the same way, and what may be beneficial for one group may induce some disadvantages in another.
- Overall, urban green space seems to provide a positive contribution to human health and well-being, but more information is required on how different sectors of society engage with green space and what the relative levels of these benefits are. Specifically, more information is required about what is meant by 'good design and management' in terms of promoting the health and well-being benefits, along with a range of other 'services'.

References

Akers, A., Barton, J., Cossey, R., Gainsford, P., Griffin, M. and Micklewright, D. (2012) Visual colour perception in green exercise: positive effects on mood and perceived exertion. *Environmental Science and Technology* 46, 8661–8666.

Astell-Burt, T., Feng, X. and Kolt, G. S. (2013) Does access to neighbourhood green space promote a healthy duration of sleep? Novel findings from a cross-sectional study of 259,319 Australians. *British Medical Journal Open* 3: e003094.

Banting, D., Doshi, H., Li, J., Missios, P., Au, A., Currie, B.A. and Verrati, M. (2005) *Report on the Environmental Benefits and Costs of Green Roof Technology for the City of Toronto*. Department of Architectural Science, Ryerson University, Ontario.

Barton, J. and Pretty, J. (2010) What is the best dose of nature and green exercise for improving mental health? A multi-study analysis. *Environmental Science and Technology* 44, 3947–3955.

Bedimo-Rung, A.L., Mowen, A.J. and Cohen, D.A. (2005) The significance of parks to physical activity and public health: a conceptual model. *American Journal of Preventive Medicine* 28, 159–168.

Bingley, A. (2013) Woodland as working space: where is the restorative green idyll? *Social Science and Medicine* 91, 135–140.

Bixler, R.D. and Floyd, M.F. (1997) Nature is scary, disgusting, and uncomfortable. *Environment and Behavior* 29, 443–467.

Björk, J., Albin, M., Grahn, P., Jacobsson, H., Ardö, J., Wadbro, J. and Skärbäck, E. (2008) Recreational values of the natural environment in relation to neighbourhood satisfaction, physical activity, obesity and wellbeing. *Journal of Epidemiology and Community Health* 62, e2.

Bowler, D., Buyung-Ali, L., Knight, T. and Pullin, A. (2010) The importance of nature for health: is there a specific benefit of contact with green space. *Environmental Evidence*: www.environmentalevidence.org/SR40.html.

Branas, C.C., Cheney, A. and MacDonald, J.M. (2011) A difference-in-differences analysis of health, safety and greening vacant urban space. *American Journal of Epidemiology* 174, 1296–1306.

Cameron, R.W.F., Blanuša, T., Taylor, J.E., Salisbury, A., Halstead, A.J., Henricot, B. and Thompson, K. (2012) The domestic garden – its contribution to urban green infrastructure. *Urban Forestry and Urban Greening* 11, 129–137.

Carney, M. (2012) Compounding crises of economic recession and food insecurity: a comparative study of three low-income communities in Santa Barbara County. *Agriculture and Human Values* 29, 185–201.

Coombes, E., Jones, A.P. and Hillsdon, M. (2010) The relationship of physical activity and overweight to objectively measured green space accessibility and use. *Social Science and Medicine* 70, 816–822.

Dadvand, P., Nieuwenhuijsen M.J., Esnaola, M., Forns, J., Basagaña, X., Alvarez-Pedrerol, M., Rivas, I., López-Vicente, M., De Castro Pascual, M., Su J., Jerrett, M., Querol, X. and Sunyer, J. (2015) Green spaces and cognitive development in primary schoolchildren. *Proceedings of the National Academy of Sciences, USA* 112, 7937–7942.

de Groot, W.T. and van den Born, R.J. (2003) Visions of nature and landscape type preferences: an exploration in the Netherlands. *Landscape and Urban Planning* 63, 127–138.

Detweiler, M.B., Murphy, P.F., Myers, L.C. and Kim, K.Y. (2008) Does a wander garden influence inappropriate behaviors in dementia residents? *American Journal of Alzheimer's Disease and other Dementias* 23, 31–45.

DeWolfe, J., Waliczek, T.M. and Zajicek, J.M. (2011) The relationship between levels of greenery and landscaping at track and field sites, anxiety, and sports performance of collegiate track and field athletes. *HortTechnology* 21, 329–335.

Donovan, G.H., Michael, Y.L., Butry, D.T., Sullivan, A.D. and Chase, J.M. (2011) Urban trees and the risk of poor birth outcomes. *Health and Place* 17, 390–393.

Donovan, G.H., Butry, D.T., Michael, Y.L., Prestemon, J.P., Liebhold, A.M., Gatziolis, D. and Mao, M.Y. (2013) The relationship between trees and human health: evidence from the spread of the emerald ash borer. *American Journal of Preventive Medicine* 44, 139–145.

Elsadek, M. and Fujii, E. (2014) People's psychophysiological responses to plantscape colors stimuli:

a pilot study. *International Journal of Psychology and Behavioral Sciences* 4, 70–78.

Elsadek, M., Jo, H., Sun, M. and Fujii, E. (2013) Brain activity and emotional responses of the Japanese people toward trees pruned using Sukashi technique. *International Journal of Agriculture, Environment and Biotechnology* 6, 465–470.

Fjørtoft, I. and Sageie, J. (2000) The natural environment as a playground for children: landscape description and analyses of a natural playscape. *Landscape and Urban Planning* 48, 83–97.

Focht, B.C. (2009) Brief walks in outdoor and laboratory environments: effects on affective responses, enjoyment, and intentions to walk for exercise. *Research Quarterly for Exercise and Sport* 80, 611–620.

Francis, J., Wood, L.J., Knuiman, M. and Giles-Corti, B. (2012) Quality or quantity? Exploring the relationship between Public Open Space attributes and mental health in Perth, Western Australia. *Social Science and Medicine* 74, 1570–1577.

Gandelman, N., Piani, G. and Ferre, Z. (2012) Neighborhood determinants of quality of life. *Journal of Happiness Studies* 13, 547–563.

Gearin, E. and Kahle, C. (2006) Teen and adult perceptions of urban green space Los Angeles. *Children Youth and Environments* 16, 25–48.

Gigliotti, C.M. and Jarrott, S.E. (2005) Effects of horticulture therapy on engagement and affect. *Canadian Journal on Aging/La Revue Canadienne du Vieillissement* 24, 367–377.

Gladwell, V.F., Brown, D.K., Wood, C., Sandercock, G.R. and Barton, J.L. (2013) The great outdoors: how a green exercise environment can benefit all. *Extreme Physiology and Medicine,* 2, 1.

Gonzalez, M.T., Hartig, T., Patil, G.G., Martinsen, E.W. and Kirkevold, M. (2010) Therapeutic horticulture in clinical depression: a prospective study of active components. *Journal of Advanced Nursing* 66, 2002–2013.

Gonzalez, M.T., Hartig, T., Patil, G.G., Martinsen, E.W. and Kirkevold, M. (2011) A prospective study of existential issues in therapeutic horticulture for clinical depression. *Issues in Mental Health Nursing* 32, 73–81.

Grahn, P. and Stigsdotter, U.A. (2003) Landscape planning and stress. *Urban Forestry and Urban Greening* 2, 1–18.

Grahn, P. and Stigsdotter, U.K. (2010) The relation between perceived sensory dimensions of urban green space and stress restoration. *Landscape and Urban Planning* 94, 264–275.

Hansmann, R., Hug, S.M. and Seeland, K. (2007) Restoration and stress relief through physical activities in forests and parks. *Urban Forestry and Urban Greening* 6, 213–225.

Hartig, T., Mang, M. and Evans, G.W. (1991) Restorative effects of natural environment experiences. *Environment and Behavior* 23, 3–26.

Hartig, T., Evans, G.W., Jamner, L.D., Davis, D.S. and Gärling, T. (2003) Tracking restoration in natural and urban field settings. *Journal of Environmental Psychology* 23, 109–123.

Haubenhofer, D.K., Elings, M., Hassink, J. and Hine, R.E. (2010) The development of green care in western European countries. *EXPLORE: the Journal of Science and Healing* 6, 106–111.

Hedblom, M., Heyman, E., Antonsson, H. and Gunnarsson, B. (2014) Bird song diversity influences young people's appreciation of urban landscapes. *Urban Forestry and Urban Greening* 13, 469–474.

Hop, M.E.C.M. and Hiemstra, J.A. (2013) Contribution of green roofs and green walls to ecosystem services of urban green. *Acta Horticulturae* 990, 475–480.

Hull, R.B. and Harvey, A. (1989) Explaining the emotion people experience in suburban parks. *Environment and Behavior* 21, 323–345.

Huynh, Q., Craig, W., Janssen, I. and Pickett, W. (2013) Exposure to public natural space as a protective factor for emotional well-being among young people in Canada. *BMC Public Health* 13, 407.

Irvine, K.N., Warber, S.L., Devine-Wright, P. and Gaston, K.J. (2013) Understanding urban green space as a health resource: a qualitative comparison of visit motivation and derived effects among park users in Sheffield, UK. *International Journal of Environmental Research and Public Health* 10, 417–442.

Jim, C.Y. and Chen, W.Y. (2006) Perception and attitude of residents toward urban green spaces in Guangzhou (China). *Environmental Management* 38, 338–349.

Kahn, Jr., P.H., Friedman, B., Gill, B., Hagman, J., Severson, R.L., Freier, N.G. and Stolyar, A. (2008) A plasma display window? The shifting baseline problem in a technologically mediated natural world. *Journal of Environmental Psychology* 28, 192–199.

Kaplan, R. and Kaplan, S. (1989) *The Experience of Nature: a Psychological Perspective.* CUP Archive.

Kaplan, S. (1995) The restorative benefits of nature: toward an integrative framework. *Journal of Environmental Psychology* 15, 169–182.

Keniger, L.E., Gaston, K.J., Irvine, K.N. and Fuller, R.A. (2013) What are the benefits of interacting with nature? *International Journal of Environmental Research and Public Health* 10, 913–935.

Korpela, K. and Hartig, T. (1996) Restorative qualities of favorite places. *Journal of Environmental Psychology* 16, 221–233.

Korpela, K.M., Ylén, M., Tyrväinen, L. and Silvennoinen, H. (2010) Favorite green, waterside and urban environments, restorative experiences and perceived health in Finland. *Health Promotion International* 25, 200–209.

Kuo, F.E. and Sullivan, W.C. (2001a) Environment and crime in the inner city. Does vegetation reduce crime? *Environment and Behavior* 33, 343–367.

Kuo, F.E. and Sullivan, W.C. (2001b) Aggression and violence in the inner city. Effects of environment via mental fatigue. *Environment and Behavior* 33, 543–571.

Lachowycz, K. and Jones, A.P. (2011) Greenspace and obesity: a systematic review of the evidence. *Obesity Reviews* 12, e183–189.

Lee, A.C.K. and Maheswaran, R. (2011) The health benefits of urban green spaces: a review of the evidence. *Journal of Public Health* 33, 212–222.

Li, Q., Morimoto, K., Kobayashi, M., Inagaki, H., Katsumata, M., Hirata, Y. and Krensky, A.M. (2007) Visiting a forest, but not a city, increases human natural killer activity and expression of anti-cancer proteins. *International Journal of Immunopathology and Pharmacology* 21, 117–127.

Li, X., Zhang, Z., Gu, M., Jiang, D.Y., Wang, J., Lv, Y.M. and Pan, H.T. (2012) Effects of plantscape colours on psycho-physiological responses of university students. *Journal of Food, Agriculture and Environment* 10, 702–708.

Lockwood, I. and Stillings, T. (1998) *Traffic Calming for Crime Reduction and Neighborhood Revitalization*. In: Institute of Transportation Engineers 68th Annual Meeting, Toronto, Ontario.

Lottrup, L., Grahn, P. and Stigsdotter, U.K. (2013) Workplace greenery and perceived level of stress: benefits of access to a green outdoor environment at the workplace. *Landscape and Urban Planning* 110, 5–11.

Maas, J., Verheij, R.A., Spreeuwenberg, P. and Groenewegen, P.P. (2008) Physical activity as a possible mechanism behind the relationship between green space and health: a multilevel analysis. *BMC Public Health* 8, 206.

Maas, J., van Dillen, S.M., Verheij, R.A. and Groenewegen, P.P. (2009a) Social contacts as a possible mechanism behind the relation between green space and health. *Health and Place* 15, 586–595.

Maas, J., Verheij, R.A., de Vries, S., Spreeuwenberg, P., Schellevis, F.G. and Groenewegen, P.P. (2009b) Morbidity is related to a green living environment. *Journal of Epidemiology and Community Health* 63, 967–973.

Milligan, C., Gatrell, A. and Bingley, A. (2004) 'Cultivating health': therapeutic landscapes and older people in northern England. *Social Science and Medicine* 58, 1781–1793.

Mitchell, R. and Popham, F. (2007) Greenspace, urbanity and health: relationships in England. *Journal of Epidemiology and Community Health* 61, 681–683.

Mitchell, R. and Popham, F. (2008) Effect of exposure to natural environment on health inequalities: an observational population study. *The Lancet* 372, 1655–1660.

Moore, E.O. (1981) A prison environment's effect on health care service demands. *Journal of Environmental Systems* 11, 17–34.

Nilsson, K., Sangster, M. and Konijnendijk, C.C. (2011) Forests, trees and human health and well-being: introduction. In: *Forests, Trees and Human Health*. Springer, Netherlands, pp. 1–19.

Ogunseitan, O.A. (2005) Topophilia and the quality of life. *Environmental Health Perspectives* 113, 143–148.

Ottosson, J. and Grahn, P. (2005) A comparison of leisure time spent in a garden with leisure time spent indoors: on measures of restoration in residents in geriatric care. *Landscape Research* 30, 23–55.

Park, B.J., Tsunetsugu, Y., Kasetani, T., Hirano, H., Kagawa, T., Sato, M. and Miyazaki, Y. (2007) Physiological effects of Shinrin-yoku (taking in the atmosphere of the forest) using salivary cortisol and cerebral activity as indicators. *Journal of Physiological Anthropology* 26, 123–128.

Park, B.J., Furuya, K., Kasetani, T., Takayama, N., Kagawa, T. and Miyazaki, Y. (2011) Relationship between psychological responses and physical environments in forest settings. *Landscape and Urban Planning* 102, 24–32.

Park, S-A, Shoemaker, C.A. and Haub, M.D. (2009) Physical and psychological health conditions of older adults classified as gardeners or non-gardeners. *HortScience* 44, 206–210.

Park, S.H. and Mattson, R.H. (2008) Effects of flowering and foliage plants in hospital rooms on patients recovering from abdominal surgery. *HortTechnology* 18, 563–568.

Parr, H. (2007) Mental health, nature work and social inclusion. *Environment and Planning* 25, 537.

Peschardt, K.K. and Stigsdotter, U.K. (2013) Associations between park characteristics and perceived restorativeness of small public urban green spaces. *Landscape and Urban Planning* 112, 26–39.

Potwarka, L.R., Kaczynski, A.T. and Flack, A.L. (2008) Places to play: association of park space and facilities with healthy weight status among children. *Journal of Community Health* 33, 344–350.

Richardson, E., Pearce, J., Mitchell, R., Day, P. and Kingham, S. (2010) The association between green space and cause-specific mortality in urban New Zealand: an ecological analysis of green space utility. *BMC Public Health* 10, 240.

Richardson, E.A., Pearce, J., Mitchell, R. and Kingham, S. (2013) Role of physical activity in the relationship between urban green space and health. *Public Health* 127, 318–324.

Rfidah, E.I., Casey, P.B., Tracey, J.A. and Gill, D. (1991) Childhood poisoning in Dublin. *Irish Medical Journal* 84, 87–89.

Roe, J. and Aspinall, P. (2011) The restorative benefits of walking in urban and rural settings in adults with good and poor mental health. *Health and Place* 17, 103–113.

Roe, J.J., Aspinall, P.A., Mavros, P. and Coyne, R. (2013) Engaging the brain: the impact of natural versus urban scenes using novel EEG methods in an experimental setting. *Environmental Science* 1, 93–104.

Schipperijn, J., Ekholm, O., Stigsdotter, U.K., Toftager, M., Bentsen, P., Kamper-Jørgensen, F. and Randrup, T.B. (2010) Factors influencing the use of green space: results from a Danish national representative survey. *Landscape and Urban Planning* 95, 130–137.

Scholz, U. and Krombholz, H. (2007) A study of the physical performance ability of children from wood kindergartens and from regular kindergartens. *Motorik* 1, 17–22.

Sempik, J. (2007) *Researching Social and Therapeutic Horticulture: a Study of Methodology.* Thrive and Loughborough University CCFR, Reading, UK.

Sempik, J., Aldridge, J. and Becker, S. (2005) *Health, Well-being and Social Inclusion, Therapeutic Horticulture in the UK.* The Policy Press, Bristol, UK.

Spring, J.A., Viera, M., Bowen, C. and Marsh, N. (2013) Is gardening a stimulating activity for people with advanced Huntington's disease? *Dementia* 13, 819-833.

Stigsdotter, U.A. and Grahn, P. (2002) What makes a garden a healing garden. *Journal of Therapeutic Horticulture* 13, 60–69.

Sullivan, W.C., Kuo, F.E. and Depooter, S.F. (2004) The fruit of urban nature. Vital neighborhood spaces. *Environment and Behavior* 36, 678–700.

Takano, T., Nakamura, K. and Watanabe, M. (2002) Urban residential environments and senior citizens' longevity in megacity areas: the importance of walkable green spaces. *Journal of Epidemiology and Community Health* 56, 913–918.

Tenngart Ivarsson, C. and Hagerhall, C. M. (2008) The perceived restorativeness of gardens – Assessing the restorativeness of a mixed built and natural scene type. *Urban Forestry and Urban Greening* 7, 107–118.

Thompson-Coon, J., Boddy, K., Stein, K., Whear, R., Barton, J. and Depledge, M.H. (2011) Does participating in physical activity in outdoor natural environments have a greater effect on physical and mental wellbeing than physical activity indoors? A systematic review. *Environmental Science and Technology* 45, 1761–1772.

Tsunetsugu, Y., Park, B.J., Ishii, H., Hirano, H., Kagawa, T. and Miyazaki, Y. (2007) Physiological effects of Shinrin-yoku (taking in the atmosphere of the forest) in an old-growth broadleaf forest in Yamagata Prefecture, Japan. *Journal of Physiological Anthropology* 26, 135–142.

Tsunetsugu, Y., Lee, J., Park, B.J., Tyrväinen, L., Kagawa, T. and Miyazaki, Y. (2013) Physiological and psychological effects of viewing urban forest landscapes assessed by multiple measurements. *Landscape and Urban Planning* 113, 90–93.

Tyrväinen, L., Mäkinen, K. and Schipperijn, J. (2007) Tools for mapping social values of urban woodlands and other green areas. *Landscape and Urban Planning* 79, 5–19.

Tzoulas, K., Korpela, K., Venn, S., Yli-Pelkonen, V., Kaźmierczak, A., Niemela, J. and James, P. (2007) Promoting ecosystem and human health in urban areas using Green Infrastructure: a literature review. *Landscape and Urban Planning* 81, 167–178.

Ulrich, R. (1990) *The role of trees in human well-being and health.* In: Proceedings of the Fourth Urban Forestry Conference. St. Louis, Missouri. American Forestry Association.

Ulrich, R.S. (2001) *Effects of healthcare environmental design on medical outcomes.* In: *Design and Health*: Proceedings of the Second International Conference on Health and Design. Svensk Byggtjanst, Stockholm, pp. 49–59.

Ulrich, R.S., Simons, R.F., Losito, B.D., Fiorito, E., Miles, M.A. and Zelson, M. (1991) Stress recovery during exposure to natural and urban environments. *Journal of Environmental Psychology* 11, 201–230.

van den Berg, A.E. and Custers, M.H. (2010) Gardening promotes neuroendocrine and affective restoration from stress. *Journal of Health Psychology* 16, 3–11.

van den Berg, A.E. and van Winsum-Westra, M. (2010) Manicured, romantic, or wild? The relation between need for structure and preferences for garden styles. *Urban Forestry and Urban Greening* 9, 179–186.

van den Berg, A.E., van Winsum-Westra, M., de Vries, S. and van Dillen, S.M. (2010) Allotment gardening and health: a comparative survey among allotment gardeners and their neighbors without an allotment. *Environmental Health* 9, 1–12.

van Dillen, S.M., de Vries, S., Groenewegen, P.P. and Spreeuwenberg, P. (2012) Greenspace in urban neighbourhoods and residents' health: adding quality to quantity. *Journal of Epidemiology and Community Health* 66, e8.

van Duijne, F.H., Kanis, H., Hale, A.R. and Green, B. (2008) Risk perception in the usage of electrically powered gardening tools. *Safety Science* 46, 104–118.

Velarde, M.D., Fry, G. and Tveit, M. (2007) Health effects of viewing landscapes – Landscape types in environmental psychology. *Urban Forestry and Urban Greening* 6, 199–212.

White, M.P., Alcock, I., Wheeler, B.W. and Depledge, M.H. (2013) Would you be happier living in a greener urban area? A fixed-effects analysis of panel data. *Psychological Science*, 0956797612464659.

Whitehouse, S., Varni, J.W., Seid, M., Cooper-Marcus, C., Ensberg, M.J., Jacobs, J.R. and Mehlenbeck, R.S. (2001) Evaluating a children's hospital garden environment: utilization and consumer satisfaction. *Journal of Environmental Psychology* 21, 301–314.

Wilson, E.O. (1984) *Biophilia.* Harvard Press, Harvard.

Yamaguchi, M., Deguchi, M. and Miyazaki, Y. (2006) The effects of exercise in forest and urban environments on sympathetic nervous activity of normal young adults. *Journal of International Medical Research* 34, 152–159.

4 Environmental Horticulture and Conservation of Biodiversity

Ross W.F. Cameron

> **Key Questions**
> - What is meant by biodiversity?
> - What pressure do species living in towns and cities face?
> - What aspects of environmental horticulture provide opportunities for urban wildlife and what pose threats?
> - What types of locations/landscapes tend to have the highest urban plant biodiversity?
> - What key traits are linked with plant/animal species that are relatively successful in surviving in urban habitats?
> - How does the management of parks and gardens affect opportunities for wildlife?
> - What plant genotypes could be used to provide a continuous resource of nectar and pollen from February through to October in temperate climates?
> - What are the key features of brownfield sites and how do these influence the species found on them?
> - What are the advantages/limitations of green roofs in providing habitat for urban wildlife?

4.1 Introduction

Environmental horticultural practices should aim to increase or maintain biodiversity. *Traditional* horticultural practices may have viewed the cultivated plants as important, but demonstrated little interest in associated plant and animal assemblies, other than to eradicate them when they were deemed in competition with, or a threat to, the 'desirable' plants. This included the use of pesticides that were indiscriminate in their targeting of both pest and benign species alike. Thankfully, plant husbandry within *environmental* horticulture takes a more holistic view. Whilst trying to maintain populations of cultivated plants, this should be achieved by minimizing the impact on other species that may use the landscape. This is not to be confused with the complete restoration of natural habitats and communities (although sometimes that is the objective), but rather to develop some form of semi-natural or artificial plant communities that still provide a range of ecological niches to local wildlife. The focus of this chapter is on the urban environment, where horticultural land management (for non-food purposes) is most prevalent, and where the pressures on wildlife are particularly challenging, through the loss and fragmentation of green space. Specific management approaches discussed may be equally relevant in rural landscapes managed by horticulturalists, for example in large country gardens and estates, but the wider landscape context may be somewhat different.

One key driver for environmental horticulture is to restore green landscapes and a degree of ecosystem function to areas that are currently highly degraded and where the biodiversity is impoverished. This does not automatically assume the use of native plant species exclusively in this process of 're-greening'. Indeed environmental horticulture has been embroiled in the debates relating to 'is native always best?', for example the arguments that non-native highly-floristic plant species can provide pollen and nectar resources to invertebrates just as well as, or in some cases better than, native plant species communities. This is not to advocate, of course, that native plant communities with high biodiversity indices should be replaced by non-native species, but that in the highly 'artificial' environment of our towns and cities, non-native species may have a significant role in improving the opportunities for wildlife. A counterbalance to this, however, is that choice of species should be aimed at minimizing the risk of introducing aggressive, invasive plant species that threaten native habitats.

4.2 A Definition for Biodiversity

Biodiversity or biological diversity is often defined as the 'totality of genes, species and ecosystems of a region' (Larsson, 2001) (Table 4.1). Most people simply understand it as the number of species present, but actually variation within a species is also an important component. The term also alludes to the fact that we somehow know what all these species are, but actually science has probably only named about 16–30% of the total species present on the globe. Nevertheless, the notion of biodiversity is a useful one as it allows some indication of the 'biological richness' a given place, location or habitat may possess.

Urban biodiversity is defined by 'the variety or richness and abundance of living organisms (including genetic variation) and habitats found in, and on the edge of, human settlements' (Müller *et al.*, 2013). Urban landscapes cover approximately 3% of the Earth's land surface and more than half of the world's human population now live in them (Wu, 2010). Urban biodiversity is the range of species associated with this landscape type and ecosystem, and essentially means the species found from the urban core out to the urban fringe of a human settlement. It typically includes habitats such as:

- Urban parks and gardens.
- Urban industrial sites – factories, business parks, wastelands, brownfield sites, vacant space, residential spaces.
- Transport corridors and networks – road verges, railway embankments, canals.

- Peri-urban agricultural land – farmland integrated with the suburbs, city farms, allotments, meadows, pastureland, arable and vegetable croplands and rangeland.
- Urban forests – woodland and forest within the urban matrix, but also street trees and remnants of more ancient woodland.
- Rock surfaces and screes – remnant vegetation and animal communities with cliffs, but also anthropogenically determined habitats such as green roofs and walls.

Prance *et al.* (2014) have gone further and termed and defined 'horticultural biodiversity'. This is the process by which diverse organisms, ecosystems and ecological processes provide economic, environmental and social benefit when interacting with managed open green space. In reality horticultural biodiversity would seem to have two roles:

1. Offering habitat to allow native species to retain or gain a foothold in an otherwise adverse environment.
2. To conserve plant species and genotypes that might otherwise be threatened, for example, the role of a botanic garden (e.g. at a species and subspecies level) and heritage gardens (often at a cultivar level).

If this is the case, then perhaps the argument should be made that horticultural biodiversity should not be excluded solely to open green space, but also should include closed green space (glasshouses, atria, conservatories etc.).

Biodiversity as a term is often used to confer a notion of richness in wildlife, at least to a lay audience. In reality, the life does not necessarily need to be 'wild'. In an urban context a zoo or botanic garden could be extremely rich in biodiversity, but there may be few native species present at all. This point needs clarifying here, because in the forthcoming sections where the term 'biodiversity' is used it refers to all species, not just native ones.

4.3 Urban Ecology and Habitats

Within the urban landscape there will be residual areas of habitat types such as woodland, grasslands, waterways and wetlands. Cities in the tropics/subtropics may still contain patches of desert, arid scrubland, heath and maquis or remnant patches of rainforest. Landscapes being transformed through

Table 4.1. Biodiversity as arranged by ecological, organism and genetic levels.

Ecological diversity	Organismal diversity	Genetic diversity
Biome	Kingdom	Population
Bioregion	Phylum	Individual
Landscape	Family	Chromosome
Ecosystem	Genus	Gene
Habitat	Species	Nucleotide
Niche	Subspecies	
Population	Population	
	Individual	

urbanization may not always have been natural/semi-natural and many cities may retain areas of agricultural land or large estate parkland. Added to these, the urbanization creates new habitats such as road verges, railway embankments and sidings, airport runways and grassed 'safe' zones, vacant plots, wastelands, cemeteries, hard surfaces including roofs and walls, parks, allotments and gardens. The value of urban habitats to wildlife conform to the same generic ecological principles/concepts as any other habitat type. A number of the concepts often used in urban ecology particularly, are briefly described in Table 4.2.

Mosaics, networks and patches

As the human population in the landscape increases, both patch density (different types of land cover and land use per km^2) and edge density (total length of all edge segments per ha) are enhanced, but connectivity between patches is reduced. Towns and villages are less spatially heterogeneous than large cities, with the former matrices having been observed to favour native mammalian and native plant species. Both patch configuration and size (i.e. the mosaic) affects the richness of remnant vegetation, with larger and more closely aligned patches helping to retain plant populations. Small patches, however, maintain their value through acting as 'stepping stones' for species movement. Indeed, Rudd *et al.* (2002) argue that small remnant patches of vegetation combined with new gardens form a habitat network in urban landscapes that is critical to the conservation of local species.

The importance of green corridors, mosaics and high connectivity within urban environments, however, may not be weighted equally across the taxonomic groups. Angold *et al.* (2006) studied the biodiversity of urban habitats in Birmingham, UK. They indicated that effective dispersal of species across the mosaic was not a limiting factor in maintaining population persistence of certain plant and butterfly taxa in their studies. Another taxa under study though, the woodland carabid species (beetles), did appear to align to some degree of geographical structuring within the city, implying that green 'woodland' networks may be more important for this group of animals. The researchers concluded that green corridors are appropriate for some species such as small and medium-sized mammals, e.g. *Arvicola amphibius* (water vole) and *Muscardinus avellanarius* (dormouse), but did not find any evidence from their field-based data to indicate that plants or invertebrates use urban greenways for dispersal. In contrast, rather than connectivity or site location being paramount, the data on carabids and butterflies illustrated the high importance of habitat quality on individual sites within the conurbation.

Promoting urban biodiversity

Environmental horticulture can be practised almost anywhere plants are cultivated, but it is more commonly associated with urban environments and as an antidote to urbanization. Urbanization is not necessarily the elimination of biodiversity, but generally it has negative impacts, including:

- The destruction of natural ecosystems.
- The fragmentation of habitats.
- The degradation of ecosystem processes.
- The alteration and modification of natural disturbance regimes.
- The introduction of non-native animal and plant species.

With the exception of the last point (as horticulture is one of the main catalysts for the introduction of non-native species), environmental horticultural practices should go some way to mitigating these other processes. Indeed, environmental horticulture needs to play a role in sustaining ecosystem services and goods, conserving biodiversity within cities and towns, endorsing sustainable design and management systems whilst also promoting education and awareness to enhance green spaces for both humans and other species.

How does urbanization affect plant and animal species habitat, composition and behaviours? The central problems are outlined in Table 4.3. Occurrence of species depends on factors such as the species' own particular traits (e.g. where it sources food) and adaptability (noise sensitive versus ability to tolerate human disturbance), population size, the history of a site or habitat, the availability of appropriate habitat and the quality and spatial arrangement of habitats. Urban landscapes are 'top-down' human-dominated landscapes, and the planning, design and management of locations often determines the diversity and abundance of species that can be found there. As with other scenarios (e.g. islands), the larger the urban green area available to wildlife (patch size), the greater the variety of

Table 4.2. Common ecological concepts and their definitions.

Ecological concept	Description
Abundance	The number of individuals representing a species in a particular ecosystem. Relative abundance is the ratio of individuals counted in one species compared to individuals in all species.
Biome	A large naturally-occurring community of flora (plants) and fauna (animals) occupying a major habitat, e.g. forest or savannah.
Biotope	Areas with uniform biological conditions (climate, soil, altitude, etc.).
Carnivore	A species that is reliant on predating and feeding on animals.
Communities	Flora and fauna including microorganisms found in a location or habitat.
Competition	The simultaneous demand for an essential common resource by two or more organisms or species and which is actually or potentially in limited supply. May relate to food, but also for the right to hold a territory or to mate. Competition can occur between individuals within a species (intraspecific) or between species (interspecific).
Competitor plant	Plants that thrive in areas of low intensity stress and disturbance and out-compete other plants by efficiently exploiting available resources. They do this with traits such as rapid growth rate, high productivity and phenotypic plasticity (allocating resources where most needed at any given time).
Corridor	A strip of vegetation or waterway used by wildlife and potentially allowing movement of biotic factors between two areas. A number of intersection patches and corridors make up a network.
Depauperate	Areas poor in species number or diversity.
Dispersal	The ability of species to move, for example from one pond to another.
Disturbance	A change (often temporary) in environmental conditions that causes pronounced changes in an ecosystem or to species dynamics. Biodiversity is often promoted by moderate levels or infrequent occurrence of disturbance.
Diversity	Differences (usually in species or genes). Alpha diversity is the amount of diversity found within a community and beta diversity is the diversity resulting from differences amongst the various communities.
Ecocline	A gradual change in the ecosystem, for example change in environmental condition as altitude increases.
Ecotone	The area between two habitat types – edge or boundary communities, for example 'woodland' edge where there may be species from the wood and from a neighbouring grassland community. Often more biodiverse than either habitat alone.
Ecotope	A small ecologically spatial unit where conditions tend to be relatively homogeneous.
Equitability	The 'evenness' of the community – how equally all species can be found in a habitat.
Food chain	A simple plot of energy flow in a community, e.g. primary producer–herbivore–carnivore.
Fragmentation	Break-up of a habitat or ecosystem into smaller parcels. Affects the number of individuals located within an area.
Food web	Describes links among many species, i.e. the combination of different food chains.
Generalist	Species that are both widespread and common because they can use many resources or are highly adaptable.
Guild	A collection of species that use similar resources in similar ways (e.g. fish-eating birds).
Habitat	The physical location or type of environment in which an organism or biological community lives or occurs.
Herbivore	An animal that relies on plants for food.
Heterogeneity	The uneven distribution of various habitat patches, land forms or concentrations of each species within an area. A landscape with spatial heterogeneity has a mix of concentrations of multiple species (ecological), or geological formations or environmental characteristics (e.g. wind, sunlight) within it. A population showing spatial heterogeneity is one where various concentrations of individuals of this species are unevenly distributed across an area, i.e. patchily distributed.
Immigration	New species or individuals entering an area.

Continued

Table 4.2. Continued.

Ecological concept	Description
Island theory	This relates to the position of vegetation stands or other habitat types in relation to one another. For example, how areas of discrete oak woodland affect each other in determining their species composition. Larger areas of woodland ('islands') will support more species. Areas of woodland further away from the others will be more isolated and will support fewer species. There is competition within an island and this can result in a turnover of the species present over time. Some species will become extinct, but other newcomers will arrive from time to time.
Matrix	Main or background ecological system, e.g. a forest matrix is a large area predominantly covered with patches of forest.
Metapopulations	A series of local populations, spatially separate from one another, but where individuals can move from one to another.
Mosaics	The patterns of patch habitats and corridors within the wider landscape matrix.
Networks	A number of inter-connecting corridors allowing different patches to be linked. Connectivity relates to how well species can use these networks to cross the landscape. Species survival tends to be higher in patches that have higher connectivity.
Niche	The habitat, limited by environmental factors (fundamental niches) and/or competitors (realized niches) where individuals of a species can survive and reproduce.
Omnivore	A species that can eat either plant or animal material.
Parasitoid	An organism that in part of its life cycle lives in or on another organism, and usually ends up killing the host.
Patch (habitat)	One of a number of areas depicting a type of habitat. A fragmented habitat may be one that has been broken up into a number of patches, by housing, road or rail networks, and where movement between patches may be difficult for wildlife.
Refuge	Area where a species is exposed to less environmental, predation or competitive pressures.
Remnant vegetation/habitat	Areas within the urban matrix that represent the type of habitat that was prominent before urban development took place – essentially 'left-over' patches.
Richness (species)	The number of different species present.
Ruderal plant	Plant species that are adapted to high intensity disturbance and low intensity stress. Such species tend to be fast growing and have rapid, short life cycles, investing heavily in seed production to ensure the next generation. Ruderals often dominate the colonization of recently disturbed land.
Saturation	The concept when all the available niches are filled by one or more species.
Scale	The size of a landscape or habitat. Can vary markedly depending on the species being considered (puddle versus ocean). Larger areas will support more species and more individuals, as there are more resources available. Larger areas also allow more individuals to possess territories, and avoid in-breeding in isolated populations.
Specialist	A species that may be very dependent on a single food source or habitat type.
Stress-tolerator plant	Plants that compete by being adapted to intense abiotic stress and low intensity disturbance. Found in adverse environments such as alpine or arid habitats, deep shade or soils which are nutrient deficient, contaminated or high/low pH. Typified by slow growth rates, long-lived leaves, high rates of nutrient retention, and low phenotypic plasticity.
Taxocene	Closely related set of species within a community (e.g. aphids).
Taxon (plural: taxa)	A group of one or more populations of an organism or organisms seen by taxonomists to form a functional or discrete unit.
Trophic levels	Subsets of species that acquire energy in similar ways (e.g. herbivores, carnivores, detritivores).
Universal adaptive strategy theory	An evolutionary theory based on the trade-off that organisms (usually plants) face when the resources they gain from the environment are allocated between either growth, maintenance or regeneration – known as the universal three-way trade-off or C-S-R (competitor – stress tolerator – ruderal) theory.

Table 4.3. Key aspects of urbanization that affect habitat provision.

Change due to urbanization	Effects on biodiversity
Loss of natural habitat or agricultural land.	Loss or degradation of previous ecosystems.
Configuration of buildings, technical infrastructure and open spaces.	Increase in non-permeable hard surfaces, roads, paving and roofs. In city centre 60% of land surface area can be covered.
Grey infrastructure such as roadways and railway lines that fragments habitats.	Inhibition of animal movement or high incidence of fatalities, causing population isolation and potential in-breeding.
Increase in built infrastructure.	Presents hazards for some species. More than 100 million birds are estimated to be killed each year by flying into windows. Alternatively, some tall buildings provide habitat, e.g. for the feral pigeon and peregrine falcon.
New built structures.	Bridges, house roofs, underpasses, overpasses, and culverts serve as nesting and roosting sites for a number of species, e.g. *Hirundo pyrrhonota* and *H. fulva*, cliff and cave swallows, respectively. In the USA, 50% of bat species use bridges as roosting sites.
Modification of the soil-moisture regimes, drier in temperate zones, but wetter desert areas due to irrigation.	Changes plant species composition to drier-adapted species in the first instance and more nutrient-competitive species in the second.
High nutrient loads in some soils, due to atmospheric pollutants, eutrophication processes and active land use change.	As above, alters the balance with respect to plant colonization strategies. Soils richer in nutrients promote competitive and ruderal 'weedy' species rather than stress tolerators.
Warmer temperature due to urban heat islands.	Enhances range of plant species that can be grown or animals that can survive in cooler temperate climate regions (but perhaps limits species in future in warm, arid climates). Can alter growing periods and other phenological aspects (e.g. lack of winter chilling that some species require for seed germination or budbreak).
Higher productivity of plant biomass due to cultivation.	Can be exploited by some herbivore invertebrates, but may cause 'pest' explosions if no natural predators present.
Abundance of food and food wastes and other resources from human activities.	Allows high populations of some generalist species to be maintained.
Higher levels of disturbance through noise and human activities.	Can alter feeding and breeding patterns.
Soil contamination (nitrogen and calcium deposition and certain heavy metals), air pollution (elevated CO_2, NO_x, aerosols, metals and ozone) and water pollution, with particular impacts.	Impacts on species compositions, particularly soil organisms, lichens, and aquatic species.
Disturbance such as removal of all vegetation, trampling, construction, mowing, radical soil change, light pollution, litter or illegal dumping, arson and vandalism.	Loss of species diversity.
Introduced non-native species of plants and animals.	Risk of invasion to surrounding semi-natural habitats.
High predation rates due to pet species.	Populations of vulnerable species such as lizards/snakes reduced. Impact on bird and small mammal breeding rates.
High proportion of habitat generalists and common plant and animal species.	Loss of biodiversity. Cityscapes, flora and fauna become more uniform across the globe.

species and number of individuals within a species, that are likely to be present.

There are gradients across the urban matrix; in general the richness of native species declines as one moves towards the city centre, whilst non-native species richness increases towards the centre of the city, or towards the older residential suburbs and away from the rural hinterland. For certain taxa such as rodents this is well illustrated. Work in Buenos Aires, Argentina, indicated that of the seven species observed, the four native rodents were dominant on sites with natural vegetation, often located at the edge of the city, whereas the three non-native species dominated in inner city shantytowns, industrial sites, and residential neighbourhoods (Cavia et al., 2009). Overall, non-natives were faring better, as their species richness increased as new habitats were created with urbanization. In contrast, the richness of native species declined as remnant natural habitats became further fragmented, isolated, or destroyed.

In the urban core approximately 30–50% of the plant species are non-native (Dunn and Heneghan, 2011). Overall plant biodiversity peaks in the suburbs where there are still relatively adequate areas of green space and associated niches for native species, due largely to a heterogeneous landscape typology (parks, woodlands, gardens, waterways), whilst the increase in garden space correlates with an abundance of non-native garden plants. Socio-economic factors play a role here too, for example more affluent citizens may spend money on bird food to augment natural supplies. In the UK, populations of *Vulpes vulpes* (red fox) are probably highest in the more affluent suburbs of UK cities than any other 'habitat' in the country, due to ready sources of food (earthworms, fallen fruit, small mammals and discarded human food), less persecution and greater tolerance from the local human population. These suburbs are typified too, by a useful balance of green open spaces to forage and secluded large gardens to rest-up in and raise cubs. Certain bird and butterfly species follow a similar pattern across the urban matrix, being closely linked to those areas where parks and large gardens are most common. A linear gradient, of course, to some degree is a simplification and heterogeneity within the city often exists because of different building types and social contexts and this alters the opportunities available for different species.

Climate impacts on the gradient affect wildlife populations too. In desert cities such as Phoenix, Arizona (USA), the 're-greening' of the landscape through irrigation and the introduction of non-native ornamental plants has boosted populations of both native and non-native birds, insects and other animals in the city compared to the surrounding (rather arid) landscape (Hope et al., 2006).

So, the extent to which the ratio of native to non-native plants is important to other species living in the urban green space remains an unresolved question. The relative value of native/non-native plants depends on the taxa in question, the size and design of the green spaces, their integration and how these spaces are managed. Surveys in Flanders, Belgium, revealed that 15 urban parks contained about 30% of the total number of wild plant species held within the region, 50% of the breeding birds, 40% of butterflies and 60% of the region's amphibians (Cornelis and Hermy, 2004).

It has been suggested that provision of greenery can ameliorate the hostility of the urban environment for wildlife, but Chong et al. (2014) argue that greenery can be either in the form of regenerating or remnant patches of natural vegetation, or as cultivated tree, shrub, and ground cover (with a high proportion of non-native species). They made comparisons of bird and butterfly species richness between natural and cultivated vegetation cover in Singapore, whilst trying to take account of traffic density (which can disturb bird behaviour through noise). As anticipated, natural vegetation cover had a positive effect on the richness of both bird and butterfly species, but it was also evident that landscapes composed of cultivated trees enhanced bird diversity too. In contrast, increasing turf grass cover had a negative effect on bird diversity. Overall, increasing the area of cultivated vegetation had a moderate positive effect on butterfly species richness, but not as dramatic as increasing natural plant cover. For butterflies (and other nectar-seeking insects), however, targeting non-native plants that are floriferous and provide nectar and pollen in abundance, can help mitigate some of the problems caused by urbanization. Chong *et al.* state that remnants of native vegetation are better for biodiversity than cultivated landscapes, but if cultivated landscapes are the only option available these should mimic the structural features of natural systems as closely as possible, for example by having tiered layers of canopies at different heights, and areas of dense undergrowth for shelter, as well as more open 'foraging habitat'.

Changes to habitat form and complexity can be problematic in urban areas, particularly those frequented or intensively managed by humans. A simplification of the forest structure through management procedures that remove old or dead trees for safety reasons, clear understorey foliage to improve sight lines and involve a general 'tidying up' of the woodland floor, actually result in the loss of niche habitats. In Australia, ground-dwelling mammals are extirpated from urban landscape not only because of habitat loss, but also due to simplification of their habitats. In these 'tidy' urban woodlands, fallen logs and branches are often removed for human safety and to reduce the risk of forest fires but these habitat components are used by small ground-dwelling mammals as protection from predation (van der Ree and McCarthy, 2005). The removal of dead wood and leaf litter increases an individual's exposure to predation, thus reducing population density and species richness. Dead wood is important for invertebrates, such as woodboring beetles or as a food source for the young of other Coleoptera (beetle) species. In the early stages of decomposition, heartwood is commonly fed upon by larvae of lesser stag beetle (*Dorcus parallelipipedus*) and rhinoceros beetle (*Sinodendron cylindricum*), forming characteristic large and convoluted galleries. All in all, it is thought that over 1700 different invertebrate species in the UK and Eire are dependent on decaying wood in order to complete their life cycles. Removal of deadwood therefore reduces a fundamental component of the food chain.

4.4 What Sort of Species Inhabit Urban Areas?

The species that live within urban landscapes tend to fall into four categories: (i) species that were present prior to, and survived, the urbanization; (ii) species native to the region that have migrated in because conditions are conducive; (iii) non-native species introduced by humans or that have colonized from somewhere else by their own means; and finally (iv) species which have no natural habitats but have evolved to adapt to agricultural, urban and industrial landscapes (anecophytes). This fourth group includes a number of species that are common to many urban conurbations across the globe; they may be 'weedy' plant species such as *Taraxacum officinale* (dandelion), birds such as *Passer domesticus* (house sparrow) or mammals, including the ubiquitous *Mus musculus* (house mouse).

Urbanization as a process creates new habitats, but often ones that non-native 'generalist' species are better at exploiting than native ones. In reality, native species vary in their ability to exploit these new habitats and niches. There are animal species that thrive living closest to mankind (synanthropic), such as the aforementioned house mouse, but also *Rattus norvegicus* (brown rat), *Columba livia domestica* (feral pigeon) and *Blattella germanica* (German cockroach); others that can tolerate the urban and suburban conditions (suburban adaptors), such as the red fox, *Accipiter nisus* (sparrowhawk), *Turdus merula* (blackbird) and *Milvus milvus* (red kite) within Europe, and *Procyon lotor* (raccoon) and *Sciurus carolinensis* (grey squirrel) in North America; and others that do not (suburban avoiders), such as *Aquila chrysaetos* (golden eagle), *Gavia immer* (northern diver), *Lepus timidus* (mountain hare). Animals that thrive in urban environments possess traits such as:

- A catholic choice of food sources (omnivorous).
- Tolerance to variation in the abiotic environment.
- Adaptability in terms of the conditions required for shelter and breeding.
- High reproductive and survival rates.
- High rates of recruitment through immigration.
- Limited pressure from competitors and/or predators.
- Habituation to human activities and an ability to cope with highly fragmented landscapes with abundant edges (Adams and Lindsey, 2009).

As indicated above, the effects of urbanization do not impact on all species evenly. Often predators at the top of the food chain (apex predators) are the most vulnerable. These species are susceptible to population declines due to habitat fragmentation and loss, reduction of landscape connectivity, and increases in road density and traffic volumes. More generalist and often smaller, better-adapted mesopredators, on the other hand, have adapted well to the highly fragmented urban landscape and have substantially increased in abundance in the absence of apex predators (a phenomenon known as mesopredator release) and due to an increase in food supply (Prugh *et al.*, 2009). Mesopredators adapted to urban conditions include *Mephitis mephitis* (skunk), the raccoon and the red fox as well as some snake species. Enhancement of populations of these species, however, shift trophic structures and is frequently detrimental to smaller prey species, for example, specialized rodents and ground-nesting birds. Domestic cats (*Felis catus*), fit within this role too and it is estimated that in the USA alone cats kill as many as 1.4–3.7 billion birds and 6.9–20.7 billion mammals per annum (Loss *et al.*, 2013).

Although the general pattern is of native species richness declining and non-native species richness increasing over time, collectively, native species can still comprise 50–70% of total species richness in a city, albeit sometimes they are found as less frequently abundant species (Kowarik, 2011). In other parts of the world, particularly vulnerable island ecosystems such as New Zealand, where European settlers retained a desire for the species common to their homeland, the proportion of non-natives is much greater, with negative consequences for many of the original endemic species.

Generating data for urban biodiversity can be surprisingly challenging. There is a disproportionately large amount of information on urban parks and woodlands, but the typology and associated biodiversity of the private garden to some extent remains a mystery, largely due to a lack of access for professional researchers. Often more than 70% of the land in urban areas is privately owned. In addition, these garden landscapes can be highly variable in their landscape typology and range of plant species incorporated. Gardens that are completely paved over may sit cheek by jowl with those that are packed with trees, shrubs, flowering herbaceous perennials and annual flowering plants that provide rich nectar and pollen sources. Sometimes though, this 'missing element' on information pertaining to domestic gardens can be accommodated through citizen science studies. In the UK, the Royal Society for the Protection of Birds (RSPB) conducts an annual survey of bird populations by asking their amateur members to record what they see in their own gardens over a one hour period in late winter, and use this to census changes in garden bird populations. Perhaps one of the most comprehensive attempts to 'fill in the gaps' and address garden biodiversity was the BUGS (biodiversity in urban garden survey) project in Sheffield, UK. Here 61 gardens were surveyed for their biodiversity value (Thompson *et al.*, 2003, 2004).

Plants

Plants with ruderal characteristics dominate many city landscapes across the globe. Examples of such species include many common 'weed' species, e.g. *Taraxacum officinale* and *T. erythrospermum* (dandelion), *Capsella bursa-pastoris* (shepherd's purse), *Plantago major* (greater plantain), *Senecio vulgaris* (grounsel), *Stellaria media* (chickweed), *Polygonum aviculare* (prostrate knotweed), *Chenopodium album* (fat-hen), as well as grasses such as *Hordeum murinum* (false barley), *Cynodon dactylon* (Bermuda grass) and *Poa annua* (annual meadow grass).

Even in terms of cultivated plants there are concerns that many cities are dominated by the same species, i.e. there is a global homogenization process at work. The popularity of aesthetically pleasing, strongly structured trees or shrubs or colourful flowering garden plants has resulted in garden and park landscapes being very similar irrespective of location around the world, or at least within similar climatic zones. The plants used in an ornamental park or garden in California, USA, may be very similar to ones used in Greece or Italy. Ignatieva (2010) analysed catalogues from a range of nurseries within the temperate zones of the USA, Russia, Germany and New Zealand, and found a high degree of commonality on the genotypes used irrespective of the location. Favoured plants across temperate zones were European deciduous trees and shrubs and some 'fashion' conifers. Plants that appeared to have universal appeal include:

- Deciduous trees species – *Acer* (maple), *Betula* (birch), *Fraxinus* (ash), *Prunus* (ornamental cherry), *Populus* (poplar), *Quercus* (oak), *Rhododendron*, *Salix* (willow) and *Ulmus* (elm).
- Evergreen tree species – *Chamaecyparis lawsoniana* (Lawson cypress and related cultivars), *Juniperus* (junipers), *Picea* (spruce), *Pinus* (pine) and *Thuja* (cedar).
- Annual and perennial flowering plant species and cultivars – *Impatiens* (busy Lizzie), *Pelargonium* (geranium), *Petunia*, *Tagetes* (marigold) and *Viola* (pansy and viola).
- Lawn grass species – *Agrostis capillaris* (common bent), *Festuca rubra* (red fescue), *Lolium perenne* (perennial rye-grass) and *Poa pratensis* (Kentucky blue grass).
- Ornamental grass species – *Cortaderia* (pampas grass), *Miscanthus*, *Pennisetum* and *Stipa*.

Parks and gardens in the tropics are less well documented but common 'global' species are likely to include: *Acacia* (wattle), *Bougainvillea*, *Casuarina*, *Croton*, *Delonix regia* (flamboyant tree), *Hibiscus rosa-sinensis*, *Jacaranda*, *Plumeria* (frangipani tree) and *Strelitzia reginae* (bird of paradise), (McCracken, 1997; Soderstrom, 2001).

In terms of non-native plants, horticulture is a principal contributor to the non-native biodiversity of urban environments. Since Neolithic times, approximately 12,000 plant species have been introduced into Europe for ornamental and cultural purposes, with about 10% of these becoming naturalized. Horticultural practices too have increased both intraspecific (within a species) and interspecific (between species) hybridization, thus swelling the numbers of non-natives further. Introduced plants seem to proliferate and become more variable too, when introduced to new areas. Gilbert (1989) noted that *Aster* genotypes (Michaelmas daisies) introduced to the UK were more variable both morphologically and in their ecological amplitude than the parent species in their native North America. Older parts of cities, especially the residential areas, have more non-native species than recently settled landscapes. Furthermore, larger urban landscapes (i.e. cities) have more non-native species than small urban landscapes (i.e. villages and towns) (Pyšek *et al.*, 2004).

Mammals

With the exception of a small number of urban adapters, such as the brown rat and house mouse, inner city zones provide few opportunities for mammals. As green space increases outside the business zone and into the suburbs though, the diversity of species can increase. In these locations, however, scale becomes important, as mammals in general need greater spatial scales than many other taxa to maintain viable population sizes. Where networks are effective, such as through the predominance of permeable boundaries between gardens (hedges), this may allow for viable populations of e.g. *Erinaceus europaeus* (hedgehog), whereas impermeable barriers (walls and fences) will not. Even so, although a relatively wide range of mammal species have been recorded in urban garden habitats and city parks, it tends to be the domain of a relatively select few, e.g. in the UK, the red fox, grey squirrel and hedgehog as well as certain bats, voles and mice.

Where extensive areas of green space exist in the suburbs and urban fringes, then opportunities do exist for larger mammal species or those requiring greater territories. Indeed, the relative lack of persecution from hunting of larger predators that occur in suburbs compared to the rural environment, means that some species may be more obvious, even if actual populations are not necessarily greater. There is increasing evidence that selected species are becoming more adapted to urban environments too (synurbization). This includes species such as *Alces alces* (elk or moose) in Sweden or Canada (Fig. 4.1), *Oryctolagus cuniculus* (rabbit), *Capreolus capreolus* (roe deer), *Meles meles* (badger), *Martes foina* (stone marten) in Western Europe and *Canis latrans* (coyote), *Lynx rufus* (bobcat), *Odocoileus virginianus clavium* (key deer) and *Tamias striatus* (chipmunk) in North America. In parts of Africa and India there is even evidence of *Panthera pardus* (leopard) entering urban situations and preying on feral dogs. Similarly in Europe, a degree of relaxation in the persecution of the *Canis lupus* (wolf) has resulted in some individuals at least crossing urban environments, if not quite settling.

Birds

Compact, densely built, busy city centres tend to be the most challenging urban environments for birds and even here non-native species richness usually decreases, due to lack of green space and other secure habitat. Nevertheless, even in these relatively inhospitable city centre locations, opportunities can arise for species adapted to take advantage of the peculiar conditions encountered. Feral pigeons thrive on the buildings and roofs that replicate their original sea-cliff homes in terms of structure and lack of ground-dwelling predators. Other cliff dwellers such as *Aeronautes saxatalis* (white-throated swift) and *Falco peregrinus* (peregrine falcon) can equally exploit tall buildings and other built structures as nesting sites. In fact, urban landscapes can be superior habitats for raptors (birds of prey) because they are free of human persecution and have high availability of abundant food, in the case of the peregrine the aforementioned pigeons. Others are more dependent on 'traditional' habitat types or vegetation groupings and species. The number of birds within the Paridae

Fig. 4.1. Surprisingly large mammals such as *Alces alces* (moose) can cohabit with humans in peri-urban environments.

(*Parus major*, *Periparus ater* and *Cyanistes caeruleus*; great, coal and blue tit, respectively) were observed to increase as the proportion of native trees proliferated across the urban forest.

Shifts in bird behaviour have been noted too. This includes urban species becoming less territorial, nesting earlier and nesting on more unusual structures (such as building ledges, telegraph poles, outhouses of domestic properties), as well as producing more than the normal one brood a year. Some may reside all year round rather than follow their traditional migration patterns. Others alter their feeding habits, such as eating seeds or other new sources of food (Blair and Johnson, 2008). A classic example of adaptation was illustrated by the blue tit, which capitalized on the popularity of milk bottle deliveries to residents' doorsteps in the UK during the 1970s and 1980s. These milk bottles were topped with metal foil, which the birds learned to peck through and access the calorie-rich cream at the top of the milk.

Remnant or residual patches of native vegetation are important to some bird species. A study of hummingbirds in the city of Campo Grande (Brazil) showed that species that were usually forest dwellers could survive in small urban fragments of cerrado (grass and tree) vegetation (Barbosa-Filho and de Araujo, 2013). The hummingbird species relied both on plant species normally pollinated by hummingbirds (ornithophilous), but also other non-ornithophilous plant species for their source of nectar. The high instances of visits to the latter plant type was thought to relate to a number of contributing factors: the scarcity of the ornithophilous species that were the traditional food source and the limited period in the year they were in flower, and the provision of nectar from non-ornithophilous species with energy values equivalent to the ornithophilous species. In essence, the hummingbirds were adapting to the reduced opportunities to feed in the urban environment, by relying on other types of flowering plants that would not normally be favoured. The number of hummingbird species recorded in the urban cerrado is comparable to that reported for well-preserved forested areas. This suggests that the small forest fragments contained within the cerrado are providing important refuge sites for those hummingbird species that manage to remain prevalent in urban areas. Similarly, the remnant vegetation may play a critical role in providing 'stepping stones' across the urban matrix for such small birds, as they attempt to reach other forested areas. Understanding the birds' requirements in detail may provide opportunities for the more cultivated 'horticultural' landscapes in Brazil to replicate these floral groups and thus afford year-round food supplies to the hummingbird species in question (Table 4.4), i.e. *Amazilia fimbriata* (glittering-throated emerald), *Anthracothorax nigricollis* (black-throated mango), *Chlorostilbon lucidus* (glittering-bellied emerald), *Eupetomena macroura* (swallow-tailed), *Heliomaster squamosus* (stripe-breasted starthroat), *Hylocharis chrysura* (gilded sapphire) and *Thalurania furcata* (fork-tailed woodnymph).

As the inter-relationships between species can be exceedingly complex, the inadvertent responses to actions are difficult to predict. Catterall (2004) in her paper entitled *Birds, garden plants and suburban bushlots: where good intentions meet unexpected outcomes* cites that spatial scaling in urban design is important in trying to conserve the small-bodied, foliage feeding birds of native *Eucalyptus* forests (the so-called neglected or forgotten foliphiles). She found that as urbanization took place in Brisbane, Australia, the remaining vegetation was composed of small remnant vegetation patches (<5 ha) or well-vegetated suburbs through the planting up of gardens and parks. Local authorities and residents encouraged the retention of small vegetated patches to provide sanctuaries for wildlife. One consequence of this fragmentation of the *Eucalyptus* forests and increase in plant diversity through the garden vegetation though was an increase in larger bird species, such as *Manorina melanocephala* (noisy miner). Increases in noisy miner bird numbers, however, correlated with decreases in the smaller foliage feeders (Table 4.5). Noisy miners drive other birds away that enter their territory using vigorous coordinated attacks and loud strident vocalizations. Indeed, targets are not limited to birds, with mammals and reptiles also sometimes wounded or killed. Catterall suggested that rather than protecting the native birds of the extensive *Eucalyptus* forests, the retention of smaller and mixed patches of vegetation (due to the fact that they included nectar-rich native plant species that were attractive to the noisy miner) had a negative effect on the desired species – a lesson that positive motives for promoting native biodiversity can have a different effect from that intended.

Amphibians

The impact of urbanization on amphibians is generally negative, although some species are intrinsically

Table 4.4. Plants species, their key characteristics and flowering times, and the number of hummingbird species visiting, mean number of open flowers per plant per day, and frequency (number of flowers visited per minute/total number of observed flowers). Shaded rows indicate those species normally pollinated by hummingbirds. (Modified from Barbosa-Filho and de Araujo, 2013.)

Plant family/species	Habit	Corolla shape	Corolla colour	Flowering time	No. of hummingbird species	Flowers per plant d^{-1}	Frequency × 1000
Bromeliaceae							
Ananas ananassoides	herb	tubular	lilac	Sep.–Oct.	2	4	390
Bromelia balansae	herb	tubular	red	Jun.–Oct.	5	9	17
Bromelia plumieri	herb	tubular	red	Dec.–Feb.	1	6	2
Dyckia leptostachya	herb	tubular	orange	Nov.	1	4	3
Bignoniaceae							
Tabebuia aurea	tree	gullet	yellow	July–Aug.	3	149	1
Fabaceae							
Camptosema ellipticum	climber	tubular	red	Jan.–Feb. & Aug.–Sep.	1	1	5
Malvaceae							
Bauhinia ungulata	shrub	brush	white	June–Aug.	3	3	14
Eriotheca pubescens	tree	open	white	Sep.	2	13	29
Luehea paniculata	tree	open	white	Aug.–Sep.	1	19	1
Sapindaceae							
Serjania ovalifolia	climber	open	white	Aug.–Sep.	1	4	3
Simaroubaceae							
Simarouba versicolor	tree	open	purple	Feb.–May	2	219	40
Styracaceae							
Styrax ferrugineus	tree	open	yellow	June–Jul.	2	25	20
Verbenaceae							
Verbena hirta	shrub	tubular	lilac	Sep.–Nov.	2	240	1
Vochysiaceae							
Vochysia cinnamomea	tree	open	yellow	Feb.–May	6	270	4
Qualea parviflora	tree	open	lilac	Oct.–Nov.	1	108	1

Table 4.5. Bird species characteristic of poorly vegetated suburb, well-vegetated suburb and native intact *Eucalyptus* forest (bushland) in the Brisbane region, Australia. Bird species classified as: A, Australian icon; E, exotic; FF, forgotten foliphiles (foliage feeding); L, large; N, new arrival; S, small; T, 'terrorist'. Bird species occurrence classed as C, Common; P, Present; O, Occasional. (Modified from Catterall, 2004.)

Scientific name	Common name	Family	Class	Poorly vegetated suburb	Well-vegetated suburb	Native *Eucalyptus* forest (bushland)
Rhipidura leucophrys	willy wagtail	Dicruridae	S N	C	O	
Passer domesticus	house sparrow	Passeridae	S E N	C	O	
Sturnus vulgaris	common starling	Sturnidae	L E N	C	O	
Streptopelia chinensis	spotted turtle-dove	Columbidae	L E N	C	O	
Grallina cyanoleuca	magpie-lark	Dicruridae	L N	C	C	
Coracina novaehollandiae	black-faced cuckoo-shrike	Campephagidae	L	C	C	O
Cracticus torquatus	grey butcherbird	Artamidae	L A		C	
Cracticus tibicen	Australian magpie	Artamidae	L A	O	C	
Corvus orru	Torresian crow	Corvidae	L A	O	P	O
Trichoglossus haematodus	rainbow lorikeet	Psittacidae	L A		C	O
Trichoglossus chlorolepidotus	scaly-breasted lorikeet	Psittacidae	L A		C	
Platycercus adscitus	pale-headed rosella	Psittacidae	L A	O	C	
Dacelo novaeguineae	laughing kookaburra	Halcyonidae	L A		C	O
Philemon corniculatus	noisy friarbird	Meliphagidae	L A		C	O
Manorina melanocephala	noisy miner	Meliphagidae	T		C	
Pardalotus striatus	striated pardalote	Pardalotidae	S FF			C
Lichenostomus chrysops	yellow-faced honeyeater	Meliphagidae	S FF			C
Melithreptus albogularis	white-throated honeyeater	Meliphagidae	S FF			C
Myzomela sanguinolenta	scarlet honeyeater	Meliphagidae	S FF			P
Pachycephala pectoralis	golden whistler	Pachycephalidae	S FF			C
Pachycephala rufiventris	rufous whistler	Pachycephalidae	S FF			P
Colluricincla harmonica	grey shrike-thrush	Pachycephalidae	S FF			C
Rhipidura albiscapa	grey fantail	Dicruridae	S FF			C
Zosterops lateralis	silvereye	Zosteropidae	S	O	O	O

better adapted to the stressors involved than others. This depends on factors such as sensitivity to environmental change, life history and the requirements at different phases of development. It is also made more complex by interspecies interactions, and the dispersal requirements of the individual species (Hamer and McDonnell, 2008). Many amphibian species too require habitats that complement each other at multiple scales, access to ponds at certain times of year, but low vegetation or even areas of forest at other times. Effective links between these are important in the urban landscape, and the advent of roads, rail networks, walls, buildings and fences disrupt these corridors and networks. Even in natural habitats the lifestyle of amphibians means they often exist in a large network of discrete metapopulations at the regional scale, and isolation of groups is exacerbated by urban infrastructure. Consequently, changes in landscape structure and complexity tend to result in the decline of amphibian species diversity. There are exceptions, however, e.g. *Rana temporaria* (common frog) has benefited from an increase in popularity of garden ponds in the UK and elsewhere in Europe, although their breeding success is optimized when there are no ornamental fish present to predate the eggs or young tadpoles. For ponds that occur in urban environments, water quality is an issue (as indeed is the case with those in agricultural landscapes), due to pollutant loads associated with fertilizers, pesticides, sediment and chemicals from storm water runoff, with de-icing salts and petrochemical products being particularly common problems where ponds are located close to roads. The fact though that urban metapopulations of amphibians may be dependent on a limited number, or even just the one pond, to breed in, increases their vulnerability, as this acts as a habitat sink and a single pollution incident can eliminate an entire population (Battin, 2004).

Other implications of urbanization include alterations in water flows, either through increased built infrastructure reducing water infiltration to the soil, or greater surface runoff; or alternatively over abstraction for domestic and industrial uses. Urbanization also has more subtle effects; it alters the period water is withheld within ponds, streams or wetlands, and this affects the dynamics between species. In Portland, USA, increased and prolonged water flows changed ephemeral wetlands to more consistent, permanent wetlands. This had a consequential change in amphibian species present, namely a transition from those species with rapid larval (tadpole) development to those with prolonged larval development (Pearl *et al.*, 2005). Permanent wetlands are colonized by fish which predate tadpoles, and species that are adapted to fish-free, temporary ponds may suffer population crashes, e.g. in the USA, species including *Ambystoma macrodactylum* (long-toed salamander), *A. tigrinum* (tiger salamander), *A. laterale* (blue-spotted salamander) and *Rana sylvatica* (wood frog) have seen populations decline as more pools have become permanent features.

Reptiles

Reptile species face similar problems to amphibians and small mammals in the urban environment. They are susceptible to accidental mortality such as road kill – especially for nocturnal species that use the heat radiating off a road to regulate their body temperatures before hunting at night. In addition, reptiles are intentionally killed because of human aversions (especially snakes, even when the species in question is harmless) as well as frequently harvested for food in some cultures. Even with these pressures, however, it is still thought that agriculture and logging pose even greater threats to reptiles than urbanization, due to the magnitude and scale of change in these areas globally (Böhm *et al.*, 2013). Again species and context are important, though.

In their study of amphibian and reptile populations in the watersheds of Georgia, USA, Barrett and Guyer (2008) noted that the richness of reptile species actually increased with urbanization, whilst amphibian species numbers declined. They hypothesized that this was because urbanization altered the conditions from a closed-canopy, shallow-water habitat (generally conducive to salamanders and frogs), to a habitat characterized by open vegetation, warmer microclimate and deeper and more open water (conditions more favourable to snakes and turtles). Thus the retention of some patches of appropriate habitat can help reptiles retain a foothold during urbanization. This is manifest in locations such as Brisbane, Australia, where the viability of reptile populations in the greater city area is strongly dependent on local habitat composition, structure and the proximity and configuration of lowland remnant forests.

Invertebrates

The impact of urbanization on invertebrates is again probably polarized depending on the taxa and species. Insects, spiders and mites all have strong influences on human society, both in positive and negative terms. They provide critical ecosystem services through plant pollination and pest control, while at the same time are unpopular due to factors including disease transmission, human discomfort (e.g. biting, stinging and sucking) and damage to horticultural, arable and forestry plants. Invertebrates can be divided into three generic groups in terms of their relationship with urban habitats, namely:

- 'Urban taxa' – principally found in urban landscapes or have their highest abundance there.
- 'Generic taxa' – found abundantly in both rural and urban locations.
- 'Rural dependent taxa' – not present, or only a low level of occurrence in urban landscapes.

Urban factors that affect invertebrates include access to water, air dynamics and thermal characteristics as well as extent of urban development. These influence the frequency of occurrence, but also trophic structures. Air movement affects a spiderling's ability to colonize new areas through its 'ballooning' activities (where a young spider releases strands of silk into the air and these are caught by the wind, potentially pulling the spiderling into the high atmosphere before eventually returning to the Earth's surface many kilometres away from its original location). Urban heat islands and the effects of climate change are raising the mean temperatures experienced within cities, thus permitting species to survive throughout the year in latitudes further north and south of the Equator than has been the case in the past. Increasingly stressful urban environments impact on plants too, which respond physiologically and subsequently alter their susceptibility to herbivores and sucking insects.

As with other animal and plant taxa, the age of the habitat or the stage it is at within its successional process influences the population of invertebrates found there (Gilbert, 1989). Comparing open, derelict spaces in cities it is evident that those that have been left unmanaged for the longest time before they are 'redeveloped' have the greatest diversity of species, and species taxa and abundance shifts from younger to older plots. (Plots that are never redeveloped of course, may eventually attain a 'stable' climax vegetation, which paradoxically may result in lower biodiversity.) Interestingly, McIntyre (2000) estimates that terrestrial arthropods (insects, spiders, mites, centipedes, millipedes and crustaceans) in urban landscapes tend to be more diverse than those in rural environments. In general, herbivore species are more abundant in cities than rural sites. Conversely, parasitoids are more abundant in rural than urban sites. A number of parasitoids are reliant on butterfly and aphid species, and their abundance correlates closely with the presence of flowering plants; as such the number of parasitoid wasps tends to decrease as flower diversity and abundance reduces. In general, as the area of impervious surfaces increases there is less space for gardens, and hence fewer parasitoid wasps present (Bennett and Gratton, 2012).

In insect assemblies in the UK, urban areas are deemed richer in bee species (Fig. 4.2) and with greater numbers of individuals recorded (Fig. 4.3) than either agricultural land or even designated nature reserves (Baldock *et al.*, 2015). The trend, however, is not repeated for true flies or hoverflies. Turrini and Knop (2015), comparing agricultural managed landscapes and urban landscapes, suggested the latter also supported a greater number of bug species and greater abundance of individuals for taxa such as bugs, beetles, leafhoppers and spiders as well as more variation within the bug and beetle species present (i.e. species adapted to different niches). In this study of the urban environment, increasing the number and size of vegetated areas present enhanced the species richness and abundance of most arthropod groups under observation. Somewhat in contrast to agricultural productive landscapes, the isolation of vegetation stands and other habitats played only a relatively limited role in determining the diversity and abundance of species within the urban ecosystems. Conversely, river and other aquatic ecosystems tend to suffer more from pollution in urban areas than their rural counterparts, and this has a negative effect on the diversity of aquatic insects and crustaceans. Overall, McIntyre (2000) believes that arthropod diversity:

- decreases with increasing air and water pollution;
- increases with the age of an urbanized area;
- of non-native species increases with the age of an urban area;
- increases with a higher proportion of vegetation incorporated into the urban matrix; and
- is also affected by the juxtaposition to native habitats as these play an important role for recruitment and dispersal into new urban habitats.

4.5 Management for Biodiversity

Management of urban green space is an important component in determining its value for wildlife and the likelihood of conserving species. Management is a factor influenced by cultural norms (e.g. the 'need' for a lawn and to keep the garden 'tidy'), educational aspects, academic/professional/community inputs and the resources at the disposal of the custodian(s). Different management styles have been compared using a range of landscape typologies in Stockholm, Sweden, to determine their effects on biodiversity and wider ecosystem services (Andersson *et al.*, 2007). Comparisons were made between allotment gardens managed by individuals, cemeteries managed by the church authorities and parks managed by the city council. Andersson *et al.* (2007) concluded that 'top-down' approaches to management were a distinct handicap compared to 'bottom-up' systems. Systems managed by individuals (i.e. bottom-up management) resulted in the greatest species diversity and abundance of insect pollinators, and promoted a different suite of seed dispersers and insectivores. The advantages of management by individuals and local groups is that the people involved will be more sensitive to specific attributes of individual sites, and observant of the need for careful management. In contrast, personnel working for a larger organization may be uninitiated in this respect, and may be simply motivated to 'just get the job done' or follow a set specification without modifying behaviours based on the realities

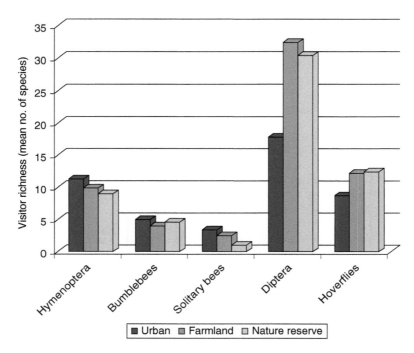

Fig. 4.2. Insect taxa richness in different landscapes of the UK. (Modified from Baldock et al., 2015.)

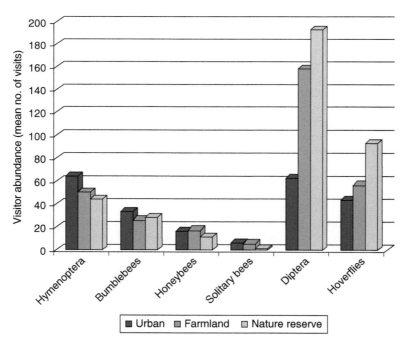

Fig. 4.3. Insect taxa abundance in different landscapes of the UK. (Modified from Baldock et al., 2015.)

on the ground. All too frequently, complaints are received each year against city council or infrastructure maintenance teams from members of the public, due to actions such as mowing areas where orchid colonies or other rare species are in flower, or where trees are damaged or killed by surrounding the base of the trunks with excavation spoil and such-like.

Invasiveness

Non-native invasive species have been cited as being the second greatest threat globally to plant biodiversity, coming second only to habitat destruction. These non-native species incorporate not only plants and animals but also fungi and bacteria, which may pose disease risks. Of the animal and plant species threatened across the globe, 57% are thought to be negatively influenced by competition or predation from non-native invasive species. These negative factors include competition for resources (light, water, nutrients), loss of genetic integrity through hybridization, alterations to wider nitrogen, phosphate or water cycles (including for example eutrophication), increased sedimentation or altering frequency and intensity of other disturbance cycles (e.g. fire). The focus in this chapter is on the threat posed by non-native invasive plants. In 2002, Pimentel *et al.* estimated that the global economic cost of invasive plants in natural areas, agriculture and gardens was in the region of £23.4 billion per annum (US$35 billion).

Horticulture needs to take its fair share of the blame here. Many plants introduced to Western Europe were done so either with the desire to increase the variety of food crops (over the last 8000 years), or more recently (over the last 500 years), to provide greater variety to landscape and garden plantings. In the USA, most woody invasive plants were introduced for horticultural purposes; one study found that 82% of the 235 woody plant species identified as colonizing natural habitats had been used in landscaping (Reichard and Hamilton, 1997). Similarly, in Australia, horticultural interests of one sort or another are blamed for approximately 60% of the naturalized flora. In New Zealand, the number of naturalized, non-native plants is now thought to be the same as the number of indigenous vascular plants (about 2500). The native flora is decimated, particularly in urban environments. In severe cases, invaders may form monocultures and completely exclude native species, such as has occurred with *Lythrum salicaria* (purple loosestrife) in wetlands of the USA. Similarly, *Eichhornia crassipes* (water hyacinth), a native of the Amazonian basin, has invaded waterways right across the tropics, being considered a major weed species in more than 50 countries. The growth rate of water hyacinth is among the highest in the plant kingdom. In some regions, this species can double its population size over 2 weeks through its rapid propagation by stolons, which allows daughter plants to spread across the water surface. By producing a dense canopy, this invasive alien shades out the native submersed aquatic plant species which are so important to invertebrate wildlife. In addition, the decaying mats of water hyacinth lower dissolved oxygen concentrations, resulting in damage to fish populations.

Not all introduced plants are necessarily invasive – indeed it is probably less than 10% – but the problem is determining which are, and which are not, potentially threatening to native species. Woody plant types and aquatic plants appear to be more invasive than terrestrial herbaceous perennial species, although the latter are considered the more problematic in some states of the USA. Alien plant species vary in their aggressiveness, ranging from benign, with little evidence of invasiveness, through to rampant and rapid colonization characteristics. Invasions may not progress linearly over time either. Some species build their populations slowly to a critical mass, after which they expand quickly to their full invasive potential. There are a number of traits common to most invasive plant species, however, including:

- rapid growth rates;
- high rates or reproduction (high flowering and fruiting abilities);
- high rates and effective seed dispersal mechanisms and high germination rates;
- tolerance to a variety of site conditions; and
- a persistent nature, which makes them hard to eradicate.

Indeed, often these are the characteristics that made them useful landscape plants in the urban environment.

Disturbed habitats tend to encourage invasion. The process of disturbance provides niches that allow aggressive species to enter the system, become established and supplant the native species. Disturbance involves not only removal of existing vegetation and soil cultivation, but also other urban processes that impact on native plant populations, such as air pollution, eutrophication, compacted soils etc. Disturbance and site degradation are not the only factors driving invasion though, the location must be conducive to the invasive species (climate, moisture availability, irradiance) and there needs to be a source of plant propagules (seeds, stems, rhizomes) for the site to be compromised.

In urban areas, parks and vacant land may be particularly vulnerable to invasion because of high levels of disturbance and close proximity to ornamental plantings, and the high frequency of ornamental landscapes and gardens within towns and cities means these locations serve as effective 'jumping off' points for invasion into natural areas. Propagules abound in heavily landscaped urban areas. Waste material from gardens or landfill sites can act as sources of contaminant. Gardeners may even plant unwanted or excess plants 'over the garden fence' rather than place them in the waste bin, due to altruistic feelings towards the plants. Commercial bird seed, and by-products from human food and waste streams can survive the 'waste-disposal process' and establish themselves around rubbish dumps and tip sites. Stormwater flooding is effective at moving aquatic and other weed propagules, and biotic factors such as birds and other wildlife disperse plant materials from one green space to another.

Eradication and control methods for invasive plants are costly and labour intensive. Both chemical (herbicides) and physical means (cutting and digging roots out) are employed in an attempt to eradicate such species. In some situations too, biological control is considered. In the UK, the non-native *Fallopia japonica* (Japanese knotweed) is a major pest along riverbanks and elsewhere, but biocontrol systems have been evaluated for controlling it. Researchers from CABI investigated both a Japanese psyllid insect (*Aphalara itadori*) and a leaf spot fungus (*Mycosphaerella polygoni-cuspidati*) as potential biocontrol agents. The insect is now licenced for release as an outdoor control agent for *F. japonica*, after many years of testing and evaluation (for effectiveness, but also to help ensure the insect itself does not become invasive). Shaw et al. (2009) found that the psyllid was able to feed off the sap of the knotweed, inhibiting its growth, but importantly, failed to breed on any of the 90 UK native plant species tested, meaning it can be specifically targeted at the invasive plant. This is the first time the release of an alien insect species to the outdoor environment was authorized within the European Union.

Control methods for invasive plants also rely strongly on education and awareness. Government officials now engage with the nursery industry, landscape professionals and gardeners to raise awareness about the potential of species to become invasive. In the latter group, it is hoped that learning about landscape plants will stimulate gardeners to take a closer interest in the choice of plants used, and give them the opportunity to make more environmentally sound decisions in their gardening practices. Before selecting a new garden plant, gardeners are being encouraged to inform themselves on issues such as:

- the plant's native range;
- how it reproduces;
- whether the plant needs a high level of maintenance to keep it in check;
- whether it is an aggressive grower;
- whether it attracts (will be dispersed by) birds or other animals; and
- whether it is an invasive anywhere, especially in areas similar to where it might be planted.

Similarly, a number of aquatic plants used in pond gardens and aquaria have been removed from sale by the horticultural industry, due to their competitiveness with native aquatic plants. Such withdrawals and similar actions are likely to increase in the future. In addition, breeding programmes for new ornamentals that focus on reducing the productive capacity of any new genotypes (e.g. releasing hybrids without viable seed) are likely to be given greater prominence in future.

Trend for planting more natives

In recent years, environmental horticulturalists have worked alongside other professionals (policy makers, landscape architects and conservation biologists) in an attempt to link landscape design with ecosystem structure and function more effectively. The prime aims are to create and restore habitats, and to reintroduce native species in urban landscapes. Where those who tend to be most involved in these processes will see the restored habitats as being 'natural', environmental horticulturalists will also embrace semi-natural habitats and the use of non-native plants to address the ecological dysfunctions.

The extent to which native and non-native plants are valued does depend on country, its location within the globe, the extent to which invasive plant species have been a problem previously and the deemed value/intactness of the existing vegetation. In Australia, for example, there is a strong movement to valuing native plantings more due to previous negative experiences of introducing alien plant (and animal) species from Europe and elsewhere, and the subsequent impact on the native flora and fauna. In Europe, where the introduction of non-native plants has taken place for at least 500 years, the attitudes, although polarized in some quarters, are in general more relaxed. In these countries there tends to be more concern with the introduction of new pathogens and pests (sometimes even to the non-native plants) rather than the introduction of new plant genotypes per se. So, for example, shipments of nursery stock plants from overseas will be scrutinized for their potential to bring in new fungi, bacteria and invertebrate pests, but little concern is expressed about the risk the plant material itself poses.

In most parts of the globe except Europe, the attitude to plantings can be summarized as having an emphasis on:

- redeveloping planting designs to take account of local climatic and historical traditions;
- re-vegetating with indigenous plants; and
- intensive site (including brownfield sites, derelict or vacant plots) management to eliminate non-native species and pests; and to increase native biodiversity whenever possible (Müller and Werner, 2010).

Within Europe the approach is less rigid, with greater emphasis on:

- reintroducing native biodiversity;
- designing with natural processes (but not always with native plants); and
- planting up as many available spaces as possible to increase biodiversity (using even very small biotopes) within the urban environment.

Over the last 10–15 years there has been a trend for landscape firms and nurseries to move away from non-native species and plant more natives (Ignatieva et al., 2008), encouraged in some instances by changes in local policies. The incorporation of native plants into park, gardens and other landscapes is seen as providing more opportunities to support food chains for native animals as well as becoming part of an integrated holistic approach to creating sustainable urban green infrastructure. There is even a new terminology to describe this landscaping style – 'biodiversinesque'. Its value is seen by some as not only providing more effective habitat and reliable food sources for native wildlife but also as a powerful visual tool to reinforce the concept of native biodiversity to the general public in everyday life (Ignatieva and Ahrné, 2013). Indeed, recent trends have not been limited to introducing native plants; insects, invertebrates and birds have also been introduced to mimic native ecosystems.

Not only are these approaches considered good for local wildlife, but there is the implicit assumption that native plant species are better adapted to the region and thus will be better suited for plantings than non-native species. This supposes, of course, that the environmental conditions in the urban environment are comparable to the habitats the species predominantly exist in. In reality, the environmental parameters required may not be collectively represented in the urban landscape, and are likely to be increasingly mismatched in future within the context of a warming climate. As Quigley (2011) states, plantings often fail because species – native or non-native – are not adapted to the urban environment, a case of simply 'the wrong plant in the wrong place'. Nonetheless, matching the 'right' native species for the right place can improve survival and enhance native species representation in urban landscapes and design. There is a trend for environmental horticulturalists to incorporate more native species into landscape and park designs.

Others challenge the 'native is always best' ideology, especially within the urban environment. This is not solely a result of environmental alterations due to urbanization and climate change either. In the UK for example, Hitchmough and Dunnett (see Chapter 1) suggest that the UK native flora is relatively poor in flowering plant species attractive to pollinating invertebrates (bees, hoverflies, butterflies), compared to analogous plant species from central Europe or North America (Hitchmough and Dunnett, 2004; Hitchmough, 2011). Salisbury et al. (2015) indicate that non-native plants may provide a particularly useful service to pollinators, by supplying nectar and pollen during late summer and autumn when the majority of the native plants have finished flowering (see Tables 4.6–4.11 for plants that provide nectar/pollen, but are not native to the UK). Hitchmough and Dunnett (2004) argue that in urban plantings that are very much in the public eye, it is important to provide 'a service' for native invertebrates and birds, whilst also ensuring these

Table 4.6. Tree genotypes providing pollen and/or nectar resources to invertebrates within different seasons in the UK. (Modified from Anon., 2015a.)

Species	Common name	Native
Spring: March–May		
Acer campestre	field maple	Y
Acer platanoides	Norway maple	
Acer pseudoplatanus	sycamore	
Acer saccharum	sugar maple	
Aesculus hippocastanum	horse chestnut	
Cercis siliquastrum	Judas tree	
Crataegus monogyna	hawthorn	Y
Ilex aquifolium	holly	Y
Malus baccata	Siberian crab	
Malus domestica	apple	
Malus floribunda	Japanese crab	
Malus hupehensis	Hupeh crab	
Malus sargentii	Sargent's crab apple	
Mespilus germanica	medlar	
Prunus avium	wild and edible cherries	Y
Prunus domestica	wild and edible plums	
Prunus dulcis	almond	
Prunus insititia	damson	
Prunus mume	Japanese apricot	
Prunus padus	bird cherry	Y
Prunus pendula f. *ascendens* 'Rosea'	flowering cherry	
Prunus persica	peach	
Prunus × *yedoensis*	flowering cherry	
Pyrus communis	pear	
Salix caprea	goat willow (male)	Y
Summer: June–Aug.		
Aesculus indica	Indian horse chestnut	
Catalpa bignonioides	Indian bean tree	
Koelreuteria paniculata	pride of India	
Robinia pseudoacacia	false acacia	
Sorbus aria	common whitebeam	Y
Sorbus aucuparia	mountain ash/rowan	Y
Tetradium daniellii	bee-bee tree	
Tilia × *europaea*	common lime	
Tilia maximowicziana	lime	
Tilia oliveri	lime	
Tilia platyphyllos	broad leaved lime	Y
Autumn: Sep.–Oct.		
Arbutus unedo	strawberry tree	
Tilia henryana	Henry's lime	

Table 4.7. Shrub and climber genotypes providing pollen and/or nectar resources to invertebrates within different seasons in the UK. (Modified from Anon., 2015a.)

Species	Common name	Native
Winter: Nov.–Feb.		
Clematis cirrhosa	Spanish traveller's joy	
Fatshedera lizei	tree ivy	
Lonicera × *purpusii*	honeysuckle	
Mahonia spp.	Oregon grape	
Salix aegyptiaca	musk willow	
Sarcococca confusa	sweet box	
Sarcococca hookeriana	sweet box	
Viburnum tinus	laurustinus	
Spring: March–May		
Berberis darwinii	Darwin's barberry	
Berberis thunbergii	Japanese barberry	
Buxus sempervirens	box	Y
Chaenomeles spp.	Japanese quince	
Cornus mas	Cornelian cherry	
Cotoneaster conspicuus	Tibetan cotoneaster	
Crataegus monogyna	hawthorn	Y
Enkianthus campanulatus	enkianthus	
Erica carnea	alpine heath	
Erica × *darleyensis*	Darley Dale heath	
Hebe spp.	hebe	
Mahonia spp.	Oregon grape	
Pieris formosa	lily-of-the-valley bush	
Pieris japonica	lily-of-the-valley bush	
Prunus incisa 'Kojo-no-mai'	cherry 'Kojo-no-mai'	
Prunus laurocerasus	cherry laurel	
Prunus spinosa	blackthorn	Y
Prunus tenella	dwarf Russian almond	
Ribes nigrum	blackcurrant	
Ribes rubrum	redcurrant	
Ribes sanguineum	flowering currant	
Salix hastata 'Wehrhahnii'	Halberd willow 'Wehrhahnii'	
Salix lanata	woolly willow (male)	Y
Skimmia japonica	skimmia	
Stachyurus chinensis	stachyurus	
Stachyurus praecox	stachyurus	
Vaccinium corymbosum	blueberry	
Summer: June–Aug.		
Aesculus parviflora	bottlebrush buckeye	
Brachyglottis 'Sunshine'	brachyglottis 'Sunshine'	
Brachyglottis monroi	Monro's ragwort	

Continued

Table 4.7. Continued.

Species	Common name	Native
Buddleja davidii	butterfly bush	
Buddleja globosa	orange ball tree	
Bupleurum fruticosum	shrubby hare's ear	
Callicarpa bodinieri var. giraldii	beauty berry	
Calluna vulgaris	heather	Y
Caryopteris × clandonensis	caryopteris	
Clematis vitalba	old man's beard	Y
Cornus alba	red-barked dogwood	
Elaeagnus angustifolia	oleaster	
Erica cinerea	bell heather	Y
Erica erigena	Irish heath	
Erica vagans	Cornish heath	Y
Erysimum 'Bowles's Mauve'	wallflower 'Bowles's Mauve'	
Escallonia species	escallonia	
Fuchsia spp. – hardy types	fuchsia	
Hebe spp.	hebe	
Hydrangea paniculata (with fertile flowers)	paniculate hydrangea	
Hydrangea anomala subsp. petiolaris	climbing hydrangea	
Hyssopus officinalis	hyssop	
Kalmia latifolia	mountain laurel	
Jasminum officinale	jasmine	
Laurus nobilis	bay tree	
Lavandula angustifolia	English lavender	
Lavandula × intermedia	lavandin	
Lavandula stoechas	French lavender	
Lavatera olbia	tree lavatera	
Ligustrum ovalifolium	garden privet	
Ligustrum sinense	Chinese privet	
Lonicera periclymenum	honeysuckle	Y
Olearia spp.	daisy bush	
Parthenocissus tricuspidata	Boston ivy	
Perovskia atriplicifolia	Russian sage	
Phlomis spp.	sage	
Photinia davidiana	photinia	
Pileostegia viburnoides	climbing hydrangea	
Potentilla spp.	cinquefoil	
Prostanthera cuneata	alpine mint bush	

Table 4.7. Continued.

Species	Common name	Native
Ptelea trifoliata	hop tree	
Pyracantha spp.	firethorn	
Rosa canina	dog rose	Y
Rosa rubiginosa	sweet briar	Y
Rosa rugosa	Japanese rose	
Rosmarinus officinalis	rosemary	
Rubus fruticosus	blackberry	Y
Rubus idaeus	raspberry	
Spiraea japonica	spiraea	
Symphoricarpos albus	snowberry	
Tamarix ramosissima	tamarisk	
Thymus spp.	thyme	
Clematis heracleifolia	tube clematis	
Elaeagnus pungens	silverthorn	
Elaeagnus × ebbingei	Ebbinge's silverberry	
Fatsia japonica	Japanese aralia	
Hedera colchica	Persian ivy	
Hedera helix	ivy	Y
Autumn: Sep.–Oct.		
Viburnum lantana	wayfaring tree	Y
Viburnum opulus	guelder rose	Y
Weigela florida	weigelia	
Zauschneria californica	Californian fuchsia	

areas remain attractive to people. For example, such areas need to be imaginatively designed, colourful and inspirational. In parallel, many ecologists have come to realize that gardens (with a high proportion of non-native plants) and not just large conservation areas, may play a critical role in offering refuge for native species in the advent of climate change by facilitating migration and seed dispersal (Goddard et al., 2010).

4.6 Environmental Horticulture in Key Urban Wildlife Habitats

A number of horticulturally important landscapes impact directly on wildlife and through careful management of these landscapes biodiversity can be enhanced. Historically of course, the concept of horticulture has been to manage areas explicitly to favour the development of some plant species over

Table 4.8. Herbaceous perennial genotypes providing pollen and/or nectar resources to invertebrates within different seasons in the UK. (Modified from Anon., 2015a.)

Species	Common name	Native
Winter: Nov.–Feb.		
Helleborus spp.	hellebore (winter flowering)	
Spring: March–May		
Ajuga reptans	bugle	Y
Arabis alpina subsp. *caucasica*	alpine rock cress	
Armeria juniperifolia	juniper-leaved thrift	
Aubrieta species	aubretia	
Aurinia saxatilis	gold dust	
Bergenia spp.	elephant ear	
Caltha palustris	marsh marigold	Y
Doronicum × *excelsum*	leopard's bane	
Erysimum 'Bredon'	wallflower 'Bredon'	
Euphorbia amygdaloides	wood spurge	
Euphorbia characias	Mediterranean spurge	
Euphorbia cyparissias	cypress spurge	
Euphorbia nicaeensis	nice spurge	
Euphorbia epithymoides	cushion spurge	
Geranium spp.	cranesbill	
Geum rivale	water avens	Y
Helleborus spp.	hellebore (spring flowering)	
Iberis saxatilis	alpine candytuft	
Iberis sempervirens	perennial candytuft	
Lamium maculatum	spotted dead nettle	
Primula veris	cowslip	
Primula vulgaris	primrose	Y
Pulmonaria spp.	lungwort	
Summer: June–Aug.		
Achillea spp.	yarrow	Y
Actaea japonica	baneberry	
Agastache spp.	giant hyssop	
Amsonia tabernaemontana	eastern bluestar	
Anthemis tinctoria	dyer's chamomile	
Antirrhinum majus	snapdragon	
Aquilegia spp.	columbine	
Argemone platyceras	crested poppy	
Armeria maritima	thrift	Y
Aruncus dioicus	goat's beard (male)	
Asparagus officinalis	common asparagus	
Astrantia major	greater masterwort	
Buphthalmum salicifolium	yellow ox eye	
Borago officinalis	borage	
Calamintha nepeta	lesser calamint	
Campanula carpatica	tussock bellflower	
Callistephus chinensis	China aster	

Table 4.8. Continued.

Species	Common name	Native
Campanula glomerata	clustered bell flower	Y
Campanula lactiflora	milky bellflower	
Campanula latifolia	giant bellflower	
Campanula persicifolia	peach-leaved bellflower	
Catananche caerulea	blue cupidone	
Centaurea atropurpurea	purple knapweed	
Centaurea dealbata	mealy centaury	
Centaurea macrocephala	giant knapweed	
Centaurea montana	perennial cornflower	
Centaurea nigra	knapweed	Y
Centaurea scabiosa	greater knapweed	Y
Centranthus ruber	red valerian	
Cirsium rivulare 'Atropurpureum'	purple plume thistle	
Crambe cordifolia	greater sea kale	
Cynara cardunculus	globe artichoke/cardoon	
Cynoglossum amabile	Chinese forget-me-knot	
Dahlia species	dahlia (open centre flowers)	
Delosperma floribundum	ice plant	
Delphinium elatum	candle larkspur	
Dictamnus albus	dittany	
Echinacea purpurea	purple coneflower	
Echinops spp.	globe thistle	
Erigeron spp.	fleabane	
Eriophyllum lanatum	golden yarrow	
Eryngium × *tripartitum*	tripartite eryngo	
Eryngium planum	blue eryngo	
Eryngium alpinum	alpine eryngo	
Erysimum × *allionii*	Siberian wallflower	
Eupatorium cannabinum	hemp agrimony	Y
Eupatorium maculatum	Joe Pye weed	
Euphorbia cornigera	horned spurge	
Euphorbia sarawschanica	Zeravshan spurge	
Ferula communis	giant fennel	
Foeniculum vulgare	fennel	
Fragaria × *ananassa*	strawberry	
Gaillardia × *grandiflora*	blanket flower	
Gaura lindheimeri	white gaura	
Geranium pratense	meadow cranesbill	Y
Geum spp.	avens	
Helenium spp.	Helen's flower	
Heliopsis helianthoides	smooth ox-eye	
Hesperis matronalis	dame's violet	

Continued

Table 4.8. Continued.

Species	Common name	Native
Inula spp.	harvest daisy	
Knautia arvensis	field scabious	Y
Knautia macedonica	Macedonian scabious	
Lathyrus latifolius	broad-leaved everlasting pea	
Leucanthemum × *superbum*	shasta daisy (open-centre flowers)	
Leucanthemum vulgare	ox-eye daisy	Y
Liatris spicata	button snakewort	
Limonium platyphyllum	broad-leaved statice	
Linaria purpurea	purple toadflax	
Lychnis coronaria	rose campion	
Lychnis flos-cuculi	ragged robin	Y
Lysimachia vulgaris	yellow loosestrife	Y
Lythrum salicaria	purple loosestrife	Y
Lythrum virgatum	wand loosestrife	
Malva moschata	musk mallow	Y
Mentha aquatica	water mint	Y
Mentha spicata	spearmint	
Monarda didyma	bergamot	
Nepeta × *faassenii*	garden catmint	
Origanum 'Rosenkuppel'	majoram 'Rosenkuppel'	
Origanum vulgare	oregano	Y
Paeonia spp.	peony	
Papaver orientale	oriental poppy	
Penstemon spp.	penstemon	
Persicaria amplexicaulis	red bistort	
Persicaria bistorta	bistort	Y
Phlox paniculata	perennial phlox	
Phuopsis stylosa	Caucasian crosswort	
Polemonium caeruleum	Jacob's ladder	Y
Rudbeckia spp.	coneflower (open-centre flowers)	
Salvia spp.	sage	
Scabiosa caucasica	garden scabious	
Scabiosa columbaria	small scabious	Y
Sedum spectabile	ice plant	
Sedum telephium	orpine	Y
Sidalcea malviflora	checkerbloom	
Solidago species	goldenrod	
Stachys byzantina	lambs' ear	
Stachys macrantha	big-sage	
Stokesia laevis	Stokes' aster	
Tanacetum coccineum	pyrethrum	
Tanacetum vulgare	tansy	Y
Telekia speciosa	yellow ox-eye	
Teucrium chamaedrys	wall germander	
Verbena bonariensis	purple top	
Veronica longifolia	garden speedwell	
Veronicastrum virginicum	Culver's root	
Autumn: Sep.–Oct.		
Aconitum carmichaelii	Carmichael's monk's hood	
Actaea simplex	simple-stemmed bugbane	
Anemone hupehensis	Chinese anemone	
Anemone × *hybrida*	Japanese anemone	
Aster spp.	Michaelmas daisy	
Campanula poscharskyana	trailing bellflower	
Ceratostigma plumbaginoides	hardy blue-flowered leadwort	
Chrysanthemum spp.	chrysanthemum (open-centre flowers)	
Dahlia spp.	dahlia	
Helianthus × *laetiflorus*	perennial sunflower	
Leucanthemella serotina	autumn ox-eye	
Salvia spp.	sage	

Table 4.9. Geophyte genotypes providing pollen and/or nectar resources to invertebrates within different seasons in the UK. (Modified from Anon., 2015a.)

Species	Common name	Native
Winter: Nov.–Feb.		
Crocus species	crocus (winter flowering)	
Eranthis hyemalis	winter aconite	
Galanthus nivalis	snowdrop	
Spring: March–May		
Crocus spp.	crocus (spring flowering)	
Muscari armeniacum	Armenian grape hyacinth	
Ornithogalum umbellatum	common star of Bethlehem	
Summer: June–Aug.		
Allium spp.	ornamental and edible onion	
Autumn: Sep.–Nov.		
Colchicum spp.	autumn crocus	
Crocus spp.	crocus autumn-flowering types	

Table 4.10. Biennial plant genotypes providing pollen and/or nectar resources to invertebrates within different seasons in the UK. (Modified from Anon., 2015a.)

Species	Common name	Native
Spring: March–May		
Lunaria annua	honesty	
Smyrnium olusatrum	Alexanders	
Summer: June–July		
Alcea rosea	hollyhock (single flowers)	
Angelica archangelica	angelica	
Angelica gigas	purple angelica	
Angelica sylvestris	wild angelica	Y
Campanula medium	Canterbury bells	
Dianthus barbatus	sweet William	
Digitalis spp.	foxglove	
Dipsacus fullonum	common teasel	Y
Eryngium giganteum	Miss Willmott's ghost	
Matthiola incana	Brompton stock	
Myosotis spp.	forget-me-not	
Oenothera spp.	evening primrose	
Onopordum acanthium	cotton thistle	
Verbascum spp.	mullein	

Table 4.11. Annual plant genotypes providing pollen and/or nectar resources to invertebrates within different seasons in the UK. (Modified from Anon, 2015a.)

Species	Common name	Native
Summer: June–Aug.		
Amberboa moschata	sweet sultan	
Anchusa azurea	large blue alkanet	
Anchusa capensis	Cape alkanet	
Ageratum houstonianum	flossflower	
Centaurea cyanus	cornflower	Y
Centratherum punctatum	Manaos beauty	
Cerinthe major 'Purpurascens'	honeywort	
Clarkia unguiculata	butterfly flower (single flowers)	
Cleome hassleriana	spider flower	
Consolida ajacis	larkspur	
Convolvulus tricolor	dwarf morning glory	
Coreopsis spp.	tickseed	
Cosmos bipinnatus	cosmea	
Cosmos sulphureus	yellow cosmos	
Cucurbita pepo	marrow/courgette	
Cuphea ignea	cigar flower	
Echium vulgare	viper's bugloss	Y
Eschscholzia californica	California poppy	
Gilia capitata	blue thimble flower	
Glebionis segetum	corn marigold	Y
Gypsophila elegans	annual baby's breath	
Helianthus annuus	sunflower (not pollen-free cultivars)	
Helianthus debilis	cucumberleaf sunflower	
Heliotropium arborescens	common heliotrope	
Iberis amara	wild candytuft	
Lavatera trimestris	annual lavatera	
Limnanthes douglasii	poached egg flower	
Linaria maroccana	annual toadflax	
Lobularia maritima	sweet alyssum	
Malope trifida	large-flowered mallow wort	
Nemophila menziesii	baby blue-eyes	
Nicotiana alata	flowering tobacco	
Nicotiana langsdorfii	Langsdorff's tobacco	
Nigella damascena	love-in-a-mist	
Nigella hispanica	Spanish fennel flower	
Papaver rhoeas	poppy	Y
Phacelia campanularia	Californian bluebell	
Phacelia tanacetifolia	fiddleneck	
Phaseolus coccineus	scarlet runner bean	

Continued

others, and to reduce the impact of herbivores. Thus, cultivated crops encouraged at the expense of 'weeds', and invertebrate species considered pests reduced or eliminated altogether. Even in environmental horticulture this may still be the case. Today, however, it may well be that non-native ornamental 'alien' plants are being eradicated in an attempt to help the establishment of native species, indeed precisely those indigenous plant species that once might have been seen to be the enemy of 'good horticultural practice'. The key areas where horticulture overlaps with biodiversity management are outlined below, although emphasis is given to those landscape typologies where the management is often led by an environmental horticulturalist or other plant-based professional, rather than professionals with a wider remit within nature conservation or an alternative discipline, such as ornithology.

Urban trees and woodland

The urban forest is considered important for supporting urban biodiversity and indeed biodiversity across a wider context. The urban forest can comprise a

Table 4.11. Continued.

Species	Common name	Native
Reseda odorata	garden mignonette	
Ridolfia segetum	false fennel	
Sanvitalia procumbens	creeping zinnia	
Scabiosa atropurpurea	sweet scabious	
Tagetes patula	French marigold	
Tithonia rotundifolia	Mexican sunflower	
Trachymene coerulea	blue lace flower	
Tropaeolum majus	nasturtium	
Verbena × hybrida	garden verbena	
Autumn: Sep.–Oct.		
Machaeranthera tanacetifolia	tansy-leaf aster	
Verbena rigida	slender vervain	
Vicia faba	broad bean	
Zinnia elegans	youth and old age	

significant percentage of a nation's tree canopy. In the USA, trees in urban counties are thought to account for nearly 25% of the country's total tree canopy cover (Dwyer *et al.*, 2000). In Flanders, Belgium, park trees provide a significant component of the region's canopy cover and this is largely composed of native species such as *Fagus sylvatica* (beech), *Acer pseudoplatanus* (sycamore), *Quercus robur* (English oak) and *Fraxinus excelsior* (ash) (Cornelis and Hermy, 2004). Trees either as individuals or as groups provide networks for species distribution and migration and act as refuges across the city landscape. Research suggests that urban forests contain a significant percentage of species that were endemic to an area before it was developed. In Guangzhou City, China, urban tree diversity was considered high (Jim and Liu, 2001). A survey of 115,000 trees in parks, university grounds and along city streets recorded over 250 different species. Native broadleaves dominated, with the top three abundant species recorded as *Ficus virens* (white fig), *Caryota mitis* (fish-tail palm), and *Melaleuca leucadendra* (paperbark). It is thought that plant species richness in Guangzhou City actually exceeds the degraded forests of the surrounding countryside (Jim and Liu, 2001). In Hong Kong, a review of the tree stock indicated 232 species present, but natives were the minority here, only contributing to 69 of the species recorded. Urban forests also may be home to endangered or conservation priority species. In Sweden, Stockholm County contains two-thirds of the red-listed species despite being densely populated, and includes rare plant species such as the fern *Dryopteris cristata* and the moss *Buxbaumia viridis* which require damp woodlands (Colding *et al.*, 2003).

Trees are an essential element to be included in the planning process but the value of native and non-native tree species still needs to be explored more deeply. Chong *et al.* (2014) suggest trees, irrespective of whether native or non-native, had a positive influence on the diversity of urban birds in Singapore by providing an extra dimension to an otherwise planar landscape. Even non-native trees provided more opportunities to local bird populations than alternative types of green space, such as grass, largely because many of these trees were tropical rainforest-adapted species and thus could still provide a number of the services to the local bird population (although biodiversity was still optimized by stands of native trees, where these are left standing). Older, veteran trees provide greater opportunities than younger, 'still extending' trees. Thus, during urban expansion or densification, priority should be given to conserving and retaining large veteran trees, or mature stands of woodland. Work in Los Angeles, USA, (Clarke *et al.*, 2013) alluded to the fact that tree species diversity was correlated with older, high-income residential areas, and that to help extend the opportunities for wildlife, new tree plantings should be focused on low income neighbourhoods to help mitigate gaps in the urban forest. Management for biodiversity, however, should maintain the existing diversity in older residential areas and encourage a wider range of species in the newer housing areas.

The actual number of systematic studies on invertebrate populations on urban trees is fairly limited. A study by Helden *et al.* (2012) compared the potential of native and non-native trees to host members of the Hemiptera (true-bugs), in Bracknell, UK. In general, native trees proved more valuable for supporting the greater number of species within Hemiptera (Fig. 4.4) and increasing the numbers of individuals found on individual trees (Fig. 4.5), although due to the variation between tree species, there was no overall statistical difference between the two groups. Indeed, some non-native trees such as *Sorbus intermedia* (Swedish whitebeam) and *Quercus rubra* (red oak) were relatively useful in providing habitat and resources for a range of bug species. The non-native *Malus* (apple) also enhanced the number of individuals that could be found

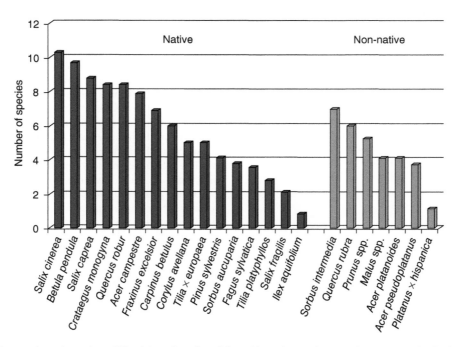

Fig. 4.4. The number of species of Hemiptera (true bugs) found in native and non-native tree species in the UK. (Modified from Helden et al., 2012.)

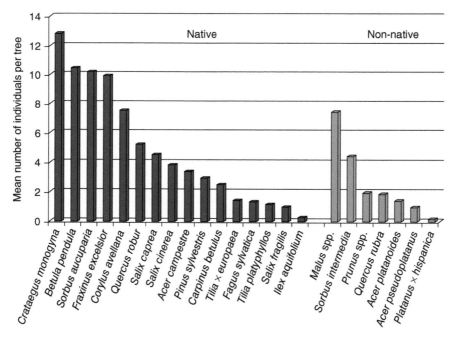

Fig. 4.5. The mean number of individual specimens (abundance) of Hemiptera (true bugs) found per tree across a range of native and non-native tree species in the UK. (Modified from Helden et al., 2012.)

within the non-native tree group. These non-native genotypes often excelled in promoting Hemiptera richness and abundance compared to 'less conducive' native species such as *Salix fragilis* (crack willow) and *Ilex aquifolium* (holly). So even when restricted to using native tree species, environmental horticulturalists may wish to consider species selection carefully as some native species may be more valuable than others. However, it should be stressed that this study just investigated one group of insects, and other taxa may have given different results.

Parks and gardens

Parks and gardens provide a novel set of habitats in the urban landscape. Depending on the socio-cultural factors and age of a city, private gardens can comprise a major component of urban green space. In the UK, private residential gardens represent up to 27% of the total land area in a city (Smith *et al.*, 2006). They deliver considerable opportunities for biodiversity, but fulfilment of this depends on design, scale and dominant management regimes. Their value is often dependent on the larger scale 'landscape ecology' framework and how they are spatially organized and fit within the wider, interconnected green space network. Indeed, their value to biodiversity is often determined by the character of adjoining habitats. Gardens promote plant species richness (although frequently through non-native species or hybrids of native species) and seem to be particularly beneficial to insect and avian species diversity through the provision of shelter, nesting sites and food.

Parks particularly, and to some extent gardens, are legacies of preceding cultural values and trends. Many urban parks (and some of the larger gardens) still retain an English landscape style promoted by the likes of Lancelot 'Capability' Brown (1716–1783) and have large expanses of grass with clumps or avenues of trees. In contrast, many urban gardens retain elements of the gardenesque style, although they can also be influenced by minimalist or naturalistic styles of gardening too. The gardenesque style (John Claudius Loudon, 1783–1843) evolved during the Industrial Revolution in Europe and the Victorian era, and was characterized by the introduction of new colourful, floristic plants, artificial designs and in some instances extravagant features such as ornate summer houses and topiary. A common theme was that the plants were very much 'on-show' and positioned and managed in such a way that each individual specimen could be displayed to its full potential in a scattered planting. The approach involved the creation of small-scale landscapes to promote beauty, variety and mystery. Typical features included island beds, winding paths, tree-planted mounds as well as open areas of mown grass. Landscape gardeners of the day often preferred the use of non-native plants that were beginning to be introduced from other parts of the world, including new genotypes from eastern Asia, North and South America. Today, urban gardens may be on a smaller scale but still many retain the legacy characterized by 'pretty', 'tidy', 'colourful' and 'beautiful' homogeneous landscapes based on non-native plants. This has implications for the species that can utilize such landscapes.

In their assessment of garden biodiversity, Thompson *et al.* (2003) found 438 plant taxa in 60 gardens in Sheffield, UK, of which two-thirds were non-native and one-third native to the UK. This was a richer collection of species compared to that found in derelict urban land (brownfield sites) or other plant communities typical of central/northern England (Fig. 4.6). Non-natives were mainly represented by species from Europe and Asia. The majority of the most-frequently encountered species were weeds, not plants intentionally planted into the garden, e.g. *Geranium robertianum* (herb Robert) or *Digitalis purpurea* (foxglove), or plants that had self-seeded from introduced species but were now acting as weeds, e.g. *Alchemilla mollis* (lady's mantle), *Aquilegia vulgaris* (columbine) and *Meconopsis cambrica* (Welsh poppy). Despite the weed species being the most common, the authors noted that there was no saturation to the species count for the garden flora, in that the number of ornamental genotypes available to gardeners is vast (theoretically at least 70,000 species and cultivars in the UK). Overall, native species richness was not correlated with garden size, but total species richness was. Garden design and management also influence the number of species present, and it was noted that through active management, more species can be maintained in a given area than otherwise would have occurred naturally (Thompson *et al.*, 2003). Others agree that gardens promote biodiversity richness, but argue that ecological functionality is not great as they are developed largely for visual appeal and do not increase trophic diversity and for the most part are not self-sustaining (Quigley, 2011).

In many tropical cities too, parks and gardens have a high proportion of non-native plants. In cities such as Bandung (Indonesia), Bangalore (India)

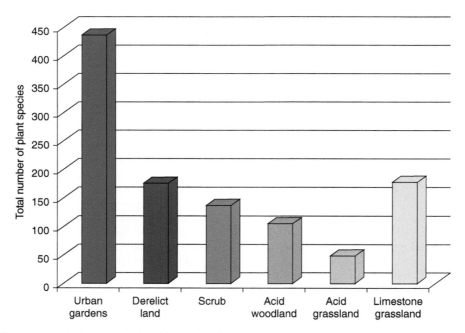

Fig. 4.6. The number of plant species found in 120 1-m² plots in a variety of habitats in the UK. (Modified from Thompson et al., 2003.)

and Rio de Janeiro (Brazil) 70–80% of woody plants found in parks may be non-native. These parks host a diverse range of plant species, but large numbers of the genotypes have been imported into the region. Domestic gardens in Hong Kong were evaluated for their diversity of tree species. The 72 species observed represented 19% of the species present in the province. Considering the relatively small area devoted to private gardens this was considered a fairly rich collection. In contrast to Western European countries, however (where tree species diversity is higher in parks and gardens than in native woodlands), the proportion was small compared to other locations surveyed. Species richness in domestic gardens was comparable to 'old and valuable' trees in Hong Kong (70 species), but considerably less than that noted for urban park trees (272 species), landscape trees in neighbourhoods around public housing (232 species) and roadside trees (149 species) in Hong Kong or, indeed, total urban trees in neighbouring Guangzhou (254 species) or Taipei (164 species). Zhang and Jim (2014) put the arboreal richness of the gardens in Hong Kong down to two factors. The tropical humid climate is conducive for a wide range of tropical tree species to be grown and, once established, garden trees have high survival and persistence rates at remarkably low population sizes, due solely to the high maintenance levels employed by gardeners and landscape professionals. Of the 72 species noted in gardens, some species were more common than others. *Juniperus chinensis* (18% of sampled trees) was dominant through its use as a garden hedge or border tree. Others were popular for their ability to provide shade (*Michelia alba*, 10% and *Dimocarpus longan*, 4.7%), flower colour (*Delonix regia* (flamboyant tree), 5.5%), or dramatic form (*Dypsis lutescens* (bamboo palm), 7.2% and *Araucaria heterophylla* (Norfolk Island pine), 4.2%). Despite the overall diversity being considered good, as with other cities there is an over-dependence on these relatively few popular species. This result was consistent with urban trees in Chicago, Mexico City and Bangkok where the top four dominant species comprise over 50% of the entire tree population.

Gardens provide particular opportunities for specific taxa and groups. The high propensity of garden ponds, for example, help maintain populations of amphibians in the UK, i.e. *Rana temporaria* (common frog), *Bufo bufo* (common toad) (Fig. 4.7) and *Lissotriton vulgaris* (smooth newt) and help counteract the loss of wetlands and farm ponds over the

Fig 4.7. *Bufo bufo* (the common toad) is generally adapted to well-vegetated gardens, but is vulnerable to degradation of pond habitat and injury from domestic pets or road traffic.

last century – a point illustrated by the fact that populations of the common frog have experienced declines in rural areas but increases in urban parks and gardens. As mentioned elsewhere, urban environments are important for bees, and gardens contribute to this. Colony expansion of bumblebee (*Bombus terrestris*) nests was greater in suburban gardens in southern England than neighbouring agricultural habitats, and the density of bumblebee nests recorded in UK suburban gardens (36 nests ha^{-1}) was comparable to that found in hedgerows and grassed pathways in rural habitats (20–37 nests ha^{-1}) (Osborne *et al.*, 2008). Similarly, in a comparison of public parks in San Francisco, USA, *Bombus* spp. were more abundant in a park within a central urban location compared to parks outside the city boundary.

The demise of bee species and other 'pollinators' is a growing concern, not only in relation to concerns for the species themselves, but also the pollinating services such taxa provide. Bees, butterflies, hoverflies and certain beetle and thrip species are important for ensuring the pollination of a wide range of food crops; the fruiting bodies of which are fundamental to human nutrition. Insect-dependent crops include *Malus* (apple), *Pyrus* (pear), *Fragaria* (strawberry), *Lycopersicum* (tomato), *Pisum* (pea) and other legumes, to name but a few. Bees also provide honey directly.

A number of initiatives in recent years have aimed to better understand pollinator populations and to encourage their viability, both in rural and urban environments. Within the UK, the Royal Horticultural Society (RHS) investigated the value of non-native plants to invertebrates, both as a source of pollen and nectar, but also as a general conducive habitat to a range of taxa. Their 'Plants for Bugs' project evaluated garden plant populations using species derived from either the UK alone, the northern hemisphere excluding the UK or from the southern hemisphere, these geographical 'communities' being used to monitor the habitat preferences of various invertebrate taxa. Each plant population were matched in their composition and form across the geographical populations, i.e. each group including similar numbers of shrubs and climbers, geophytes, herbaceous perennials as well as genotypes of ornamental grasses. Data is still being collated from the 4-year project, but findings from this project and other parallel studies, in addition to the extensive records and experience of RHS entomologists, gardeners and beekeepers, has enabled the RHS to provide a comprehensive list of plants that are useful for pollinating insects (Tables 4.6–4.11). This list has now been branded 'RHS Perfect for Pollinators' (Fig. 4.8) and has been endorsed by the England National Pollinator Strategy. This allows both gardeners to attain (and nurseries to promote) those genotypes particularly useful to pollinating insects, irrespective of whether the plants are 'native' or not (Anon., 2015a). Other studies have taken similar approaches and information is now available to gardeners in North America as to what garden plants should be planted to encourage hummingbirds to visit (Table 4.12).

Urban parks are valued for their ecosystem services and positive aesthetical and social values, being the 'green lungs' of many cities, but they are also useful refuges for wildlife. The historic city parks of Europe, for example, provide a network of vegetation types, ranging from the highly ornamental to almost near naturalistic. Along with gardens, the larger urban parks are characterized as one of the most heterogeneous green spaces in urban ecosystems, and often demonstrate a range of habitat type and high vegetation diversity. Typical components include ponds, grassland (short and meadow-like), woodland, shrub communities, stand-alone specimen trees and ornamental annual and perennial borders. They often contain and support the preservation and conservation of endangered and rare taxa (Kümmerling and Müller, 2012). European parks are based largely on indigenous flora supplemented with alien ornamentals introduced since the 16th century; the reliance on native species for the framework has not been mirrored, however, in countries

Fig. 4.8. The Royal Horticultural Society in the UK is promoting plants that provide nectar and pollen to insects through their 'Perfect for Pollinators' awareness campaign. Image courtesy of the RHS.

that were European colonies. In European colonies, non-native species imported from the colonizing country were used rather than indigenous flora in the original park design. Parks therefore, particularly those outwith Europe, can be a source for plant invasions through their widespread use of non-native plants (Dehnen-Schmutz et al., 2007).

Many urban parks were criticized in the 1970s and again in the 2000s for being overly dominated by turf grass, which required excessive maintenance through gang mowing, resulting in a high use of fossil fuels. A consequence of which was that the term 'green deserts' was coined to describe park landscapes. Since then, many park authorities have attempted to turn a higher proportion of their closely mown turf over to meadow-type landscapes with varying degrees of success. Increasing the length of the grass (height of cut) has been associated with increasing plant richness. Falk (1980) observed that intensively managed (i.e. mown, fertilized and irrigated) lawns had 50% fewer species than the less intensively managed lawns. Intensively managed and close-cut lawns promote grass species at the expense of broadleaved forbs (herbaceous plants). For example the grasses *Festuca arundinacea* (tall fescue), *Poa pratensis* (Kentucky bluegrass) and *Cynodon dactylon* (Bermuda grass) occupied 90% cover in an intensively managed lawn, but this dominance dropped to 70% in the less intensively managed counterparts. The non-grass species also altered in their abundance and domination, with a transition from *Trifolium repens* (white clover) to *Digitaria ischaemum* (crabgrass) as the intensity of the management decreased.

The area of land within urban parks has been linked to diversity in bird species, with larger areas increasing richness. However, the number of sub-habitats, the complexity of these habitats and their ability to complement each other and hence fulfil the entire ecological needs of a species also play key roles. Biological factors such as size of home range and specific interactions between species are critical too.

Table 4.12. Examples of garden plants attractive to hummingbirds in the USA. (Modified from Anon., 2015b, c.)

Species	Common name
Aesculus californica	California buckeye
Aesculus pavia	red buckeye
Aesculus × carnea	red horse chestnut
Aesculus × carnea 'Briotii'	ruby red horse chestnut
Agapanthus orientalis	African lily
Asclepias tuberosa	Indian paintbrush
Buddleia alternifolia	butterfly bush
Caesalpinia gilliesii	poinciana
Caesalpinia pulcherrima	poinciana
Callistemon citrinus	red bottlebrush
Campsis radicans	trumpet vine
Cephalanthus occidentalis	buttonbush
Chilopsis	desert willow
Citrus spp.	citrus species – orange, lemon etc.
Cuphea ignea	cigar plant
Erythrina herbacea	cardinal spear
Hamelia patens	firebush
Hibiscus spp.	swamp mallow
Jatropha integerrima	peregrina
Justicia californica	chuparosa
Malus floribunda cultivars	crabapple
Kalanchoe blossfeldiana	kalanchoe
Lantana spp.	lantana
Lonicera sempervirens	honeysuckle
Odontonema strictum	firespike
Rhaphiolepis indica	Indian hawthorn
Schlumbergera bridgesii	Christmas cactus
Strelitzia reginae	bird of paradise

More intensively managed parks tend to be a negative component, but the term 'intensively managed' needs to be carefully defined and designated here. Herbaceous flower borders are 'heavily managed' but also are useful nectar 'hotspots' within parkland landscapes, attracting many pollinating insects.

Bird populations were monitored in relation to landscapes within Israeli urban parks and the degree of management imposed (Fig. 4.9, and see Table 4.13 for definitions of management imposed). Although intensively managed landscapes decreased the richness of the species present, areas that were managed at a moderate intensity were deemed good as bird habitats and comparable to unmanaged areas. Overall, the moderately managed areas were richer habitats for all the categories of birds identified (urban adapter, urban exploiter, alien and locally rare).

Practical aspects to garden and park management that environmental horticulturalists can employ to aid wildlife include the following:

- Reduce lawn mowing to allow greater plant diversity and for more prostrate forbs (herbaceous, non-grass species) to flower.
- Avoid the use of pesticides, or utilize more benign, more targeted or organic alternatives.
- Provide water features, ponds and wet/damp areas. For example damp mud from drying puddles is used by *Delichon urbicum* (house martin) to build their nests in the eaves of houses.
- Create heterogeneous plant groups to mimic natural habitats – canopy, shrub, and ground layers of vegetation. Introduce features that provide shelter, e.g. hedges, compost heaps, log piles and areas of long grass. As most gardens are part of a mosaic, not all 'habitat' types will need to be incorporated within a single garden.
- Increase the number of trees and the amount of foliage.
- Encourage at least some trees/shrubs to grow to their natural height, and promote a canopy structure 2 m in height.
- Although perhaps unsightly, dead tree trunks can be left and then used to support more attractive features such as climbing *Rosa* (rose), *Clematis* or *Lonicera* (honeysuckle).
- Leave cut logs and branches to support wood-boring invertebrates and fungal microorganisms.
- Provide an area of the garden/park that is not intensively managed.
- Retain areas that are not excessively 'tidy'.
- Ensure the majority of the landscape is composed of soft rather than hard landscape materials (plants and exposed soil, rather than concrete and paved slabs).
- Introduce nectar and pollen rich flowering plants – create floral rich plant communities using annuals, biennials or herbaceous perennials.
- Promote complementary flowering plant types, e.g. night-scented, rosette, panicle and tubular forms, continuity of flowering and long flowering periods, seasonal variety and flowers that produce edible seeds.
- Generally, single flower forms are more advantageous than doubles or multiples, because the nectaries and stamens will not have been sacrificed in the breeding process for more petals, or any nectaries that are present will be more easily accessible.
- Promote a proportion of plants that tolerate predation effectively, e.g. *Rosa* cultivars that are

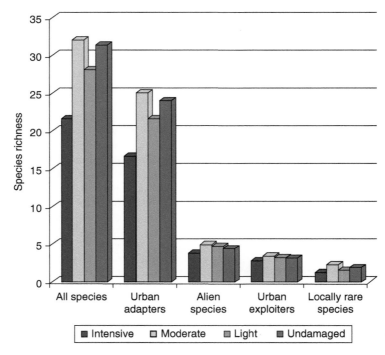

Fig. 4.9. Bird species richness in an urban park, Tel Aviv, Israel. (Source: Shwartz et al., 2008.)

Table 4.13. Description of the management regimes and habitat in Yarkon Park, Tel Aviv. (Modified from Shwartz et al., 2008.)

Management regime	Management Intensity[a]	Habitat description	Plant understorey	Irrigation	Human activity[b]
Intensive	++++	Open grass lawn, scattered trees	Mainly exotic	Yes	High
Moderate	+++	Gardens, orchards, agricultural farm	Partly exotic	Yes	Moderate
Light	++	Syrian ash, pine, tamarisk coppices, fallow fields with annuals	Mainly native	No	Low
Unmanaged	+	*Eucalyptus* groves, fallow fields, annual ground cover	Mainly native	No	Low

[a]Landscaping maintenance operations such as grass cutting, pruning, fertilizing levels range from high (++++) to low (+).
[b]As estimated by the quantity and development of pathways (paved versus unpaved), number of people observed eating or playing within a 100 m radius during five sampling minutes in each location (means: 5 for low, 17 for moderate and 37 for high) and the number of waste bins (relative rankings of high, moderate, low).

not unduly damaged by aphids. Use alternative species/genotypes if plants are highly susceptible to herbivores.
- Certainly for gardens in Western Europe, plant provenance does not appear to be of paramount importance in supporting wildlife, because many herbivores select species within a genus or family, regardless of original provenance.
- Use companion planting to protect crops and other desirable plants. These companion plants either produce volatiles that inhibit pest insect species from visiting the locality, or alternatively attract predatory insects that prey on the pests.
- Provide habitat and nesting sites for birds, small mammals and invertebrates.

- Consider 'green' infrastructure solutions – rain gardens and swales rather than underground pipes to draw water off.
- Leave dead plant parts intact until late in the dormant season, as these provide overwintering shelter and potential food sources for wildlife.
- 'No tillage' – i.e. avoiding or reducing the amount of soil digging – is favourable to some, but not necessarily all species. Although there is a wide range of responses among different species, evidence from agricultural systems would suggest most organism groups have greater abundance or biomass in 'no-till' than in conventional tillage systems. This is particularly so for the larger soil organisms as they are more sensitive to digging and other soil interventions than the microfauna. Population changes may not just be due to physical disturbance of the soil, but also to the addition and burial of organic matter such as crop residues or compost, and the changes in soil chemistry, water and temperature this may bring on.
- Invest in bird and bat nest boxes and place these in suitable locations within the landscape.

Community gardens and allotments

Although still understudied, these urban landscapes seem valuable for wildlife, due to the heterogeneous nature of the landscape and variety of vegetation forms found within them. Many community gardens encompass areas of bare open ground but also vegetated areas through both the food crops present and rougher areas of ruderal/shrub cover. The food crops themselves mimic different tiers of natural vegetation: orchard fruit = tree layer, fruit bush = shrub layer and vegetables and herbs = ground layer vegetation. These cropping plants are often augmented by herbaceous flowering plants, as well as many native wild plants, at the less managed edges of the gardens. Despite representing highly managed plant communities, Lin et al. (2015) consider urban agricultural sites to exhibit high levels of biodiversity, and indeed, that they often exceed the biodiversity of other green space areas within the city. They attribute this largely to the varied vegetation structure, increased diversity of native plants and the reduction in impervious surfaces compared to other urban land forms; features they note that also contribute to important urban ecosystem services such as pollination, pest control, and climate resilience.

Although allotments are managed for home food production, a high proportion of allotment holders in Yorkshire, UK, expressed a strong interest in urban biodiversity irrespective of the age or management style of their plots (Turnbull, 2012). Pitfall traps were used across 42 plots to assess the invertebrates living on the soil surface. The data showed a trend for abundance actually increasing from rural to urban plots, with Coleoptera (beetles), Oniscidea (woodlice) and Araneae (spiders) constituting almost 80% of species caught. To some extent, taxa abundance varied with management styles too, although not always in a consistent manner. When plots were split based on either traditional (highly managed) or a less intensive, wildlife-friendly management style, molluscs (snails and slugs) and woodlice were noted to be more abundant on the wildlife-friendly plots, whereas beetles were more abundant on the traditional ones. Spiders, Opiliones (harvestmen) and myriapods (millipedes and centipedes), on the other hand, showed no significant difference. The majority of species surveyed were generalists (flexible opportunists) when ranked within their taxa, and whilst spider populations fitted with the concept of intermediate levels of disturbance increasing biodiversity, the same was not true for the woodlice and beetle species present. Although there was a trend for increases in species diversity along the rural–urban gradient, individual sites also strongly influence biodiversity, further highlighting the importance of site history and management (Turnbull, 2012). A key point to note from this data is that rural areas dominated by intensive agricultural practices may themselves be relatively poor in supporting a diverse range of species.

Recent increased interest in converting derelict, brownfield land and using it for food production led to one study examining the implications of these transitions on invertebrate biodiversity (specifically arthropod generalist predators) (Gardiner et al., 2013). This study in Ohio, USA, showed that the abundance of Coccinellidae (ladybirds) and Syrphidae (hoverflies), and the activity of Carabidae (predatory ground beetles), Formicidae (ants), and Lycosidae (wolf spiders) were roughly equivalent among the vacant plots and the urban allotment sites. The conversion of brownfield sites to allotments, however, had a negative effect on abundance levels of Dolichopodidae (long-legged flies) and the density levels of Linyphiidae (small spiders) and Opiliones (harvestmen). In contrast, the abundance of Anthocoridae (flower bugs) and the number of active Staphylinidae (rove beetles) were greater within the urban allotments compared to the derelict land. Overall, although allotments and community gardens may impact on biodiversity, the response appears to be very taxon-dependent, and as long as allotment sites are carefully managed, they can be positive locations for urban wildlife.

Wetlands and ponds

It is frequently quoted in garden books and magazines that if you wish to introduce animal life to your garden, construct a pond or some other water feature. Open water is normally a rare resource in city business districts and even the outlying suburban belts may have few natural water features remaining. Therefore, the provision of new ponds and bog gardens, as well as careful management of existing natural water courses are key components in drawing wildlife into cities. Despite the fact that many towns and cities developed precisely because they were located on a river, this rarely means the natural character of the river has been preserved. Rivers have been disconnected from their natural flood plains, channelized and culverted-over with concrete – all of which are highly detrimental to aquatic life. Thankfully, these approaches are beginning to be reversed in the more enlightened cities at least, as the value of blue and green infrastructure begins to become more widely recognized (see section 2.5 on sustainable urban drainage systems).

In a survey of the larger aquatic invertebrates (macroinvertebrates) it was shown that garden ponds tended to be less rich than the traditional field ponds in rural locations (Hill and Wood, 2014). These researchers found 44 taxa from garden ponds (10 of which were unique to garden ponds) but over three times that number of taxa in field ponds. They, however, noted the value of garden ponds in providing habitat for certain taxa such as dragonflies and damselflies (Odonata) due to their high mobility at the adult phase and thus relative ease in colonizing new ponds. Additional features such as fountains and artificial waterfalls allowed a number of taxa normally defined as specialists of natural fast flowing water, to colonize these ponds. In contrast, frequent cleaning and dredging of ponds was detrimental to some of the taxa under observation. Overall, urban garden ponds afford value at the landscape scale, due to their relative abundance and diversity, but the biodiversity that is held within them is still strongly determined by their management.

To enhance the value of urban ponds, again a number of practical activities should be encouraged:

- The creation of shallow edges to allow terrestrial species to drink safely and for amphibious species to enter/exit the pond easily.
- The placement of stones, logs, marginal plants or other 'refuge' features at the pond edge to provide some protective cover for species that use the shallows.
- Ensure the middle of the pond is dug to a depth of at least 0.8–1 m as this will enable some water to remain unfrozen during cold winters (ensure such ponds, however, are not accessible to small children).
- Include substrate at the base of the pond as this provides a habitat for invertebrates and amphibians to burrow in and shelter from predators, but this needs to be formed from nutrient-poor material, as excess nitrogen and phosphorus will quickly cause eutrophication problems by encouraging excessive algal blooms.
- As urban ponds will be slow to colonize naturally with aquatic plant species, these can be introduced, but care is required to avoid the use of invasive species, most notably *Azolla filiculoides* (fairy moss), *Crassula helmsii* (New Zealand pygmyweed), *Eichhornia crassipes* (water hyacinth), *Hydrocotyle ranunculoides* (floating pennywort), *Lagarosiphon major* (curly waterweed), *Myriophyllum aquaticum* (parrot's feather) and *Pistia stratiotes* (water lettuce). Balanced planting of marginal plants in the shallows, oxygenating species for the pond bottom and floating species such as *Nymphaea* spp. (water lilies) being advocated to help establish the pond quickly. Floating species exclude light, thus helping to counteract excessive algal blooms.
- Avoid introducing ornamental fish, as most are superb predators of invertebrates and amphibian tadpoles.
- Encourage invertebrates such as *Daphnia* and *Cyclops* spp. as they are algal grazers and often important in the food chain of garden ponds.
- Remove sediment and sludge to stop excess silting of the pond, although doing so too frequently can impair invertebrate populations. Pond sludge and dead vegetation can be left at the side of the pond to allow the more mobile invertebrates to attempt to migrate back to the pond, before being finally removed and composted at a location away from the aquatic environment. Cleaning

ponds with detergents or other chemical agents is not encouraged.
- The design and construction of larger ponds or mosaics of ponds can encourage less abundant species, e.g. in the UK amphibians such as *Triturus cristatus* (great crested newt) or *Natrix natrix* (grass snake).

Roads, railways and verges

Road and rail traffic do not tend to mix well with wildlife! Road traffic can have a significant toll on populations of some vertebrate species, particularly slow moving creatures, including toads and frogs. Wider and busier roads have been shown to be problematic for both bird and butterfly species (Chong *et al.*, 2014). Birds appear to be disturbed by the noise and vehicle movement, and even the air turbulence caused problems for butterflies. Providing trees that overhang the edge of roads, and building roads where trees can be incorporated into the middle reservation are deemed positive interventions to help aerial animals traverse roads and hence migrate across the wider urban matrix. There are some advantages too for humans, with vegetation planted down the middle of highways reducing glare from headlights from on-coming traffic. This needs to be balanced, however, with the potential for more serious injuries to drivers/passengers when cars collide directly with trees.

Low-growing woody vegetation, i.e. shrubs and hedgerows, immediately in the vicinity of roads (at least those with heavy traffic volumes) has a negative effect on bird populations, through increasing the number of fatalities due to collisions with road vehicles. Urban specialists such as *Turdus merula* (blackbird), *Passer domesticus* (house sparrow) and *Prunella modularis* (dunnock), which have low lines of flight, are placed in danger when flying from one patch of shrubbery to another across a roadway. Those that tend to patrol along hedgerows and ditches at 2–3 m above the ground such as *Tyto alba* (barn owl) are also particularly vulnerable where such features lead up to roadways. Careful design of roadside vegetation can offset some of these problems through avoidance of 'desire lines' that transect directly across a road. For example, staggering and extending the distance between patches of shrubbery at either side of a road will encourage some species to fly higher, and the placement of vegetation further away from the road will improve the line of vision, thus helping birds perceive any on-coming traffic. Conversely, vegetation can be designed to encourage terrestrial animals to cross roads via wildlife tunnels and green bridges (Fig 4.10).

Roadside verges are underutilized nature reserves. Through better management these could encourage more pollen- and nectar-rich flowering forbs that are beneficial to insects. Different mowing management treatments have been evaluated in the Netherlands (Noordijik *et al.*, 2009). Mowing of roadside verges twice per year with the grass cuttings (hay) being removed, increased the number of flowering plant species and the total number of inflorescences observed compared to other treatments imposed. This had positive effects on the abundance of insects and the frequency at which flowers were visited. Although mowing twice a year with hay removal was most beneficial in terms of management interventions, there was still a complete absence of flowering plants immediately after the mowing. Any extension of this or similar schemes, therefore, should encourage rotational cutting in different areas along the roadway to ensure some flowering plants are available to the local invertebrate populations continuously throughout the entire summer.

Brownfield sites, vacant plots and wastelands

Brownfield sites, vacant plots and wastelands are not areas that would necessarily normally fall under the jurisdiction of a horticultural office, but the value of such brownfield sites needs to be considered in the wider context of urban green space and biodiversity conservation; not least because horticulturalists will often be called on to help in the 'development' of ex-industrial 'brownfield' sites. The reality is that these 'waste' areas are frequently quite species-rich and provide opportunities for a spectrum of species, including those native to the area, or from agri-rural environments as well as ruderal, non-native species (Kelcey and Müller, 2011). Moreover, as they are very atypical to other urban open/green spaces they offer habitat for numerous rare and endangered species. Previous land activities and their legacy often dictate the vegetation dynamics and wider biological significance of such sites. Rubble left over from demolished buildings results in patches denuded of vegetation and a prevalence of dry sandy or stony soils and their accompanying warm localized microclimates, conditions conducive to species such as ground beetles *Cicindela campestris* (green tiger beetle), *Cicindela sylvatica* (heath tiger beetle) and solitary

Fig. 4.10. Encouraging longer, coarser vegetation at entry points to green bridges and wildlife tunnels can encourage small mammals to use such routes to traverse road and railway lines.

bees such as *Andrena barbilabris*. Brownfield sites also attract stress-tolerator plant genotypes, such as orchids, due to the residual contamination from chemicals associated with the previous industrial processes. Nutrient-poor soils with high metal content enhance the competitiveness of plants with stress-tolerator traits over ruderal and competitor species. Not all brownfield sites have nutrient poor soil, however, and the heterogeneity of patches with different nutrient pools adds to the biodiversity. Indeed, soil fertility is often key in determining species dominance in these patches. In Western Europe, woody plants such as *Populus tremula* (aspen) dominate on poor sites, whereas *Sambucus nigra* (elderberry) and *Salix caprea* (goat willow) dominate the moderately fertile sites. Tredici (2010) describes this suite of species occupying vacant plots rather protractedly as a 'cosmopolitan assemblage of early-successional, disturbance-tolerant species that are pre-adapted to the urban environment'.

Brownfield sites are under socio-economic pressures due to redevelopment for housing and industrial uses. The impact of this on their biodiversity is poorly understood. For example, it is unknown how much open area needs to be preserved and indeed whether conservation is possible without completely excluding economic development (Kattwinkel *et al.*, 2011). Simulation studies using species distribution models for plants, grasshoppers, and leafhoppers demonstrated that overall dynamic land use (i.e. the 'cycle' of construction and development of new building on some sites, but the demolition of buildings with plots left vacant on others) supports urban biodiversity in terms of species richness and rarity. From this modelling it was apparent that setting aside brownfields before redevelopment for a period of on average 15 years supported the highest conservation value. As such, Kattwinkel *et al.* recommend integrating the concept of 'temporary conservation' into urban planning when dealing with industrial and business areas. This concept requires habitat to be destroyed by redeveloping brownfield sites and converting them to built-up structures, but simultaneously creating new open spaces due to abandonment of urban land uses at other locations. They claim this maintains a spatio-temporal mosaic of different successional vegetational stages ranging from pioneer to pre-forest communities. Such approaches enable urban populations of some unique species to remain viable, over the long term.

Conventional walls and roofs

Built infrastructure, although an adverse environment for most species is not without life. Hard surfaces dominate due to the high density of buildings, but these vary in their construction and integrity. Modern city-centre buildings constructed largely of glass or metal provide limited opportunities for wildlife, unless specific interventions relating to green roofs or walls are implemented. On more traditional brick and stone surfaces, on the other hand, non-vascular taxa such as lichens and moss can gain a foothold, even if the surface is relatively smooth. Non-mortared structures and those buildings, walls, pavement or roadways that are beginning to wear, provide niches for vascular plants. Compared to newer walls, older walls and mortar tend to have more plant species because they have weathered more, have had time to neutralize alkaline salts, and accumulate some organic material in crevices. Nevertheless, these remain challenging environments due to the relative limited availability of soil and water, and the extreme temperatures experienced. As temperature and moisture profiles strongly influence the potential for life, the predominant climate determines the range of species present. Maritime climates with high rainfall and low fluctuations in temperature promote a wider diversity and abundance of wall flora, and relatively large plants such as ferns and even *Buddleia davidii* (butterfly bush) can be commonplace on crumbling wall systems. Patterns of plants types found in walls and other hard surfaces have been classified (Lundholm and Richardson, 2010; Francis, 2011). In Atlantic and Central Europe hemicryptophytes (perennials with their buds at or near the soil surface) dominate, whilst in Mediterranean Europe, chamaephytes (woody species with resting buds at or near the soil surface) are more common. In India, on the other hand, therophytes (annual species) tend to be better adapted, whereas in Israel phanerophytes (woody species with resting buds above the soil surface) were observed to be the dominant group colonizing walls.

Green roofs and walls

One of the key drivers for green roofs (see Chapter 10) has been habitat provision for wildlife. The opportunities for wildlife are influenced by factors such as the design of the green roof, substrate composition and depth, degree of exposure the roof experiences, water availability, the wet/dry cycles of the substrate, and the location with respect to other nearby green spaces. Most green roofs are effectively ecological islands, with continuity of habit between them generally not feasible. Green walls may allow some connection with roofs and thus potential for green networks and corridors, but the vegetation type is often very different, and this still makes the assumption that the migrating animal species can climb! Species that can fly or are able to propel themselves via wind currents such as fungal spores, lightweight plant seeds, or young 'parachuting' spiders will be able to colonize new roofs. Human-assisted introductions are possible, as are inadvertent introductions, e.g. via substrate or soil imported on to the roof. In one of the small green roofs at the University of Sheffield, UK, a frog has taken up residence and was possibly brought in with a damp vegetation mat or suchlike.

Brenneisen (2006) argues that natural soils and substrates are important in colonizing a roof system. The adaptation of spider and beetle fauna to natural soil and other substrates such as sand and gravel from riverbanks seemed to be a key factor to success in a number of Swiss green roofs. This research showed that near-natural habitats can be established on roofs, with different roofs in Switzerland partially replicating certain microhabitats. This included the establishment of invertebrate populations associated with riverbanks, rock scree and debris, alpine and dry grassland habitats. Roofs with restricted drainage have been used to replicate the conditions typical of wet or dry meadows, heathland and acidic moorland. Even the inclusion of ponds and shallow moisture-retentive areas can encourage marsh and wetland species.

Local governments in different regions of both Switzerland and Germany have stipulated that new buildings of a certain size, and which have a flat roof, are legally obliged to put a green roof system in place. This has been a catalyst for the expansion of the green roof industries in these countries. In Switzerland, as a means to promote biodiversity, roofs exceeding 500 m^2 in area also need to be composed of appropriate natural soils from the surrounding region and must be contoured to provide varying depths of substrate to support different plant and invertebrate communities. One of the oldest green roof complexes in Switzerland is the Wollishofen water plant in Zurich, which was originally installed in 1914 to keep the water within the building cool. These roofs were composed of native soils and plant species, and unlike the surrounding agricultural land, which has subsequently become intensively managed, have succeeded in retaining their complex and diverse plant communities. This green roof has now become famous due to its role as a refuge for 175 recorded plant species, including a number of Red Data book (i.e. endangered) terrestrial orchids. A number of orchid species have been found growing on this and other European green roofs (Table 4.14; Fig. 4.11). Unlike many modern green roofs, the Wollishofen roof had local soil placed over a layer of gravel, resulting in a substrate that alternates between high water-retention and much drier periods, thus providing conditions not dissimilar to those found in semi-natural habitats such as wet meadows and moorland. Such conditions were important factors in conserving the typical local and regional biodiversity on these green roofs (Brenneisen, 2003).

A number of researchers and land managers have advocated that green roofs can play an important role in preserving wild flower species that are under pressure in the wider rural environment, due to loss of habitat, soil nutrient enrichment and increased competition from more aggressive species or changes in land management practices; in the latter case, for example, the loss of traditional hay meadows to silage production – the paradox being that urban green roofs may facilitate the conservation of rare plant species normally associated with the traditional or rural, agrarian landscape. The fact that green roofs are a highly artificial system, but where key factors such as soil nutrient levels or moisture can be manipulated, means there are opportunities to recreate important ecological niches that have become rare in natural or agro-ecosystems. Working in the Mediterranean climate of Italy, Benvenuti (2014) monitored plant population dynamics on a semi-extensive 200 mm-deep substrate roof, with irrigation being used during establishment and the more severe dry periods. He found that by planting communities of plants that flowered throughout the season (Table 4.15) continuous food supplies could be provided to a number of different insect taxa (Figs 4.12 and 4.13). The adverse environment encountered on the roof to some extent could be countered by a reliance on geophytes that survived from one year to the next due to their dormant phase aligning with the driest and hottest mid-summer periods. In conclusion, Benvenuti considers that wildflower roofs may contribute to ecological corridors across the city matrix and partially substitute for the loss of green space at ground level. The relatively 'unutilized' potential of roof space should be given greater priority in future to help address biodiversity loss in city centres.

One of the initiatives behind the advent of green roofs in the UK in the 1980s was the desire to recreate brownfield habitat that was being lost at ground level due to the economic 'boom' that took place during that decade. This resulted in vacant ex-industrial spaces within cities being developed for new office blocks and housing. Brownfield sites are important habitat for certain taxa of invertebrates, but also one European bird species, namely *Phoenicurus ochruros* (black redstart). This species, perhaps more than any other, has become synonymous with the ecological green roof movement. Within the UK, the black redstart is at the edge of its northern distribution, and so having a resident breeding pair using your green roof as feeding habitat is seen as the ultimate symbol of success. As more brownfield sites were lost to development, the greater the pressure became to initiate green roofs as alternative habitat for black redstarts.

The key criteria that green roofs can provide for birds are water, food, shelter and roosting opportunities. Some species may also choose to nest on a green

Table 4.14. European orchid species observed on green roofs.

Orchid species	Common name
Dactylorhiza fuchsia	common spotted orchid
Dactylorhiza incarnata	early marsh orchid
Dactylorhiza majalis	western marsh orchid
Dactylorhiza sambucina	elder-flowered orchid
Epipactis palustris	marsh helleborine
Gymnadenia conopsea	fragrant orchid
Listera ovata	twayblade
Ophrys apifera	bee orchid
Ophrys sphegodes	early spider orchid
Orchis mascula	early-purple orchid
Orchis militaris	military orchid
Orchis morio	green-winged orchid
Platanthera bifolia	lesser butterfly-orchid

Fig. 4.11. A number of 'iconic' orchid species have been found growing on green roofs including left *Ophrys apifera* (bee orchid) and right *Dactylorhiza fuchsia* (common spotted orchid).

Table 4.15. Plant communities evaluated by Benvenuti (2014) in Pisa, Italy, for potential use on green roofs, utilizing some increasingly rare wild plants and determining their potential to yield pollen and nectar sources to native insects.

Plant community	Species	Common name
Early flowering	*Anemone hortensis*	broad leaved anemone
	Crocus vernus	spring crocus
	Iris chamaeiris	dwarf bearded iris
	Narcissus tazetta	paperwhite daffodil
	Ornithogalum umbellatum	star of Bethlehem
Spring flowering	*Anthemis arvensis*	corn chamomile
	Centaurea cyanus	cornflower
	Chrysanthemum myconis	corn marigold
	Nigella damascena	love-in-a-mist
	Tuberaria guttata	spotted rock-rose
Summer flowering	*Anthirrhinum majus*	snapdragon
	Calamintha nepeta	lesser calamint
	Consolida regalis	forking larkspur
	Dianthus carthusianorum	carthusian pink
	Scabiosa columbaria	small scabious
Late flowering	*Allium carinatum*	keeled garlic
	Colchicum autumnale	meadow saffron
	Leontodon tuberosus	tuberous hawkbit
	Scilla autumnalis	autumn squill
	Sternbergia lutea	yellow autumn crocus

Environmental Horticulture and Conservation of Biodiversity

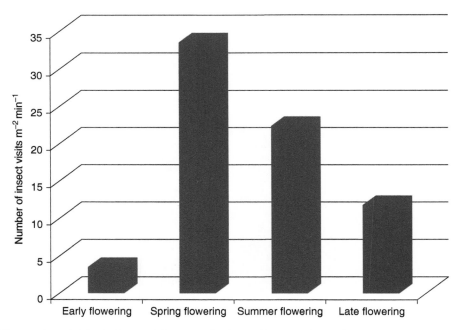

Fig. 4.12. The number of visits by insects (all taxa studied) per m² in one minute observation periods in plant communities representing different potential green roof flora in Italy. (Modified from Benvenuti, 2014.)

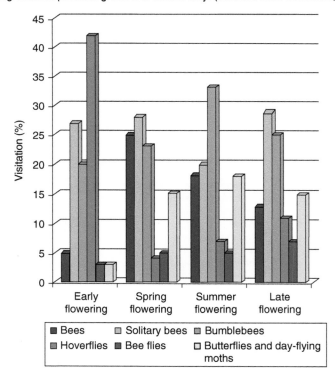

Fig. 4.13. The proportion of different insect taxa visiting green roof plant communities as determined by different flowering times during the year. (Modified from Benvenuti, 2014.)

roof depending on their normal requirements, for example the gravel substrates employed may replicate the stony beaches that species such as terns, gulls and ringed plovers normally utilize (Table 4.16). Food sources commonly include berries, seeds and invertebrates attracted in by the vegetation and substrates. As highlighted above, however, not all green roofs have the same ecological value for feeding or breeding birds. The opportunities presented for colonizing bird species vary depending on roof type, vegetation cover and plant species composition, presence of water, space (territory) available and levels of maintenance (potential disturbance).

Food is perhaps the main reason that certain bird species are attracted to green roofs – a point that may explain the roofs in highly urbanized areas being more frequently visited (as there are few alternative sources of food) than those in rural areas or at the edge of the city (where there is more choice) (Dunnett and Kingsbury, 2004). Cantor (2008) observed that those bird species that relied on the plants or invertebrate species on roofs were present more frequently than more common urban species which did not appear to have an ecological niche associated with the roofs. Green roofs may also provide a secure nesting or roosting site, and the fact that they are generally located in urban habitats may actually mean that there are fewer predators around compared to more traditional rural nesting sites (Gering and Blair, 1999). Large numbers of young of ground nesting birds such as *Vanellus vanellus* (northern lapwing), *Sterna hirundo* (common tern) or *Larus canus* (common gull) are predated on by *Vulpes vulpes* (red fox), *Mustela nivalis* (weasel), *Mustela erminea* (stoat) and *Neovison vison* (American mink), yet these predators will be almost universally absent from tall roofs at least. Larger plants on roofs facilitate roosting, by providing physical shelter against adverse weather, with roofs having additional advantages such as enhanced night-time temperatures (due to heat loss from buildings); extreme cold being a major cause of calorie consumption and thus a significant killer of small birds, or reducing their fitness to breed in the subsequent spring. Shrubs and small trees are particularly favoured by song birds (passerines) and their presence on roofs can both widen the range of species visiting and enhance the numbers of individuals within a species found at a given location.

Although water is not an absolute requirement, its presence on a green roof may reduce the number of visitations a bird needs to make to water bodies at ground level, thereby reducing its chances of being predated upon, and saving it energy. Surface or standing water is thus often a useful addition to roofs that are designed to provide habitat for birds. This does not need to be a dedicated pond, but impermeable areas where puddles can form or where irrigation water can be allowed to accumulate will be readily exploited by any birds present. Some bird species are adroit in attaining water from their food sources or even drinking dew; however, the provision of water baths and water stations may be an alternative means to ensure water is supplied effectively and regularly.

There is still relatively little information available as to how important green roofs may be in providing

Table 4.16. Bird species that have been reported to breed on green roofs (Sources: Burgess, 2004; Brenneisen, 2003, 2006; Baumann, 2006; Anon., 2010; Fernandez-Canero and Gonzalez-Redondo, 2010.)

Species	Common name
Anas platyrhynchos	mallard
Anthus pratensis	meadow pipit
Alauda arvensis	sky lark
Branta canadensis	Canadian goose
Carduelis chloris	green finch
Charadrius dubius	little ringed plover
Charadrius hiaticula	ringed plover
Charadrius vociferus	kildeer
Columba livia domestica	feral pigeon
Columba palumbus	wood pigeon
Corvus corone	carrion crow
Cyanistes caeruleus	blue tit
Falco peregrinus	peregrine falcon
Falco tinnunculus	kestrel
Fringilla coelebs	chaffinch
Galerida cristata	crested lark
Haematopus ostralegus	Eurasian oystercatcher
Larus canus	common gull
Motacilla alba	pied wagtail
Muscicapa striata	spotted flycatcher
Oenanthe oenanthe	northern wheatear
Parus major	great tit
Passer domesticus	house sparrow
Passer montanus	tree sparrow
Phoenicurus ochruros	black redstart
Phylloscopus trochilus	willow warbler
Pica pica	magpie
Rissa tridactyla	kittiwake
Sterna hirundo	common tern
Turdus merula	blackbird
Vanellus vanellus	northern lapwing

a network of green spaces across the city matrix, and whether they assist birds in migrating across urban conurbations. Potentially they could become important 'refuelling stations' for some species as they conduct their biannual migrations or even just move from one green space to another looking for new territories. Even if certain species do not nest on green roofs, appearances on them can become commonplace as they search out food. The green roof at the Ford Motor Company's Dearborn construction plant, Michigan, USA, is frequented by *Contopus cooperi* (olive-sided fly catcher) and *Charadrius vociferus* (killdeer). Keeping with the motor industry theme, Burgess (2004) observed on the green roof at the Rolls Royce factory in West Sussex, UK, bird species as diverse as *Corvus frugilegus* (rook), *Larus canus* (common gull), *Charadrius hiaticula* (ringed plover), *Motacilla alba* (pied wagtail), *Turdus philomelos* (song thrush) and *Carduelis cannabina* (linnet). Green roofs in Portland, Oregon (USA) have been recorded hosting *Cyanocitta cristata* (blue jay), *Passer domesticus* (house sparrow), *Selasphorus rufus* (rufous hummingbird), *Corvus brachyrhynchos* (American crow), and *Hirundo rustica* (barn swallow). Green roofs may provide these generic services of food, shelter and water, but can also be designed to mimic habitats within the urban area to specifically benefit relatively rare or endangered species.

Green walls, particularly green façade types, provide nesting and foraging habitat for urban-adapted birds. Research in north Staffordshire, UK, compared bird activity on or within a 10 m radius of green walls and compared this to similarly unclad walls (Chiquet *et al.*, 2013). Birds were not encountered on bare walls, although they were observed on nearby roofs and vegetated areas. Birds, however, were more abundant in areas with green walls. On the green walls themselves, the birds' activity was always restricted to the upper half of the wall vegetation. Cladding green walls with dense foliage appears to offer some habitat opportunities to birds, and may help supplement urban green infrastructure in those locations where urban density restricts the opportunities for trees and other ground-covering vegetation (Chiquet *et al.*, 2013). In suburban situations the use of wall climbers such as *Hedera helix* (ivy), *Clematis montana*, *Passiflora caerulea* (Passion flower), *Hydrangea petiolaris* and *Parthenocissus quinquefolia* (Virginia creeper) on houses or garden structures are known for their ability to afford secluded nesting habitat.

Birds may be drawn to green facades as a resource for food. Observational studies by Matt (2012) suggested that green façades had a higher abundance and diversity of arthropods than non-vegetated building façades. Vegetated walls were shown to contain 16–39 times more arthropods per m^2 than equivalent bare walls. Invertebrate species found included herbivores, predators, parasitoids and detritivores, indicating a range of niche opportunities within the wall systems. Arthropod abundance and richness were most strongly correlated with habitat availability and increased with density of the foliar canopy.

Although green roofs provide better opportunities for wildlife than bare roofs, some have criticized the implications that green roofs, particularly extensive roofs with shallow substrates, are an alternative to habitat provision at ground level. Therefore, the justification that a new building can be placed on a green field site and a replacement habitat simply placed on the roof of the building should be challenged. The disadvantages associated with green roofs are not solely focused on their accessibility for some species (such sites being difficult for numerous terrestrial species); also the soil temperatures may be supra-optimal, or they may retain inadequate moisture levels to let some plant and animal species survive. Earthworms, for example, struggle to thrive on many green roofs, as there is not the depth of soil required to retreat to cooler, moister conditions in summer. Each green roof may be too small to support individual populations of any one species, and it is difficult to replicate the scale of habitat that some species require. In comparing green roofs and ground habitats in Basel, Switzerland, Brenneisen (2006) noted that the areas available for colonization at ground level covered several hectares and were thus in a different order of magnitude to that of green roofs, which may only cover between a hundred and a few thousand square metres.

4.7 Urban Biodiversity and Humans

Urbanization is a significant threat to biodiversity and contributes to the physical, geographical and emotional separation of people from nature (Strohbach *et al.*, 2009). In addition to removing and fragmenting habitat, urbanization presents an additional growing problem for biodiversity conservation, through increasingly isolating half of the world's population from the experience of nature. The separation of people from nature is likely to be an increasingly important environmental issue, as it fundamentally influences the way people value nature and their willingness to conserve it. A lack of knowledge leads to a lack of empathy. A vibrant, observable and easily accessed urban biota can help address this trend and improve people's understanding of the natural world, although perhaps not always providing a comprehensive awareness.

The ability of citizens to understand and connect with urban biodiversity was studied by Shwartz *et al.* (2014), where they intentionally enhanced the diversity of flowers, birds and insect pollinators in small public gardens in Paris and observed people's responses. Species diversity and abundance was increased by providing flower meadows and introducing bird nest-boxes. Residents were interviewed before and after the interventions. Results showed respondents expressed a strong preference for a rich diversity of species (excluding insects) and interestingly related this diversity to feelings of well-being in the gardens. There was no indication, however, that they actually noted the change in the diversity of species before and after the habitat interventions. Respondents underestimated species richness and only noticed the changes in native flower richness in those locations where the interventions were explicitly advertised and where public engagement activities were organized. Shwartz *et al.* claim that their results highlight a 'people–biodiversity paradox' as there is a mismatch between people's perceptions of biodiversity and actual awareness of it. Further studies are needed to explore the role that urban biodiversity plays in people's daily lives and the importance of this interaction for raising public support for general conservation policies.

Conclusions

- Urban biodiversity is the variety or richness and abundance of living organisms (including genetic variation) and habitats found in and on the edge of human settlements.
- Urbanization tends to diminish and fragment green space, resulting in limited resources for remaining populations of plants and animals, and inhibiting movement (reduced connectivity) between one area and the next for some species/taxa.
- Other problems associated with urban environments include: direct injury due to vehicles and buildings, limited access to soil and altering hydrological patterns, temperature profiles and soil nutrients, more disturbance through physical means and noise, greater predation due to pet species and competition from introduced 'alien' species.
- A key driver for environmental horticulture is to restore green landscapes and a degree of ecosystem function to areas that are currently highly degraded and where the biodiversity is impoverished. This may include utilizing non-native plant species as well as native, where appropriate to do so.
- Occurrence of species within urban habitats depends on factors such as a species' key traits and adaptability, population size, the history of a site or habitat, the availability of appropriate habitat and the quality and spatial arrangement of habitats.
- Typically, in city centres 30–50% of the plant species are non-native. Plant biodiversity though tends to peak in the suburbs where there are still relatively adequate areas of green space and associated niches for native species, due largely to a heterogeneous landscape typology (parks, woodlands, gardens, waterways), whilst the increase in garden space correlates with an abundance of non-native garden plants.
- Environmental horticulturalists need to champion the urban forest concept as urban forests (garden, street and park trees as well as areas of dedicated woodland) are important for supporting urban biodiversity and allowing species to move across the urban landscape.
- The design and management of garden space will become increasingly important as the effects of urbanization and climate change place pressure on native animal and plant populations.
- Carefully managed gardens, allotments, parks and even brownfield sites are important refuge sites for urban wildlife, and have the potential to support greater biodiversity and numbers of individuals of certain species/taxa than agriculturally managed areas in the rural environment.
- Green roofs and green walls can help certain taxa in the urban environment, but will not necessarily replace terrestrial green space lost to urbanization on a like-for-like basis. Soils and substrates may be limited in volume and in their ability to remain sufficiently moist, the vegetation mantle may be relatively thin, and some species may have difficulty accessing such 'habitats'. Future research should evaluate the mechanisms by which green roofs and walls can be better planned, designed

and managed to promote biodiversity and other ecosystem services.

- Urban biodiversity will become increasingly important in enabling a largely urban human population to access and understand nature. Environmental horticulturalists can help 'bridge the gap' between traditional, highly manicured green space and more ecologically robust spaces, whilst ensuring such places remain highly relevant to the human population.

References

Adams, E. and Lindsey, K.J. (2009) *Urban Wildlife Management*, 2nd edn. CRC Press, Georgia, USA.

Andersson, E., Barthel, S. and Ahrné, K. (2007) Measuring social-ecological dynamics behind the generation of ecosystem services. *Ecological Applications* 17, 1267–127.

Angold, P.G., Sadler, J.P., Hill, M.O., Pullin, A., Rushton, S., Austin, K. and Thompson, K. (2006) Biodiversity in urban habitat patches. *Science of the Total Environment* 360, 196–204.

Anon. (2010) Habitat action plan – Green Roofs. University of Sheffield, Green Roof Centre, www.thegreenroofcentre.co.uk/Library/Default/Documents/Sheffield%20Green%20Roof%20HAP%20Feb%202010_634159246532552600.pdf (accessed 22 January 2015).

Anon. (2015a) RHS Perfect for pollinators plant list. Royal Horticultural Society, www.rhs.org.uk/science/pdf/conservation-and-biodiversity/wildlife/rhs_perfectfor-pollinators_plantlist-jan15 (accessed 22 January 2015).

Anon. (2015b) Hummingbirds of Florida. Universities of Florida Extension Services, edis.ifas.ufl.edu/uw059 (accessed 25 January 2015).

Anon. (2015c) Plants for hummingbirds. Oregon State University extension services, extension.oregonstate.edu/gardening/plants-hummingbirds (accessed 25 January 2015).

Baldock, K.C., Goddard, M.A., Hicks, D.M., Kunin, W.E., Mitschunas, N., Osgathorpe, L.M. and Memmott, J. (2015) Where is the UK's pollinator biodiversity? The importance of urban areas for flower-visiting insects. *Proceedings of the Royal Society of London B: Biological Sciences* 282, 20142849.

Barbosa-Filho, W.G. and de Araujo, A.C. (2013) Flowers visited by hummingbirds in an urban Cerrado fragment, Mato Grosso do Sul, Brazil. *Biota Neotropica* 13, 21–27.

Barrett, K. and Guyer, C. (2008) Differential responses of amphibians and reptiles in riparian and stream habitats to land use disturbances in western Georgia, USA. *Biological Conservation* 141, 2290–2300.

Battin, J. (2004) When good animals love bad habitats: ecological traps and the conservation of animal populations. *Conservation Biology* 18, 1482–1491.

Baumann, N. (2006) Ground-nesting birds on green roofs in Switzerland: preliminary observations. *Urban Habitats* 4, 37–50.

Bennett, A.B. and Gratton, C. (2012) Local and landscape scale variables impact parasitoid assemblages across an urbanization gradient. *Landscape and Urban Planning* 104, 26–33.

Benvenuti, S. (2014) Wildflower green roofs for urban landscaping, ecological sustainability and biodiversity. *Landscape and Urban Planning* 124, 151–161.

Blair, R.B. and Johnson, E.M. (2008) Suburban habitats and their role for birds in the urban–rural habitat network: points of local invasion and extinction? *Landscape Ecology* 23, 1157–1169.

Böhm, M., Collen, B., Baillie, J.E., Bowles, P., Chanson, J., Cox, N. and Cheylan, M. (2013) The conservation status of the world's reptiles. *Biological Conservation* 157, 372–385.

Brenneisen, S. (2003) Ökologisches Ausgleichspotenzial von extensiven Dachbegrünungen – Bedeutung für den Arten – und Naturschutz und die Stadtentwicklungsplanung. Doctoral dissertation, Institute of Geography, University of Basel, Switzerland.

Brenneisen, S. (2006) Space for urban wildlife: designing green roofs as habitats in Switzerland. *Urban Habitats*, 4, 27–36.

Burgess, H. (2004) An assessment of the potential of green roofs for bird conservation in the UK (Master's research report). University of Sussex, Brighton, UK.

Cantor, S.L. (2008) Green roofs in sustainable landscape design. Norton and Company, New York.

Catterall, C.P. (2004) Birds, garden plants and suburban bushlots: where good intentions meet unexpected outcomes. In: Lunney, D. and Burgin, S. (eds) *Urban Wildlife: More Than Meets the Eye*. Royal Zoological Society of New South Wales, Mosman, pp. 21–31.

Cavia, R., Cueto, G.R. and Suárez, O.V. (2009) Changes in rodent communities according to the landscape structure in an urban ecosystem. *Landscape and Urban Planning* 90, 11–19.

Chiquet, C., Dover, J.W. and Mitchell, P. (2013) Birds and the urban environment: the value of green walls. *Urban Ecosystems* 16, 453–462.

Chong, K.Y., Teo, S., Kurukulasuriya, B., Chung, Y.F., Rajathurai, S. and Tan, H.T.W. (2014) Not all green is as good: different effects of the natural and cultivated components of urban vegetation on bird and butterfly diversity. *Biological Conservation* 171, 299–309.

Clarke, L.W., Jenerette, G.D. and Davila, A. (2013) The luxury of vegetation and the legacy of tree biodiversity in Los Angeles, CA. *Landscape and Urban Planning* 116, 48–59.

Colding, J., Elmqvist, T., Lundberg, J., Ahrné, K., Andersson, E., Barthel, S., Borgström, S., Duit, A., Ernstsson, H. and Tengö, M. (2003) The Stockholm Urban Assessment (SUA-Sweden). In: *Millennium Ecosystem Assessment Sub-Global Summary Report*, Stockholm.

Cornelis, J. and Hermy, M. (2004) Biodiversity relationships in urban and suburban parks in Flanders. *Landscape and Urban Planning* 69, 385–401.

Dehnen-Schmutz, K., Touza, J., Perrings, C. and Williamson, M. (2007) The horticultural trade and ornamental plant invasions in Britain. *Conservation Biology* 21, 224–231.

Dunn, C.P. and Heneghan, L. (2011) Composition and diversity of urban vegetation. In: Niemeliä, J. (ed.) *Urban Ecology: Patterns, Processes and Applications*. Oxford University Press, Oxford, pp. 103–134.

Dunnett, N. and Kingsbury, N. (2004) *Planting Green Roofs and Living Walls*. Timber Press, Portland, USA.

Dwyer, J.F., Nowak, D.J., Noble, M.H. and Sisinni, S.M. (2000) *Connecting people with ecosystems in the 21st century: an assessment of our nation's urban forests*. USDA Forest Service. General Technical Report PNW-GTR-490.

Falk, J.H. (1980) The primary productivity of lawns in a temperate environment. *Journal of Applied Ecology* 17, 689–695.

Fernandez-Canero, R. and Gonzalez-Redondo, P. (2010) Green roofs as a habitat for birds: a review. *Journal of Animal and Veterinary Advances* 9, 2041–2052.

Francis, R.A. (2011) Wall ecology: a frontier for urban biodiversity and ecological engineering. *Progress in Physical Geography* 35, 43–63.

Gardiner, M.M., Burkman, C.E. and Prajzner, S.P. (2013) The value of urban vacant land to support arthropod biodiversity and ecosystem services. *Environmental Entomology* 42, 1123–1136.

Gering, J. and Blair, R. (1999) Predation on artificial bird nests along an urban gradient: predatory risk or relaxation in urban environments? *Ecography* 22, 532–541.

Gilbert, O.L. (1989) Allotments and leisure gardens. In: *The Ecology of Urban Habitats*. Springer, Netherlands, pp. 206–217.

Goddard, M.A., Dougill, A.J. and Benton, T.G. (2010) Scaling up from gardens: biodiversity conservation in urban environments. *Trends in Ecology and Evolution* 25, 90–98.

Hamer, A.J. and McDonnell, M.J. (2008) Amphibian ecology and conservation in the urbanising world: a review. *Biological Conservation* 141, 2432–2449.

Helden, A.J., Stamp, G.C. and Leather, S.R. (2012) Urban biodiversity: comparison of insect assemblages on native and non-native trees. *Urban Ecosystems* 15, 611–624.

Hill, M.J. and Wood, P.J. (2014) The macroinvertebrate biodiversity and conservation value of garden and field ponds along a rural-urban gradient. *Fundamental and Applied Limnology/Archiv für Hydrobiologie* 185, 107–119.

Hitchmough, J. (2011) Exotic plants and plantings in the sustainable, designed urban landscape. *Landscape and Urban Planning* 100, 380–382.

Hitchmough, J. and Dunnett, N. (2004) Introduction to naturalistic planting in urban landscapes. *The Dynamic Landscape*. Taylor and Francis, pp. 336.

Hope, D., Gries, C., Casagrande, D., Redman, C.L., Grimm, N.B. and Martin, C. (2006) Drivers of spatial variation in plant diversity across the central Arizona-Phoenix ecosystem. *Society and Natural Resources* 19, 101–116.

Ignatieva, M. (2010) Design and future of urban biodiversity. In: Müller, N., Werner, P. and Kelcey, J.C. (eds) *Urban Biodiversity and Design*, 1st edn. Wiley-Blackwell, Oxford, pp. 118–144.

Ignatieva, M. and Ahrné, K. (2013) Biodiverse green infrastructure for the 21st century: from 'green desert' of lawns to biophilic cities. *Journal of Architecture and Urbanism* 37, 1–9.

Ignatieva, M., Meurk, C.D., van Roon, M., Simcock, R. and Stewart, G.H. (2008) *How to Put Nature into our Neighbourhoods: Application of Low Impact Urban Design and Development (LIUDD) Principles, with a Biodiversity Focus, for New Zealand Developers and Homeowners*. Landcare Research Science Series no. 35. Manaaki Whenua Press.

Jim, C.Y. and Liu, H.T. (2001) Patterns and dynamics of urban forests in relation to land use and development history in Guangzhou City, China. *The Geographical Journal* 167, 358–375.

Kattwinkel, M., Biedermann, R. and Kleyer, M. (2011) Temporary conservation for urban biodiversity. *Biological Conservation* 144, 2335–2343.

Kelcey, J.G. and Müller, N. (2011) *Plants and Habitats of European Cities*. Springer Science and Business Media.

Kowarik, I. (2011) Novel urban ecosystems, biodiversity and conservation. *Environmental Pollution* 159, 1974–1983.

Kümmerling, M. and Müller, N. (2012) The relationship between landscape design style and the conservation value of parks: a case study of a historical park in Weimar, Germany. *Landscape and Urban Planning* 107, 111–117.

Larsson, T.-B. (2001) Biodiversity evaluation tools for European forests. *Ecological Bulletin* 50. Blackwell Science, Oxford, UK.

Lin, B.B., Philpott, S.M. and Jha, S. (2015) The future of urban agriculture and biodiversity-ecosystem services: challenges and next steps. *Basic and Applied Ecology* 16, 189–201.

Loss, S.R., Will, T. and Marra P.P. (2013) The impact of free-ranging domestic cats on wildlife of the United States. *Nature Communications* 4, 1396.

Lundholm, J.T. and Richardson, P.J. (2010) Habitat analogues for reconciliation ecology in urban and industrial environments. *Journal of Applied Ecology* 47, 966–975.

Matt, S. (2012) Green façades provide habitat for arthropods on buildings in the Washington, DC metro area. Doctoral dissertation, The University of Maryland, Maryland, USA.

McCracken, D.P. (1997) *Gardens of Empire: Botanical Institutions of the Victorian British Empire*. Leicester University Press, London & Washington.

McIntyre, N.E. (2000) Ecology of urban arthropods: a review and a call to action. *Annals of the Entomological Society of America* 93, 825–835.

Müller, N., Ignatieva, M., Nilon, C.H., Werner, P. and Zipperer, W.C. (2013) Patterns and trends in urban biodiversity and landscape design. In: *Urbanization, Biodiversity and Ecosystem Services: Challenges and Opportunities*. Springer, Netherlands, pp. 123–174.

Müller, N. and Werner, P. (2010) Urban biodiversity and the case for implementing the convention on biological diversity in towns and cities. In: Muller, N., Werner, P. and Kelcey, J.G. (eds) *Urban Biodiversity and Design*. Wiley-Blackwell, Oxford, pp. 3–33.

Noordijk, J., Delille, K., Schaffers, A.P. and Sýkora, K.V. (2009) Optimizing grassland management for flower-visiting insects in roadside verges. *Biological Conservation* 142, 2097–2103.

Osborne, J.L., Martin, A.P., Shortall, C.R., Todd, A.D., Goulson, D., Knight, M.E. and Sanderson, R.A. (2008) Quantifying and comparing bumblebee nest densities in gardens and countryside habitats. *Journal of Applied Ecology* 45, 784–792.

Pearl, C.A., Adams, M.J., Leuthold, N. and Bury, R.B. (2005) Amphibian occurrence and aquatic invaders in a changing landscape: implications for wetland mitigation in the Willamette Valley, Oregon, USA. *Wetlands* 25, 76–88.

Pimentel, D., Lach, L., Zuniga, R. and Morrison, D. (2002) Environmental and economic costs associated with non-indigenous species in the United States. In: Pimentel, D (ed.) *Biological Invasions: Economic and Environmental Costs of Alien Plant, Animal and Microbe Species*, Taylor & Francis, Oxford, pp. 285–306.

Prance, G.T., Dixon, G.R. and Aldous, D.E. (2014) Biodiversity and green open space. In: Dixon, G.R. and Aldous, D.E. (eds) *Horticulture: Plants for People and Places*, Volume 2. Springer, Netherlands, pp. 787–816.

Prugh, L.R., Stoner, C.J., Epps, C.W., Bean, W.T., Ripple, W.J., Laliberte, A.S. and Brashares, J.S. (2009) The rise of the mesopredator. *Bioscience,* 59, 779–791.

Pyšek, P., Chocholousková, Z., Pyšek, A., Jarošík, V., Chytrý, M. and Tichý, L. (2004) Trends in species diversity and composition of urban vegetation over three decades. *Journal of Vegetation Science* 15, 781–788.

Quigley, M.F. (2011) Potemkin gardens: biodiversity in small designed landscapes. In: Niemelä J. (ed.), *Urban Ecology: Patterns, Processes and Applications*. New York, Oxford University Press, pp. 85–91.

Reichard, S.H. and Hamilton, C.W. (1997) Predicting invasions of woody plants introduced into North America. *Conservation Biology* 11, 193–203.

Rudd, H., Vala, J. and Schaefer, V. (2002) Importance of backyard habitat in a comprehensive biodiversity conservation strategy: a connectivity analysis of urban green spaces. *Restoration Ecology* 10, 368–375.

Salisbury, A., Armitage, J., Bostock, H., Perry, J., Tatchell, M. and Thompson, K. (2015) Enhancing gardens as habitats for flower-visiting aerial insects (pollinators): should we plant native or exotic species? *Journal of Applied Ecology* 52, 1156–1164.

Shaw, R.H., Bryner, S. and Tanner, R. (2009) The life history and host range of the Japanese knotweed psyllid, *Aphalara itadori* Shinji: potentially the first classical biological weed control agent for the European Union. *Biological Control* 49, 105–113.

Shwartz, A., Shirley, S. and Kark, S. (2008) How do habitat variability and management regime shape the spatial heterogeneity of birds within a large Mediterranean urban park? *Landscape and Urban Planning* 84, 219–229.

Shwartz, A., Turbé, A., Simon, L. and Julliard, R. (2014) Enhancing urban biodiversity and its influence on city-dwellers: an experiment. *Biological Conservation*, 171, 82–90.

Smith, R.M., Warren, P.H., Thompson, K. and Gaston, K.J. (2006) Urban domestic gardens (VI): environmental correlates of invertebrate species richness. *Biodiversity and Conservation* 15, 2415–2438.

Soderstrom, M. (2001) *Recreating Eden: a Natural History of Botanical Gardens*. Vehicule Press, Montreal, Canada.

Strohbach, M.W., Haase, D. and Kabisch, N. (2009) Birds and the city: urban biodiversity, land use and socioeconomics. *Ecology and Society* 14, 31.

Thompson, K., Austin, K.C., Smith, R.M., Warren, P.H., Angold, P.G. and Gaston, K.J. (2003) Urban domestic gardens (I): putting small-scale plant diversity in context. *Journal of Vegetation Science* 14, 71–78.

Thompson, K., Hodgson, J.G., Smith, R.M., Warren, P.H. and Gaston, K.J. (2004) Urban domestic gardens (III): composition and diversity of lawn floras. *Journal of Vegetation Science* 15, 373–378.

Tredici, P.D. (2010) Spontaneous urban vegetation: Reflections of change in a globalized world. *Nature and Culture* 5, 299–315.

Turnbull, S. (2012) Epigeal invertebrates of Yorkshire allotments: the influence of urban-rural gradient and management style. Doctoral dissertation, University of Hull, UK.

Turrini, T. and Knop, E. (2015) A landscape ecology approach identifies important drivers of urban biodiversity. *Global Change Biology* 21, 1652–1667.

van der Ree, R. and McCarthy, M.A. (2005) Inferring persistence of indigenous mammals in response to urbanisation. *Animal Conservation* 8, 309–319.

Wu, J. (2010) Urban sustainability: an inevitable goal of landscape research. *Landscape Ecology* 25, 1–4.

Zhang, H. and Jim, C.Y. (2014) Contributions of landscape trees in public housing estates to urban biodiversity in Hong Kong. *Urban Forestry and Urban Greening* 13, 272–284.

5 Landscape Trees, Shrubs and Woody Climbing Plants

ROSS W.F. CAMERON

Key Questions
- Trees and shrubs – what is the difference?
- What roles do trees, shrubs and climbing plants play in the landscape?
- What are the different methods of woody plant propagation?
- What are the constraints on plant viability/quality during production and establishment?
- What are the advantages and disadvantages of using clonally produced material in the landscape?
- What are the different sizes of trees that are available from the commercial nursery stock industry?
- How does tree size affect planting, establishment and subsequent management?
- What threat do weeds pose to woody plants?
- What is an urban forest?

5.1 Introduction

Trees are a key component in the landscape and provide the dominant feature in many of the world's biomes, being the climax vegetation type where appropriate temperature and water availability facilitate their growth (Table 5.1). Within the landscape they not only determine many of the characteristics of the ecosystem, but from a human perspective are also valued for providing 'a sense of place', possessing functional and cultural or spiritual value (Fig. 5.1). They are dominated by dicotyledonous species (gymnosperms – the conifers – and angiosperms – the broadleaves), although there are a number of notable monocotyledonous species too, some with important economic value, e.g. *Phoenix dactylifera* (date palm). The subdivision between trees and shrubs tends to be a rather imprecise one, but shrubs are generally considered to be smaller than trees, with multiple branches derived from nodes near the base of the plant rather than a single dominant stem (trunk). Ecologically, shrubs tend to play a role as understorey plants (the shrub layer) in woodland, or as effective pioneering species when habitats are in transition, e.g. 'scrub' vegetation on 'unmanaged' chalk downland. Shrubs can dominate too, where ecological pressures such as limited water availability, frequent fires or browsing activity by herbivores inhibit true tree development (e.g. Mediterranean biomes such as the chaparral of North America or the maquis of southern Europe have a high preponderance of woody shrub and woody sub-shrub species). Woody climbing plants or vines are found in woodland settings or in rocky escarpments, where their growth habit allows them to compete for light (irradiance) by growing up other plants or cliff faces.

Although trees, shrubs and woody climbers were originally exploited as a source of food, firewood, timber, fibres and herbal medicines, they have also been planted to improve the aesthetics of landscapes, both rural and urban. They possess a range of aesthetic qualities; these being defined as form, colour (hue), texture and line within the landscape. They are used architecturally to provide 'canopies' or 'walls' of vegetation, and screen off unsightly objects or provide privacy. In addition to performing as living sculpture in their own right, woody plants 'soften' the surrounding architecture, by acting as a foil to adjacent buildings or improving the texture of the building surface (e.g. green walls/facades). Trees and shrubs are used to define boundaries, roadways and other access routes, and specimen trees have a long tradition as landmark icons.

© R.W.F. Cameron and J.D. Hitchmough, 2016. *Environmental Horticulture: Science and Management of Green Landscapes* (R.W.F. Cameron and J.D. Hitchmough)

Table 5.1. Forest biomes and their key characteristics.

Forest biome	Annual precipitation range (mm)	Range of mean annual temperature (°C)	Features
Boreal (taiga)	100–2000	–7–5	Dominated by slow-growing gymnosperms, particularly conifers.
Temperate	350–2500	0–18	Dominated by deciduous 'hardwood' species. Leaves abscised before winter.
Temperate rainforest	2000–3400	5–20	Has both fast-growing gymnosperms and deciduous hardwood species.
Tropical seasonal	1300–2750	18–28	Dominated by deciduous 'hardwood' species. Leaves abscised before dry season.
Tropical rainforest	2400–4500	18–27	Dominated by evergreen 'hardwood' species.

Fig. 5.1. Trees add character to a landscape, but the landscape can also dictate the character of trees. In this location in Derbyshire, UK, a boulder field on a moorland has restricted the access and movement of browsing livestock, enabling a forest to establish. The wind-prone location and boulder-strewn land surface have combined to alter the form of the trees and create these gnarled, contorted trunks and branches. Such iconic trees often have a strong resonance with local people.

Increasingly, woody plants are being valued not just for their aesthetics, but also for a range of other functional purposes (see Chapters 2 and 3). Trees and shrubs modify microclimates through shading, evaporative cooling and reducing air movement by acting as wind breaks. Landscape practitioners now advocate their wider use in regulating building temperature. They can be used to divert sound waves, filter sunlight, reduce glare or block sources of light pollution, such as street lamps or car headlights. Woodlands, forests, parklands and even street and garden trees contribute to reducing precipitation intensity and rates of runoff, thereby reducing the likelihood of flash flooding. As some societies become more conscious of their environmental footprint, trees are increasingly valued as a habitat and food resource for wildlife – standing dead specimens often providing as valuable a resource as living ones.

In urban locations, trees are used to align streets and avenues, as individual specimens or small groups in parkland or gardens and in mass plantings within urban forests. Genotype selection will depend on aspects such as final size, natural tree form (e.g. fastigiate or compact), or the ability of the genotypes to respond to management techniques such as stem training, coppicing or pollarding. As pressure for space increases within high density urban environments, there is a trend to plant species with a small final height and breadth. As a consequence, planting of smaller species and cultivars has become widespread, for example in Europe, *Prunus* (ornamental cherries and almonds, typical final height 8–12 m), *Malus* (crab apple, 10–14 m), *Sorbus* (rowans, 8–16 m) and *Pyrus* (ornamental pears, 10–16 m) are being preferred to larger growing species such as *Platanus* × *acerifolia* (London plane, 40–44 m), *Quercus* (oaks, 25–35 m), *Fagus* (beech, 35–38 m) or *Acer platanoides* (Norway maple, 25–28 m).

Shrubs too, reflect their natural ecological niche by being utilized as the understorey of new woodland plantings, but are also exploited in roadside plantings, and are used as screens, barriers, ground cover and to stabilize soil on slopes. This is in addition to providing structure and interest in shrubberies and mixed ornamental borders. Aesthetic attributes largely rely on foliage colour and form, as well as a multitude of genera grown notably for their flowers (e.g. *Rosa*, *Rhododendron*, *Hydrangea*, *Hypericum*, *Hibiscus*,

Callistemon, Lagerstroemia, Ceanothus, Helianthemum), fruiting berries (e.g. *Cotoneaster, Berberis, Viburnum*) or scent (*Philadelphus, Syringa, Daphne*). Some of these genera are popular due to the wide range of species used in cultivation; in *Rhododendron* for example, there are prostate alpine species such as *R. campylogynum* (0.025–0.1 m tall) and others that are semi-trees like *R. arboreum* (about 20 m tall). Crossing different species (hybridization) has increased the range of garden/landscape genotypes further. This no more so than for *Rosa*, of which there are in excess of 3000 cultivars currently in commerce – covering every hue of colour in their flowers, with the exception of the true blue or the 'fabled' black.

Woody climbing plants tend to fall into two categories: large vigorous types that can cover a building wall or act as ground cover over a slope or embankment (*Hedera, Parthenocissus, Wisteria, Monstera, Bougainvillea*) or climbers that are more limited in their extension growth and are often associated with container planting or better quality cultivated soil (e.g. *Jasminum, Clematis* hybrids, *Lonicera*; Fig. 5.2).

The number and variety of woody plants used within urban public space can vary significantly. Within Europe, most cities have 50–80 street trees per 1000 inhabitants, but density can be as low as 20 street trees per 1000. Although there are geographical trends, the strategies and priorities within different local authorities also has a strong influence; Grenoble in France had 90 street trees per 1000 inhabitants compared to only 20 in Nice, despite the cities only being 200 km apart (Pauleit *et al.*, 2002). There is also a broad correlation between forest cover within a country and the prevalence of urban trees. Tree cover in cities within the UK, the Netherlands, Denmark and Iceland is relatively low (10 m^2 canopy cover per person), compared to Finnish cities (140 m^2 cover per person), reflecting the total woodland in each country; a similar pattern also being observed between different states of the USA (Nowak *et al.*, 1996). Climate determines species richness, with a trend for more natural diversity of plant (and animal) species as latitude decreases. This is not necessarily reflected in the use of landscape plants, however, as this is dictated by climate but also cultural factors, such as private ownership of gardens, interest in gardening and the strength of garden industries in different counties. In mild temperate climates such as Western Europe and Northwest USA, which are conducive to the successful cultivation of plants from a wide range of different biomes, the range of garden plants can be extensive. This is due to the use of a large proportion of non-native imported species and the development of new genotypes bred/selected for their aesthetic traits. In the UK, the Royal Horticultural Society lists 70,000 ornamental plants available to the garden trade, which include woody and non-woody genotypes but excludes varieties of bedding plants (due to their extensive range and relatively limited longevity in the marketplace). The range of plants used in public open spaces tends to be much less than that of private gardens (see section 5.8 on urban forests).

5.2 Woody Plant Production

Trees, shrubs and woody climbers are propagated by seed, cuttings and grafting onto rootstocks. A few commercially important species are also propagated via micropropagation (tissue culture), due to being difficult to produce successfully from the first three methods. A large component of the ornamental woody plant genotypes available are derived from either selected mutations (sports) or cross-bred 'cultivated' genotypes, and these will not reproduce 'true to type' via seed (i.e. sexual) propagation. Clonal (asexual) propagation via cuttings, grafting or micropropagation is then utilized as a means to build up stock of these genotypes.

Sexual propagation

Sexual propagation is used where genetic variability is desirable and a lack of uniformity does not impact on the plant performance or design.

Fig. 5.2. Woody perennial climbers such as *Lonicera* 'Mandarin' (honeysuckle) help screen unsightly walls or fences. Climbing plants can provide colour, scent, habitat and food sources for wildlife, as well as be used as thermal screens and to modify microclimate.

Seed

The use of seed for the propagation of woody perennials relates most to the production of woodland or hedgerow specimens, particularly where native species are being utilized, or even demanded. For woodland grants in the UK for example, landowners may be required to plant young trees produced from local provenances of seed (genotypic forms common to that geographical area). Seed-propagated specimens are predominantly used for woodland planting or large landscapes contracts, such as mass roadside plantings. Seed is used much less as a means of propagation for plants in more formalized urban areas such as street boulevards, parks, or even for those genotypes that are retailed through garden centres, largely because the latter are composed of a selected clone of a particular species (desired for either its uniformity of growth, or enhanced aesthetic characteristics such as unusual flower/foliage colour or form).

Seed is collected from known 'superior' specimen trees, or where large volumes are required from 'seed orchards'. These are collections of parent trees planted together for the specific purposes of producing high volumes/quality progeny. Such orchards may include selections that are of a particular form of phenotype or have enhanced disease tolerance characteristics.

For seeds to germinate successfully, a number of factors need to come together:

- Individual seeds must possess enough carbohydrate, fat and protein reserves (for this reason large seeds frequently have better germination rates compared to smaller seeds).
- The dormancy breaking requirements have been satisfied, that is, (i) sufficient moisture has been imbibed by the endosperm to activate growth, and (ii) the seed coat has been weakened enough to allow the radicle (first root) and hypocotyl (first 'storage' leaves) to emerge.
- Temperature and irradiance spectra are optima for the given species.
- Moisture content of the soil is appropriate for further development, without impeding oxygen diffusion to the rapidly respiring tissues.

Dormancy covers a number of factors constraining seed germination. Physical dormancy may be due to too hard a seed coat. This is overcome by physical abrasion with sandpaper or a file, or exposure to acidic or warm water solutions; a process known as scarification. These techniques mimic the processes that would occur in nature before a seed coat was weakened, i.e. freezing and thawing stress, action of soil microorganisms, passage through the digestive tract of mammals/birds or action due to fire. Even if the integrity of the seed coat is compromised, the embryo itself may not germinate. This is due to endogenous dormancy as controlled by phytohormones that act as inhibitors of growth. Exposure to moisture and chilling (stratification) is required to deactivate the action of these hormones (e.g. degradation of abscisic acid) and thus facilitate normal cell division and expansion.

In contrast, ectodormancy usually refers to the situation where seeds remain viable, but external environmental conditions are not conducive for germination (temperature, irradiance, humidity). As a further challenge, some woody plants enter secondary dormancy if water has been imbibed but external conditions do not promote growth. This is again controlled hormonally, and further treatment with chilling (or in some species, irradiance) is essential. Artificial exposure of seed to treatment with gibberellic acid (GA3 – a phytohormone that promotes cell expansion and growth) can also overcome secondary dormancy.

Controlled germination is feasible for many woody plants through the use of temperature schedules, scarification or the application of hormonal treatments. Various protocols exist that encompass the requirements of different species. These are widely used in commercial production and facilitated by the use of refrigerated cabinets and glasshouses. In nature, of course, germination is much more sporadic, but natural re-seeding is still a valid technique in certain landscape redevelopments – for example, allowing woodland regeneration after mature trees have been felled.

Vegetative propagation

Vegetative propagation allows specific genotypes to be 'cloned' and ensures their 'trueness to type', thus providing a degree of predictability on final size, form and colour of foliage or flower. Various techniques exist for landscape plants (Table 5.2).

Cuttings

The technology involved in rooting cuttings of woody plant species varies considerably. The simplest method is to insert cut stem sections in the ground

Table 5.2. Vegetative propagation techniques.

Technique/facility	Methods
Hardwood cuttings (semi- or fully-lignified stem sections, with no leaves present)	
Field hardwood cuttings	Stems inserted into field soil – often used for species with pre-formed root initials, that generate new root systems readily, e.g. *Salix* spp.
Protected hardwood cuttings	Basal heat applied to encourage callus and root formation at the base of cut stems. Systems such as 'Malling bin' used, where stems are bundled and placed on heated sand beds covered with bark. Bark/sand strata keep cuttings moist whilst allowing extra water to drain. Basal temperatures of 18–21°C encourage rooting, but aerial temperatures are kept cool and light excluded to avoid premature bud development and shoot extension.
Softwood cuttings (un- or semi-lignified stems, with leaves present)	
Sun tunnel	This is a 'mini-polytunnel' with polythene placed over low hoops, and cuttings inserted into the ground or raised bed. The polythene helps retain humidity, with cuttings hand-watered intermittently.
Shade box	Box placed over soil, pit or raised bed and covered with palm leaves or open hessian. Used in the tropics where natural humidity is high, and shading material helps reduce irradiance intensity and minimizes stress on the leafy cuttings.
Contact polythene	Cuttings inserted into sand beds or rooting media held within shallow trays. Polythene sheet draped over the cuttings is in direct contact with the upper leaves. Contact with polythene reduces transpirational water loss from the leaf. Sheets pulled back temporarily at times of watering. Often used in combination with glasshouse shading to avoid 'leaf scorch'.
Open mist	Cuttings in sand beds, trays or pots and placed under mist nozzles. Mist frequency controlled by either computer (e.g. based on irradiance integrals), timer or 'wet-leaf' system. In the latter system an electrical signal is maintained by moisture on the artificial leaf and when this dries out it triggers nozzles to spray, thus re-wetting the cuttings but also reinstating the electrical circuit on the artificial leaf. Growing medium is a free-draining, 'open' texture to avoid waterlogging due to frequent misting.
Enclosed mist	As above, but enclosed within polythene tents that also help raise and maintain humidity, whilst still providing intermittent leaf wetting.
Fog	Controlled in a similar manner to mist. Finer water particles though help in the maintenance of humidity, without excessively wetting the rooting media. Fog also disperses incoming solar irradiance, indirectly providing a form of shade.
Root cuttings	Procedures vary depending on the diameter of the roots being harvested. For species with thin roots, sections 25–50 mm long are cut and placed longitudinally on an 'open' medium within trays. These are covered with polythene or glass to provide heat and humidity. For species with thicker, fleshier roots, sections (50–75 mm long) are placed vertically into the growing medium, ensuring polarity is correct. The upper part of the root cutting is conventionally given a horizontal cut, while the basal end is cut at an angle to bestow a point at the base, thus aiding insertion into the medium. New shoots regenerate from the cut root sections.
Layering	This involves the bending or laying down of a shoot, pegging it at ground level and covering with earth. The act of bending the shoot alters phytohormone flow, and 'earthing up' with soil encourages roots to form on the stem. Once these develop the original shoot can be cut, severing it from the mother plant. In due time the resultant plant is lifted from the soil and potted on. Air layering follows a similar principle, but in this case the rooting medium is packed around a stem and held within a polythene bag. Again exposure to the dark, moist medium encourages root formation and the rooted stem section can be cut from the original plant.

Continued

Table 5.2. Continued

Technique/facility	Methods
Micropropagation/tissue culture (apical meristems, but also other tissues such as pollen or ovaries)	This involves the use of aseptic *in vitro* systems. Small pieces of plant tissue (explants) are sterilized, and then placed on a culture medium. Explants are placed in liquid or agar-based medium containing a cocktail of chemical components (nutrients, growth regulators, vitamins, carbohydrates). The composition of which determine how the explants develop subsequently. For example, the agar medium may be used to proliferate new shoots from explants, or form an undifferentiated cell, 'callus' culture. Transfer of these tissues to a new medium encourages further cell transformations, for example inducing root formation on shoot sections, or the formation of somatic embryos (seed-like structures). These final tissue cultures are subsequently weaned off the *in vitro* systems and form young plantlets. In this way many thousands of new plants can be generated from just a single explant.
Grafting	The fusion of two separate tissue sections, normally from two distinct genotypes. Most commonly shoot sections (scion) are grafted on to another plant which has a developed root system (rootstock). The shoots of the rootstock are then removed (headed back) allowing the scion wood to develop and form the new branch framework. Grafting is used on woody plants to improve or control the vigour of the scion growth, e.g. *Rosa* (roses) are grafted onto rootstocks that impart vigour to the flowering scions, while *Malus* (apple) may be grafted onto rootstocks that constrain growth and encourage earlier fruiting in the life of the plant (precocity).
Chip budding	Used when the plant is not actively growing as in spring before bud-burst or in late summer when extension growth has ceased. The scion bud is removed with about 20–30 mm of stem attached. This is placed into a notch made on the rootstock of equivalent size, and tied in with budding tape. The rootstock is not normally cut back until the union is complete, e.g. the following spring.
T-budding	Used on active growing stem sections 5–25 mm diam. A 'T'-shaped cut is made on rootstock and existing budwood from scion (bud with approximately 40 mm of cambial wood) is inserted into the 'T' and tied in.
Patch budding	A regular section of bark is removed from the rootstock and an equivalent sized piece of bud wood excised from the scion. This is then 'patched-on' and tied in.
Whip graft	This is used for small diameter material (e.g. 5–15 mm diam.) and involves the rootstock being given a complete slanting cut, of which a further cut is inserted in the middle, leaving a small flap (tongue) of tissue. Similarly, a slanting cut is made in the scion, with a further insertion to provide a corresponding flap. The stock and scion are slipped together with the flaps interlocking. The intersection is then tied tightly and sealed in warm wax to minimize moisture loss.
Splice graft	As above but without the flaps (tongues).
Side graft	There are various forms of these, but all involve a vertical cut in the rootstock, with a wedged shaped piece of scion wood slipped into the aperture. This is then tied and waxed, with the rest of the rootstock above the union being cut off. Side veneer grafts involve longer sections of scion wood (e.g. 40–50 mm). These are often used for grafting young conifers and other evergreens.
Cleft graph	Cleft graphing involves the horizontal cutting of the rootstock and then inserting a vertical split across the cut section. Small sections of scion budwood are then slipped into the aperture, ensuring cambium contact between scion and rootstock. The whole area is then sealed with wax.
Wedge graft	This is similar to the cleft graft, but vertical notches are removed from the rootstock and the scion section gently tapped into place to allow a tight fight between cambial tissues.
Bark graft	This is similar to the wedge graft, but loose flaps of bark around the vertical cuts are nailed back in place to help apply pressure around the scion/rootstock interface.

Continued

Technique/facility	Methods
Topworking	This is a special type of grafting where the crown is reduced on mature trees and complete cross-sectional cuts made across the branches leaving stumps. Scion sections are then grafted on to these stumps and develop into branches of the new scion genotype. Opportunities exist for more than one scion genotype to be carried by the rootstock, so increasing the varieties of e.g. cherries or apples derived from the 'one tree'.

Table 5.2. Continued

and encourage latent, pre-formed root initials to expand and develop. This occurs in *Salix*, *Populus*, *Ribes* and *Forsythia* (propagated as hardwood leafless cuttings in winter). For species that do not naturally develop pre-formed roots then the synthesis of new adventitious roots is required. These are generated from undifferentiated cells usually formed after wounding (such as occurs at the base of an excised stem section). In commerce, this is most commonly achieved by propagating new plants from leafy softwood cuttings, although the levels of success are very much dictated by genotypic and environmental factors.

With these leafy softwood cuttings, creating a propagation environment that supports the cutting long enough to allow roots to form is critical. In practical terms this means providing an environment that facilitates leaf photosynthesis whilst minimizing water loss from the cutting, yet also ensuring the medium the cuttings are inserted into is free-draining and does not become anaerobic. Tolerance to the stresses experienced during propagation and the inherent rhizogenic potential of different genotypes affects the ability of cuttings to form roots. Some species have reputations as being 'difficult-to-root', whereas others are categorized by nursery managers as 'easy-to-root'. In reality, the levels of stress a cutting experiences depends on the interactions between the different environmental components.

So propagation environments that have relatively high irradiance (and hence are advantageous due to their ability to maximize photosynthesis) may need to provide higher humidity to minimize moisture loss from the leaves. This is accomplished by enclosing cuttings within polythene tunnels or alternatively fogging the cuttings with fine water droplets (10–50 μm diameter). High irradiance levels in a protected structure of glass or polythene correspond to enhanced temperatures, therefore additional leaf cooling and wetting through misting (water droplets of 50–100 μm diameter) may be warranted. Conversely, propagation benches that are more heavily shaded may be able to keep cuttings viable simply by draping a polythene sheet over the canopy of the cuttings (contact polythene) to help minimize water loss. The drawback of such systems, however, is that the suboptimal light reduces photosynthetic capacity, and cuttings may take longer to root, or may be more prone to rotting due to low carbohydrate levels. Over and above these 'trade-offs' associated with irradiance levels, systems that both retain high humidity and keep leaves cool through the deposition of water droplets enhance rooting success. Fog and polythene-enclosed mist systems for example are useful for optimizing rooting in challenging subjects by providing both these key factors. In contrast, open mist has more variable humidity levels with negative effects for some species (Fig. 5.3). The majority of shrubs and woody climbers are propagated via softwood cuttings, and although cutting preparation procedures are similar throughout, idiosyncrasies do arise; e.g. *Clematis* cuttings are prepared with the base derived from the former internodal section of the parent stem, whereas in most other species the base of the cutting conventionally comprises a former leaf/bud node – the nodal section considered to have greater cell regeneration potential.

A secondary factor affecting 'ease of rooting', however, is the level and action of phytohormones either within the cutting or applied to the cutting (i.e. auxin-based rooting powders and solutions). For cuttings of difficult-to-root species the generation/movement of endogenous auxins (often synthesized in regions of active cell division – developing shoot tips, new leafs etc. and then exported to the base of the cutting where root initiation occurs – a process known as polar auxin transport) is critical. For this reason rooting potential in difficult species is optimized when cuttings are derived from the actively growing shoots in early summer, with a very rapid drop-off in the ability to root as shoot extension rates slow, and resting apical buds begin to form.

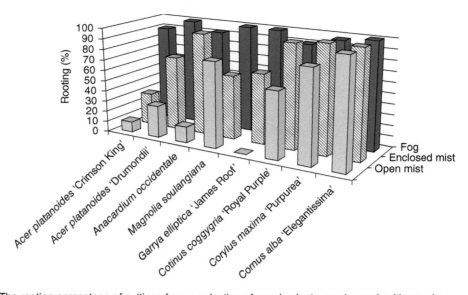

Fig. 5.3. The rooting percentage of cuttings from a selection of woody plants species and cultivars when propagated in either open mist (open columns), polythene enclosed mist (striped columns) or fog (solid columns) propagation environments. Note the variation in tolerance to open mist between genotypes. (Modified from: Harrison-Murray et al., 1998; Saranga and Cameron, 2007.)

A study by Cameron et al. (2001) demonstrated the importance of the developing branches within the cuttings for root generation. *Cotinus coggygria* (smoke bush) stock plants were manipulated to encourage branching on the developing shoots and were then excised as cuttings. Rooting was highest when the developing branches were left intact on the cutting (Fig. 5.4) as these were providing a source of endogenous auxin. Placing a chemical inhibitor to auxin movement within the cutting, either on the branches or the stem, reduced rooting. Conversely, where the developing branches were removed and subsequent rooting percentages were low, adding exogenous auxin to either the cut branch stumps or the base of the cutting tended to increase rooting once again.

As cuttings are prone to stress, nursery managers will apply fungicides to the crop, particularly to control *Botrytis cinerea*, although their use is not always absolutely necessary. Very heavy wetting through frequent fogging/misting for example may inhibit fungal spore germination and development.

The rooted cutting is known as a 'liner' or 'steckling' and will require weaning from the high humidity environment it was rooted under. This is easily accomplished by rolling up the polythene covers around the propagation bed and/or reducing the frequency of misting/fogging by a small proportion each day for about one week. Once weaned, liners are left in their propagation modules to establish an effective root system, and once a small intact root system is formed (rootball) they can be potted on into small pots (e.g. 90 mm diameter). These are then grown on under protection, or placed on container beds outdoors to reach marketable size. Landscape plants typically produced via leafy softwood cuttings include: trees such as *Chamaecyparis*, *Magnolia*, *Paulownia*, *Thuja* cultivars; shrubs such as *Buddleia*, *Buxus*, *Cistus*, *Cotoneaster*, *Euonymus*, *Forsythia* and *Viburnum* cultivars; and climbers such as *Clematis*, *Hedera* and *Lonicera* cultivars.

Grafting

Grafting is the union of a favourable genotype (scion) onto another plant (rootstock). Once a successful union between the two plants has taken place, the stem and branches of the rootstock plant are removed (headed-back), allowing the scion genotype solely to exploit the resources derived from the root system. There are various types of graft-union and mechanisms for grafting (Table 5.2).

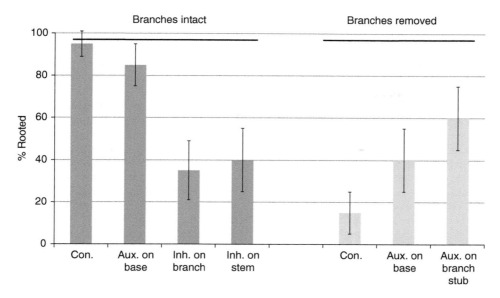

Fig. 5.4. Rooting (%) in branched cuttings of *Cotinus coggygria* 'Royal Purple', when branches are left intact (Con.), or removed and treated with auxin (Aux.) or an auxin inhibitor (Inh.). Bars, SE. (Modified from Cameron *et al.*, 2001.)

Ornamental deciduous trees are commonly produced under field conditions when a single bud from a scion is grafted (budded) onto the stem of the rootstock at approximately 100–150 mm above ground level. Common landscape trees such as *Acer platanoides* 'Crimson King', *Robinia pseudoacacia* 'Frisia', *Sorbus* cultivars (rowans/whitebeams) e.g. 'Joseph Rock', *Prunus* cultivars (flowering cherries and almonds) e.g. 'Fragrant Cloud', *Malus* cultivars (crab apples) e.g. 'Red Jade', *Betula* cultivars (birches) e.g. 'Grayswood Ghost' etc. are all produced this way. Budding normally takes place in August in the northern hemisphere, when the rootstock stem has started to lignify and there is sufficient warmth for the bud cells to fuse with the cambium of the rootstock.

In contrast, a side veneer graft carried out in winter is the preferred method for grafting selected cultivars of conifers. Rootstocks are grown in pots and allowed to go dormant in winter. After exposure to low temperature (5–10°C) for at least 6 weeks the rootstock is brought into a cool greenhouse. These are left for a few days to activate new root growth before grafting takes place. The plant should not be watered at this time though, as excessive sap rise through the xylem can push the graft apart after 'tying-in'. Notable ornamental conifers including *Picea pungens* 'Hoopsii', *Picea omorika* 'Pendula', *Pinus strobus* 'Blue Shag', *Pinus sylvestris* 'Glauca Fastigiata' and *Pinus radiata* 'Aurea' are usually propagated this way.

Growing-on

Once plants have been successfully propagated they are grown-on for a period at the nursery before being sold. Retailers specify the size and shape of plants they require, as well as stipulating other parameters (the crop specification), e.g. minimum number of flower buds present, pot colour, a peat-free growing medium etc. These specifications need to be achieved and a relatively uniform crop provided by the nursery in as speedy and as economical manner as possible. Profit margins for commercial nurseries rely strongly not only on volume of sales, but also on minimizing wastage and reducing production times. Hence, the newly propagated liners are often grown-on under protection (glasshouse or polytunnels) to optimize growth and aid root establishment, with plants being frequently pruned to improve their branching habit and retain crop uniformity. With woody climbers this stage of production is associated with tying and training to a framework of bamboo canes or the like. Not only does this make the plants more presentable, but aids handling and avoids the stems/flowers being damaged. Chemical growth regulators may also be applied during production for similar objectives or to encourage flower formation on the young stock. (Depending on species, a premium is paid for plants just coming into flower as this enhances sales impact, or it may be part of the pre-agreed specification.)

During nursery production, plants may be potted on 2–3 times and the final pot/plant size reflects the labour and other resources involved plus the space/time taken to grow the crop; large stock being considerably more expensive than smaller specimens of the same species. Depending on geographical location, the growing media that is used during production can vary. In countries such as the UK, peat has traditionally been the favoured medium due to its capacity to hold both high volumes of oxygen and water, possess a good cation exchange capacity and be relatively light to use and transport. Concerns over carbon dioxide (CO_2) emissions from peat extraction and the loss of peat bog habitat, however, is putting pressure on the commercial use of peat in horticulture (in the UK at least) and alternative media are being sought and researched. Shredded bark, composted wood chips or sawdust, green (waste) compost, coir, sand and loam and many other locally sourced organic-based media are also used across the globe as components in potting substrates.

Shrubs and woody climbers tend to be grown in containers throughout (container or pot grown), but trees (and roses) are more frequently budded and grown-on in the field and either sold as bare root stock or put into containers before selling (containerized). Field-grown specimens that are larger than these are lifted from the soil and their root systems wrapped up in burlap (hessian or similar material) in a process known as 'rootball and burlap'. A large proportion of the root system can be lost during field-lifting and nursery teams will often undercut their stock some months prior to lifting. This is accomplished to encourage a fibrous root system to form in close proximity to the stem. Undercutting with a subsoil blade severs the primary roots; a process that initiates secondary (side) roots to form. During the lifting operations, bare root stocks are particularly susceptible to root desiccation, and these should be kept damp and placed in a cold store, before further packing and shipping.

Trees are sold based either on their height (younger specimens) or their stem girth (older specimens where the circumference is measured at breast height). Within the UK there is a size classification (BS 3936 – *Part 1: Nursery Stock Specification for Trees and Shrubs*; Anon., 1992). In addition, certain terminology helps describe the tree age and form (Table 5.3).

5.3 Retail and Markets

Ornamental trees, shrubs and woody climbers tend to have two main market arenas: garden centres that sell plants to the gardening public, and the landscape (or amenity) sector which provides nursery

Table 5.3. Nursery tree classification within the UK.

Terminology	Description
Seedling	A 1-year-old or younger tree, sold by height, e.g. 40–60 cm.
Transplant	Trees started in a seed bed and undercut (primary roots severed by a subsoil blade) at either 1 or 2 years of age. These can be left *in situ* to grow on or be transplanted to a new bed. Sold with information relating to undercutting or transplanting and plant height.
Whip	Trees produced from seed or cuttings, with a central stem and little or no side branching. Sold by height.
Feathered whip	Larger than a whip, with side branches (feathers). Sold by height.
Maiden	Similar to a whip but produced via budding or grafting.
Feathered maiden	A maiden tree with side branches, usually 1, sometimes 2 years older than a maiden.
Standard trees	Tree with girth (circumference) measured at breast height (gbh).
Light standard	6–8 cm gbh
Standard	8–10 cm gbh
Select standard	10–12 cm gbh
Heavy standard	12–14 cm gbh
Extra heavy standard	14–16cm gbh
Advanced heavy standard	16–18 cm gbh
Semi mature	18+ cm gbh

stock lines to landscape architects or their contractors. Within both sectors product price can vary considerably depending on more subtle divisions within each. Garden plants, for example, may be retailed through 'state of the art' garden centres as a component of an extensive range of garden products on sale, or alternatively may be sold by large multiple retailers such as DIY stores and supermarkets (Fig. 5.5). Those nurseries that specialize in supplying the 'landscape'/amenity sector may grow plants to a specific contract, or provide a more limited and predictable range of plants. Conversely, they are more likely to produce greater volumes of any single line to supply the needs of their customers for large landscaping contracts, such as a new housing development. In recent years though, a number of these nurseries have set up their own, 'cash and carry' outlets which have a greater range of plants, but are still retailed at wholesale prices to landscape or garden designers. In addition to these two main retail routes, the traditional outlet, and still a significant component of the market, are those nurseries that grow their own lines and sell direct to the public. This may be either through their own retail area or via the internet. Some of these nurseries will specialize in particular groups of plants, e.g. *Rhododendron*, *Rosa*, *Clematis*, conifers etc., to develop their own unique market niche.

Sustainable production

As with other sectors, concerns over the environmental footprint of woody plant production is driving research and new initiatives to improve performance and sustainability. This includes investigations into:

- new growing media, including the use of recycled products;
- a reduction in pesticide use and a drive for greater integrated pest management and biocontrol approaches;
- more efficient use of water and the development of water recycling technologies;
- applying fertilizers to meet crop requirements rather than routine applications, irrespective of need; and
- attempting to reduce energy use, through alternative heat sources and low-energy crop lighting, e.g. the use of low-energy LED systems (see also Chapter 8).

5.4 Establishment

Mature trees are large significant landscape features, not readily moved, so site selection for planting young trees is critical. Provision of adequate space is key, but consideration also needs to be made for proximity to buildings, roads and other infrastructure, the impact

Fig. 5.5. Plants retailed through garden centres and DIY stores are required to be grown to set specification of size and form, and supplied with barcode and information label.

of the trees on sight lines, future shading traits and potential change in land use. A number of factors require consideration in selecting the appropriate species, particularly climatic conditions (temperature, wind exposure – including coastal 'salt-laden' winds, humidity), soil structure and chemistry (drainage and water availability, gas exchange, nutrient levels, pH, organic matter content, presence of phytotoxic elements) and biotic factors such as the likely presence of pests or pathogens (Table 5.4). Susceptibility to

Table 5.4. Factors to consider when choosing tree and shrub species.

Factor	Considerations for trees and shrubs
Climatic	
Temperature	Summer temperatures too low to support appropriate shoot growth and development. Frost and freezing mid-winter temperature below that which genotype can tolerate – 'hardiness zone' information provides preliminary guidance on species selection. 'Unseasonal' frost and damage to tissues such as developing shoots or flowers. Sufficient winter chilling is required to release buds from endodormancy – poor and uneven bud development can occur if winter temperatures are too high.
Wind	Physical damage due to strong wind, e.g. newly emergent leaves, or stem/branch structure. Exposure due to strong or excessively cold wind – tissue desiccation. Coastal, salt-laden winds may induce osmotic and phytotoxic stress, e.g. tip burn on leaves.
Precipitation/soil moisture availability	Frequency of drought or flooding events (see points on soil structure).
Solar irradiance	Supra-optimal irradiance may cause photoinhibition, UV wavelengths inducing direct cell damage. Tree canopy cover or planting in the shade of a building or wall may be a prerequisite for some shade-adapted species. Cultivars with gold/yellow foliage are particularly susceptible to excessive irradiance.
Soil	
Particle size and chemical composition	Determines soil type (sand, silt, clay) and degree of organic matter content, which will in turn affect physical, chemical, hydraulic and biotic factors.
Structure	Defines drainage, water holding and aeration characteristics – roots require free-draining soils, with pores to retain moisture and provide root aeration. Soils with high bulk density will impede root extension.
Fertility	Provide macronutrients – nitrogen, phosphorus and potassium – as well as essential minor nutrients.
pH	Needs to be suitable to avoid direct damage to roots, but also availability of key nutrients.
Toxicity	Soils/substrates reclaimed from ex-industrial sites may contain chemical contaminants that are phytotoxic to plants, e.g. mercury, cadmium, chromium, residues of oil etc.
Biotic	
Pests or pathogens	Likelihood of damaging infestations of pests or the presence of pathogenic fungi, bacteria, viruses.
Weeds	If management of site is to be limited, then species need to be competitive and resilient against weed competition.
Mycorrhizae	The presence of these fungi may aid establishment of certain woody plants through their symbiotic action.
Wildlife	Wildlife conservation may be a consideration, with species chosen for their ability to support invertebrates, or provide food for birds and mammals.
Aesthetic	Meet the requirements outlined with respect to colour, form, line and texture. Provide interest and beauty at specific, or for prolonged, periods. This may include spring blossom, autumnal foliage colour, or interesting bark colour/texture in winter.
Physical	Ability to provide shade, wind abatement, shelter from rain or solar glare, cooling influence, noise and aerial pollutant barrier, stabilize soil etc.
Social and economic	
Functional qualities	Privacy and seclusion, complement buildings, provide scale to landscape and sense of place, provide timber or food.
Management	Heavy and long-term investment required, training, staking, pruning etc.

pests and pathogens has resulted in rapid declines in the popularity of certain tree species. *Aesculus hippocastanum* (horse chestnut) is no longer recommended for planting in Western Europe due to its susceptibility to *Pseudomonas syringae* pv. *aesculi* (bleeding canker) and *Cameraria ohridella* (horse chestnut leaf minor). Chalara dieback in ash (*Hymenoscyphus pseudoalbidus*; also known as *Chalara fraxinea*) has caused an almost complete collapse in the market for ash trees (*Fraxinus excelsior*) in the UK, and even crops with phytosanitary certificates are unsaleable due to the perceived threats. Even in circumstances where a whole species is not under threat, horticultural knowledge is important in making prudent decisions. Rosaceous species should not be planted on soils that previously held the same/similar species due to a phenomenon known as 'rose replant disease'. The precise cause of this is unknown, but may be a combination of a build-up of pathogens in the soil and a depletion of certain micronutrients. Replacement plantings of the same/similar species often demonstrate signs of poor vigour and the dieback of roots.

Trees and shrubs are categorized by their preferences or tolerances to climatic and other environmental factors which aids in appropriate plant selection. Climate change and associated changes in biotic populations, however, are beginning to alter what is suitable for specific locations (including the concept that native species may not always represent the optimal or best-adapted species choice).

Failure of young trees and other woody plants to establish successfully in the landscape is a common problem. This is due to a variety of reasons, including:

- Lack of water during the establishment phase – not watered immediately after planting or during dry periods in the first year.
- Competition from weeds for (largely) water, but also nutrients and light.
- Transplant shock (technically a check in growth due to poor plant acclimation when moved from one environment to another). In practical terms most land managers define it as an imbalance of the root to shoot ratio. Young trees may lose a high proportion of the fine roots due to lifting from the field, root trimming or desiccation when bare root stocks are left in warm or low humidity conditions. Once planted such plants struggle to acquire sufficient water to support top-growth (leaf and stem development). Typical symptoms are that the tree does not grow or growth is slow over a number of years, as the roots and shoots re-establish homeostasis; a period where the tree is susceptible to competition.
- Herbivore damage (rabbits/voles/deer etc. eating young trees or stripping their bark).
- Vandalism.
- Poor maintenance – particularly injury to the bark of young trees due to strimmer damage (used to keep weed growth down around the base of the tree).
- The use of de-icing salts for roads and pedestrian areas.
- Limited 'root run' due to restricted space – small containers or planting pits.
- Trenching of the soil by utility companies, cutting roots and causing soil compaction.

Where trees and other woody plant are located has a strong influence on their establishment rates and subsequent success, with trees in urban environments probably experiencing the greatest range of stress factors. As such, urban trees are considered to have slower growth rates than their rural counterparts (Quigley, 2004), but that final size is strongly influence by their ecological characteristics. Early- and mid-successional type species (often light demanding, e.g. *Liriodendron* (tulip tree) and *Gleditsia* (honey locust)) can reach their final height and girth dimensions in urban situations, whereas late-successional types (often shade-tolerant species such as *Quercus* (oak)) regularly fail to attain their final potential size. As well as environmental and social influences, establishment rates vary strongly based on the level of management at and after planting. Regional variances are apparent in this respect, with 220 out of every 1000 Danish urban street trees needing to be replaced within 10 years of planting, compared to only 100 in Finland and Iceland. In general, park trees have higher survival rates than street trees, although local problems, e.g. vandalism in some parks, still induce significant losses. Costs associated with planting and establishing trees also vary between countries, with some spending less than €250 (£176) per tree and others in excess of €2000 (£1400) per tree. Within Europe there was a gradient from north to south in terms of the number of genera or even species used in streetscapes, with northern cities often relying on limited genotypes. In contrast, Mediterranean cities and towns utilize a wide range of species, including frost-sensitive subjects such as *Citrus* and palms (Pauleit *et al.*, 2002).

Size/age of transplant

In more prestigious or iconic planting schemes there is a preference to plant large specimens to provide an instant effect. The downside is that larger plants may take longer to establish successfully, especially if there has been root damage due to lifting (e.g. via the ball and burlap method). In temperate regions (e.g. northern USA) it may take 5 years for a 100 mm diameter tree to regenerate its root system entirely, whereas after transplanting a larger tree (250 mm diameter) the root volume may not be fully restored for up to 13 years (Watson and Himelick, 1997). Root regeneration rates, however, tend to be faster in warmer climates.

It is often quoted that young trees once successfully established will outgrow more mature specimens, but there are conflicting reports as to whether this is true (Watson, 2005). So although many reports indicate that planting younger trees is advantageous, others such as Struve et al. (2000) point out that in reality young specimens are still unlikely to outgrow their larger counterparts. A key element is how much root has been lost in the transplanting process. For example, large container-grown material (with minimal root damage?) may establish and grow more effectively than smaller plants (Thetford et al., 2005). Other factors too can influence the relative benefits between larger or smaller transplanted trees, including the species chosen, provenance of the genotypes, type of root system (predominant tap root versus fibrous roots), seasonal and other environmental conditions at time of planting, previous husbandry and the physiological state of the tree, as well as root to leaf canopy ratios and relative rootball to soil backfill volume (size of hole).

Soil cultivation and amelioration techniques vary significantly depending on the size of tree being planted. For young whips, planting into the parent soil simply involves inserting a slit or notch into the soil and carefully teasing in the plant roots, whereas a large standard tree requires major soil excavations and backfilling with quality topsoil. For planting large trees, the planting 'pit' is recommended to be 2–3 times the diameter of the rootball and backfilled with a good quality soil of a known standard (e.g. in the UK complying with British Standard BS 3882). It is important that the backfilled soil integrates with the parent soil to ensure movement of moisture through the profile and to avoid proliferating roots potentially having to bridge an air gap where the two substrates meet (for example, if there is soil/substrate shrinkage due to dry periods). More mature plants usually require greater investment in staking, tying and irrigation compared to smaller, younger stock. For some of the large grades of trees, tying and securing may involve both above- and below-ground techniques.

Soil conditions

Droughty soils and soils prone to compaction and waterlogging influence survival and establishment. For environmental horticulturalists there is a trade-off with how often woody plants should be watered post-planting, and the labour costs associated with doing so. Similarly, in arid areas, there may be considerations with respect to conserving water whilst meeting the needs of newly planted specimens. Some species can be classified by their water-use characteristics, with, for example, *Salix matsudana* (Chinese willow) and *Tilia cordata* (small-leaved lime) having high water use, *Platanus × acerfolia* (London plane) and *Fraxinus pennsylvanica* (green ash) moderate water use and *Acer platanoides* (Norway maple) low water-use traits (Montague et al., 2004).

Providing water may not only be important in ensuring the survival of recently planted specimens, but has longer term influences. Differences in the root systems of *Acer rubrum* due to irrigation regimes imposed over the first 24 weeks after planting were still evident 5 years after the treatments terminated (Gilman et al., 2003). Regular irrigation during establishment pays dividends in terms of growth. Trees that were irrigated frequently during the post-planting stage had larger trunk cross-sectional areas (175 cm^2) compared to those watered less frequently (137 cm^2). This was attributed to greater root numbers in the surface soil layers. In contrast, the infrequently watered trees had a higher proportion of their roots at a deeper level (30% of roots were at least at 300 mm below the surface). Reducing irrigation once plants have become established is often considered a positive influence as it encourages root proliferation at deeper depths and hence can help secure water supply, should surface layers dry-out during drought events.

Pre-conditioning plants on the nursery through controlled moderate drought stress has been considered as a technique to improve their resilience once planted out in the landscape. (They should, however, still be well watered in the container before planting as well as watered-in on planting.) Drought pre-conditioning can be carried out in the nursery 3–4 weeks preceding sale, once plants have reached their specified height. For woody plants destined to be planted out in the autumn, reduced irrigation can be introduced in August or September. It is important that irrigation is not switched off completely, but rather plants are weaned onto lower moisture availability, thereby inducing physiological changes that enhance acclimation to drought (Fig. 5.6). Such changes may include: more effective or prolonged closure of the stomata; reductions in cell water content and increases in solute (osmotic) potential; induction of cells that are smaller in size and increases in cell wall elasticity; alterations in leaf and xylem water potential; and modification to endogenous hormone concentrations, e.g. increased concentrations of abscisic acid.

In an attempt to minimize failure rates with bare root seedlings after planting-out, roots can be dipped into clay slurry or hydrogels, or packed in damp peat to help minimize root desiccation. Even relatively short periods of exposure to air can reduce viability, i.e. as measured in minutes. *Pseudotsuga menziesii* (Douglas fir) seedlings exposed to warm air (32°C) dropped from 100% survival after 15 min exposure to only 60% and 50% survival after 60 and 120 min exposure, respectively (Hermann, 1967). Overall, the effect of hydrogels either used as a root dip or used at the time of planting appears mixed. Apostol *et al.* (2009) indicated that applying hydrogels to roots of *Quercus rubra* (red oak) reduced cell leakiness and promoted more rapid budbreak in the spring after planting (both evidence of reduced physiological stress on the roots). After planting though, there was no apparent advantage associated with this treatment as there was no difference with non-treated plants in terms of shoot length, dry weight, root volume, net photosynthesis or stomatal conductance. Similarly, others have found no advantage

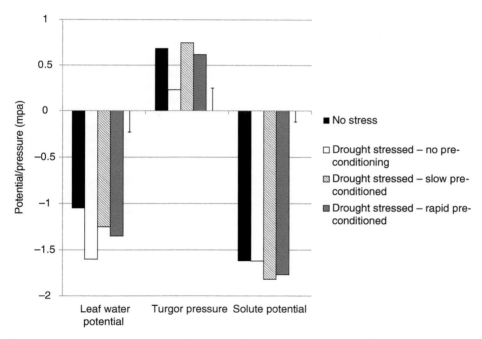

Fig. 5.6. The effect of pre-conditioning (controlled reduced irrigation in nursery) in *Forsythia* 'Lynwood' when exposed to a subsequent drought stress. Data shows leaf water potential, turgor pressure and solute potential in: control plants ('no stress', solid bar) and drought-stressed plants with either no previous treatment ('no pre-conditioning', open bar), or previous gradual ('slow pre-conditioning', striped bar) or rapid ('rapid pre-conditioning', shaded bar) exposure to reductions in water availability on the nursery. Bars, LSD. Note how pre-conditioned plants retain higher values for leaf water and turgor pressure and lower values for solute potential than non-pre-conditioned specimens, indicating greater adaptation to drought. (Modified from Cameron *et al.*, 2008.)

(e.g. Heiskanen, 1995), whereas when Specht and Harvey-Jones (2000) incorporated gel into the growing medium, they noted increased water uptake, stomatal conductance and increased biomass in *Quercus coccinea* (scarlet oak).

Drought is not the only stress to reduce transplant survival, or delay effective establishment. Tolerance to waterlogging, wind, heavy metals, low nutrient availability, salinity etc. are also strongly influenced by genotype.

Urban soils

Urban soils are frequently impoverished and this may be particularly problematic to trees and other woody plant species. The reduced lifespan of urban trees compared to trees in more rural settings has been attributed to poor soil structure, lack of water and pollutants prevalent in the urban environment. In surveying urban soils in Hong Kong, Jim (1998) identified that they were a difficult medium for plant growth due to limited space for root development and interference from the abundant underground utility lines, pipes and cables. Surface and sub-surface soil layers are commonly compacted (both unintentionally by e.g. vehicle movements, and intentionally in attempts to stabilize building foundations) resulting in restricted root spread and limiting the movement of air and water to roots. Soil 'sealing' through the use of impermeable roadways and pavements similarly impairs the movement of water and gases from the atmosphere to the rhizosphere (Fig. 5.7). The presence of stones, boulders, construction rubbles, mortar, bricks, old paving, old and extant foundations, and other obstructions often frustrate planting efforts and render many sites unusable. To add to the problems, most soils are depleted of organic matter with insufficient quantities of the major nutrients required for vigorous plant growth. Release of carbonate from the calcareous construction waste tends to increase pH, making soils more alkaline and making it more difficult for plants to absorb phosphorus and micronutrients such as iron, copper, zinc, boron, cobalt and manganate. There may be other soil contaminants such as lead that directly impair growth. Additional injury to planted specimens can be induced by trenching and tunnelling through the rhizosphere via the actions of utility companies. Many urban soils have lost their natural soil horizons and instead have artificial horizons composed of poor quality infill material.

Fig. 5.7. Soil sealing is an issue in urban areas with high proportions of road tarmac and concrete inhibiting the movement of water and oxygen into the soil. Here the location around the tree has been paved with block pavers, where gaps between individual blocks allows for the ingress of water and gas exchange. Note also the slit drain on the road to allow runoff water to be directed towards the tree roots.

To some degree, soils can be 'designed' to help street and plaza trees perform well, whilst minimizing the damage to the physical infrastructure such as pavements and roadside kerbs. 'Structured' soils are an aggregate mix of gravel and soil that encourage deeper and more prolific tree root development. They tend to have a high porosity that aids aeration and supplies an effective water holding capacity after rain – large pores draining while small pores can hold water against gravity. Such materials are becoming popular when back-filling tree planting pits (Fig. 5.8). Structural soils not only allow for tree root development, but are structurally strong enough to enable paving stones and other surface coverings to be positioned firmly on top. Planting pits themselves are useful to designate a specific volume for root growth, and impair roots from causing displacement of pavement slabs or kerbstones. The downsides of structured soils, however, may

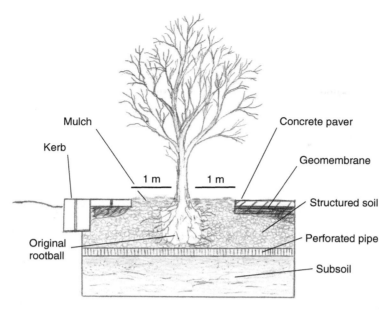

Fig. 5.8. Tree pit with structured soil incorporated.

relate to the open structure and high aggregate composition not retaining enough moisture during very dry periods or indeed, not retaining nutrients as effectively as organic or clay-based soils (which tend to have a higher cation exchange capacity).

Polypropylene cells systems are sometimes used in the root zone, again to strengthen soil weight bearing capacity and protect roots from compressive forces. Rather than being applied in isolation around individual trees, the combination of structured soils and cells can facilitate the formation of 'root zone corridors' placed linearly along roadways and pavements. This process thereby enhances the total volume of soil accessible to the roots of any one tree. In other locations where soil compaction is a problem and high soil bulk density inhibits root growth, pneumatic tools are used to expand the soil with compressed air. This creates fractures and fissures in the soil which allow roots to explore further through the soil profile thereby expanding the overall area for root development.

Mycorrhizae

Mycorrhizae are soil fungi that have a symbiotic relationship with plants, and through attachment to roots (ectomycorrhizae) or even integration into the plant root cells (endomycorrhizae) effectively extend the plant's ability to access nutrients and water from the soil (the surface absorbing area can be increased 100–1000-fold). In return the mycorrhizae derive energy from photosynthates directed towards the plant roots. Through the interchange of nutrients, water and hormones, the presence of mycorrhizae stimulates plant growth and accelerates further root development. They also improve resilience against soil-borne pathogens and to environmental stresses such as drought, salinity and high pH. The endomycorrhizal fungi exchange nutrients with plants via finely branched arbuscules that form within the plant cell and are known as arbuscular mycorrhizae (AM) (glomeromycetes). Rather than arbuscules, some species form sac-like vesicles within the host cell and are denoted as vesicular arbuscular mycorrhizae (VAM). Arbuscular mycorrhizae are the oldest in evolutionary terms and the most widespread mycorrhizal fungi. It is thought that more than 90% of all plant species on the planet form symbiotic associations with these fungi. In contrast, ectomycorrhiza are most commonly associated with tree interactions, forming relationships with species in the Betulaceae, Fagaceae, Pinaceae, Rosaceae and Salicaceae. These fungi (mainly Basidiomycetes) form a 'Hartig net': a network of hyphae held within the extracellular spaces of the root and that in time form a protective sheath around the root.

Plant/mycorrhizae relations are a normal part of soil ecology, but in soil that has been disturbed by human activity, the quantity of mycorrhizae decreases drastically. Following disturbance of forest soils, levels of colonization from AM are reduced by almost 50% compared to undisturbed soil (Jasper *et al.*, 1991). So typically, damaged urban soils may be bereft of the appropriate mycorrhizal partners (Fig. 5.9). Stabler *et al.* (2001) found differences in mycorrhizal populations associated with woody plants between urbanized residential landscapes and neighbouring undisturbed, natural desert landscapes, with less colonization in the residential areas evident as long as 10 years after such sites had been landscaped.

Low fungal propagule numbers in damaged soils may result in poor colonization by mycorrhizae, and present subsequent difficulties in establishing mycorrhizal-dependent woody plant species. Even when colonization can occur, the rates are reduced. Trees growing in urban areas are associated with diminished AM and ectomycorrhizal colonization compared to more rural environments (Bainard *et al.*, 2011). This may be due to these environments having a lower mycorrhizal fungal propagule abundance in the first instance, or the fungi having an impaired ability to infect the tree roots, compared to their rural counterparts. A number of factors could account for this: the highly disturbed and modified (pH, nutrient content, pollutants, lack of aeration) nature of urban soils; relatively lower proportions of host plant species present in urban environments; and greater competition from non-arbuscular mycorrhizal species. Previous studies have shown that mycorrhizal associations with plants can be negatively influenced by environmental stress, regardless of mycorrhizal abundance (Klironomos and Allen, 1995).

Compounding the fact that urban/degraded or disturbed soils may have lower levels of mycorrhizae present, or reduced ability to propagate and colonize woody plants, plants grown on nurseries may themselves have limited mycorrhizal associations. Container plants grown-on with the 'luxury' of supra-optimal levels of nutrients and water are thought to provide little opportunity for their symbiotic mycorrhizae to colonize. Moreover, those mycorrhiza that do occur naturally in the nursery are not capable of surviving in urban soils (Weber and Claus, 2000). This then leads to the scenario of plants which have been grown under ideal conditions being transplanted out to harsh field conditions, but with little opportunity for adaptation (acclimation) and without the presence of their 'natural' mycorrhizal support network.

For landscape specimens, mycorrhizae can be introduced at one of two periods. This may be either within pots during the production phase, or used to inoculate the planting hole at the time of transplanting into the soil. Both approaches may only achieve limited success, however, if conditions are not conducive to the fungi. As such, results from research are mixed. Drought tolerance was increased in young specimens of *Acer*, *Quercus* and *Tilia* during the nursery phase when mycorrhizae were introduced (Fini *et al.*, 2011). Although inoculation did not enhance shoot growth, it provided several physiological benefits such as the maintenance of less negative leaf water potential, higher apparent carboxylation rate, higher RuBP regeneration, and higher quantum yield of photosystem II under water shortage. In essence, inoculation with mycorrhizae appeared to enhance the trees' ability to extract water or to use it more efficiency. In other studies, the benefits of inoculation during the nursery stage have been less tangible, e.g. in oak species such as *Quercus palustris*, *Q. coccinea* and *Q. virginiana* (Gilman, 2001; Martin *et al.*, 2003).

Planting depth and protection

One area often overlooked in practice is planting depth. It is vital that the tree is planted so that its root collar (the section of tissue that provides the interface between stem and roots) is aligned to the surface of the soil. Amazingly, it has been estimated that 75–93% of professionally planted trees may have their root collars buried by soil or mulch (Wells *et al.*, 2006). Burying a plant too deep interferes with the roots' capacity to uptake water, nutrients and oxygen, thereby delaying establishment. In the worst cases it may induce transplant failure. Although in their studies replication rates were relatively low ($n = 6$), Arnold *et al.* (2007) observed that survival rates were reduced to between 66% and 50% in *Lagerstoemia*, *Fraxinus*, *Nerium* and *Platanus* when root collars were planted 76 mm below soil level. Similarly, *Prunus* × *yedoensis* suffered 50% failure when root collars were planted at either 150 or 310 mm below the soil surface, although 100% survival was recorded in *Acer rubrum*. *Acer*, however, had increased incidence of root girdling (roots crossing and constraining the stem and other roots) when planted at the greater depths.

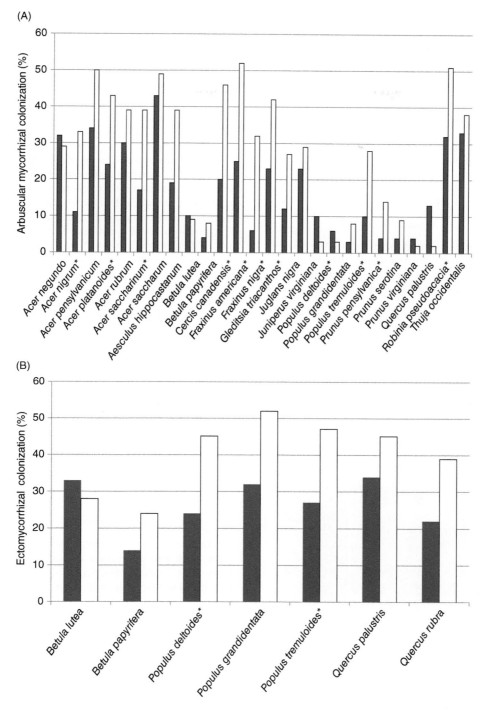

Fig. 5.9. Colonization of (A) arbuscular mycorrhizae and (B) ectomycorrhizae in tree species in either urban (solid bar) or rural (open bar) environments. Species marked with * show statistical differences between environments. (Modified from Bainard et al., 2011.)

Young trees need to be protected from the local wildlife too. Tree guards are commonly used to protect the stems of saplings from herbivores such as rabbits and deer and are considered to provide a favourable microclimate to the young tree, although the extent to which the latter is true has been challenged (Fig. 5.10). Strapping and other physical barriers are also utilized to protect bark from being stripped. Tree shelters are widely used to protect young trees, for which there is strong evidence of improved survival rates (Bellot *et al.*, 2002).

5.5 Maintenance

Weed control

Weed control is required for effective establishment of newly planted trees and shrubs. Weeds compete with landscape plants for moisture, nutrients and, where growth is excessive, for light. Competition for water is greatest in drier climates and where soils have limited water holding capacity. Removing weeds improves tree establishment under such circumstances (Fig. 5.11), although factors that aid weed growth, such as the incorporation of fertilizer, can result in even greater competition (Davis *et al.*, 1999). Davies (1985) states that to be effective, weed control must eliminate competition below the soil surface, i.e. favouring tree root development over the root growth of the weeds. Control methods such as cultivation, herbicides or mulching are most effective, whereas cutting weeds above ground level are ineffectual over the long term. Studies on pine (*Pinus taeda* and *P. elliottii*) demonstrated that removal of competing weeds stimulated new growth in recently planted specimens just as effectively as any fertilizer treatment (Neary *et al.*, 1990). In *Quercus ilex*, optimum establishment was achieved when weed control was combined with ventilated tree shelters in areas where weed pressure was high Cerrillo *et al.* (2005). The importance of removing weed competition to aid young trees cannot be overstated.

Traditionally, weed control in amenity areas would involve herbicides. However, legislation and the cost of producing new agrichemicals for relatively niche roles is now limiting the number of active ingredients available for use in urban areas. In the UK, only glyphosate is now widely available for eliminating perennial weeds. Other countries have banned completely the use of herbicides in amenity settings (e.g. Denmark and Sweden: Kristoffersen *et al.*, 2008) due to concerns relating to groundwater contamination.

Fig. 5.10. Opaque, wrap-around tree guard used to protect vulnerable stems from rabbits and hares. Other wider tree guard types can enclose the foliage too, reputably increasing temperature and humidity, whilst still permitting the transmission of irradiance.

Mulches

Mulches are ground coverings that suppress weed growth as well as restrict moisture evaporation from the soil surface (Fig. 5.12). Some mulch materials help increase water infiltration and reduce soil surface temperature (e.g. by 2.2–3.3°C: Skroch *et al.*, 1992); Box 5.1 gives some further benefits (and drawbacks) in mulch use. Mulches are classified as organic (pine needles, bark, chipped wood, etc.) and inorganic (usually polyethylene or polypropylene woven fabrics, although pebbles and other aggregates are employed in some high value landscapes). Choice depends on aesthetic values, cost, durability and availability of the materials as well as effectiveness. Polyethylene and polypropylene mulches are considered more effective than organic

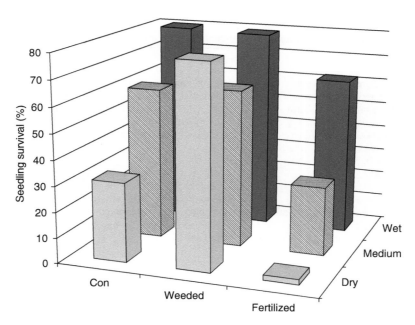

Fig. 5.11. Seedling survival of *Quercus ellipsoidalis* grown in dry (open bar), medium (shaded bar) or wet (solid bar) soils with weeds present ('con'), removed ('weeded') or present but with nitrogen fertilizer ('fertilized'). (Modified from Davis et al., 1999.)

Fig. 5.12. An organic mulch of bark chippings, being used to suppress weeds in a recently planted shrub border.

materials in suppressing robust weeds, but are unattractive and other mulch materials are often placed on top. Of the organic mulches, pine bark is popular due to its attractive surface as well as possessing a high durability. Organic mulches reduced total weed counts by 50% compared to control plots, and underlying organic mulches with polyethylene resulted in complete control (Skroch et al., 1992).

Landscape Trees, Shrubs and Woody Climbing Plants

Box 5.1. Potential benefits and drawbacks associated with the use of mulch materials.

Benefits	Drawbacks
Weed suppression. Can eliminate solar irradiance thereby reducing weed germination, but some may also form physical barriers to weed growth. Some mulches are considered to have allelopathic properties, inhibiting weed germination by phytochemical means.	Additional costs, including labour associated with placement of the mulch. Cost/benefit ratios can be positive, if plant survival is higher, or there is a lower requirement for subsequent irrigation and there are lower costs associated with weeding and pest/pathogen control.
Reduced evaporation.	Organic mulches may have contaminants present such as arsenic or formaldehyde, e.g. if wood materials have been previously treated with preservatives.
Temperature modification, e.g. insulation against frost.	Wood chips and other organic mulches if not properly composted before use may contain pathogen spores or weed seeds.
Inorganic mulches such as gravel and pebbles can help keep soil cool in summer.	
Resistance against compaction/compression forces.	Black plastic mulches may heat the soil – a possible advantage in winter, but a disadvantage in summer.
	Even dark organic mulches may raise local temperatures, resulting in greater evapotranspiration and water loss in the adjacent plants.
Reducing precipitation runoff and improving infiltration to soil.	Some plastic or fabric mulches interfere with water infiltration and movement.
Protect soils from wind and water erosion.	Wind may blow mulch materials off the soil.
Rapid root growth and hence stabilization of new plants.	Subsequent removal of fabric mulches can cause root damage, as fine surface roots are removed too.
Increased survival rates in young transplants.	Bark, wood and certain other organic mulches may 'immobilize' soil nitrogen due to the microbial process associated with their degradation. Research suggests though that this may be very localized in the soil/mulch interface, and would, for example, not interfere with the ability of deeper roots of woody plants to access nitrogen (Chalker-Scott, 2007).
Reduce 'rainsplash' back into the plant foliage, thereby reducing potential for pathogens (*Pythium*/*Phytophthora* spp. to infect leaves).	Mulches may accommodate non-pathogenic microorganisms, such as the 'dog vomit' slime mould (*Fuligo septica*), which detracts from the aesthetic value of the mulch.
Microbial populations associated with organic mulches can be competitive or predatory on plant pathogens, thus reducing disease incidence.	Certain mulch materials may be flammable and add to the risks in fire-sensitive locations.
Other mulches may possess chemicals that are toxic to pathogenic organisms or inhibit microbial activity – e.g. thujaplicin from *Thuja* spp. wood or bark chips.	
Such chemicals may also repel invertebrate pests.	Mammals, particularly rodent species utilize plastic and fabric sheets for shelter. Some of these species may damage tree/shrub stems.
Some mulch materials have a positive aesthetic value.	Plastic/fabric materials are often considered unsightly.
Market for waste materials, including recycled wood, rubber and glass.	The sustainability of some materials may be questionable, either derived from petrochemical products, or sourced and transported from distant quarries.

Billeaud and Zajicek (1989) evaluated a range of organic mulches and showed that deeper mulches are more effective at controlling weed growth but also tend to lock up soil nitrogen, decrease soil pH and reduce the visual quality of the landscape plants. A good compromise was shallow organic mulches (e.g. 50 mm) laid over a polypropylene weed barrier. Others believe the role of mulches in immobilizing nitrogen is overstated, and any effect of immobilization is very localized to a few millimetres around the mulch/soil interface.

Pruning

Pruning is carried out on urban trees to improve the aesthetic appearance of specimens as well as reduce hazards to people and property. Pruning corresponds to a number of different strategies. 'Structural pruning' is encouraged on relatively young trees to ensure the continued development of a leading shoot (the 'leader' that eventually forms the main trunk), an 'open' branch structure with even spacing of the lateral branches initiated from the main stem, and to remove any poorly-angled branches that may later misshape the tree or cause splitting of the trunk. This structural pruning is important as it improves the integrity of the whole structure, enabling trees to have greater resilience against future storms. In older trees, 'crown cleaning' is employed to remove dead, diseased or overcrowded branches, and 'crown thinning' aims to increase air movement and irradiance penetration through the tree canopy. Where lower branches may interfere with vehicle or pedestrian movement, or where there is a requirement to provide more room for buildings and other urban infrastructure, these may be removed via 'crown raising'. 'Crown reduction' reduces the overall size of the canopy by shortening the length of lateral branches. This can be done to ensure the trees remain in scale to surrounding buildings or to avoid entanglement in overhead power or telephone wires. Pruning of large, mature trees is often limited to the removal of dead or hazardous limbs, and even here care is required to avoid pathogens and insects penetrating the wound site. Branches and limbs should not be cut flush to the main trunk, as a small section of remaining branch stub (collar) ensures cambial tissues can help form a protective covering around the heartwood, thus securing the wound site.

Shrubs are also pruned, sometimes for form enhancement, but more often to promote flower production. Rose pruning is probably the classic example, and the bane of many a novice gardener. The key to shrub pruning, however, lies closely with understanding where and when flowers are initiated. Some species predominantly initiate flower buds on the developing shoots, but these need to be physiologically mature enough to support flower induction. For this reason these plants should be pruned immediately after flowering, thereby encouraging new vegetative shoot extension, which matures sufficiently to form flower buds that then overwinter and provide a full flower display the following year. *Philadelphus* spp. fall into this category. In these sorts of shrubs, selective hard pruning too of some of the old thick stems will encourage the promotion of younger vigorous growth from the base, keeping the plant rejuvenated and providing the mainframe structure for the future.

In contrast, the entire specimen of *Buddleia davidii* (butterfly bush) or bush (hybrid-T) roses can be pruned to the base, as flowers develop in the actively extending stem growth of the following season. Such annual severe pruning is also used when the aesthetic characteristics of the plant can be improved. Hard pruning of trees such as *Paulownia tomentosa* (foxglove tree) or *Catalpa bignonioides* (Indian bean tree) results in larger leaves than normal and provides a striking feature; likewise hard pruning of colourful stemmed *Cornus* (dogwood) and *Salix* (willow) varieties results in more vivid colouring associated with the new young shoots. Annual severe pruning, however, can tax the vigour of individual specimens, so the presence of a nutrient-rich, moisture-retentive soil becomes more critical.

Pruning of woody climbers and vines is no less complex. If there is one genera that challenges *Rosa* in causing confusion in the minds of gardeners and horticulturalists, it is *Clematis*. Here pruning strategies vary strongly with genotype and to some extent depend on where (how high up) and when one wishes the plant to flower. Pruning strategies include:

- No pruning. Early-blooming species and cultivars, including the *C. montana* and *C. alpina* types, e.g. *C. montana* 'Elizabeth'. (Group 1).
- Light pruning, removing weak and dead stems only, in late winter. After flowering, a few shoots may be hard pruned to encourage regrowth and rejuvenation from the base of the plant. These *Clematis* types flower on more 'mature wood' and comprise the large-flowered cultivars that flower in early summer on short shoots developing from the previous year's growth. In some cultivars, however, there can be some later, late

summer flowering too, just as the current season's growth begins to mature. Included within this group is the iconic *Clematis* 'Nelly Moser' with its pink-and-white barred flowers. (Group 2).

- Severe pruning to basal buds in late winter. This stimulates new, strong vegetative growth from the base of the plant and flowers form on the current season's growth. Includes species such as *C. viticella*, *C. texensis* and *C. orientalis* and their derived cultivars, as well as larger flowered hybrids such as *C.* 'Jackmanii' and *C.* 'Ernest Markham'. (Group 3).

Effective pruning of other climbers, such as *Wisteria*, require some understanding of their physiological development too, and reference to quality 'gardening' books can be invaluable in helping to implement appropriate pruning strategies for professional and 'amateur' horticulturalists alike.

Hard pruning to ground level is also accomplished on some tree species through the process known as coppicing. This is not done every year, but perhaps every 5–7 years, for example with *Corylus* (hazel) coppiced woods. This is a legacy of traditional woodland management for the production of fence stakes, bean poles and weaving material. Today, however, it is usually carried out to temporarily open up the forest floor to light and encourage flowering in the forest herb layer. It is a key management strategy to encourage a more biologically diverse piece of woodland (a wider range of flowering forbs, but also associated invertebrates that rely on them). In a similar manner, some trees will be cut back on a regular basis, but 2–3 m from ground level (pollarding) to constrain the size of the tree crown (e.g. in space-restricted streetscapes) and provide more irradiance at ground level.

Frequently, landscape shrubs are pruned to regular shapes such as squares and rectangles; where this fits well with the geometrical patterns of nearby buildings/other urban infrastructure (Fig. 5.13) or is in keeping with historical landscapes such as parterres, this is appropriate. All too often though it is used within a mind-set to keep things 'neat and tidy', and which results in landscapes that look constrained or incongruous with the plants species that have been chosen. Moreover, such approaches limit the heterogeneity within the planted landscape, thus reducing the opportunity to provide much-needed 'habitat' for urban birds and invertebrates.

5.6 Right Plant, Right Place

Horticulture, and gardening in particular, always has had an element of 'challenge' in it. This is one of

Fig 5.13. Pruning shrubs into symmetrical shapes often results in poor results, and reflects a lack of imagination in the management of urban greenery. But here the extent of the process – providing an additional horizontal plane at a level higher than the ground – is used to good effect.

the attractions of the discipline, to grow something that others cannot, or to create unique designs that express different ideas or concepts. Whether it is the immense effort that goes into growing the biggest and the best for the local flower show, or the challenge of developing a new colour break or form of flower, these aspects add much of the intrigue and fascination involved with the subject. A garden, almost by definition is an area where a range of plants grow that would not normally occur there naturally, or at least where the local species are actively manipulated to favour certain plant genotypes over others. So, in essence, the idea of using plant species that might not quite fit the local conditions has always been a driving force in horticulture, with the skill of the cultivator being to adapt the conditions to suit the plant. As a consequence, tropical plant species are grown in (expensive to construct and run) glasshouses at 50° north and 40° south of the Equator, *Rhododendron* and other acid-loving ericaceous species are plunged into peat beds in gardens over alkaline chalk, bog gardens are created on sandy free-draining sites and heavy clay soils are backfilled with gravel to facilitate the growing of species that require 'sharp' drainage!

Life is made much easier, however, if the plant species used in a set location have some adaptation to the conditions prevalent on the site. As such, a number of the concepts within environmental horticulture relate to using, on a more frequent basis, species well-fitted to the conditions presented, and not necessarily attempting to alter radically the conditions themselves. This does not automatically mean that only locally adapted species need be used, but rather that considerations should be given to the types of plants that are likely to be successful, including species from other continents or biomes.

An understanding of a species ecophysiology (where it comes from and what it is adapted to), or even just a quick inspection of its morphology can help provide some indications of the conditions it will tolerate.

Wind tolerance

Trees and shrubs that tolerate strong winds or can resist desiccation from drying winds often have identifiable traits. This includes waxy or needle-like leaves that help restrict water loss from the leaf. Pine and spruce trees are useful for exposed windy sites due to these characteristics, for example, *Pinus nigra* subsp. *maritima* (Corsican pine), *P. sylvestris* (Scots pine) and *Picea sitchensis* (sitka spruce). Similarly, broadleaf species may include *Quercus ilex* (holm oak) or *Escallonia* spp. (particularly good on more temperate coastal situations) due to the thick cuticles on their leaves. At a lower scale, the strap-like leaves of sub-shrubs such as *Calluna vulgaris* (heather), *Erica* spp. (heath), *Lavandula* spp. (lavender) and *Rosmarinus* spp. (rosemary) help retain moisture. Some species avoid the wind by growing close to the ground in a hummocky (e.g. *Pinus mugo* and its cultivars) or prostrate form (e.g. *Salix arctica* or *Juniperus squamata* 'Blue Star'). Other species may grow tall but can resist the wind by either possessing thin branches that allow the wind to filter through, or have an ability to bend and be flexible in the wind (e.g. *Betula* spp., *Salix* spp. and *Pinus flexilis*). In addition, species as typified by *Acer pseudoplatanus* (sycamore) are expert at tolerating wind, cold and even salt spray in combination.

Wet soils and flooding tolerance

Some locations are more prone to water accumulation than others and planting should take account of this. Swales and rain gardens may be designed to capture runoff water and hold it temporarily in an attempt to avoid urban drainage systems being overloaded. Other planted areas may just naturally be at the base of a hill or slope and receive all the natural runoff and percolation of water from the surrounding landscape. Alternatively, many urban green spaces are actually located along riverbanks, on floodplains or on other areas prone to natural flooding. These locations are seen by planners as 'sacrificial land', i.e. areas that should not be developed on, precisely because they are a relief space to allow flood water to accumulate rather than inundate nearby buildings and houses. Finally, certain clay soils may also be prone to wetness, even when on higher land due simply to poor drainage capacity.

Many species in the plant kingdom are well-adapted to wet soils and periods of flooding, but many others are not and careful selection is required for such sites. Riparian (river-side) woody plants usually do well in such locations, and species such as *Alnus glutinosa* (alder), *Populus nigra,* (black poplar), *P. deltoides* (cottonwood), *Betula nigra* (river birch), *Salix* spp., *Taxodium distichum* (swamp or bald cypress) and *T. ascendens* are ideal for their ability to tolerate 'out and out' flooding. Other species are more likely to be damp woodland species in terms of their ecophysiology and are good at coping with wet clay soils rather than prolonged periods of inundation per se. These include small trees such as *Amelanchier canadensis* (service berry), *Aronia arbutifolia* (red chokeberry), *Styrax japonicus* (Japanese snowbell), *Halesia carolina* (little silverbell), *Carpinus betulus* (hornbeam), *Corylus avellana* (hazel), *Sambucus nigra* (elder) and *Populus tremula* (aspen); larger trees such as *Fraxinus excelsior* (ash), *Cercidiphyllum japonicum* (katsura), *Nyssa sylvatica* (black tupelo), *Davidia involucrata* (handkerchief tree) and *Liquidambar styraciflua* (sweetgum), as well as shrubs such as *Ribes sanguineum* (flowering currant), *Euonymus europaeus* (spindle), *Cornus stolonifera* (dogwood), *Rubus hispidus*, *Viburnum opulus* (guelder-rose), *Hydrangea* spp., *Vaccinium corymbosum*, (highbush blueberry), *Kalmia polifolia* and *Myrica gale*.

The severity of effects due to waterlogging depends on the extent to which plant species themselves can adapt to the prevalent conditions, but also on aspects such as season of flooding (temperature and growth phase), duration of flooding, depth of waterlogging, soil type and degree of microbial activity as well as the condition of the flood water (e.g. degree of deoxygenation). Most plant species are more prone to waterlogging injury when temperatures are higher (i.e. summer) as this accelerates respiration and oxygen deficiency within the soil (Fig. 5.14).

The paradox with waterlogging is that the foliar parts of many plants actually experience 'physiological drought', as root cells fail to maintain the transpiration stream. This results in water no longer effectively being transported into and up the xylem tissues. There are a number of stress factors that impair plant development during waterlogging, but a lack of oxygen in the root zone alters respiration with less energy available for cells to regulate water uptake. Oxygen diffuses through air much more effectively than it does water, so when soils become saturated diffusion rates decrease dramatically, leading to oxygen starvation and anaerobic conditions at the root soil interface. Flooding (or more appropriately in physiological terms 'anaerobism') affects a number of processes in the transpiration stream, not only water absorption into cells but also root hydraulic conductance and stomatal opening. Not all plants respond in the same way, however, leading to some debate as to what precisely the stress factors are and how plants respond to them. This is accentuated by the fact that not all species demonstrate

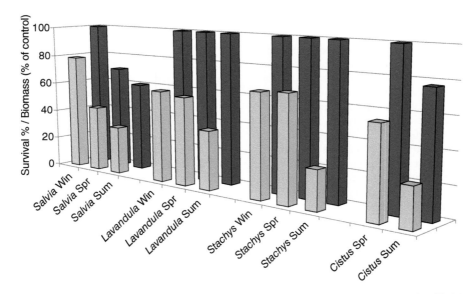

Fig. 5.14. Effects of waterlogging and seasonal (temperature) on four Mediterranean plant species (*Salvia officinalis*, *Cistus* × *hybridus*, *Lavandula angustifolia* and *Stachys byzantina*). The effect of 18-day waterlogging on survival (solid bar) and biomass (shaded bar) when flooded in semi-controlled, glasshouse conditions in winter ('Win'), spring ('Spr') or summer ('Sum'). Survival rates decrease in *Salvia officinalis* and *Cistus* × *hybridus* during warmer periods. *Lavandula angustifolia* and *Stachys byzantina* survive all waterlogging events but biomass is reduced during flooding in summer. (Modified from King *et al.*, 2012.)

the symptoms of water deficit during flooding. One theory is this may be due to subtle effects relating to soil oxygen and CO_2 partial pressures as well as differences between plant species. High CO_2 levels that accumulate during flooding due to respiration from both soil microflora and roots, may be transformed to carbonic acid (H_2CO_3). This in turn is transported to root cells and acidifies the cytoplasm. This is thought to inhibit aquaporin activity; these are the pores by which water moves between one cell and the next. Tournaire-Roux et al. (2003) claim it is this transformation of CO_2 to H_2CO_3 and the resultant acidification of the cell cytoplasm that inhibits root hydraulic conductivity under anaerobic conditions. Unfortunately, many studies on waterlogging document oxygen depletion, but do not monitor for CO_2 accumulation and activity. As such it is still difficult to say whether loss of water uptake is directly due to a lack of oxygen or alternatively a build-up of CO_2.

Others argue that stomatal closure (and hence transpiration) is mediated by root-derived chemical signals (Jackson and Ricard, 2003) and although the role of the stress hormone abscisic acid (ABA) is implicated, the exact nature of the possible chemical signals is still unknown. A further mechanism may be that phytotoxic compounds generated in the roots during anaerobic conditions are responsible for interfering with water absorption and movement within the roots, for example ethanol, acetaldehyde (ACC), or lactic acid (Kamaluddin and Zwiazek, 2001). Irrespective of the mechanisms involved, the consequences are usually the same – significant and drastic reductions in growth in non-adapted species.

So how do some species seem to tolerate waterlogging better than others? This may be due partly to different biochemical responses to the effects outlined above, but other physiological factors/alterations help some species cope with anaerobism. This includes the formation of airways (aerenchyma) through the stem to the roots that allow oxygen to diffuse down the plant internally, and/or the development of surface roots that can access oxygen from the atmosphere more readily. Indeed, genera such as *Salix* have pre-formed root initials present in their stems, allowing them to send out new roots quickly when water levels change. These new 'surface' roots then provide plants with the water and nutrients they need, and compensate for the older, now dysfunctional roots located deeper down the soil profile in the anaerobic zone.

Some of the differences in tolerance between species are explained by such factors. When reviewing those trees suitable for urban soils prone to waterlogging, Smith *et al.* (2001) noted that *Corymbia maculata* (spotted gum) and *Platanus orientalis* (oriental plane) were able to induce new roots when waterlogged, and to recover quickly after waterlogging. In contrast *Platanus × acerfolia* (London plane) did not induce roots whilst under stress, but recovered well on draining, whereas the least well adapted, *Lophostemon confertus*, was unable to initiate roots both during and sometime after waterlogging. Although riparian species generally have the best adaptations to waterlogging, it can be surprising which other species possess favourable characteristics and tolerances. *Corymbia*, for example, grows in well-drained, rocky locations in eastern Australia, not habitats normally prone to flooding. Similarly, certain drought-adapted 'Mediterranean' sub-shrubs such as *Lavandula* and *Salvia* possess some adaptation to waterlogging (King *et al.*, 2012). One key factor in waterlogged soils is the absence of oxygen (anaerobism) for root respiration. Providing low oxygen conditions (hypoxia) prior to the full removal of oxygen (anoxia), however, increased root survival in *Salvia*, indicating the potential to adapt to low oxygen environments (Fig. 5.15).

Drought tolerance and xeriscaping

Drought-tolerant plants tend to be easier to identify than those with tolerance to flooding. Typical clues include the presence of succulent, pubescent (hairy), small or narrow leaves, or leaves that are coloured to reflect the incoming irradiance (blue, grey or silver). The absence of leaves altogether may also be a feature. Some tropical species shed their leaves during the dry season to minimize water loss. In a similar manner, species such as *Eucalyptus*, adapted to warm, dry climates, have chlorophyll in their stems and bark which contributes to photosynthesis when the leaves' ability to fix CO_2 is compromised. Other adaptations may include deep tap roots or conversely fine surface roots (the latter being used in desert situations to capture moisture from light rains or dew that forms overnight, and where the water does not penetrate deep into the soil profile). There are exceptions to these rules too, of course. The conventionally-shaped, broadleaved, rapid-growing shrub *Cotinus coggygria* is surprisingly drought tolerant.

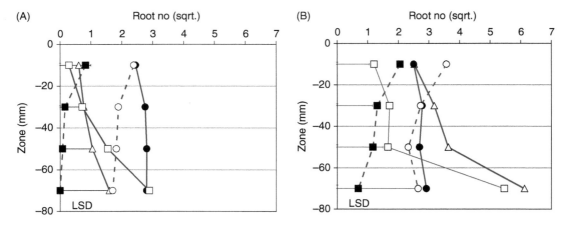

Fig. 5.15. Number of new roots per plant (square root transformed) recorded in different zones down the root profile: (A) after exposure to gaseous treatments (day 10) and (B) after recovery (day 20) in *Salvia officinalis*. Treatments: control (□), anoxia (■), hypoxia (Δ), prolonged hypoxia (●) and hypoxia/anoxia (○). Bars, LSD. (Source: King et al., 2012.)

Xeriscaping is a special form of landscape design, where landscaping and horticultural strategies are designed to minimize water use. It is used in those situations where soil moisture supplies are insufficient, and/or irrigation water is in short supply or too expensive to be used to support non-drought-adapted landscape plants. Depending on location, soil type and choice of plant species, water use can be reduced by one-third to two-thirds if xeriscaping principles are adopted when designing, planting and maintaining a landscape. Obviously there are variations within this theme, ranging from entire desertscapes with cacti and other succulents, to plants representing Mediterranean biomes, including trees, seasonal flowering shrubs, sub-shrubs, geophytes and herbaceous perennial species (forbs). Within continental and even temperate climatic zones too, there are some woody species better adapted to drier, low water conditions. Typical woody plant species useful in xeriscapes are outlined in Boxes 5.2 and 5.3.

5.7 Pests and Pathogens

Many woody plants are now planted to encourage biodiversity in urban landscapes (see Chapter 4), and in a robust ecosystem, plants are armed with a range of defences to tolerate pest and pathogen pressure. This, combined with the more complex food chains that are associated with natural/semi-natural landscapes, ensures that herbivorous invertebrates themselves are predated upon, thus regulating numbers attacking the plants. As such, plant populations for the most part are resilient enough to tolerate a degree of pressure from herbivorous animals, pathogenic fungi, bacteria and viruses, although individual plants may succumb from time to time. In natural woodland settings a large percentage of deadwood actually represents a 'healthy' ecosystem.

In ornamental and other urban situations the story is more complex. Unsightly damaged trees and shrubs are not often appreciated for their aesthetic value. Dead trees and particularly dead branches (limbs) that might fall are considered a danger to the public, and so need managing; ideally individual trees should not become so stressed that they die prematurely anyway. Urban environments do not represent entire natural ecosystems and a limited choice of trees and shrubs, including the use of monocultures (e.g. an avenue composed of one tree species alone) actually encourages pest and pathogen pressure on particular species. The use of clonal genotypes, with no variation in genetic resistance to herbivores or pathogens, also increases the susceptibility of the tree and shrub stock. Abiotic (environmental stress) exacerbated by urban conditions and climate change can reduce plant defences and increase susceptibility to pest/pathogen attack. The movement of plant material through increased global travel, the import of planting stock from other regions/countries and perhaps climatic changes encouraging wider species distributions are also resulting in trees/shrubs being exposed to new species of pathogens or pests, or new pathovars (races) of existing pathogens. In these scenarios tree and

> **Box 5.2.** Trees with some tolerance of dry soils and potentially useful for xeriscapes (depending on climatic zone).
>
> | | | |
> |---|---|---|
> | *Acer grandidentatum* | *Fraxinus pennsylvanica* | *Pinus ponderosa* |
> | *Acer negundo* | *Ginkgo biloba* | *Prunus armeniaca* |
> | *Acer tataricum* | *Gleditsia triacanthos* | *Pyrus ussuriensis* |
> | *Ailanthus altissima* | *Juniperus* spp. | *Quercus gambelii* |
> | *Catalpa speciosa* | *Ilex* spp. | *Quercus ilex* |
> | *Cedrus deodara* | *Pinus aristata* | *Quercus macrocarpa* |
> | *Celtis occidentalis* | *Pinus edulis* | *Robinia pseudoacacia* |
> | *Crataegus ambigua* | *Pinus mugo* | *Styphnolobium japonicum* |
> | *Crataegus crus-galli* | *Pinus pinea* | *Zelkova* spp. |

> **Box 5.3.** Shrubs and sub-shrubs with some tolerance of dry soils and potentially useful for xeriscapes (depending on climatic zone).
>
> | | | |
> |---|---|---|
> | *Amorpha canescens* | *Cowania mexicana* | *Prunus besseyi* |
> | *Artemisia* spp. | *Elaeagnus* spp. | *Rhus glabra cismontana* |
> | *Atriplex canescens* | *Ephedra equisetina* | *Rhus trilobata* |
> | *Berberis* spp. | *Fallugia paradoxa* | *Rosa woodsii* |
> | *Buddleia alternifolia* | *Fendlera rupicola* | *Rosmarinus officinalis* |
> | *Caragana arborescens* | *Forestiera neomexicana* | *Rubus deliciosus* |
> | *Caryopteris* × *clandonensis* | *Genista* spp. | *Shepherdia argentea* |
> | *Ceanothus* spp. | *Hippophae rhamnoides* | *Symphoricarpos* spp. |
> | *Cercocarpus* spp. | *Holodiscus dumosus* | *Salvia* spp. |
> | *Cistus* spp. | *Ligustrum vulgare* | *Syringa vulgaris* |
> | *Chamaebatiaria millefolium* | *Lavandula angustifolia* | *Viburnum lantana* |
> | *Chrysothamnus* spp. | *Potentilla* spp. | |

shrub stock may have limited defences and whole populations can be, and are from time to time, decimated.

Possibly the most infamous case of trees succumbing to a new pathogen is that of the elm. Populations of elms (of different species and hybrids) in both Europe and North America have been killed by the ascomycete pathogens of the genus *Ophiostoma*, e.g. *O. novo-ulmi*, (the activities of which are generally known as Dutch elm disease). These ascomycete fungi are spread through the action of bark-burrowing beetles such as *Hylurgopinus rufipes*, *Scolytus multistriatus* and *S. schevyrewi*. In the UK the numbers of mature elms killed over the last century exceeds 25 million, resulting in radical changes to the character of the UK landscape, notable through the loss of large iconic hedgerow trees. Similarly, cities such as New Haven, USA, which relied strongly on elms for street and park trees, lost much of their character within a few years.

New epidemics following the pattern of Dutch elm disease are a concern for environmental horticulturalists. The threats vary depending on the dominant tree species used in different parts of the world and potential exposure to new pathogens (Table 5.5) or pests (Table 5.6). In terms of landscape trees, most documentation relates to Europe or North America, although Australia and New Zealand also document threats (and impose strict regulations on plant imports to help protect native species).

Over and above these potentially devastating pests and pathogens, many woody plants will host a range of invertebrates and microorganisms that will cause infestations or disease symptoms when conditions are conducive. For example this includes the powdery mildews (e.g. *Erysiphe*, *Podosphaera*, *Microsphaera*, *Sawadaea* spp.), root rots (e.g. *Pythium* and *Phytophthora* spp.), bootlace fungus (*Armillaria* spp.), bacterial cankers (*Pseudomonas* spp.) and pests such as aphids (e.g. *Myzus*,

Table 5.5. Examples of potentially damaging pathogens of woody plants in temperate climates.

Pathogen species	Description
Ceratocystis fagacearum (oak wilt)	Spreads through the shipment of infected wood or through insect activity and between one tree and its neighbour via natural root grafts (fusion of two separate roots).
Ceratocystis platani (plane wilt)	Very damaging to *Platanus* spp. (up to 80% of trees affected in some parts of France). Enters through wounds, causing cankers and eventual death. Originating in southeastern USA, it has spread throughout urban populations of *P.* × *acerfolia* (London plane) in the cities of the eastern USA.
Chalara fraxinea (ash dieback)	An aggressive fungal disease of *Fraxinus* spp. (ash) which causes crown death, leaf wilting and dieback of branches.
Cryphonectria parasitica (chestnut blight)	Currently a localized, but highly damaging, disease of *Castanea sativa* (sweet chestnut).
Dothistroma septosporum (Dothistroma needle blight or red band needle blight)	A fungal pathogen of pine that can cause significant defoliation and subsequent death in species such as *Pinus nigra* subsp. *maritima* (Corsican pine), *Pinus contorta* (lodgepole pine) and *Pinus sylvestris* (Scots pine).
Erwinia amylovora (fireblight)	Affects species in the Rosaceae. Affected areas appear blackened, shrunken and cracked, as though scorched by fire. Primary infections are established in open blossoms and tender new shoots and leaves in the spring when blossoms are open.
Fusarium circinatum (pine pitch canker)	Affects *Pinus* spp. and *Pseudotsuga menziesii* (Douglas fir). Spores are spread by wind/rain and can enter trees or seedlings through existing wounds or the roots. Originally found in Mexico but has spread to North, Central and South America, Europe and to parts of Asia and Africa. Causes bleeding infections that can encircle branches, exposed roots and trunks. The wood beneath the infection site is saturated with resin and becomes a characteristic honey colour.
Mycosphaerella dearnessii (brown spot needle blight)	In North America it causes serious growth check to seedlings and young trees, and has rendered Christmas tree plantations unsaleable. In Central Europe has recently been found affecting several pine species.
Phytophthora austrocedrae	Confirmed as the cause of dieback and deaths of *Juniperus* (juniper), a priority conservation species, in northern England, UK, in 2011. This pathogen had previously been almost solely associated with *Austrocedrus chilensis* (Chilean cedar) trees in South America.
Phytophthora kernoviae	Confirmed only in UK, Ireland and New Zealand, and only in a very few trees, so far. The fact that it seems to be able to infect a number of trees such as *Fagus* (beech) and *Quercus* (oak) spp., as well as woodland understorey species such as *Vaccinium* (bilberry) and *Rhododendron* spp., makes it a concern, particularly for historically important gardens with rare or unusual species.
Phytophthora lateralis	Infects and usually kills *Chamaecyparis lawsoniana* (Lawson cypress) trees.
Phytophthora ramorum (ramorum disease/sudden oak death)	A fungus-like organism which attacks a large range of woody plants. *Larix decidua* (European larch), an economically important silviculture species is a key host, and large numbers have had to be felled.
Puccinia psidii (Eucalyptus/guava rust)	A pathogen with a wide host range in the Myrtaceae, which includes about 3000 tree and shrub species – many of which are of great economic and conservation significance. It is native to parts of South America, but also now occurs in parts of North and Central America. The pathogen causes disease symptoms in young shoots, flower buds and developing fruit depending on the host plant. Highly susceptible trees may be malformed or killed outright. Growth rates of infected trees are diminished.
Acute oak decline	A condition affecting *Quercus* in parts of the UK, in which bacteria, including one species previously unknown to science, are believed to be involved.

Metopolophium, *Phorodon*, *Rhopalosiphum*, *Aphis* and *Elatobium* spp.), capsid bugs (e.g. *Plesiocoris* and *Corythucha* spp.), scale insects (e.g. *Lecanium*, *Aulacaspis*, *Saissetia* spp.), spider mites (e.g. *Tetranychus*, *Panonychus*, *Oligonychus* spp.) and caterpillars of various moth and butterfly species (Lepidoptera) amongst many others.

A number of strategies can be employed to minimize infection of woody plants in the landscape, or control infections/infestations depending on the

Table 5.6. Examples of potentially damaging pests of woody plants in temperate climates.

Pest species	Description
Agrilus anxius (birch borer)	European and Asian *Betula* (birch) spp. are highly susceptible. Characteristic D-shaped exit holes, but only once adults have left.
Agrilus planipennis (emerald ash borer beetle)	A major pest in North America able to infect *Fraxinus pennsylvanica* (green ash), *F. americana* (white ash), *F. nigra* (black ash) and *F. quadrangulata* (blue ash). Larvae feed on inner bark leading to girdling, leaf yellowing, branch dieback and eventual tree mortality. Symptoms of infestation similar to birch borer.
Anoplophora chinensis (Chinese longhorn beetle)	Can spread long distance via movement of infested living (particularly potted plants) or sawn wood. Attacks base of trunk of suitable hosts which include wide range of broadleaved trees, e.g. *Acer, Cotoneaster, Platanus, Malus, Prunus, Salix, Rosa* spp. First sign is usually a round exit hole. Small indentations from female chewing and oviposition may be the only other external sign. Die-back of foliage during early attack phase with sustained attacks resulting in tree mortality.
Anoplophora glabripennis (Asian longhorn beetle)	A wood-boring insect that can cause extensive damage across a range of broadleaf tree species Characteristic die-back of foliage during early phases. Beetles develop galleries within the wood and exit holes weaken the integrity of infested trees. Larvae are considered to be the most damaging as they tunnel through the cambium tissues. Now one of the most destructive non-native insects in the USA; it and other woodboring pests cause an estimated US$3.5 billion in annual damages.
Cameraria ohridella (horse chestnut leaf miner)	Originating in Macedonia the species has spread rapidly throughout Europe. The larvae mine within the leaves of *Aesculus hippocastanum* (horse chestnut) and up to 700 visible tunnels have been recorded on a single leaf under favourable conditions. Severely damaged leaves shrivel and turn brown by late summer, well before natural abscission.
Dendroctonus micans (great spruce bark beetle)	Present throughout much of the Eurasian region. In addition to spruce such as *Picea abies* (Norway spruce) also attacks *Pinus sylvestris* (Scots pine) and *Abies* spp. (firs). Beetles infect the root and stems of trees and breed in the bark; where the larvae then feed creating tunnel galleries and undermine the viability of the whole tree.
Dendrolimus pini (pine tree lappet moth)	This moth can cause extensive defoliation of pines and other conifer trees in parts of its native range in Europe and western Asia.
Thaumetopoea processionea (oak processionary moth)	Capable of defoliating *Quercus* spp. over several seasons, leading to dieback, long-term decline and mortality. Larval hairs cause severe health problems for people and animals (skin rash, eye and throat irritation, allergic reactions).

pathogen/pest in question. Careful vigilance can help offset problems associated with new diseases/pests to a specific area. This includes restricting imports of timber or live plants, using 'plant passports' (certificates of plant health) and imposing quarantine conditions to any imported material. Disease/pest outbreaks can be monitored through raising awareness within horticultural/landscape professional circles or with the general public, e.g. notices in entrances to urban woodlands.

Once a pathogen is present, then cultivation regimes can help avoid disease. This includes avoiding drought stress in landscape plants or minimizing insect damage (because, for example, viruses can be transmitted from plant to plant via aphids or other sap-sucking invertebrates). A number of soil pathogens build up high spore levels or other propagules when their host plant has been present for a long time. Subsequently, planting the same species again in the parent soil results in poor growth and disease symptoms ('specific replant disease/disorder'). A case in point is that *Rosa* spp. (roses) should not be planted in soil that recently grew *Rosa* before, similarly for *Prunus* (cherries and plums) or *Malus* (apples).

Chemical pesticides have been the traditional means of controlling pests and diseases, and are still used to a high degree. In landscape situations, however, there is increasing reluctance to spray chemicals where the public may be present, or where more than the target pest species is harmed (non-selective pesticides). Nevertheless, in some countries copper and other metal-based chemicals are used, especially in nursery situations, e.g. fungicides such as chlorothalonil, Bordeaux mixture, benomyl and maneb. In mature trees, trunk injections are used to control pests and pathogens, e.g. imidacloprid or thiamethoxam have been used to help control Asian and Chinese longhorn beetles (*Anoplophora glabripennis* and *A. chinensis*). Where a number of individuals of the same species are grown together, it may be worth investing in pheromone traps; these emit insect sex pheromones, attracting attacking insects to the traps rather than the food source trees. In a similar manner, physical barriers can be put in place to stop insects accessing the vulnerable part(s) of the plants, e.g. sticky or cloth bands may be put around the trunks of trees to stop flightless insects from climbing up into the canopy. Where labour allows, manual removal of pests is feasible; tent-forming caterpillars can be removed via hand or even vacuuming. For localized infections, selective removal of infected branches may help stop the spread of a pathogen. Similarly, radical action such as the felling and removal of entire stands of trees may be necessary for stopping the spread of the more virulent pathogens or invasive pest species. When managing infected specimens, tools and machinery need to be cleaned and sterilized afterwards to avoid cross-contamination onto healthy stock. There may be some benefit of treating wounded areas and cut stumps with fungicide, but the value of bitumen-based wound sealants is unproven, and may even inhibit the natural callus and wound sealing processes.

Biocontrol or integrated pest management (IPM) techniques are attempted in woody plant populations prone to specific disease or pests. To some extent in outdoor situations this is accomplished by encouraging natural predators of the pest species through habitat management, for example providing more ground-cover vegetation, or companion plants that offer habitat or alternative food sources to predatory species. In some circumstances the use of specific biocontrol agents is justified, for example spraying inoculations of bacteria (*Bacillus thuringiensis*) or nematodes (e.g. *Steinernema kraussei*) onto insect larvae, where these parasitic organisms invade the host's body and kill them. Parasitoid wasps do something similar by laying their eggs within the body of the pest species. A longer term strategy for woody plant biosecurity, of course, is to select or breed for genotypes that have greater tolerance to the specific problematic pest or pathogen.

5.8 Urban Forests

Urban forest is a term used to denote the tree stock in towns and cities and incorporates street, park and garden trees as well as areas of woodland (natural, semi-natural or intentionally planted) within urban conurbations. Urban forestry was defined by Johnston and Hirons (2014) as 'the planned, systematic and integrated approach to the planting, maintenance and management of trees and woodland in and around urban areas'.

Traditionally, urban trees were valued for their aesthetic qualities, although wider recognition of their ecosystem service role is increasingly used as the rationale underpinning careful maintenance and new plantings (see Chapters 2 and 3). Urban forests can be a fundamental component of promoting a city's character or a 'sense of place' (Fig. 5.16). Many of the world's great cities and conurbations can be identified with their tree stocks; seen for example in the use of iconic London plane (*Platanus × acerfolia*) in the city squares of London, UK, or large imposing specimens of *Dipterocarpus alatus* in the streetscapes of tropical cities such as Hoi Chi Minh in Vietnam. Such tree stocks, however, are under pressure from environmental factors or continued densification of the urban environment. Street trees particularly can be exposed to environmental stress associated with:

- excessive heat (urban heat island effect);
- drying winds (canyon effect);
- excessive shade from nearby buildings;
- lack of water reaching their roots through the extensive use of non-permeable paving materials;
- aerial pollutants;
- phytotoxicity from road de-icing salts;
- a lack of root development due to soil compaction or containment within limited soil volumes;
- poor nutrient availability through a lack of natural nutrient cycling; and
- root damage from tunnelling and trenching activities as new cables and pipes are lined out under pavements and roadways.

Species that form large specimens are particularly under threat as perceptions associated with a lack of city-centre space mean that they may not be selected during the planning of inner city redevelopment, with smaller stature species selected in preference. Large and older specimens, however, are thought to provide more benefits (Fig. 5.17). Well maintained trees in appropriate locations may live for 100 years or more, but life expectancy of many street trees post-establishment is thought to be only 19–28 years (Roman and Scatena, 2011), with failure at the establishment phase perhaps reducing this mean figure to as little as 12 years in some localities. Diversity in city trees is also a concern, especially in the context of promoting resilience against future climatic changes and biotic factors. In central and northern Europe, at least 250 landscape plant species are thought to be used in city locations, but actually less than five genera may account for 50–70% of all street trees in many of these cities. Similarly, surveys in Argentina demonstrated that five species alone accounted for 86% of city tree stocks. Paradoxically, in temperate climates the urban heat island effect (where city centres can be up to 11°C warmer than surrounding rural districts) can increase the range of species grown, as well as encouraging better rates of growth and earlier flowering/fruiting. The lack of late season frost is a bonus for species such as *Magnolia*, where flower blooms avoid being blighted by freezing injury, as might occur more commonly in colder, rural localities.

The threats to city trees can be alleviated significantly by careful planning, design and management of urban green spaces. Integrating green space within the main planning process allows space to be created for appropriate tree species and these components to be linked to other key objectives on, for example, flood mitigation, noise abatement, thermal comfort and biodiversity. Such integrated approaches require dialogue between different disciplines and specialisms, with environmental horticulturalists, landscape architects and

Fig 5.16. Urban forests soften the geometric outlines of city built-infrastructure and help promote a positive sense of place.

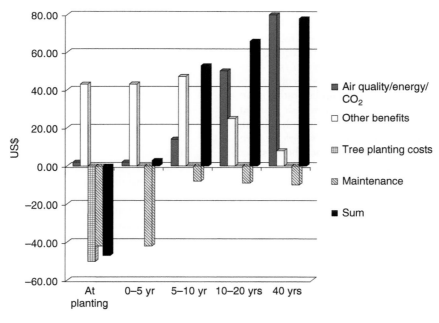

Fig 5.17. The beneficial effects of trees tend to increase with age, with larger older specimens delivering greater ecosystem services. Fiscal values per tree are calculated in relation to the annual benefits derived in terms of improving air quality, building energy reduction or CO_2 sequestration, other indirect benefits (e.g. increase in property value) and the costs associated with planting and maintenance. The 'sum' denotes the net average annual fiscal loss or gain associated with the tree. (Modified from Armour *et al.*, 2012.)

arboriculturalists working alongside planners, engineers, architects, community groups, utility companies and others with an interest in the urban form. Ideally, this takes place from an early stage in the development process.

Management of the tree stock is dependent on professionally trained tree officers and arboriculturalists. Their role is to inspect and identify trees for any early indications of pathogens or pest problems, improve aesthetics through appropriate tree choice and structural pruning, and ensure that the urban tree stock remains safe to the public and nearby properties. This will include implementing selected tree limb removal or aiding tree crown stabilization through bracing techniques.

Traditional expertise in these matters is being augmented by newer technologies, such as GIS systems and aerial photography. Imaging systems can aid estimations of urban forestry canopy cover and even species composition, with data being downloaded to centralized databases. Similarly, evaluations of city trees now employ systems that help integrate information from a variety of sources, and enable tree officers to provide an aesthetic or economic value to the trees within their jurisdiction. Such systems vary in their complexity. The Helliwell system is entirely based on expert judgement, focuses on visual amenity value and has very little requirement for field data to be collected. The Capital Asset Value for Amenity Trees (CAVAT) process identifies the wider benefits of trees to communities rather than purely visual amenity, but does not provide detailed benefit and cost requirements. On the other hand, i-Tree requires data collected from a sample or even a complete inventory of the neighbourhood street trees and interlinks this to information relating to the local community and existing management procedures (e.g. current management costs, local population size, energy use) before providing cost/benefit data outputs.

In tandem with developing these highly technical approaches to understanding the tree population, effective community engagement is paramount too. Citizens who identify and care for their neighbourhood trees are much more likely to engage with preservation of existing stocks and support replacement strategies and new additional plantings. Indeed, such community groups have a significant role in aspects such as species choice and planting efforts, and underpin initiatives such as the Community Forest Schemes in the UK.

Conclusions

- Urban form is strongly influenced by the presence of woody plants, with scale and context dictating the types of plants used. Trees provide a strong structural background to many avenues, parks and gardens, whereas shrubs and climbers are effective as screens or to promote visual interest at a smaller scale. These plant types contribute significantly to ecosystem service delivery within the cityscape.
- The diversity of woody plant species in the urban environment is greatest in private gardens, with diversity decreasing with more civic-based planting schemes, and in those landscapes within the public realm. Some cities are reliant on a limited range of genotypes for their urban tree stock numbers, and numbers of individual trees per citizen vary radically.
- Woody plants are propagated using a variety of techniques, but tree propagation is dominated by seed production or grafting – field-budding being particularly prevalent for popular species such as *Malus* and *Prunus*. Clonal production of shrubs tends to be dominated by propagation from cuttings.
- The production of ornamental varieties of woody landscape plants supports a large and diverse nursery trade industry, with scheduled production to meet market demands. Large numbers of 'hardy' woody plants will start life within protected or semi-protected facilities and will be grown under intensive management regimes before finally being planted in the landscape. This has implications for the management of resources such as water, energy, nutrients, pesticides and substrates such as peat. Research into the sustainable production of such plants is a high priority.
- Location within the landscape will determine species choice following the rule 'right plant, right place'. Plant selection should be based on: final size; aspect and exposure; temperature range; soil type, structure and pH; moisture availability; predisposition to and prevalence of significant pests and pathogens; aesthetic merit; and the overall objectives of the planting design.
- Establishment is a critical phase in the lifespan of woody plants, with the greatest plant failures occurring at this post-planting stage. Key factors to account for include provision of adequate water and reducing competition from

- weeds until plants become fully established in the parent soil.
- Urban soils are particularly challenging to plants, as those around built structures are frequently heavily compacted and contain physical and chemical contaminants. They are typified by high bulk densities, limited organic matter and poor structure that impedes movement of water and air. Such soils need amelioration, or the construction of special root zone features, such as spacious tree planting-pits to help ensure the longevity of the landscape plants.
- Weed, pest and pathogen control are important in ensuring survival of woody plants, but in addition their presence/absence is linked to the fundamental component of aesthetic quality, although this can depend on context (being generally less tolerated in highly formalized or designed plantings). Increasingly, mulches are employed to help combat weed pressure, with reduced emphasis on chemical herbicides. Biocontrol measures can be effective in dealing with pests. Vigilance is also part of the armoury though, for example, to ensure serious pathogens do not spread from one area to another.
- 'Urban forest' is the term used to denote the tree population and other woody plants found in and around urban conurbations. Urban forestry is tree management within the entire urban area, not only the management of individual trees, and takes account of social, economic as well as environmental factors associated with trees.

References

Anon. (1992) *BS 3936-1: Nursery Stock. Specification for Trees and Shrubs*. British Standards Institution (BSI), London.

Apostol, K.G., Jacobs, D.F. and Dumroese, R.K. (2009) Root desiccation and drought stress responses of bareroot *Quercus rubra* seedlings treated with a hydrophilic polymer root dip. *Plant and Soil* 315, 229–240.

Armour, T., Job, M. and Canavan, R. (2012) *The Benefits of Large Species Trees in Urban Landscapes: a Costing, Design and Management Guide* (C712). CIRIA, London.

Arnold, M.A., McDonald, G.V., Bryan, D.L., Denny G.C., Watson W.T. and Lombardini L. (2007) Below-grade planting adversely affects survival and growth of tree species from five different families. *Arboriculture and Urban Forestry* 33, 64–69.

Bainard, L.D., Klironomos, J.N. and Gordon, A.M. (2011) The mycorrhizal status and colonization of 26 tree species growing in urban and rural environments. *Mycorrhiza* 21, 91–96.

Bellot, J., De Urbina, J.O., Bonet, A. and Sánchez, J.R. (2002) The effects of treeshelters on the growth of *Quercus coccifera* L. seedlings in a semiarid environment. *Forestry* 75, 89–106.

Billeaud, L.A. and Zajicek, J.M. (1989) Influence of mulches on weed control, soil pH, soil nitrogen content, and growth of *Ligustrum japonicum*. *Journal of Environmental Horticulture* 7, 155–157.

Cameron, R., Harrison-Murray, R., Fordham, M., Wilkinson, S., Davies, W., Atkinson, C. and Else, M. (2008) Regulated irrigation of woody ornamentals to improve plant quality and precondition against drought stress. *Annals of Applied Biology* 153, 49–61.

Cameron, R.W.F., Harrison-Murray, R.S., van Campfort, K., Kesters, K. and Knight, L.J. (2001) The influence of branches and leaf area on rooting and development of *Cotinus coggygria* cv. Royal Purple cuttings. *Annals of Applied Biology* 139, 155–164.

Cerrillo, R.M.N., Fragueiro, B., Ceaceros, C., del Campo, A. and de Prado, R. (2005) Establishment of *Quercus ilex* L. subsp. *ballota* [Desf.] Samp. using different weed control strategies in southern Spain. *Ecological Engineering* 25, 332–342.

Chalker-Scott, L. (2007) Impact of mulches on landscape plants and the environment. A review. *Journal of Environmental Horticulture* 25, 239.

Davies, R.J. (1985) The importance of weed control and the use of tree shelters for establishing broadleaved trees on grass-dominated sites in England. *Forestry* 58, 167–180.

Davis, M.A., Wrage, K.J., Reich, P.B., Tjoelker, M.G., Schaeffer, T. and Muermann, C. (1999) Survival, growth, and photosynthesis of tree seedlings competing with herbaceous vegetation along a water-light-nitrogen gradient. *Plant Ecology* 145, 341–350.

Fini, A., Frangi, P., Amoroso, G., Piatti, R., Faoro, M., Bellasio, C. and Ferrini, F. (2011) Effect of controlled inoculation with specific mycorrhizal fungi from the urban environment on growth and physiology of containerized shade tree species growing under different water regimes. *Mycorrhiza* 21, 703–719.

Gilman, E.F. (2001) Effect of nursery production method, irrigation, and inoculation with mycorrhizae-forming fungi on establishment of *Quercus virginiana*. *Journal of Arboriculture* 27, 30–39.

Gilman, E.F., Grabosky, J., Stodola, A. and Marshall, M.D. (2003) Irrigation and container type impact red maple (*Acer rubrum* L.) 5 years after landscape planting. *Journal of Arboriculture* 29, 231–236.

Harrison-Murray, R.S. Knight, L. and Howard B. (1998) *Strategies for Liner Production Designed to Achieve High Quality Container Plants*. HDC HNS55. Horticultural Development Council, UK.

Heiskanen, J. (1995) Physical properties of two-component growth media based on *Sphagnum* peat and their implications for plant-available water and aeration. *Plant and Soil* 172, 45–54.

Hermann, R.K. (1967) Seasonal variation in sensitivity of Douglas-fir seedlings to exposure of roots. *Forest Science* 13, 140–149.

Jackson, M.B. and Ricard, B. (2003) Physiology, biochemistry and molecular biology of plant root systems subjected to flooding of the soil. In: *Root Ecology*. Springer, Berlin, Heidelberg, pp. 193–213.

Jasper, D.A., Abbott, L.K. and Robson, A.D. (1991) The effect of soil disturbance on vesicular–arbuscular mycorrhizal fungi in soils from different vegetation types. *New Phytologist* 118, 471–476.

Jim, C.Y. (1998) Urban soil characteristics and limitations for landscape planting in Hong Kong. *Landscape and Urban Planning* 40, 235–249.

Johnston, M. and Hirons, A. (2014) Urban trees. In: Dixon, G.R. and Aldous, D. (eds) *Horticulture: Plants for People and Places*, Volume 2. Springer, Netherlands, pp. 693–711.

Kamaluddin, M. and Zwiazek, J.J. (2001) Metabolic inhibition of root water flow in red-osier dogwood (*Cornus stolonifera*) seedlings. *Journal of Experimental Botany* 52, 739–745.

King, C.M., Robinson, J.S. and Cameron, R.W. (2012) Flooding tolerance in four 'Garrigue' landscape plants: implications for their future use in the urban landscapes of north-west Europe? *Landscape and Urban Planning* 107, 100–110.

Klironomos, J.N. and Allen, M.F. (1995) UV-B-mediated changes on below-ground communities associated with the roots of *Acer saccharum*. *Functional Ecology* 9, 923–930.

Kristoffersen, P., Rask, A.M., Grundy, A.C., Franzen, I., Kempenaar, C., Raisio, J., Schroeder, H., Spijker, J., Verschwele, A. and Zarina, L. (2008) A review of pesticide policies and regulations for urban amenity areas in seven European countries. *Weed Research* 48, 201–214.

Martin, T.P., Harris, J.R., Eaton, G.K. and Miller O.K. (2003) The efficacy of ectomycorrhizal colonization of pin and scarlet oak in nursery production. *Journal of Environmental Horticulture* 21, 45–50.

Montague, T., Kjelgren, R., Allen, R. and Wester, D. (2004) Water loss estimates for five recently transplanted landscape tree species in a semi-arid climate. *Journal of Environmental Horticulture* 22, 189.

Neary, D.G., Rockwood, D.L., Comerford, N.B., Swindel, B.F. and Cooksey, T.E. (1990) Importance of weed control, fertilization, irrigation, and genetics in slash and loblolly pine early growth on poorly drained spodosols. *Forest Ecology and Management* 30, 271–281.

Nowak, D.J., Rowntree, R.A., McPherson, E.G., Sisinni, S.M., Kerkmann, E.R. and Stevens, J.C. (1996) Measuring and analyzing urban tree cover. *Landscape and Urban Planning* 36, 49–57.

Pauleit, S., Jones, N., Garcia-Martin, G., Garcia-Valdecantos, J.L., Rivière, L.M., Vidal-Beaudet, L. and Randrup, T.B. (2002) Tree establishment practice in towns and cities. Results from a European survey. *Urban Forestry and Urban Greening* 1, 83–96.

Quigley, M.F. (2004) Street trees and rural conspecifics: Will long-lived trees reach full size in urban conditions? *Urban Ecosystems* 7, 29–39.

Roman, L.A. and Scatena, F.N. (2011) Street tree survival rates: Meta-analysis of previous studies and application to a field survey in Philadelphia, PA, USA. *Urban Forestry and Urban Greening* 10, 269–274.

Saranga, J. and Cameron, R. (2007) Adventitious root formation in *Anacardium occidentale* L. in response to phytohormones and removal of roots. *Scientia Horticulturae* 111, 164–172.

Skroch, W.A., Powel, M.A., Bilderback, T.E. and Henry P.H. (1992) Mulches: durability, aesthetic value, weed control and temperature. *Journal of Environmental Horticulture* 10, 43–45.

Smith, K.D., May, P.B. and Moore, G.M. (2001) The influence of compaction and soil strength on the establishment of four Australian landscape trees. *Journal of Arboriculture* 27, 1–7.

Specht, A. and Harvey-Jones, J. (2000) Improving water delivery to the roots of recently transplanted seedling trees: the use of hydrogels to reduce leaf and hasten root establishment. *Forest Research* 1, 117–123.

Stabler, L.B., Martin, C.A. and Stutz, J.C. (2001) Effect of urban expansion on arbuscular mycorrhizal fungal mediation of landscape tree growth. *Journal of Arboriculture* 27, 193–202.

Struve, D.K., Burchfield, L. and Maupin, C. (2000) Survival and growth of transplanted large- and small-caliper red oaks. *Journal of Arboriculture* 26, 162–169.

Thetford, M., Miller, D., Smith, K. and Schneider, M. (2005) Container size and planting zone influence on transplant survival and growth of two coastal plants. *HortTechnology* 15, 554–559.

Tournaire-Roux, C., Sutka, M., Javot, H., Gout, E., Gerbeau, P., Luu, D.T. and Maurel, C. (2003) Cytosolic pH regulates root water transport during anoxic stress through gating of aquaporins. *Nature* 425, 393–397.

Watson, G.W. and Himelick, E.B. (1997) *Principles and Practice of Planting Trees and Shrubs*. International Society of Arboriculture, Champaigne, Illinois.

Watson, W.T. (2005) Influence of tree size on transplant establishment and growth. *HortTechnology* 15, 118–122.

Weber, G and Claus, M. (2000) The influence of chemical soil factors on the development of VA mycorrhizas of ash (*Fraxinus excelsior* L.) and sycamore (*Acer pseudoplatanus* L.) in pot experiments. *Journal of Plant Nutrition and Soil Science* 163, 609–616.

Wells, C., Townsend, K., Caldwell, J., Ham, D., Smiley, E.T. and Sherwood, M. (2006) Effects of planting depth on landscape tree survival and girdling root formation. *Arboriculture and Urban Forestry* 32, 305.

6 Herbaceous Plants and Geophytes

James D. Hitchmough

> **Key Questions**
> - What are the distinguishing features of both herbaceous plants and geophytes?
> - What are the different spatial arrangements that are used to plant herbaceous specimens in contemporary landscapes?
> - What factors contribute to plant robustness or resilience, and why are these important when selecting herbaceous and geophyte genotypes?
> - Why are many herbaceous plants susceptible to mollusc damage particularly? Where are these pressures the greatest?
> - Describe the different plant/pot sizes that might be used in a contracted planting of herbaceous species.
> - What are ideal soil conditions for planting herbaceous species and geophytes?
> - Why is it important to control weeds in herbaceous plant communities and what are the mechanisms that might be used?

6.1 Introduction

In the last decade in particular herbaceous perennials and geophytes have become more important in public designed landscapes, as part of a latent public appetite for more colour and seasonal change that is being articulated more forcefully. These life forms do, however, present significant challenges in public landscapes; many species are dormant for part of the year and, unlike shrubs, are often subject to severe damage from slugs and snails. In addition, because they are potentially capable of (particularly in herbaceous plants) rapid increase in canopy volume within a single growing season, less well-fitted species can quickly be lost from plantings through competition. Integrating these ideas into the design and management of plantings of herbaceous plants and geophytes is the focus of this chapter.

6.2 Distinguishing between Herbaceous Plants and Geophytes

Both of these major life forms share two important characteristics: (i) They do not manufacture lignin; they are not woody. (ii) Their growth is therefore based on a cycle of active shoot development, followed by the abscission of the above ground parts as the plants enter a period of dormancy or at least quiescence. This cycle is made possible by an underground or surface perennation organ that stores the carbohydrates made in the previous growth cycle as fuel to initiate the next cycle. Beyond these two characteristics, it becomes increasingly difficult to find commonality.

With herbaceous plants a number of different life forms have evolved based on the typical duration of longevity. This chapter deals largely with perennials, i.e. those herbaceous plants that have at least three growth cycles and in most cases many more before they die. There are, however, also annual herbaceous plants in which there is only one cycle of growth, culminating in massive seed production (see Chapter 8). There are also biennial herbaceous plants (e.g. some *Digitalis* species – foxgloves), as well as monocarpic herbaceous plants that share the traits of biennials, but in which the growth period prior to the final and fatal flowering is stretched out from 4 to 7 years. Monocarpism is perhaps most obviously found in herbaceous plants in the Sino-Himalayan region, e.g. in *Meconopsis* (Himalayan poppy).

The herbaceous life form is found in dicots, monocots and pteridophytes (ferns). In Western European horticultural practice most of the plants referred to as herbaceous plants are typically dicots. Ecologists refer to dicot herbaceous plants as 'forbs'; this is a convenient term for distinguishing between, for example, non-grassy, and grass-like herbaceous plants, such as sedges and true grasses. The latter two groups are often referred to by ecologists as graminoids, again a useful cover-all term.

The term 'geophyte' is used here in this text rather than the rather more common horticultural term 'bulb' or 'bulbous', to get around the problem that many of the plants generically referred to as bulbs, technically are not. Bulbs have a very distinctive perennation structure; familiar to anyone who has ever chopped up that most ubiquitous of bulbs, the culinary onion. The bulb consists of concentric rings of compressed leaves. This growth form is the norm in the Alliaceae (e.g. *Allium*), and the Amaryllidaceae (e.g. *Crinum, Cyrtanthus, Fritillaria, Narcissus, Scilla*). Corms, which unlike bulbs have no obvious internal structure, are essentially a bag of carbohydrates, with a root meristem at the base and a shoot meristem further up the corm. Corms are the norm in geophytic members of the Iridaceae (e.g. *Crocus, Gladiolus, Iris, Ixia, Watsonia*). Finally there are tubers, of which the culinary potato is the most familiar. Again they are a bag of carbohydrate with growth points (roots and shoots) distributed across the exterior, although in some tubers (e.g. *Cyclamen*), the location of the shoot and root meristem is clearly defined as top or bottom. Tubers typically differ from corms in being asymmetric, often fleshy and sometimes creeping; they are found in *Anemone, Cyclamen* and *Tropaeolum*, amongst others.

In an ordered world it would be very easy to distinguish between herbaceous perennials and geophytes: herbaceous perennials would have a growth cycle of 6–8 months and geophytes would have a shorter growth cycle of 3–4 months. In the real world there are what most people would consider herbaceous perennials, such as *Incarvillea*, which have 4-month growth cycles; and there are corm-forming evergreen geophytes such as *Dierama, Dietes*, certain *Moraea* species and *Watsonia*, some of which are in active growth for 6–8 months. Some botanical authors place *Kniphofia* (Asphodeliaceae) in the geophyte camp, whereas most horticulturalists would be much more comfortable seeing these species as herbaceous perennials.

In essence, all of these classifications are a human attempt to create systematic order on part of the natural world in which order is rather erratically distributed. What is more important is developing an understanding of how these different life-forms interact with cultivation practice.

6.3 Patterns of Growth in Herbaceous Plants and Geophytes

Herbaceous plants

Due to the fact that herbaceous plants and geophytes do not make wood, the patterns of growth by which canopy expansion takes place is much more varied than in trees and shrubs. Understanding these patterns is very important when using herbaceous plants, as they determine the outcome of competition between different species in plantings, and in particular the capacity of one species to eliminate another. The following are frequently observed:

1. Gradual shoot expansion but slow canopy expansion.

In this growth form, the subterranean or soil surface shoot meristems expand slowly to form a concentric ring. Annual additions to the radius of the shoot ring may be less than a centimetre in very slow-growing, small species, for example *Scabiosa columbaria, Geum coccineum* or *Pulsatilla vulgaris*. As a result, the shoot tissues produced from these buds only expand approximately the same limited distance each year; total canopy extension is very gradual and predictable, with the plants forming neat clumps. Mature plants of *P. vulgaris* will, for example, rarely occupy a volume of spaces wider than 400–500 mm. Where species with this growth habit are very long lived, as is the case for some *Hosta* species, this leads to a very large increase in the space occupied by the foliage between planting and maturity; the canopy radius (but not the ring of subterranean shoots) of *H. sieboldiana* reaches 1.2 m across or more at maturity.

There are also qualitative differences in how canopy expansion proceeds in species with this growth strategy. In some species buds that continue to produce shoots are retained in the centre of the clump, as for example in the case of *Epimedium* or *Omphalodes cappadocica* maintaining a dense clump in perpetuity. In other species the buds in the centre of the clump are abandoned, resulting in the development of doughnut-like canopies with an empty centre.

2. Gradual shoot expansion but seasonally rapid canopy extension.

This growth form is common in many *Geranium* species, e.g. *G. psilostemon* and *G. wallichianum*, *Oenothera macrocarpa*, *Persicaria amplexicaule*, and *Potentilla* hybrids, amongst others. It is relatively common in the herbaceous plants popular in traditional border plantings, where a clump of each species was often separated from its neighbours by a large space that was subsequently filled by the expanding leaf canopy as the plants grew by mid-summer. This growth habit does, however, in mixed plantings at closer spaces lead to the elimination of slower, shade-sensitive species by taller herbaceous plants with this growth trait.

3. Endless spread by below-ground or above-ground shoots.

This growth pattern varies in terms of the extent of spread possible. At the conservative end of the spectrum there are sheet-forming plants such as *Ajuga reptans*, *Phlox stolonifera*, *Tiarella cordifolia* (above-ground rooting stems) and *Ajuga genevensis* (below-ground rooting stems). These species tend to form in time open, clonal patches rather than a 'smooth' continuous unbroken carpet of foliage, although because they do root as they grow they can theoretically colonize large areas. The next step up in scale comes with species such as *Geranium macrorrhizum*, which slowly forms long permanent closed carpets. At the extreme end of the sheet-forming spectrum are *G. procurrens* and *Lamiastrum galeobdolon*, which expand their canopies by over 1 m each year. This growth pattern is also found in species with tall leafy stems, e.g. *Lysimachia clethroides*, *Euphorbia griffithii*, and *Macleaya cordata*. These species form tall clonal patches and eliminate everything that is not highly shade tolerant or summer dormant as they spread.

4. Slow canopy expansion but with new individuals established by self-sowing.

Many of the species that use most of their photosynthates to produce large amounts of biomass (growth forms 2 and 3 above) have relatively limited seed production, and only occasional self-sown seedlings. Because plants with slow canopy expansion are often associated with unproductive habitats in which there is incomplete vegetation cover, these species tend to have higher levels of seed production and self-sowing because there are opportunities for this strategy. Species that have a higher than average capacity to do this include: *Berkheya purpurea*, *Dianthus carthusianorum*, *Dracocephalum rupestre*, *Origanum vulgare*, *Penstemon barbatus*, *Polemonium reptans*, *Scabiosa columbaria* and *S. ochroleuca*.

Geophytes

The scale of difference in terms of canopy expansion tends to be reduced in geophytes, but it is still possible to identify important differences:

1. Solitary individuals; little or no vegetative expansion.

This is the case in species that do not readily produce corm, bulb or tuber offsets. With the exception of some tulips, plants with this habit tend to be uncommon in mainstream environmental horticulture use, but potentially are important in amateur horticultural circles. Some examples are *Brunsvigia*, *Eremurus*, *Iris latifolia*, *I. reticulata* and *Romulea* spp.

2. Gradual but slow expansion through vegetative offsets.

This clump-forming growth habit occurs because of offset formation, and is common in some species within *Narcissus*, *Eucomis*, *Gladiolus*, *Galtonia*, *Nerine* and *Watsonia*.

3. Solitary individuals with slow offset formation with new individuals established by self-sowing.

Where these species are relatively large with abundant foliage this can be a problem in plantings where there is insufficient maintenance to remove seed heads prior to seed dispersal. For example, where self-seeding is not checked, plants such as *Allium × hollandicum* eventually produce 'Allium forests' leading to the elimination of low-growing herbaceous plants. This habit is also highly developed in *Galtonia candicans* and *Tulipa sprengeri*. It is also found in much smaller-growing species such as many *Chionodoxa*, *Crocus* and *Cyclamen*, and *Narcissus* such as *N. bulbocodium*.

6.4 The Role of Herbaceous Plants and Geophytes in Designed Landscapes

Whilst many of these plants make a substantial contribution to planting schemes through their foliage, the main reason for using them is for their flowers. It is impossible to know precisely when herbaceous plants and geophytes were first cultivated in gardens, although wall paintings from the classical civilizations suggest that some of the more spectacular species, such as *Lilium candidum*, have a very long history of cultivation. It seems highly likely that initially the range of flowers used were largely restricted to native and near-native species, as in the case of *L. candidum* in the Mediterranean, but some species were clearly being moved substantial distances along trade routes, before 1000 CE, for example tulips from central Asia. Geophtyes are easy to move around (when done at the right time of year) because of their desiccation-resistant storage organ. Many of the herbaceous plants depicted in tapestries in the Middle Ages actually exist in the real world, as opposed to solely in the artists' imaginations. Some of these species appear to have been planted in gardens during this period. The range of plants expands rapidly from the Renaissance onwards as world trade increases. The first tulips to arrive in Vienna via the Ottoman Empire in 1512 (Culver-Campbell, 2001) caused a horticultural 'storm'.

There is evidence that many herbaceous plants and geophytes were planted into close cut lawns – a practice known as enamelling, sometimes in parterre-like compartments; a sort of 'cultural meadow' (Woudstra and Hitchmough, 2000). This style is also represented from 1800 in central European gardens such as Průhonice in Prague. Flower gardens are also presumed to have existed, however, given their relatively ephemeral nature, evidence is limited. Research by Laird (1999) has shown that even during the height of the English Landscape Movement, flower gardens continued to be in vogue, often at the rear of grand houses, and this tradition continues to develop from *c*.1820, as the herbaceous border, often set within a formal architectural structure. In the latter part of the 19th century, however, there is also a more naturalistic school of use, pioneered by writers and gardeners such as William Robinson (who wrote *The Wild Garden* published in 1870) and his contemporaries in Germany (Woudstra and Hitchmough, 2000). From then on the 'cottage garden', in which herbaceous plants are used in naturalistic arrangements, becomes popular. Very few of these new ideas filter through to public green space, where herbaceous planting is normally restricted to herbaceous borders and geophytes to spring formal bedding or the use of *Crocus* and *Narcissus* (daffodils) in mown grass. Probably the most dramatic shift in the use of herbaceous perennials in public landscapes occurs in Germany from 1950 onwards, through a more 'ecologically-inspired' view of planting herbaceous perennials and geophytes to form naturalistic plant communities (Hansen and Stahl, 1994). Through the institution of Garden Festivals, these plant communities were developed to a high level of sophistication, using a mixture of native and non-native species, with semi-natural vegetation types such as the steppe (a dry grassland extending through Eurasia) and prairie (a summer growing North American equivalent) as key inspirations.

These ideas are then re-fashioned through plantsmen/plantswomen designers such as Oudolf (Oudolf and Kingsbury, 2014), and applied to prestige urban landscape projects on a grand scale, as for example in the High Line in New York. Together with the ideas and practices from the Sheffield School of Dunnett and Hitchmough (Dunnett and Hitchmough, 2004) this leads to a situation in the second decade of the 21st century where herbaceous planting becomes highly desirable at least as an aspiration in public planting, although in practice it remains relatively uncommon.

6.5 Contemporary Options for Herbaceous and Geophyte Planting in Public Landscapes

Nearly all the herbaceous planting now practised in public landscapes embodies two characteristics that were not much considered in the UK prior to 2000; the application of (i) ecological and (ii) functional ideas.

These ecological ideas often address plant selection (discussed later in this chapter) in relation to the environment of the planting site, whilst functional ideas pertain to, for example, how the plants might be used and managed to reduce weed invasion and to fulfil a role beyond the purely aesthetic. Frequent roles are supporting biodiversity and managing rainwater runoff (Dunnett and Clayden, 2007). Incorporating these ideas into design results in various forms of planting. The planting styles that arise from this can be classified in different ways, for example in terms of spatial arrangement in the three dimensions of space, and then in terms of the locations and/or conditions in which planting might be used. With the exception of evergreen geophytes such as *Dierama*, the extended dormancy period of most geophytes means that within sustainable forms of plant use (as opposed to ephemeral planting displays), they normally have to be planted in conjunction with plants that retain a structure and presence during the geophyte dormancy period. Typically this might involve seasonally unmown grass and meadows, herbaceous planting or planting beneath deciduous shrubs and trees. These approaches are discussed later in this chapter.

6.6 Spatial Arrangements for Herbaceous Planting

Block-based planting

This is the standard 19th century style of planting that formed the basis of most planting (herbaceous and shrub) throughout the subsequent 20th century. Plants are placed in groups (blocks) of the same species or cultivar to form island monocultures which are then juxtaposed with other species/cultivars in adjacent blocks. The size of block can vary from the small, say five to ten plants, up to hundreds or thousands depending on the scale of the landscape. This approach reached its apotheosis in the USA in the work of James Van Sweden and Wolfgang Oehme in the latter part of the 20th century (Oehme and Van Sweden, 1991). On a much smaller scale it was also the style adapted in the early work of Piet Oudolf (Oudolf and Gerritsen, 2013). The advantage of this approach is that it is simple to conceptualize, making it easy to plant; each species can be set at the spacing that best suits, i.e. perhaps highly vigorous species at 600 mm centres, very slow species at 250 mm and so on. It also looks highly designed and intentional, so it provides a strong sense of 'cues to care', that is, people respond more positively to it precisely because they think it is actively managed and maintained (rather than 'unkempt' nature). It is relatively easy to maintain, as anything that invades the monoculture is clearly either a weed or a volunteer from elsewhere in the planting and can be readily identified and removed. Research in the Netherlands has investigated plant spacing and the effect this has on establishment and subsequent resistance to weed invasion. Planting small plants (plugs or 9 cm pots) at high densities, e.g. at 200 mm apart, produces a very dense sward that is potentially very resistant to weed invasion. Using the right plant selection, and implementing sensitive maintenance, such plantings can look good for 10–15 years or even longer; for example the Oudolf designs at Pensthorpe Natural Park in Norfolk, UK.

The negatives of this planting style are largely aesthetic; unless the blocks repeat they potentially look uncomfortable, because they have a clear sense of grain or direction that may be at odds with surrounding spatial form. Each block can only look really attractive when in flower, so only one flowering event can be generated from each m² of planting because of the monocultural basis. In structural terms this form of planting is simple; there is only ever one layer per unit area (Fig. 6.1). The scientific

Fig. 6.1. Block planting and the relatively simple layer structure associated with each species block.

evidence suggests that this style of planting may not be as useful to invertebrates as habitat with a more complex structural form. This form of planting also suffers on sites when there is a gradation of soil and other environmental factors across the blocks, resulting in some plants or patches of plants within the block performing poorly or even dying. Because there is only one species present, block plantings have the lowest capacity to respond positively to changing site conditions.

Repeating plantings of small blocks or individuals

This term covers a diversity of approaches, in which the main difference is that the number of plants forming each group is reduced, allowing the creation of vegetation that is more complex per unit area than traditional block planting.

The number of plants forming each block

Within this genre plant number per 'block' can range from, for example, 5 to 1. In essence, the smaller block size means that there is more opportunity for a greater range of species per unit area and the block form will be much less apparent. The smaller block allows more repeats to be used, further breaking down the grain of the planting from that which is evident in large blocks.

When the block is typically composed of only one plant with potentially a different species or cultivar at each planting location, this has often been referred to as matrix planting, although there is little consistency in the application of these terms. Because it is essentially impossible to show individual plants on most planting plans, matrix plantings are typically designed as a planting mix with only the key visual dominants (the biggest plants typically) shown on a plan, and then often only to inform the judgement of the planters about intended spatial distribution. These mixes are planted at a uniform spacing, typically around nine plants per m^2, although this varies depending on soil productivity, with density being decreased as soil productivity increases and vice versa. The key design decisions in creating these mixes are: how many species and which species are to be used to create the plant community and how many of the typical nine planting positions in each m^2 should be allocated to each species? The judgements in making these decisions are essentially exactly the same as for any other type of planting. Key factors are the visual and physical degree of plant diversity required, the surface topography of the resulting planting, i.e. how tall, how low, how bumpy and the duration of the flowering season required.

The number of layer structures to be built in

Block planting is by definition a mono-layer within each block, and the larger the block the more this is the case. In semi-natural 'wild' vegetation mono-layers are uncommon except on very productive soils where one or perhaps two large, fast-growing species dominate and inhibit the establishment of additional species. On less productive sites more mixed communities are the norm because there are always a diversity of species that can invade; some are short ones which will be shade tolerant, whereas many of the tall species will require full sun.

What are the advantages and disadvantages of building in more than one layer above each m^2 of ground? Most plantings using the matrix approach effectively have two or three layers. The lower or ground layer in a two-layer system will generally consist of a spring, vernal growing layer of species that are associated with woodlands and woodland edges. These plants start to grow very early in the year, and tolerate being shaded from late spring/early summer onwards by the leaf canopies of taller, summer and autumn flowering herbaceous species. Typical examples would be *Primula vulgaris* and *P. elatior*, *Ajuga reptans*, *Tiarella cordifolia* and *T. wherryi*, *Omphalodes verna* and *Saxifraga stolonifera*. By having this structure the ground is covered by leaves (many of these species are evergreen or at least semi-evergreen) in winter, hence reducing winter weed colonization; they also look more attractive in winter than the frequently winter-dormant taller species and produce a dramatic spring/early summer flowering display in their own right. They then disappear under the emerging shoots of the later flowering species, the first of which flower in mid-summer and continue through to autumn, giving a 6–8 month period of flowering. The key factor for the summer and autumn flowering species is that they must be unattractive to slug and snail grazing as the shoots push through, as these herbivores will be more abundant where a vernal layer is present rather than absent. Species that most tolerate this elevated spring grazing pressure include: *Aster novae-angliae*, *A. umbellatus*, *Aconitum* spp., *Euphorbia* spp., most grasses,

Leucanthemum maximum, *L. serotina*, *Lythrum salicaria*, *Sanguisorba* spp., *Veronica longifolia* and *Veronicastrum*. Because the understorey species are deeply shade tolerant and often have some capacity to become summer dormant, the presence of complementary species that subsequently form the upper canopy layer and thus intercept much of the light, do not pose a problem for this system.

Designing and creating matrix-type plantings is clearly more challenging than block monocultures, and this system also raises more questions for maintenance, particularly with identification of what is a desirable planted specimen and what is a weed that should be removed. What are the advantages of this planting style that might make these additional challenges worth considering? In aesthetic terms matrix type plantings provide the capacity to produce a succession of flowering events from each single m^2 of planting, perhaps 5–6 months rather than 2 weeks. It also allows spring flowering species to be used, many of which are very unattractive in summer and autumn, with the taller emergent species visually masking their decline. Because of the repeating pattern, when a species present at a density of say 1–2 plants per m^2 comes into flower, the effect generated is of the whole area of the planting turning this colour. Matrix-like plantings potentially generate much greater levels of drama for longer than with block planting in which only a small area is typically flowering at any point in time. On the functional level, these matrix-like plantings have much greater capacity to fix themselves when soil or other environmental gradients are inherent (as they always are) in a planting. Whilst the aim of planting design is to try to get the placement of plants 'right' so, for example, the species in question thrive, this is often much more difficult in practice due to the difficulty of knowing just how tolerant a species is of a given set of environment conditions. Matrix plantings are more forgiving in terms of getting these decisions 'wrong'. For example: where plantings run from full sun to under newly planted trees, as the shade generated by the trees increases, the most shade-tolerant species/cultivars in the herbaceous layer will be preferentially advantaged and the more shade-intolerant species disadvantaged. Thus, providing both shade- and sun-tolerant species within the planting design allows the planting mix to respond to changing conditions even in the absence of intelligent maintenance.

The ratio of plants of later flowering species (often the tallest) and shorter summer-flowering species needs to be monitored to assure that the tall, later flowering species do not eliminate the summer-flowering species though shading. For plantings involving nine planting sites per m^2, typically two would be a tall autumn species (but not the same species in each m^2), three would be summer flowering (but not the same species in each m^2), and four would be spring flowering.

Three-layer systems involve the same principles. If the upper two emergent layers are kept relatively sparse, sufficient light will be available to the ground level species that are relatively shade intolerant. For this to work the ratio of low to medium to tall plants needs to be in the order of 10:2:0.5. In addition to the species listed previously, species such as *Ajuga genevensis*, *Geum rivale/coccinea* cultivars, *Heuchera* cultivars and species, *Omphalodes cappadocica* and *Pulsatilla vulgaris* can be used, with the final choice dependent on soil moisture levels and productivity. Emerging species that have basal foliage but naked flowering stems are preferred for the two upper layers, in order to prevent casting too much shade and hence eliminating shade-intolerant species in the lower layers. Species that form clumps are preferred too, as these reduce the likelihood of growing over the crowns of the emergent species in spring, casting shade and leading to their extirpation.

Incorporating geophytes into herbaceous and other plantings

Irrespective of the nature of the plantings that geophytes are to be added to, the universal principles that govern success are that the growth cycle of the geophytes must be sufficiently compatible with the host vegetation to allow them to be able to photosynthesize adequately. Where this does not occur, the geophytes will gradually stop flowering and ultimately decline and disappear. Some combinations that allow this to happen are: winter or spring growing/flowering bulbs in winter deciduous herbaceous planting; the same geophtyes beneath deciduous shrubs and trees; summer growth geophytes in herbaceous planting with a lower foliage canopy than the geophytes. Plantings composed of evergreen herbaceous plants are more challenging unless the geophyte foliage is taller than the canopy of the former. Evergreen trees only work with species adapted to tolerate the low levels of light and moisture stress during their growth period (very few) and evergreen shrubs are generally incompatible with geophytes unless the canopies of the latter admit unusually high amounts of light.

6.7 Plant Selection

Robustness in herbaceous plants and geophytes

In practice, the criteria on which an assessment of plant robustness might be made are typically (i) its capacity to do what was expected of it in terms of growth and flowering, and (ii) its persistence on a site at relatively low levels of care. But why are some plants demonstrably much more robust than others? Whilst it would be elegant to assume that all plants are of equivalent robustness, when presented with their own specific ecological (niche) conditions and when cultivated in gardens and other highly-managed landscapes, some species are clearly better adapted than others. Some plants just refuse to give up no matter how badly treated they are, and this seems to occur across a range of conditions. In many cases this robustness is derived from self-evident aspects of their biology: large, fast-growing but long-lived plants with multiple growth points and/or colonizing rhizomes and leaves and growth points that are unpalatable to slugs and snails. Very small, slow-growing plants with few growth points and high palatability to herbivores are unlikely to be performing well in 10 years' time.

In many cases there is a relationship between soil productivity and longevity in herbaceous plants, with longevity, particularly in short-lived species, declining as soil productivity increases. Species such as *Achillea*, for example, are relatively long-lived in their habitats without recourse to division and other cultural practices. In cultivated soils with much higher nutrient and water levels, however, these species fall apart relatively quickly, presumably because their pre-programmed cycles involved in longevity are completed far more quickly.

Planting species in the 'wrong' habitat often has a very negative effect on robustness, and plant 'fitness' for a particular environment is also extremely important, if sometimes a less obvious factor, in determining robustness. Key elements in this fitness is the ecological range or amplitude of a species in their habitat. This includes many factors, for example tolerance of stress in terms of moisture, light and nutrients. The amplitude of a species in response to these ecological factors depends on a mixture of evolutionary history and chance. Some species demonstrate much broader amplitude to these sorts of ecological factors, i.e. they might grow perfectly well in sun or shade, or from wet to dry soil, than would be expected from their actual habitats in the wild. Other species can only be robust within a much more narrow range of possibilities, for example bog primulas such as *Primula florindae* are extremely robust in cool summer climates when grown in saturated/near-saturated soil (as in the peat-dominated Flow Country of northern Scotland, UK, where this species is now naturalized), but lacks resilience and is short-lived in dry soil. Sometimes these interactions are complex; *Salvia nemorosa* is a continental, dry-summer species that can be robust in maritime wet-summer climates providing herbivory from slugs in spring/summer is controlled. Where this is not the case, this species is eaten out sooner or later where rainfall exceeds a threshold value of rainfall frequency. In geophytes an important factor in robustness is the degree of soil wetness during the dormant periods. Many geophytes evolved in parts of the world with dry summers and the resting organs are often subject to damage from fungal pathogens in summer moist soil. *Tulipa* is an important example of this, although species and cultivars are very variable in their response, see for example Wilford (2006) on persistence in the different wild-occurring *Tulipa* species. Relationships between plant robustness and fitness are shown diagrammatically in Fig 6.2.

Despite the implication in the horticultural literature that fitness operates at the species level, this is rarely the case. Because ecological amplitude is a product of the mosaic of local ecological conditions under which individuals and populations of a species are constantly exposed to natural selection, fitness is defined at this latter level. A compounding factor in this is that within a population in a given site, different individuals, because of within-population genetic variation, may be much fitter in a given set of environmental conditions than others. Studies by the author (unpublished) evaluating hundreds of South African species show that even within poorly fitted populations, there are often a small percentage of individuals (typically <5%) that are demonstrably much better fitted. For every 100 seeds of a species sown there are a handful that will not be killed or damaged by the factors that lead to the failure of the other 95% of the seedlings.

Across the entire natural distribution of a species the very different climatic and edaphic conditions

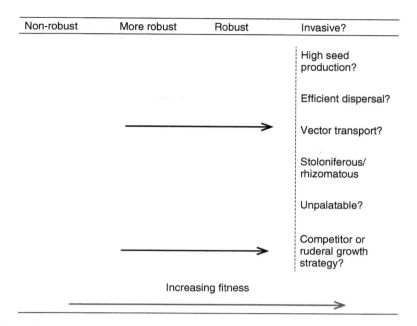

Fig. 6.2. The relationship between increasing plant fitness for a planting site and its behaviour in terms of robustness, i.e. ease of cultivation. To become invasive, plants need more than just high fitness; they need to possess at least some of the traits in the right hand column.

inevitably result in hugely varying levels of intrinsic fitness to a given set of conditions. Currently cultivated commercial populations of *Gladiolus cardinalis* are derived from plants originally collected from 1000–1400 m altitudes in the mountains of western South Africa; however, populations extend to 2000 m, with the latter being much more tolerant of severe winter cold and increasingly less tolerant of summer heat. This particular species is for example almost unable to be grown in the Mediterranean climate of Cape Town, South Africa, even though it is just 50 km from its native mountain habitats. For most horticultural practitioners, outside of the world of alpine plants (where scholarly appreciation of the habitat of the plants in relation to their appearance and behaviour is cherished), there is very limited understanding of how fitness might differ depending on the highly localized geographical, altitudinal and habitat origin of the particular plant population.

The nursery production industry has rarely consciously utilized these processes. It has, however, simply selected the seedlings that grew best, or even just survived, thereby typically increasing 'fitness' in the process. Cultivation typically increases fitness (at least for the urban or garden environment) from one generation to the next from the point of introduction from the wild through these selection processes. This is assuming new, much less-fit genes are not incorporated by intentional or unintentional plant breeding. In some cases fitness is also increased by a change in ploidy levels in seedlings.

The practical outcome of these discussions is that plant selection in terms of the species, cultivars, and other subspecific variants of a species is an incredibly important process in terms of getting the best long-term fitness possible. Plant selection protocols for landscape planting should deal with two key ideas in addition to the standard morphological and phenological questions about flower colour and timing:

1. How close is the habitat type of the planting site in relation to that in the wild?
2. How close climatically is the planting site environment to that in which the plant is derived in the wild?

Habitat types are typically shaped by interactions between two key factors: the degree of soil moisture and the degree of shade (Table 6.1). Steppe plants,

Table 6.1. Fitness is relative rather than absolute and changes according to interaction between overall climate and planting locations within these environments. The table shows how groups of species become more or less fitted when placed in different planting combinations of soil moisture and light.

Condition	Highly fitted	Moderately fitted	Poorly fitted
Wet, shady	Astilboides tabularis Carex pendula Matteuccia struthiopteris Petasites hybridus Primula sieboldii Primula vulgaris	Aconitum carmichaelii Aruncus dioicus Cimicifuga simplex Geranium sylvaticum Geum rivale Primula pulverulenta	Allium christophii Euphorbia rigida Origanum vulgare Penstemon strictus Pulsatilla vulgaris Salvia nemorosa
Wet, intermediate	Aconitum carmichaelii Aruncus dioicus Cimicifuga simplex Geranium sylvaticum Geum rivale Primula pulverulenta	Aster divaricatus Cyclamen hederifolium Geranium macrorrhizum Lathyrus vernus Tellima grandiflora Trachystemon orientalis	Anemone sylvaticum Buphthalmum salicifolium Centaurea scabiosa Geranium pratense Hemerocallis flava Knautia arvensis
Wet, sunny	Camassia leichtlinii Hesperantha coccinea Persicaria bistorta Primula pulverulenta Succisa pratensis Trollius europaeus	Astilboides tabularis Carex pendula Matteuccia struthiopteris Petasites hybridus Primula sieboldii Primula vulgaris	Aster divaricatus Cyclamen hederifolium Geranium macrorrhizum Lathyrus vernus Tellima grandiflora Trachystemon orientalis
Dry, shady	Aster divaricatus Cyclamen hederifolium Geranium macrorrhizum Lathyrus vernus Tellima grandiflora Trachystemon orientalis	Allium christophii Euphorbia rigida Origanum vulgare Penstemon strictus Pulsatilla vulgaris Salvia nemorosa	Camassia leichtlinii Hesperantha coccinea Persicaria bistorta Primula pulverulenta Succisa pratensis Trollius europaeus
Dry, intermediate	Anemone sylvaticum Buphthalmum salicifolium Centaurea scabiosa Geranium pratense Hemerocallis flava Knautia arvensis	Camassia leichtlinii Hesperantha coccinea Persicaria bistorta Primula pulverulenta Succisa pratensis Trollius europaeus	Aconitum carmichaelii Aruncus dioicus Cimicifuga simplex Geranium sylvaticum Geum rivale Primula pulverulenta
Dry, sunny	Allium christophii Euphorbia rigida Origanum vulgare Penstemon strictus Pulsatilla vulgaris Salvia nemorosa	Anemone sylvaticum Buphthalmum salicifolium Centaurea scabiosa Geranium pratense Hemerocallis flava Knautia arvensis	Astilboides tabularis Carex pendula Matteuccia struthiopteris Petasites hybridus Primula sieboldii Primula vulgaris

for example, are typically best adapted to soils that are relatively dry in summer and often in winter too, with a requirement for full sunlight. Prairie plant communities are associated with sunny open sites with often substantial summer rainfall. Woodland edge communities are composed of species that tolerate sun and moderate shade and often low soil moisture levels. Plant selectors need to undertake research as to the conditions in which a plant is known to grow in the wild to answer this first question. Climatic similarity/dissimilarity interacts with habitat types to provide either further restrictions or alternatively more freedom of choice.

Climatic similarity is driven by three key factors: (i) overall patterns of precipitation/evaporation, (ii) altitude, and (iii) latitude. Plants adapted to grow in soils that are, however, only wet in winter and spring for example, typically cope badly in soils that are wet all year round and so are not ideally fitted. Many locations, such as South Africa and North America, have winter rainfall on western sides and summer rainfall on eastern sides, with a mix of both in the middle. European horticulture literature on plants from South Africa typically ignores this, and sees South Africa as a climatically homogeneous place with potential catastrophic consequences. Altitude interacts with precipitation patterns; typically as altitude increases precipitation increases and even in Mediterranean climates the rain-free season becomes much less severe or even non-existent. Plants from really high altitudes are often much less tolerant of high summer temperatures, but far more tolerant of severe winter minimum temperatures. Latitude interacts with altitude. The most common origin of plants in UK gardens is temperate Asia, often at latitudes much more southerly than the UK. For example, western Sichuan and Yunnan in China are at approximately 38–40°N – more southerly than the UK – but the plants here are naturally occurring at approximately 3000–5000 m. Notions of plant fitness currently accepted will no longer be valid in the future as the impacts of climate change take hold; e.g. in 50 years' time, the currently slightly winter-cold limited *Gladiolus cardinalis* is likely to be a ubiquitous, widespread species in the UK.

Structural form in herbaceous/geophyte plant selection

The orientation of leaves and stems of herbaceous plants in space and time has a major effect on their performance and how they can be used in plantings, and is a major factor to consider in plant selection. It is possible to think about these plant form issues from both a functional and aesthetic perspective.

From a functional perspective one of the most important foliage factors is the duration of leaf retention. Herbaceous plants with evergreen leaves typically inhibit weed invasion during the winter months more than deciduous species, although herbaceous plants with dead foliage that is very slow to decompose can also be very effective, for example *Iris sibirica*. In multi-layer plantings it is highly desirable to have as many of the ground layer species having evergreen or semi-evergreen foliage as possible; this trait is relatively common in many slow-growing, highly shade-tolerant woodland species, such as *Tiarella wherryi*, *Tellima grandiflora*, *Asarum* and *Epimedium* spp.

The distribution of the foliage mass in space has a major effect on the capacity of a species to out-compete or to be out-competed by other herbaceous plants. As herbaceous plants grow, their leaves and stems cast increasing amounts of shade, which potentially has a deleterious effect on lower-growing neighbours. The species that are most sensitive to competition for light when grown alongside other herbaceous plants are those that are shade intolerant with either mat-like (e.g. *Thymus*, *Oenothera macrocarpa*), or rosette-like foliage (e.g. *Verbascum phoeniceum*, *Salvia argentea*). In mixed communities, low-canopied species generally need to have only occasional tall species adjacent to them if their longer-term survival is not to be compromised.

Ironically, the most problematic herbaceous plants for mixing with shade-intolerant smaller species are generally those that are historically the most popular, namely those with tall leafy stems. In these the architecture normally involves erect stems furnished with leaves from ground level to the inflorescence. A classic example would be the almost shrub-like *Helianthus* 'Lemon Queen'. Leafy stem morphologies are least problematic when the vertical leafy stems form a narrow, tightly packed bundle, as in *Veronicastrum virginicum* or *Andropogon gerardii*.

An alternative morphology involves what is known as the naked stem. In these the foliage is normally restricted to a relatively low clump with tall leafless flowering stems that do not significantly restrict irradiance from reaching the lower canopies. Species with these growth forms include *Echinacea pallida*, *Ferula communis*, *Silphium terebinthinaceum*, and *Stipa gigantea*. Many geophytic genera have this architecture too, for example *Agapanthus*, *Allium*, *Dierama*, *Bulbinella*, *Eremurus*, *Galtonia* and *Watsonia* (Fig. 6.3). The naked stem architectures also have a particularly useful role aesthetically in that they do not block views through plantings, allowing the designer to create far more contrasts of colour and form as a result.

6.8 Phenology

Due to the fact that herbaceous plants and geophytes are often used primarily to provide colour via their flowers, it is important in plant selection to know when a particular species or cultivar commences flowering and when it typically ends. This information is often poorly documented in the horticultural literature. In gathering this information it is very useful to use a spreadsheet in which time is split up into weekly columns that

Fig. 6.3. Species where there is a low clump of leaves at the base and a tall flowering stem. This type of arrangement allows irradiance to reach ground level and avoids mutual shading. Plants with these growth forms include *Echinacea pallida, Ferula communis, Silphium terebinthinaceum* and *Stipa gigantea*. Many geophytic genera have this architecture too, e.g. *Agapanthus, Allium, Dierama, Bulbinella, Eremurus, Galtonia* and *Watsonia*.

can be coloured to represent, line by line of the spreadsheet, the flowering period of each proposed species.

Phenology is also very important in relation to when foliage is produced and then declines, both in herbaceous perennials and geophytes. Herbaceous plants that come into growth late in the growing season are often associated with warmer habitats in which a relatively high temperature soil/air threshold must be exceeded to initiate growth; some examples include *Incarvillea, Oenothera macrocarpa, Gladiolus* (South African summer rainfall species), *Roscoea, Ruellia* spp., *Schizachyrium scoparium, Zantedeschia albomaculata* and *Zinnia grandiflora*. There are, however, some species native to Western Europe that also respond in this way, for example *Eryngium maritimum*. These plants are often problematic in terms of weed colonization in that they do not occupy space effectively until June or July. They are also prone to heavy shading and therefore extinction when planted on the shady side of taller, earlier to emerge herbaceous plants. On the other hand, they can often be combined in interesting ways with species that emerge or flower very early in the year to create plantings in which there is little competition for space in the first part of the growing season.

6.9 Attractiveness to Invertebrates

Whilst most of the nature conservation literature suggests that nativeness is the key factor in whether herbaceous plants are attractive to native insects of their region (e.g. Tallamey, 2007), current scientific research suggests that this is only a part of the picture and in some cases not important at all (Smith *et al.*, 2006; Owen, 2010). Many insects within the Palearctic fauna, which stretches from Ireland to Japan and from the Arctic circle south to the Himalayas, share a long evolutionary overlap with the plants of this vast band of the Earth's surface, thus increasing likely suitability. There is, however, also much serendipity in these relationships. Some species of plants are hyper-attractive to pollinating invertebrates with which they share no evolutionary history. Hanley *et al.* (2014) in a UK study, found that one of the most attractive plants to native bees was a *Hebe* from New Zealand. Papachristou (pers. comm.) found no difference in pollinating insect visits to six species of native and non-native Dipsaceae. Non-native species such as *Echinacea* and *Buddleia* are clearly extremely attractive to native invertebrate pollinators. Research is currently in progress in the UK to obtain a more reliable clear measure of flower attractiveness to generalist insect pollinators. This appears to relate to factors such as the flower morphology, sugar concentration in the nectar, the palatability of pollen and so on. Sugar concentration is easily measured and provides a convenient ranking system to determine the 'value' of the plant to invertebrate pollinators, in this respect at least. Until this work is further developed, observation and the literature, much of which is very fragmentary, has to be used. Trials conducted by the Royal Horticultural Society (RHS), UK, now make observations on the relative attractiveness of different species and cultivars to invertebrates. In the *Sedum* Evaluation Trial that concluded in 2006 surprisingly large differences were noticed in the attractiveness of different *S. autumnale* × *S. telephium* cultivars to insects.

The plants that are generically poor for generalist native pollinators are those in which the reproductive organs have been converted into double flowers, but beyond that it is difficult to generalize. Because most Western European native species flower in spring or early summer, plants that flower in late summer and autumn are particularly important for generalist pollinators, as they extend the season that food sources are available.

6.10 Palatability to Molluscs

Herbaceous plants and geophytes are particularly sensitive to slugs and snails because in most species they have to push their new growth through the soil, creating feeding opportunities for slugs and snails that are not offered by shrubs. Because slugs, and to a lesser degree snails, feed nocturnally, the impact of these herbivores is completely underestimated in the horticultural literature, with the exception of the most highly palatable genera such as *Delphinium*. Even in *Delphinium* palatability varies considerably at the species level. Many Asian *Delphinium* such as *D. bulleyanum* are remarkably unpalatable, the same is also true of the North American *D. glaucum* (Table 6.2). The patterns in herbaceous plants as a whole do not mirror the situation in *Delphinium*. In species derived from the northern hemisphere palatability tends to be very high in genotypes from the eastern and central USA. Here molluscs are not an important part of the herbivorous fauna and therefore many of the herbaceous species have not expended resources on developing the physical and chemical weaponry (hairs on leaves or silicon crystals) to deter grazing. Hitchmough and Wagner (2011) investigated the relative palatability of approximately 30 North American prairie plant species to slugs and snails, finding most species susceptible to mollusc predation, especially if exposed to high grazing pressure at the seedling stage. By contrast (*Delphinium elatum* aside), most European and Eurasian species are much less palatable, with temperate Asian species intermediate, although within each of these very broad groupings there are still distinct contrasting palatabilities between species. Other than the spin of the evolutionary dice, many meadow species of high altitude are often very palatable because the climate precludes high mollusc densities (as do cold continental regions such as Eurasia) and hence effective defences are often absent. In the European Alps the distribution of palatable species such as *Arnica montana* are known to be determined by slug densities (Breulheide and Scheidel, 1999).

6.11 Plant Establishment

Specification, plant procurement, production systems

Procuring herbaceous plants during the dormant season is more problematic than with woody

Table 6.2. Herbaceous and geophyte genera and their relative palatability to molluscs.

Highly palatable	Intermediate	Unpalatable
Arnica	*Allium*	*Aconitum*
Asclepias	*Aster*	*Agapanthus*
Cacalia	*Bulbinella*	*Astilbe*
Delphinium	*Camassia*	*Crocosmia*
Dracocephalum	*Crinum*	*Crocus*
Echinacea	*Dianthus*	*Euphorbia*
Monarda	*Dodecatheon*	*Filipendula*
Ratibida	*Eremurus*	*Geranium*
Rudbeckia	*Eucomis*	*Helleborus*
Salvia	*Galtonia*	*Iris*
Trillium	*Liatris*	*Kniphofia*
	Narcissus	*Limonium*
	Primula	*Linum*
	Ranunculus	*Penstemon*
	Solidago	*Potentilla*
		Sanguisorba
		Succisa
		Trollius
		Tulipa

plants, both in terms of judging the quality of plants that are dormant and confirming precise identification. On projects in which the plant material is procured via tenders issued by the landscape contractor, there is often a high level of undisclosed substitution either with or without the knowledge of the contractor. The net result of this process is that by the time identification can be confirmed it is often mid-summer and there is a general reluctance amongst all parties to reject and re-plant plantings. This leads to plantings that at best may look completely different from what was designed and at worst fail to work as intended.

Procuring through contractors or nurseries with whom a personal relationship exists is a valuable means to minimize this malpractice. Buying herbaceous plants and geophytes unseen from any nursery where a previous relationship of understanding has not been built up, or where a large proportion of the plants may be grown by sub-contractors is probably best avoided. Another approach is to commission nurseries to 'contract grow' material where timescales between production and planting in the landscape make this feasible. Procuring plant material during the growing season where this is possible has the advantage in that it is much easier to confirm identification/quality.

Selecting the nursery product

Most herbaceous plants and evergreen/semi-evergreen geophytes are grown in containers, allowing planting (with irrigation post-planting during spring to autumn) to occur throughout the year. Across the range of native and non-native herbaceous plants in the market, the size of containers can range from 200 cm^3 plugs, to 9 cm diameter, 1, 2 and 3 litre pots. As a general principle, plugs are best restricted to sites with reasonable control of planting conditions and reliable post-planting maintenance. The growing media volume is so small in a plug that drying out post-planting or plugs being buried in large planting holes leads to a high level of mortality under average site conditions. Plants grown in 9 cm and 1 litre pots are preferred on most sites because they are large enough to handle and plant easily, large enough to withstand suboptimal planting practice, and the cost per unit is insufficient to inhibit high density planting. Pots of 2–3 litres are expensive and problematic in terms of the size of the holes that need to be dug, and the level changes that result from the spare excavated backfill.

An alternative to container-grown stock is the bare root transplants, which are still grown by field-production specialist nurseries. These restrict plantings to autumn through to spring, but when correctly handled often produce very good results. The standard bare root herbaceous plant product is typically the same price as a 9 cm pot specimen, but the size of the plant, and the number of buds etc., is much greater in the bare ground product.

With the exception of evergreen species, the most widely planted geophytes are nearly always procured as dry bulbs, corms and tubers in late summer and early autumn. Top-quality bulb growing companies specify the grading size on sale; larger bulbs tend to perform more robustly. Procuring and planting geophtyes as early as possible has a very significant effect on performance; bulbs planted at the tail end of the season in, say, late October in the northern hemisphere are likely to be desiccated and often with roots or shoots protruding that will be broken on planting.

Timing of planting

In landscape practice this is often determined by factors other than what is biologically best for establishment, however when it is possible to do so planting should be in early autumn (e.g. in the UK planting in September is often good as soils are not too wet, so reducing inadvertent site compaction as a result of the planting process), allowing root establishment prior to winter dormancy. Plants planted in autumn are typically much larger in the following year than those planted 5 months later in spring. Exceptions to the autumn planting rule of thumb are species that may suffer damage in their first winter, such as *Begonia*, *Dierama*, *Hedychium*, *Moraea*, *Roscoea* and *Watsonia*.

Planting protocols

When planting between spring and autumn, ensuring the root balls are soaked prior to planting is important, as most herbaceous plants and evergreen geophytes are grown in organic matter composts based on peat, green waste or bark, and these materials shrink and are hydrophobic following drying. Once planted, dry root balls are almost impossible to re-wet, leading to stress and in some cases high levels of mortality. Many nursery-grown herbaceous plants come with a weed flora growing on the top of the pots, that then seed post-planting and contaminate the planting mulch. Better quality herbaceous plants will have less of this but it is useful to specify in all cases that the upper 2–3 cm of the pot compost is manually removed prior to planting.

Planting is easiest and quickest when the soils into which planting is taking place are relatively loose and non-compacted. This can be achieved by either cultivating the soil surface a few days prior to planting, or in some cases by covering planting sites for a couple of weeks prior to planting with tarpaulins to keep the soil drier. The latter allows planting to go ahead in-between rainfall events. The general principle is to make a hole (trowel planting) that is large enough to take the root ball and then to compress the edges of the hole inwards to create close contact between the soil phase and the growing medium phase to assist in water transfer. When in doubt herbaceous plants should be planted slightly deeper, i.e. the surface of the root ball slightly below the finished soil level, rather than above. This deeper planting allows new roots to emerge from the base of the above-ground parts, and often reinvigorates planting stock.

The actual depth herbaceous plants are planted at will also need to take into account the depth of mulch that is applied on top of the soil. This is most problematic with herbaceous plants whose foliage is evergreen and arranged in a basal rosette as, if this gets covered with mulch, it can lead to the death of the foliage and ultimately the plant. For herbaceous plants with upright stems and high 'shoot thrust', depth is not a major problem.

Various recommendations for planting depth for geophytes are available in many bulb catalogues. As many (but not all) of these species have contractile roots which physically drag the corm or bulb more deeply into the soil, many geophytes place themselves at the depth that they prefer themselves, so if in doubt land managers should err on the side of a planting depth that is relatively shallow. In their wild habitats, many geophytes are found at greater depths in the soil than would be expected from the size of the bulb/corm or tuber, e.g. Mediterranean *Crocus* corms are often found at depths of 100–200 mm.

Deep planting is often useful for bulbs and corms that are sensitive to occasional extreme winter cold. In most cases the most susceptible tissues to chill or cold injury are at the basal plate of the bulb or corm from which roots are initiated. By placing this deeply in the soil, the minimum temperature experienced in winter is moderated by the very high thermal inertia of soil. Geophytes that might benefit from being treated in this way include *Bulbinella*, *Crinum*, *Eucomis*, *Gladiolus* (South African species), *Ornithogalum* (South African species) and *Watsonia*.

Mulching

In most designed landscapes, plantings are only maintainable when some form of mulch is employed on a regular basis to blanket the weed seed bank in the soil, and present a 'quick to dry' surface that is hostile to in-blown weed seed. The critical input in year one of herbaceous planting is to prevent blown-in weeds establishing and producing seed whilst the planting is relatively open. If plantings can be kept largely weed-free during this period, year two and subsequent maintenance is greatly reduced, as the canopy of the planted material provides much more competition for light. Horticulturally trained staff are sometimes indifferent to these principles, seeing weeding as about doing maintenance to a timetable rather than breaking the life cycle of weed development. Subsequent maintenance to manage weeds is much greater where weeds are first allowed to self-seed before removal.

The least effective materials for mulching are colloidal organic materials such as composted green waste; whilst these are effective for blanketing weed seed banks and even shading out newly germinated weed seedlings on the layer below, they are an excellent germination surface for blown-in seed. Garden management organizations such as the National Trust in the UK, which generally have vast amounts of composted arisings to use, often mulch their herbaceous plantings twice a year to get around this problem.

In general, however, coarse granular materials such as composted bark (particles larger than 15 mm), pea gravel, and grit/very coarse sand are highly effective mulches. The advantages of the mineral material is that they are more persistent. They are also relatively hostile environments for slugs, compared to composted organic material that may actually increase local slug densities. The disadvantage is that they need to be applied in deep layers to be effective; a minimum of 100 mm, but ≥150 mm is better as it makes it more difficult for seedling weed roots to reach the moist underlying soil prior to dying of moisture stress. Surface scarification of gravel-type mulches on a warm dry day is very effective in terms of eliminating cohorts of blown-in annuals and biennials. Planting in deep mulches of sands as a deliberate high-stress strategy to minimize weed colonization in naturalistic plantings is covered in Chapter 7.

6.12 Longer-Term Maintenance

Weed control

As has previously been discussed, the use of mulches is key to making herbaceous plantings manageable when maintenance resources are limited. These are most cost-effective when applied in winter (e.g. January/February in the UK) and spread over the entire surface of the planting. This is required very occasionally with mineral mulches, but with organic debris mulches such as bark are probably required every other year. Species with basal rosettes or those that are in leaf in January are likely to be damaged by this process. In addition to mulches the characteristics of the plants and planting design employed are also very important. High density plantings (>9 plants per m^2), with dense basal foliage, preferably evergreen, greatly inhibit weed colonization from blown-in seed. However, the essential quality of many herbaceous plantings and summer-growing geophytes is the loss of canopy cover between late autumn and spring that facilitates weed invasion.

The weed species that invade herbaceous planted landscapes tend to reflect the nature of those plantings and the forms of management used. In the first year after planting, the weeds are mainly those with small seed that are well distributed by wind etc. Many of these species are ruderals, such as *Poa annua* (annual meadow grass), *Cardamine hirsuta* (hairy bittercress) and *Epilobium* (willow herb) hybrids. They typically come in with the planted specimen as weeds on the tops of pots as well as blown or carry-ins and are themselves very quick to set seed. The 'invasion' rapidly gains momentum from a few weeds to many, and is often greatest between autumn and winter when the site is most open and the surface of mulches are permanently moist. These invasion events are predictable and can be managed by either burning-over the surface in March or applying a contact herbicide that defoliates but does not kill established planted perennials.

Flash burning using either paraffin or biofuel flame guns or propane-fuelled burners used for road repairs are also very effective at killing the cohorts of annual winter weeds and, in doing so, breaking the invasion cycle. In UK conditions this is particularly effective when undertaken in April just as or before the leaf emergence of the planted species, as the canopy then closes very quickly after burning, before a second wave of weed germination can take place. When undertaken on a dry day in March/April after the dead landscape plant material has been strimmed-off and removed from the site, there is little or no smoke, so it is readily undertaken in urban areas. Many plants are tolerant of being burnt-over at this time of year. Few herbaceous plants that are winter dormant will be damaged, species that are evergreen or semi-evergreen and/or have most of their growth points at or above the soil surface (e.g. *Fragaria* species) are more likely to suffer damage, although there is currently little data available on this beyond the experience of the author. A list of species that the author has found to withstand burning are given in Box 6.1.

Some species are highly tolerant of burning even when in full growth. Hitchmough (unpublished) has burnt North American prairie species when in leaf in May and nearly all recovered very quickly, most producing another set of leaves within 14 days. Much more research, however, needs to be undertaken on this aspect of management.

One downside of using propane and similar based hydrocarbons to burn off material is that even if the volume of greenhouse gases released is relatively small, the perceptions alone may not be one of sustainable management. A less carbon-intensive approach (see Hitchmough and de la Fleur, 2006) is to overspray plantings with a contact herbicide in March/April. This method is essentially a non-thermal equivalent to burning, in that the death of tissues is largely restricted to the point of contact with the chemical. Many of the historically most widely used contact herbicides such as diquat are highly toxic and potentially problematic in public spaces. Recently, vinegar (acetic acid) has been registered in the UK as a

---

Box 6.1. Species that are tolerant of burning-over in March/April.

Agapanthus campanulatus and *inapertus* types
Berkheya spp.
Dianthus carthusianorum
Dierama (delays flowering, not to be undertaken every year)
Incarvillea spp.
Kniphofia (most)
Oenothera macrocarpa
Pulsatilla vulgaris
Polemonium reptans
Rudbeckia fulgida with overwintering leaf rosette

contact herbicide following its successful use for a number of years in Scandinavia. Trials are currently underway by the author's research group at the University of Sheffield, UK, into the efficacy of vinegar used in this way.

Even when managed 'disturbances' such as spraying and burning are employed there will always be some establishment of perennial weed species that tolerate these regimes. The most common examples of these species in the UK are grasses, and in particular: *Elymus repens* (couch grass), *Agrostis stolonifera* (creeping bent), *Holcus lanatus* (Yorkshire fog), *Ranunculus repens* (creeping buttercup), *Taraxacum officinale* (dandelion), *Urtica dioica* (nettle) and *Calystegia sepium* (bindweed). In the early stages of colonization all of these plants can be managed satisfactorily by careful physical removal. This is particularly so where deep gravel mulches are used as the roots do not snap off and remain in the medium when the plant is removed. In the longer term some of the above species require different approaches. Glyphosate traditionally has been the main landscape herbicide used to control such weeds, however, its use (certainly in non-agricultural locations) is now being reviewed in a number of countries. Where it remains available, stoloniferous grasses can either be painted with a glyphosate-based, translocated herbicide, or over-sprayed with a graminicide-specific herbicide to reduce risk to adjacent species. Most graminicides used in herbaceous plantings are 'off-label' applications, so what is available varies from year to year, and from country to country. For example, fluazifop-P-butyl and clethodium are currently approved for amenity or woodland vegetation in some EU countries but not others. The advantage of graminicides is the entire planting can be sprayed in April/May to remove or severely damage the entire grass weed cohort. Tussocks of planted ornamental grasses can be covered with builders' buckets. Evidence from the author's research and practice to date shows that these graminicides are highly selective, with no evidence of any non-grassy weed species suffering severe damage from their application.

Of the likely broadleaved invaders, *Calystegia* (bindweed) and *Rumex* spp. (docks) can generally only be controlled effectively by painting the shoots with a glyphosate herbicide applied via a hand weed wiper or brush. Commercial gel-based formulations are now available too. To prevent cross plant transmission via leaf leakage it is desirable to enclose the treated foliage of voluminous weeds such as bindweed in tubes designed to protect tree seedlings from rabbits until absorption is complete.

Controlling long-term community development

Weed management is a critical part of controlling the composition of the community, but so also is controlling the growth of ornamental species that are planted to prevent over-dominance. Even with much experience and very detailed ecological knowledge it is difficult at the design stage to accurately predict which species will become over-dominant. This process is context specific; it depends on which other plants are present, and how much counter-competitive pressure is exerted. It also depends upon the moisture and nutrient status of the site. Very often, competitive dominants that emerge in plantings are those species that are both productive and the least palatable to slugs and snails.

Managing these processes is frequently quite challenging as the natural inclination is to often let the community establish its own dynamics, rather than to have to make decisions about what species/specimens to remove and how many and when. The degree to which such decisions need to be made depends upon the context and human expectation. In the highly visited steppe-prairie plantings at the RHS Wisley Gardens, UK, designed by the author, it took 5 years for dominance to become a major issue. By 2012, however, the biomass produced by *Aster* species, combined with a very wet summer, became so large as to reduce the photosynthesis of *Echinacea pallida*, the mainstay of the summer display. The shade generated by the large biomass of *Aster* supported a significant population of slugs that grazed the *E. pallida*, exacerbating its decline. This interaction between herbaceous plants/geophytes and slug grazing is little recognized but extremely important. This chain of events necessitated the removal of up to 50% of the *Aster* and the planting of new lower species into the gaps to allow recovery of *E. pallida*.

As in the example given, it is necessary for managers to set priorities on which species are sacrosanct and which are not, and to make decisions to achieve these priorities. This is not easy when many of the species planted are self-seeding, resulting in a chaotic and constantly changing plant community.

6.13 Manipulating Attractiveness

In many contemporary plantings of herbaceous plants and geophytes, design is used to minimize maintenance, by avoiding species which lodge (bend over) badly after heavy rain, or possess unattractive spent inflorescences or post-flowering leaves. In such contexts, it may be necessary to remove untidy plant parts either for reasons of aesthetics or to reduce excessive self-seeding.

Conclusions

- Herbaceous plants and geophytes have become more popular recently for use in urban landscapes, due to strong aesthetic features associated with flower colour and plant form.
- Their relatively low-growing stature, however, poses problems in terms of resisting stress factors associated with mollusc herbivory and weed competition.
- Both herbaceous perennial plants and geophytes are non-woody in nature, and survive periods of abiotic stress by storing reserves within underground or ground-level perennation organs. For geophytes this tends to be specialized storage organs such as bulbs, corms or tubers.
- Both herbaceous perennials and geophytes have evolved strategies to ensure survival in mixed plant communities. These are useful traits when exploiting these plant types within environmental horticulture, either when planted as individual specimens or when allowed to propagate themselves through off-sets and seed.
- In recent years, the increase in popularity of herbaceous plants and geophytes within the public realm has been driven largely by ecological and functional philosophies.
- Such landscapes are becoming increasingly popular due to their ability to attract invertebrates, especially those with a high public acceptance (e.g. bees and butterflies).
- The key to successful planting in public space is understanding individual species requirements, promoting an appropriate balanced 'community' and taking decisive action to reduce both weed pressure and interspecies competitive effects within the planting.
- Climatic factors (timing and degree of wetness/dryness, but also predominant temperature regimes), soil nutrient status and biotic pressure (e.g. from slugs and snails or degree of canopy shading) are important criteria to consider when choosing species for any particular designed landscape.

References

Bruelheide, H. and Scheidel, U. (1999) Slug herbivory as a limiting factor for the geographical range of *Arnica montana*. *Journal of Ecology* 87, 839–848.

Culver-Campbell, M. (2001) *The Origin of Plants*. Transworld Publishers, London.

Dunnett, N. and Clayden, A. (2007) *Rain Gardens: Managing Water Sustainably in the Garden and Designed Landscape*. Timber Press, Portland, Oregon.

Dunnett, N. and Hitchmough, J.D. (2004) *The Dynamic Landscape, Design, Ecology and Management of Naturalistic Urban Vegetation*. Taylor and Francis, London.

Hanley, M.E., Awbi, A.J. and Franco, M. (2014) Going native? Flower use by bumblebees in English urban gardens. *Annals of Botany* 113, 1–8.

Hansen, R. and Stahl, F. (1994) *Perennials and their Garden Habitats*. Oxford University Press, Oxford, UK.

Hitchmough, J.D. and de la Fleur, M. (2006) Establishing North American Prairie vegetation in urban parks in northern England: effect of management practice and initial soil type on long term community development. *Landscape and Urban Planning* 78, 386–397.

Hitchmough, J.D. and Wagner, M. (2011) Slug grazing effects on seedling and adult life stages of North American Prairie plants used in designed urban landscapes. *Urban Ecosystems* 14, 279–302.

Laird, M. (1999) *The Flowering of the Landscape Garden, English Pleasure Grounds, 1720–1800*. University of Pennsylvania Press, Philadelphia.

Oehme, W. and Van Sweden J. (1991) *Bold Romantic Gardens: the New World Landscape of Oehme and Van Sweden*. Acropolis Books, New York.

Oudolf, P. and Gerritsen, H. (2013) *Dream Plants for the Natural Garden*. Francis Lincoln Press, London.

Oudolf, P. and Kingsbury, N. (2014) *Planting a New Perspective*. Timber Press, Portland, Oregon.

Owen, J. (2010) *Wildlife of a Garden, a 30 Year Study*. RHS Publications, London.

Robinson, W. (1870) *The Wild Garden*. J. Murray, London.

Smith, R.M., Warren, P.H., Thompson, K. and Gaston, K.J. (2006) Urban domestic gardens (VI): environmental correlates of invertebrate species richness. *Biodiversity and Conservation* 15, 2415–2438.

Tallamey, D. (2007) *Bringing Nature Home: How You Can Sustain Wildlife with Native Plants*. Timber Press, Portland, Oregon.

Wilford, R. (2006) *Tulips: Species and Hybrids for the Gardener*. Timber Press, Portland, Oregon.

Woudstra, J. and Hitchmough, J.D. (2000) The enamelled mead: history and practice of exotic perennials grown in grassy swards. *Landscape Research* 25, 29–47.

7 Semi-Natural Grasslands and Meadows

JAMES D. HITCHMOUGH

> **Key Questions**
> - What factors have driven the recent interest in 'urban meadows' and meadow-like plant communities?
> - What environmental conditions in nature tend to promote grasslands and forb-rich communities?
> - What factors increase plant diversity in managed grassland communities?
> - How can more colour be introduced into an existing area of grass?
> - How do mowing regimes affect plant competition and thus the species composition of meadows?
> - How would you ideally design and manage a highly diverse, species-rich floristic meadow community, if you were starting from scratch?

7.1 Introduction

Meadows have a special place in human history and development within Europe and elsewhere, and often invoke romantic notions of the countryside. This is especially so if the meadow communities are full of flowering dicotyledonous plants that bestow swathes and mosaics of bright colour. Perhaps no other landscape type signifies lost innocence and the desire to return to a bygone 'golden' era as much as the meadow landscape. Similarly, the demise of traditional hay meadows are a metaphor for the wider, rapid changes in agricultural practices that have occurred over the last 200 years, and indeed, for the generic loss of biodiversity within the countryside. Prairies have a similar symbolism in North America. The fact that meadow plant communities were themselves often a byproduct of practical necessity that underpinned rural living, tends to get lost with time and the more recent trends that romanticize country life. Indeed, without the need to provide livestock with fodder throughout the winter months, many of these biodiverse, flower-rich communities would not have existed at all in past-times. Nevertheless, the emotional ties humans have for meadows and other grassland types are strong. The aim here, however, is not to recount the history of the rural meadow per se, but rather to discuss why these and similar 'meadow-like' landscapes are gaining popularity within an urban context.

7.2 The Role of Grass-Based Plant Communities in Urban Spaces

Flower (forb)-rich meadow communities occur where some form of abiotic stress inhibits vegetation succession to trees and shrubs. This may be cold, fire, drought or grazing pressure. In many of these scenarios grass species dominate, but are augmented by dicotyledonous species, many of which produce brightly coloured flowers to attract pollinating invertebrates. As such, these landscapes can have a high aesthetic appeal. Many meadows are semi-natural landscapes in that the florally-rich plant communities are encouraged through human activity, such as grazing of domestic livestock, cutting for hay or even flooding to improve the quality of pasture land (traditional wet-meadows).

Many meadow landscapes have been lost over the last century due to more intensive agricultural practices, such as the prairie systems of North America, where only <5% of the original area is thought to remain (with perhaps <1% of the tall-grass prairies left). Similarly, in countries such as the UK, changes in agricultural practices have seen the demise of colourful flowering meadows to be replaced by near-monocultures of *Lolium perenne* (perennial ryegrass) as more productive pasture, or by arable crops. These activities, however, have acted as a catalyst not only to preserve what rural meadow systems still remain, but also to try and replicate typical traditional meadow communities within urban settings. Meadow making, particularly in urban situations, is becoming increasingly popular as an alternative to more intensive horticultural landscapes, for which the resources to manage are increasingly unavailable. Meadow-like vegetation has become seen as a sort of 'magic bullet'– colourful and attractive to people, good for supporting biodiversity and yet also relatively inexpensive to create and manage.

In the UK, in those locations where dense trees are absent, grass-dominated landscapes tend to be the mainstream, much more so for example than in continental Europe. This is due to lower levels of summer moisture stress, mild winters that facilitate grass dominance during this period and high levels of slugs and snails that reduce the competitiveness of many grassland broadleaves. Because of the presence of silicon in the leaves, grasses are generally less palatable to slugs than are forbs (O'Reagain, 1989).

Within UK urban public space the dominant vegetation type is gang-mown grass, a gang being a number of individual cylinder (reel) mowing units working in combination behind a tractor. The dominance of mown grass is largely due to its multifunctional characteristics, but also because it is generally the cheapest vegetation surface in terms of maintenance, at least within a moist maritime climate that is conducive to grass growth (Cobham, 1990). Mown grass is cheapest to manage in large spaces, without obstacles interfering with the action of the multiple gangs. In other parts of the world mown grass is more difficult to maintain and hence often retains a cachet of exclusivity, particularly where water is short, as in southern Europe, California or the Gulf States. In the UK, turf is ubiquitous and for most people it maintains a sort of neutral constant presence. Because it is almost never irrigated or fertilized or treated with pesticides (in general public space) it attracts little of the environmental opprobrium that it does, for example, in much of the USA (Herbert-Bormann *et al.*, 2001).

On the negative side, mown grass, because it is used so extensively often to the exclusion of all other urban vegetation in UK green space, is both relatively limited in terms of the experiences it can offer users (it is often seen as rather boring) and is also quite limited in its biodiversity value (Smith *et al.*, 2015). Debates have raged about the capacity of mown grass to store carbon (Townsend-Small and Czimczik, 2010), and whilst it does incorporate carbon into the soil from grass clippings post-mowing, these levels are similar to the carbon stored from alternative management options such as managing a meadow. All in all, from a sustainability perspective, the UK probably has much more gang-mown grass than can rationally be justified and hence there is interest in how the mown-grass estate could be managed differently.

7.3 Changing the Mowing Regime to Create a Spring Meadow

In this approach the grass is mown as normal with a break between January/February and May in the northern hemisphere to allow plants that can flower during this time to do so. In long mown grass the main species to take advantage of this are likely to be *Taraxacum* spp. (dandelion) (Fig. 7.1), *Bellis perennis* (lawn daisy), *Ranunculus repens* (creeping buttercup) and *Plantago lanceolata* (lawn plantain), all of which can create a very attractive effect. In addition it is possible to add early spring flowering bulbs to the sward to supplement and lengthen the flowering season (see section 7.4 below).

7.4 Changing the Mowing Regime to Occasional Flail Cutting

In situations where the work rate of gang mowing is slowed by topography or obstacles such as trees and edges, there is a trend in some public open spaces to switch from gang mowing at 30 cuts per annum to cutting once or twice a year using a flail mower, often in winter. This treatment has similar, or in some cases, slightly reduced costs to gang-mown turf (Cobham, 1990).

Characteristics of occasional flail-cut grass

Flails macerate the cut grass into lumps which are left on the cut surface. As a result, as the cut grass decomposes the nutrients present in it are recycled back into the soil, and hence on highly fertile clay or loam-based soils there is no diminution of soil productivity with flail cutting. What this means is that, except on especially infertile sandy soils, the plant community produced by flailing will shift from rhizomatous-stoloniferous grass and rosette-forming forbs such as *Bellis* and *Taraxacum* to tall,

Fig. 7.1. *Taraxacum* spp. (dandelions) producing a dramatic spectacle in sporadically mown turf.

coarse, tussock-forming grasses such as *Dactylis* and *Arrhenatherum*. The shading and competition provided by the latter grass taxa leads to the loss of forbs and a subsequent drop in plant diversity. This management practice is often 'sold' to residents on the basis that cessation of regular cutting will encourage wild flowers to develop but this does not normally occur on productive soils because the dense growth of tussock grasses is extremely hostile to the establishment of meadow forbs. Even when seedling forbs do establish after cutting of the grassland in winter, come spring the proportionately massive leaf canopies of the grasses typically cause the elimination of most of the established seedlings.

Anecdotal evidence as well as some research studies (Junge *et al.*, 2009) suggests that the resulting tall coarse grassland communities are very disliked by the public in strongly urban settings; hence the pejorative term 'rough grassland'. Whilst plant diversity is likely to decline under this regime, animal biodiversity is likely to increase (Humbert *et al.*, 2012). Rough grass is an excellent habitat for sap-sucking insects, and beetles, spiders, slugs and snails. It is also good habitat for many small mammals and amphibians, like frogs and toads.

Diversifying flail-cut grass

Geophytes

Since natural colonization of rough grass by attractive native forbs is likely to be incredibly slow or even impossible where these plants are very distant from the site, techniques that speed up this process are a priority. One group of plants (largely not native) that are well suited to cultivation in this vegetation type are geophytes such as *Narcissus* and *Crocus* and other species with similar late winter growing habits (Lloyd, 2004) (Fig. 7.2 and Table 7.1). Flailing grass in late autumn/early winter inhibits the formation of tussock structures, thus allowing even small geophtyes to project their foliage into the light and photosynthesize adequately to increase and spread, both vegetatively and sometimes also from seeds.

These species are relatively easy to establish because they compete for resources such as light and nutrients

Fig. 7.2. Bulbs such as *Narcissus* spp. (daffodils) are robust enough to survive in rough grass, as long as their foliage is allowed to die back naturally before the first cut of the grass.

when the grasses are at their least competitive, i.e. they utilize a window of reduced competition within the grassland. Despite this, the higher the standing biomass of the flailed grass, the more attention needs to be given to selecting more vigorous geophytes with taller leaves, such as taller growing *Narcissus* and *Camassia*. Small geophytes such as *Crocus* are more successful in less productive grasslands. Planting needs to be undertaken in autumn for best establishment and hence cutting the grassland in September in the year of planting is recommended, and mowing it in November (in the northern hemisphere) to ensure competition for light is limited in this first year. Commercial bulb mixes for mass planting into mown grass have been developed in the Netherlands and are readily available commercially.

Forbs

Irrespective of whether these are native or non-native, forbs are more difficult to establish because their growth period generally coincides with that of the grasses, and hence there is no separation of competition in time as is the case with the geophytes. The critical factor of establishment in these highly competitive grass swards is whether the proposed forbs are potentially 'ecologically fit' for the grassland in question, for example as determined

Table 7.1. Geophyte genera suited to cultivation in flail-cut grass.

Galanthus	snowdrop
Crocus	crocus
Narcissus	daffodil
Hyacinthoides	bluebell
Camassia	quamash

by their palatability to slugs and snails, and the height and shade tolerance of their leaves.

Fitness of species at a planting site is determined generally by how similar the climate and the soil conditions (in terms of wetness–dryness and fertility) are to those that the plants evolved in. In the UK for example, this does not mean that only species native to Britain will grow well in grasslands in the UK, but rather that successful species are likely to be drawn from places in the world where due to altitude and other factors, similar habitat conditions can be found, such as the mountains of Europe or the Caucasus – geographically distant but supporting many species well-fitted to the UK. *Lychnis chalcedonica* for example, is found in Russia in wet grasslands along streams, but is very tolerant of growing in rough grass in the UK, as is *Geranium psilostemon* from northern Turkey and *Euphorbia palustris* (Hitchmough, 2009) from fens and boggy grassland in Europe. Most North American species are not very tolerant of these grasslands because they are prone to being highly palatable to slugs and snails (Hitchmough and Wagner, 2011).

Species with taller leaves that project into the light are in general much easier to establish in rough grass than are species with basal leaves, which are generally heavily shaded by surrounding tussocky grasses. Exceptions to this are rosette-forming species with high levels of shade tolerance that grow actively during the winter months, for example *Primula veris* and *P. vulgaris*. *Geranium* species are particularly tolerant of growing in rough grass because of their capacity to continue to elongate their leaf petioles so that their leaves are always in the sun, even in very rank grassland.

Species that have demonstrated competitiveness in grasslands managed on a flail cutting regime are shown in Table 7.2.

7.5 Changing from Gang Mowing to a Meadow Regime

This approach is less common than flail cutting because the need to cut and remove the herbage from the site increases the cost over that of flail cutting. The advantage of this regime, however, is that over long periods of time (typically in excess of 10 years) the annual removal of the peak standing biomass in summer gradually reduces the productivity of the grass sward, creating opportunities for natural colonization of native forbs, plus more reliable establishment of forbs through planting or over-seeding

Table 7.2. Forb species capable of growing successfully in rough grass swards cut in late autumn or winter.

Wet to moist soil types	Moist to dry soil types
Euphorbia palustris (marsh spurge)	*Geranium pratense, G. psilostemon*
Iris sibirica (Siberian iris)	*G.* × *magnificum* (cranesbills)
Lychnis chalcedonica (Maltese cross)	*Hemerocallis* (day lily) (vigorous cultivars)
Lythrum salicaria (purple loosestrife)	*Malva moschata* (musk mallow), *M. alcea*
Persicaria bistorta (common bistort)	*Papaver orientale* (oriental poppy), *P. bracteatum* (Iranian poppy)
Primula vulgaris (primrose)	*Primula veris* (cowslip)
Veronicastrum virginicum (Culver's root)	*Sanguisorba officinalis* (great burnet)

(Wells, 1980; Marrs, 1985). It is, however, important to be realistic about what can be achieved through meadow cutting alone. It will not lead to highly floristically diverse meadows in the short term where soil productivity is high and where there are few if any meadow forbs growing adjacent to the site. In the absence of the latter, a situation known as seed limitation prevails; forbs do not establish simply due to the fact that there is no seed of these species present. Meadows can therefore remain forb species-poor for decades or longer under these circumstances. Most urban sites are essentially islands separated from more biodiverse patches of land from which seed rain might derive. Nor can much if any faith be placed in the idea that desirable forbs might recruit following a switch to a meadow regime. Most of these species do not persist or do not persist for long within the 'seed bank' of the soil. To adapt an adage from the great Westerns: the Cavalry is definitely not just around the corner!

The costs associated with cutting the sward in summer as hay vary substantially depending on the scale of the operation and the machinery available (Cobham, 1990). Much of the green-space network of Stockholm, Sweden, is dominated by meadow-like grasslands (floristically relatively poor) and are managed by cutting in August followed by baling using large-scale agricultural equipment. The actual management costs of this operation per unit area are unknown but are likely to be substantially lower than the 4 hours per 100 m^2 typically required for hand cutting with a strimmer and brushcutter followed by manual raking and removal. There is an extensive range of cut and bale equipment on the market. Some of the machinery is designed for small-scale meadow cutting in the mountains of Europe, and often involves self-propelled reciprocating blade mowers and small-scale baling machines.

Timing of cutting has a major impact on competitive relationships between species in the meadow, and ultimately which species are dominant and subordinate. When dealing with grass-dominated systems (gang-mown grass allowed to grow long) in which the forbs present are initially mostly low-growing creeping or rosette-forming species, cutting early in the summer, typically late June or early July in the northern hemisphere, is likely to be the most appropriate strategy to maximize forb dominance. Cutting at this time checks the physical dominance of the tall grasses such as *Arrhenatherum* and *Dactylis* that are likely to be shading the community, allowing light back into the system soon enough to limit the extirpation of light-demanding low forbs. Implementing this regime during the first 5–10 years is going to be beneficial for moving towards a more diverse rather than tall and rank meadow. On really fertile soil a second autumn cut and cuttings removal in addition to an early spring mow (no cutting removal) to 'tidy up' is advantageous.

This regime is also likely to be effective in terms of aiding the establishment of species in the sward by checking the degree of shading from tall species (in the first instance mainly grasses). This regime will, however, continue to disadvantage the development of forb species with taller leafy stems (for example *Malva moschata* or *Knautia arvensis*). To favour these species a delayed cut is required, for example in August or even September.

These cutting regimes are only a few of the options with meadow management. It is important to recognize that traditional hay meadow management was designed to maximize hay yield for animal feed and maximizing plant biodiversity or the abundance of a given species was purely coincidental. Although it involves two hay cuts per annum, a regime in which the meadow is cut in late May and then again in October is worth considering for high profile sites as it pushes the main flowering peak back into summer, and reduces meadows looking very 'strawy' and unattractive during this period. The author used this approach in the wildflower meadows at the London Olympic Park, UK, to great effect. Again this system tends to check the dominance of tall grasses, and removes the flowers and developing seeds of the grasses. In practice, environmental horticulturalists need to experiment with individual sites in question, as local conditions can influence species competitiveness. Effective managers will observe the changes, and judge whether the cutting regime favours the desired species and is preferred by those people using the landscape.

The species that cope with summer hay-cut grassland are generally different to those that establish well in summer-uncut grasslands (Table 7.3).

Establishing native or non-native forbs in grasslands managed via a summer to early autumn cutting regime is not particularly easy, especially when the productivity of the soil is high. The dominant problem is competition with the grasses both for water and nutrients (below-ground competition) and for light (above-ground competition). Grasses typically have higher growth rates than many forbs (see Hitchmough *et al.*, 2008) and this is particularly

Table 7.3. Native UK/Western European forb species well suited to summer/early autumn cut hay meadows.

Centaurea nigra (common knapweed)	
Centaurea scabiosa (greater knapweed)	Preference for/tolerance to drier soils
Galium verum (lady's bedstraw)	
Geranium pratense (meadow cranesbill)	
Hypochaeris radicata (cat's ear)	
Leontodon autumnalis (autumn hawkbit)	
Leucanthemum vulgare (ox-eye daisy)	
Lotus corniculatus (birdsfoot trefoil)	
Malva moschata (musk mallow)	
Primula veris (cowslip)	
Prunella vulgaris (primrose)	
Ranunculus acris (meadow buttercup)	Preference for/tolerance to wetter soils
Sanguisorba officinalis (great burnet)	Preference for/tolerance to wetter soils
Scabiosa columbaria (small scabious)	
Stachys officinalis (betony)	Preference for/tolerance to drier soils
Succisa pratensis (devil's-bit scabious)	Preference for/tolerance to wetter soils

problematic when small, relatively young forbs have to compete with established grasses. The second factor is predation from slugs on forbs. Because many grasses have siliconized leaves, slugs selectively graze on forbs rather than grasses (Murray *et al.*, 1996), and the defoliation associated with this further tips the competition balance in the direction of the grasses.

The gap between the competitive capacity of established grasses and young forbs (technically referred to as competitive asymmetry) can theoretically be narrowed by trying to establish forbs as container-grown plants. Historically these have often been grown in small plug cells for this purpose, but the use of large pot-grown material, such as 9 cm pots, results in large plants that take longer to be defoliated by slugs and have more capacity and carbohydrate reserves to push leaves into the light than do plugs. The species that establish most reliably in meadow-like swards as small plants (e.g. plugs) are those that have a high degree of shade tolerance and also unpalatability to slugs (Davies *et al.*, 1999). *Primula veris* is one such species and there is much visual evidence of how effective plugs have been for this species on motorway verges over the past 15 years. Another species that is probably effectively established in this way is *Succisa pratensis* but there is relatively little research on this. Planting in autumn (e.g. October in the northern hemisphere) after the final hay cut or winter tidy-up mow, is probably the most opportune time to establish pot-grown forbs as the soils are still warm enough to get some root establishment into the soil profile. Planting in spring increases the risk of roots not establishing before the organic growing medium transferred from the pot dries out and shrinks away from the surrounding soil, leading to a high likelihood of failure.

The problem with planting is that it inevitably involves relatively few plants per unit area of grassland. Even with locally native species, existing grass swards are very heterogeneous both below and above ground in terms of the degree of competition; some plants will fail in one planting location that would succeed 50 mm away. Serendipity, or – to use the ecological term – 'stochasticity', is important in all ecological processes but is a major issue in establishment in meadows. This competitive heterogeneity is only rarely visually evident, and is a challenging idea to environmental horticulturalists, who generally manipulate environments and competition to avoid the unpredictability that comes with heterogeneity. The more individuals that are planted, the greater the likelihood that some will be placed in microsites where competition is insufficient enough to permit establishment. To achieve establishment of a wide range of species it is generally necessary to plant every year over a long period of time; to see this process as ongoing rather than a one-off. This will statistically greatly increase the likelihood of positive outcomes in a process that is largely uncontrollable. As a result, this process is ideally undertaken by community groups, supported logistically by the managing authority.

There has been some research on whether it is possible to increase successful establishment of forbs in grassy swards by providing a competition-free space (e.g. Davies *et al.*, 1999) – normally referred to in ecological terms as a 'gap' – around each planted specimen. These gaps can be made either mechanically by stripping off small patches of turf or more economically by spraying out a circle or square of grass with a glyphosate-based herbicide 8 weeks before planting. Work on trees shows that the creation of competition-free gaps is immensely helpful to establishment (Riginos, 2009); trees grow much bigger in year one and cast more shade and compete for water more effectively, and gradually these advantages translate into the balance of competitive dominance switching from the herbaceous ground layer to the trees. This is not the pattern that is generally observed in planting forbs in grass-dominated swards. In these, providing a gap significantly improves the amount of forb biomass made in the first year, but then in the second year, when the gap generally closes due to grass reseeding or vegetative encroachment from the edge, forb biomass is either static or declines (Hitchmough, 2009). In the third year, if the forbs are (i) well-fitted to the environment they are planted into and (ii) placed in a microsite which allows them to establish, they may then begin to increase their biomass. Where this is not the case, most of the planted individuals either remain very stressed and dwarfed, or simply die (Davies *et al.*, 1999). This fitness for the site can often be thought of in terms of soil moisture regimes, for example *Ranunculus acris*, *Succisa pratensis* and *Lychnis flos-cuculi* are species designed to compete in moist to wet soils, whereas *Salvia pratensis*, *Centaurea scabiosa* and *Scabiosa columbaria* are plants that are most competitive in dry soils. In horticultural terms all of these species will grow satisfactorily under low levels of competition in an average garden soil, but when subjected to competitive pressure, as in a meadow sward, the range of tolerance shrinks back to that defined by the species' evolutionary niche; i.e. in the above examples, wet soils or drier soils.

Fitness is also much affected by growing season air temperatures, sometimes directly, but also indirectly. *Salvia pratensis* is a continental European species adapted to warm summers, which has outlier populations in the most continental (climatically speaking) parts of the southern UK. In northern parts of the UK, it will grow seemingly satisfactorily but is often eliminated by the greater slug densities associated with cooler, moister sites. On very sandy or gravelly soils outside its natural range it may perform well due to reduced densities of molluscs, but in competitive grassy swards this is unlikely to be the case.

For species that are well-fitted to the site, the likelihood of successful establishment is not significantly improved in the longer term by providing a competition-free gap at planting. It is a case of what will be will be. Again whilst the reasons for this are in logical terms clear, it is quite challenging to horticultural thought that a technological approach cannot override a background ecological process. Greater understanding of these interactions can be found in Hitchmough (2000, 2009).

The statistical likelihood of overcoming the problem of serendipity, i.e. being placed in the right microsite, clearly increases with the number of seedlings planted, providing the species used are sufficiently fit for the meadow environment under consideration. Establishing forbs from sowing seed *in situ* potentially involves very large numbers of seedlings and is therefore at least notionally an appealing way to diversify grass-dominated swards.

To maximize the chances of success, sowing needs to be undertaken in autumn, immediately after cutting the grass off as close to the soil level as possible, and then heavily scarifying the surface to ensure seeds are in physical contact with mineral soil. Some UK meadow forbs (e.g. *Ranunculus acris*) are typically autumn germinators (Grime *et al.*, 1996) and will potentially germinate in the months post-sowing. Many species, however, have some type of dormancy, either mechanical (e.g. hard seed coats) or physiological, which inhibits germination. In many species this dormancy is effectively overcome by sitting in cold wet soil over winter and experiencing thawing and freezing cycles, with germination occurring mainly in March to May in the northern hemisphere.

Once these seedlings germinate and emerge, however, the challenge is to reduce their loss through competition for light and water with the surrounding established grasses. The more fertile the site, the more productive will be the grasses, and the greater the mortality of the seedlings. Practices such as mowing in spring to defoliate grass and maintain light access to the soil level seem sensible practices, but because cutting may cause lateral grass shoot growth this may actually negate many of the expected benefits (Hitchmough *et al.*, 2008). As a result, high levels of mortality are to be expected and in productive grasslands this over-sowing practice will only very slowly lead to species establishment and diversification. As with planting, over-sowing will be most successful when carried out every year, until good sized populations of the desired species have been established. At that point seed rain from within the swards will become much more intense, and over-sowing will no longer be required.

One strategy that may be useful to establishment in northern Europe is to reduce productivity of the sward by establishing the hemiparasite *Rhinanthus minor*. This is an annual species, which after germination attaches via haustoria onto the roots of the most common species in a grassland. In a grass-dominated meadow these will mainly be grasses, however, *R. minor* is far from grass specific; it will also attach onto forbs where these are abundant. Indeed, legumes such as clover are often preferred as hosts (Jiang *et al.*, 2008). This is a process of statistical chance; species that are common get parasitized more because there is a greater chance of an individual being next to a germinating *R. minor*.

The hemiparasite 'borrows' carbohydrates and water from the host plants, although the amount of biomass it then produces is less than would have been the case if the hosts had kept their own water and carbohydrates. The net result is therefore a reduction in the standing biomass of the meadow, in effect a reduction in total productivity as might occur with meadow on much less fertile soils. In the author's experience, these drops in biomass are often large; in one experiment (Hitchmough, unpublished) *R. minor* reduced dry biomass from approximately 800 g m^{-2} to around 375 g m^{-2}. So *R. minor* can potentially be very helpful, however, the difficulties in establishing the species are several fold. Firstly, *R. minor* seed is quite expensive, and the quality of much of the seed of this species on sale in the native wildflower industry is often poor, so the cost of establishing each *R. minor* is relatively high. Secondly, *R. minor* requires a long period of chilling to be able to germinate (typically in excess of 190 days), so given that it typically germinates very early in the year (often in February), it needs to be sown in late summer or early autumn.

The final problem with *R. minor* is that it itself is often subject to competitive elimination by shading when sown into highly productive meadow communities. Hence *R. minor* does not always do what is expected of it. However, it is worth considering, especially as cycles of cutting and removal of meadow biomass begin to reduce the productivity of the sward. There are many other hemiparasites that could be used in meadows in other parts of the world, although the evidence is that they do not reduce productivity in the same way as *R. minor*. The genus *Pedicularis* has its main distribution in Western China but its species are found everywhere in the northern hemisphere, and include many that are highly attractive. In western North America *Castilleja* is another major hemiparasite genus, again with spectacular floral structures.

7.6 Creating 'Meadow' Communities from Scratch

Establishing meadow-like communities from scratch has a number of advantages and disadvantages over working with existing grasslands. In contrast to the latter, it provides the potential to create heavily forb-dominated, even grass-free communities that are florally dramatic within two years. Research into public attitudes to wildflower meadows confirms that the more flowers there are, the more attractive the meadow (Lindemann-Matthies and Bose, 2007). In the urban political arena this can be used to obtain the support of a broad section of the community and contrasts with the 'glacially' slow approach of working with existing grasslands. Creation from scratch allows a much more bespoke approach. The capacity to manage grass (at least temporally) out of areas to be sown allows a much greater range of species, both native and non-native, to be established. This allows the meadow to be designed to flower for longer, or at times of the year when grass-dominated meadows are potentially very unattractive. The approach can also be used to better target resources (such as nectar and pollen) to invertebrates at times when these are in short supply.

On the negative side, creating meadows from scratch involves significant capital costs for site preparation, seed/plants, sowing and initial maintenance, which are largely additional to the costs of working with existing grasslands. In addition, its success requires a workforce with the capacity to understand the process and its practical deliverance. When dealing with new projects in conjunction with construction work, many of these costs will be expended in any case, whatever the landscape design envisaged, and indeed even the most ambitious sown meadows will appear to be relatively inexpensive compared to most of the alternatives, such as conventional decorative planting, roll-out turf or hard surfaces.

7.7 Choice of Plant Community

Most people have romantic notions of meadows, envisaging purely native herbaceous plant communities in the countryside. By the early 21st century, outside of a few national parks and areas of outstanding natural beauty, this is something of a bygone world in the intensive agricultural landscape of many developed countries. Designed, sown meadows offer a far more extensive range of possibilities ranging from those based on semi-natural meadows drawn from the native country or elsewhere in the world, through to those completely synthetic-designed communities of native and exotic, or indeed entirely exotic, species.

Meadows

The term 'meadow' is most usefully applied to vegetation that is regularly cut for hay every year, normally between summer and autumn. After the clearing of forest, generally well underway in many temperate parts of the world by the Bronze Age, grazing of the resulting herbaceous understorey alternated with cutting for hay for winter-feed. This resulted in the classic 'lowland hay meadow' model. Lowland (and many upland) meadows are therefore a semi-natural agricultural vegetation, based on a reordering of native plant species through human management. In addition to these, there is another type of meadow found at high altitudes, often above the tree line. These 'alpine meadows' are natural and exist without cutting as the summer growing season is too cool for woody plants to invade; unlike agricultural meadows they cannot become anything else without a major shift in the climate. Typically, alpine meadows occur on unproductive soils, in high rainfall climates, throughout the mountains of the world. Classic examples are found in the Rocky Mountains of North America, the Alps of Europe and Caucasus, Himalaya and other mountains chains in Asia (Fig. 7.3). They are often grazed by wild ungulates but are otherwise stable.

Steppe

The next type of meadow-like vegetation is that of steppe (Fig 7.4), a word derived from Russian but with its precise origins remaining obscure. Unlike meadows, steppe landscapes are not regularly cut for hay although they are often grazed by domestic livestock. The general conception of steppe is a dry grassland associated with continental climates, i.e. with alternating hot, relatively dry summers and very cold winters, with often large diurnal temperature fluctuations even during the summer. The most characteristic

Fig. 7.3. An alpine meadow at 4000 m in western Sichuan, China; a climate similar to that of the UK.

Fig. 7.4. Steppe communities well-adapted to the continental climate of Romania.

genus of steppe, certainly in Eurasian and in some southern hemisphere steppe biomes is the spear, or feather, grass (*Stipa* spp.). Although associated with large tracts of continental Eurasia, naturally occurring steppe is also found in Eastern Europe, parts of central Spain, and southern central France. Some steppe landscapes are historically burnt every spring to remove dead overwintering plant material and produce a new, more palatable growth for livestock.

Prairie

The third major meadow-like grassland type is prairie, which is typically considered to be restricted to

North America but which has similarities with tall grassland and forb communities in other parts of the world, such as eastern South Africa, and parts of China and Asia. Prairie appears to be a semi-natural vegetation, in which the species associated with clearings and edges of woodland have been reordered into a distinctive plant community through the combination of grazing by bison, deer and antelope in combination with regular aboriginal spring burning. Some plant geographers see prairie as a component of steppe, which makes sense if one considers the very dry end of the prairie continuum, but not at all when one looks at the moist (mesic) to wet end of the spectrum, where prairie may be 2 m tall and essentially have an enclosed canopy and huge biomass.

Probably the main issues to bear in mind when making decisions on what type of meadow-like community to create is the environmental conditions of the site and the resources available for management. To be successful there is a fundamental need to match the ecological requirements of the proposed species with that of the site conditions. If, for example, it is a post-industrial site around a new building where the soils are composed of crushed concrete and rubble, then this would be excellent for a very dry meadow, steppe vegetation or a very dry prairie. Conversely, if the site is on very good, highly productive soil, species characterized by large biomass, such as tall leafy stem meadow species, or prairie, should be preferred. In contrast, smaller-growing stress-tolerating species associated with dry unproductive sites will quickly be eliminated by the most vigorous species sown or planted as part of the community, and in the longer term by invading species from the outside. Responding thoughtfully to productivity and moisture gradients is the bedrock for creating sustainable meadow-like vegetation.

All meadow-like plant communities in the UK ultimately invade with tall weedy, native grasses such as *Dactylis* and *Arrhenatherum*, unless active management is put in place to prevent this. The seed of these species is dispersed by small mammals and is widely distributed sooner or later. If it is anticipated that there are not the resources to prevent this process taking place then it is best advised to use species and communities that are more tolerant of invasion by these and other native grasses. Tolerance of these grasses is best developed in forb species that have co-evolved with what are cool season C_3 (spring growing) grasses, i.e. species native to maritime Western Europe (Table 7.4). These species have deployed chemical defences to be relatively unpalatable to slugs and snails. So, whilst their growth is reduced by competition for light with grasses, the increase in shading and humidity – conducive to molluscs – caused by the greater cover of grasses will not typically result in their being eaten out by the slugs and snails. North American prairie forbs would not fare so well in this situation, having not needed to evolve such defences in their native habitat.

7.8 Designing a Seed Mix

In choosing meadow seed, one option is to buy 'off the peg' seed mixes, which are readily available particularly for native species, and to a lesser degree native and non-native species (such as Pictorial Meadows™ products in the UK). If using the conventional native wildflower meadow products, it is strongly advised that the grass component of the seed mix is omitted (or at least reduced to <10% of total). The native meadow mixes in the UK are still nearly always based on the work of the Institute of Terrestial Ecology from the 1970s: 80% grass seeds (by weight or seed numbers) and 20% forb seeds. These recommendations came about by wishing to 'green-up' sown areas quickly and also because grass seed is cheap and forb seed is more expensive. The problem with these mixes is that they tend to create a grass-dominated community from the outset, rather than the forb-dominated one most people want to see. The latest generation of Pictorial Meadows™ sowing mixes have been developed in conjunction with the author, Nigel Dunnett and other members of the University of Sheffield's (UK)

Table 7.4. Species that tend to 'hold their own' in grass-dominated plant communities in the UK.

Species native to Western Europe	Exotic species
Centaurea nigra	*Buphthalmum salicifolium*
Galium mollugo	*Euphorbia palustris*
Geranium pratense	*Lychnis chalcedonica*
Knautia arvensis	*Malva alcea*
Lythrum salicaria	*Papaver orientale*
Malva moschata	*Veronicastrum virginicum*
Primula veris	
Primula elatior	
Primula vulgaris	
Sanguisorba officinalis	

Department of Landscape to overcome this problem, at least for UK conditions.

If environmental horticulturalists wish to produce a bespoke sowing mix, they need to develop a list of the species to include, and then try to imagine how the various elements will compete with one another. Based on this they will need to calculate approximately how many seedlings of each species they would wish to establish in each m^2. In most commercial seed mixes this is little more than a very crude guess! A more precise way is to use a spreadsheet in which the numbers of seed found in a gram weight is ascertained and correlated with estimates of likely percentage field emergence. For example, the author has identified typical values for almost 1000 species, which are typically between 10% and 30%. The final piece of information needed to work out how much seed of each species to put in a seed mix is the target density: for example, is the aim to have 1, 5 or 20 seedlings of each species per m^2? Once this is determined a formula needs to be added to the final column that then integrates these three pieces of information to calculate the number of grams of seed for each m^2 to be sown within the mix. This process is described in detail in Hitchmough (2004). Examples of seed sowing rates with respect to the desired number of plants per m^2 are given for a number of different communities/environmental scenarios (Tables 7.5–7.9).

The species-rich turf vegetation, of which an example is given in Table 7.5, has been developed to a high degree by Smith *et al.* (2015) through planting plugs. Whilst this is a very expensive treatment, as a concept it has considerable merit, but probably needs to be achieved by sowing seed *in situ* to be commercial applicable to most public open spaces.

7.9 Seed Management and Establishment

Site preparation for sowing

An important reason why many wildflower meadows in urban areas disappoint or just completely fail, is because they are normally sown into cultivated soil that contains both viable fragments of vegetative weeds and/or a numerically huge bank of viable weed seeds. Control of perennial standing weeds should be undertaken at least 4 months before the proposed autumn sowing, to allow a minimum of two cycles of spraying glyphosate during the active growing season of the weeds or grasses currently occupying the site. One application rarely gives the necessary level of control. Cultivation post-herbicide cycles is only required when the soil is genuinely heavily compacted.

Cultivation provides the disturbance cues that lead to mass germination and emergence from within this weed seed bank, leading (on fertile soils) to the death of many of the slow-growing sown species as a result of shading from the fast-growing weedy species that dominate the soil weed seedbank. To gain more reliable establishment of species, it is necessary to switch off the germination of these weed seeds. This is most readily achieved by spreading a 75–100 mm layer of sowing mulch that does not contain a weed seed bank, on top of the existing soil surface. This layer inhibits germination

Table 7.5. Seed mixes used for species-rich lawn/spring meadow community at the London Olympic Park, UK, 2012.

	Target plants (number m^{-2})	Weight of seed required (g m^{-2})
Grasses		
Agrostis vinealis	50	0.011
Agrostis capillaris	50	0.011
Anthoxanthum odoratum	30	0.067
Carex flacca	20	0.111
Cynosurus cristatus	50	0.104
Festuca rubra	50	0.167
Lolium perenne	50	0.333
Trisetum flavescens	50	0.067
Total grasses	350	0.871
Forbs		
Achillea millefolium	50	0.042
Bellis perennis	50	0.031
Cardamine pratensis	50	0.123
Galium verum	50	0.132
Hypochaeris radicata	50	0.208
Leontodon autumnalis	50	0.179
Leontodon hispidus	50	0.278
Lotus corniculatus	50	0.500
Plantago lanceolata	50	0.125
Primula veris	50	0.263
Prunella vulgaris	50	0.250
Ranunculus bulbosus	50	0.952
Thymus polytrichus	50	0.125
Trifolium pratensis	50	0.333
Total forbs	700	3.542

Table 7.6. Seed mixes used for hay meadow on moister slopes at the London Olympic Park, UK, 2012.

	Target plants (number m^{-2})	Weight of seed required (g m^{-2})
Forbs		
Achillea millefolium	5	0.003
Agrimonia eupatoria	1	0.200
Betonica officinalis	10	0.250
Centaurea scabiosa	3	0.133
Deschampsia cespitosa	5	0.005
Festuca ovina	20	0.077
Galium mollugo	5	0.016
Galium verum	15	0.036
Geranium pratense	5	0.303
Geranium sanguineum	3	0.200
Knautia arvense	5	0.083
Leucanthemum vulgare	10	0.017
Linaria vulgaris	10	0.006
Malva moschata	5	0.125
Origanum vulgare	15	0.100
Primula veris	15	0.079
Prunella vulgaris	10	0.042
Ranunculus acris	10	0.250
Sanguisorba officinalis	5	0.179
Succisa pratensis	5	0.238
Trifolium pratense	1	0.006
Total forbs	160	2.41

Table 7.7. Seed mixes used for hay meadow on drier slopes at the London Olympic Park, UK, 2012.

	Target plants (number m^{-2})	Weight of seed required (g m^{-2})
Forbs		
Calamintha nepeta	10	0.01
Campanula glomerata	10	0.01
Centaurea scabiosa	10	0.44
Daucus carota	10	0.05
Echium vulgare	5	0.08
Festuca ovina	10	0.04
Galium verum	20	0.05
Leontodon hispidus	10	0.11
Leucanthemum vulgare	10	0.02
Linaria vulgaris	5	0.00
Lotus corniculatus	5	0.03
Malva moschata	5	0.13
Origanum vulgare	20	0.01
Primula veris	20	0.14
Prunella vulgaris	10	0.04
Salvia pratensis	5	0.05
Scabiosa columbaria	20	0.29
Thymus polytrichus	20	0.04
Total forbs	205	1.74

from the soil seed bank below not only by reducing diurnal temperature fluctuation and increasing CO_2 levels, but also by causing the death of many small-seeded weeds as they try to emerge through the layer to reach the light. Any material that is relatively low in nutrients and which can store some water and supply this to germinating seeds can be used; final choice comes down to cost, availability and its moisture supplying properties. Two materials that are particularly widely available in urban areas are sand, and composted green waste. Sand is heavy to move around but it creates a very well-drained surface, which on a large scale is hostile to slugs, and also reduces productivity to some degree. Green compost (composted green waste) is lighter but depending on its C:N ratio may increase productivity. With sand, irrigation is required to ensure good germination in the absence of substantial and regular spring rain, whereas germination on compost is normally more reliable without irrigation. The flip side to compost is that it is a great receptor site for the germination of blown-in weeds during the first year. Where it is possible to do so, and particularly on sites where visual expectations of the meadow vegetation will be very high, temporary irrigation between March and June in the northern hemisphere is essential for success. It is the combination of severe spring moisture stress and competition with weeds from the soil seed banks that causes poor performance in many sown perennial meadows.

Sowing practice

Uniform distribution is very important in creating successful meadows because without it, patches with few seedlings present will always be invaded by weeds because of the lack of competition pressure from the sown species. Similarly, areas of high density lead to competitive elimination of sown seedlings by a process known as self-thinning (see Hitchmough *et al.*, 2004).

Table 7.8. USA prairie seed mix used at Oxford Botanic Gardens, UK.

	Target plants (number m^{-2})	Weight of seed required (g m^{-2})
Forbs		
Agastache rupestris	1	0.0056
Amorpha canescens	1	0.0101
Asclepias tuberosa interior form	5	0.1563
Aster oblongifolius	1	0.0023
Echinacea pallida	2	0.0400
Echinacea paradoxa	5	0.1250
Echinacea purpurea 'Prairie Splendor'	4	0.0533
Erigeron glaucus 'Albus'	2	0.0027
Eryngium yuccifolium	2	0.0294
Geum triflorum	5	0.0196
Helianthella quinquenervis	1	0.0011
Liatris aspera	5	0.0455
Liatris scariosa 'Album'	5	0.0549
Mirabilis multiflora	1	0.0114
Oenothera macrocarpa var. incana	5	0.0980
Penstemon barbatus 'Coccineus'	4	0.0211
Penstemon cobaea	3	0.0364
Phlox pilosa	5	0.0490
Rudbeckia maxima	0.25	0.0091
Rudbeckia missouriensis	3	0.0100
Ruellia humilis	1.5	0.0556
Silphium laciniatum	0.33	0.0314
Silphium terebinthinaceum	0.33	0.0413
Solidago speciosa	2	0.0016
Stokesia laevis 'Omega Skyrocket'	1.5	0.0682
Total forbs	66	0.762

Given the substantial difference in the physical size of seeds in sowing mixes (from 50 to 20,000 seeds per gram), in most cases the most successful way to establish meadows is by hand sowing. This sounds very slow but a team of one person sowing and two people mixing the seed can result in typically 2000–3000 m^2 per day being sowed. Given that sowing rates (forbs only) are generally about 1–1.5 g m^{-2}, it is necessary to sow with a bulking carrier. The best material for this (because it is light and very visible on a sand surface) is generally sawdust, although unless there is a sawmill close by, this is sometimes challenging to procure in volume. Sowing in two passes produces the required uniformity. The calculated seed required for the area is halved. One half is then mixed with the number of square metres of the area to be sown in terms of handfuls of sawdust. For example, if the area is 1000 m^2, 1000 handfuls of sawdust are placed in a container (often clean wheelbarrows). Hand sizes differ, but a 15-litre builder's bucket typically holds around 100 handfuls. The seed is then carefully mixed with the sawdust; doing this by hand in wheelbarrows is generally easier than using cement mixers. As it is only possible to mix about four builders' buckets maximum in a wheelbarrow, a number of barrows are required and the seed for the first pass must be divided as evenly as possible between these. For inexperienced sowers, three string lines at 1 m intervals are laid across the long axis of the site and the sower carrying a bucket of seed and sawdust walks down the lanes sowing the seed with a swinging arm action. The entire area is sown in this way, moving the string lines as required across the site. The process is then repeated as a second pass with the remaining half of the seed. The seed is then raked into the sand with a landscape rake. Where disturbance by local wildlife or domestic animals is anticipated (e.g. foxes or cats) and no irrigation is possible, covering the sown area with an open weave jute erosion matting appears to improve germination but adds about £1.00 m^{-2} (US$1.50 m^{-2}) to the cost of the process, so it is most likely to be used on prestige, relatively small-scale sites.

Post-sowing maintenance

Species have different temperature requirements for germination, so germination is staggered across quite a long temporal window. Woodland species often germinate at very low temperatures; *Primula elatior* and *P. vulgaris* for example may germinate in February under UK conditions, whereas heat-demanding prairie species such as *Asclepias tuberosa* do not emerge until June. In the main, however, seedlings of most species emerge in April and May in the northern hemisphere. The main factor that maximizes field emergence (the percentage of seed sown that results in a seedling) is freedom from soil moisture stress, hence irrigation during April and May has a very large effect on 'success'. Irrigation should only involve relatively small amounts at each irrigation

Table 7.9. Steppe-like mix used at Oxford Botanic Gardens, UK.

	Target plants (number m^{-2})	Weight of seed required (g m^{-2})
Allium senescens	8	0.1778
Aster oblongifolius	0.2	0.0005
Astragalus centralpinus	0.5	0.0431
Campanula persicifolia 'Grandiflora'	2	0.0013
Dianthus carthusianorum	2	0.0080
Dianthus carthusianorum 'Ruperts Pink'	5	0.0227
Dracocephalum argunense 'Fuji Blue'	2	0.0400
Echinops ritro	1	0.1111
Eryngium maritimum	1	0.1667
Eryngium planum 'Blaukappe'	1	0.0147
Euphorbia polychroma	0.5	0.0172
Euphorbia nicaeensis	3	0.0606
Galium verum	3	0.0048
Hyssopus officinalis var. aristatus	3	0.0141
Incarvillea delavayi 'Bees Pink'	1	0.0333
Incarvillea zhongdianensis	1	0.0250
Inula ensifolia	3	0.0125
Laserpitium siler	0.2	0.0222
Limonium latifolium	3	0.0150
Linum narbonense	3	0.0791
Malva alcea 'Fastigiata'	0.33	0.0047
Marrubium supinum	1	0.0070
Papaver orientale 'Brilliant'	1	0.0011
Paradisea lusitanica	1	0.0500
Perovskia atriplicifolia	1	0.0061
Pulsatilla vulgaris	2	0.0333
Salvia nemorosa 'Blaukonigin'	1	0.0056
Scabiosa lachnophylla	2	0.0152
Scabiosa ochroleuca 'Moon Dance'	3	0.0182
Scutellaria baicalensis	3	0.0174

Table 7.9. Continued.

Sedum telephium 'Emperors Waves'	5	0.0020
Silene schafta 'Persian Carpet'	5	0.0095
Teucrium chamaedrys	2	0.0182
Veronica incana	5	0.0018
Total forbs	77	1.0654

event – enough to saturate the sowing mulch but not much more. Even with sand or compost sowing mulches, some weeds will emerge, particularly where the spread layers are too thin (layers shallower than 50 mm are not very effective at preventing weed emergence) or where sand/compost has been stored in builders' yards surrounded by weedy vegetation that distributes its seeds onto the heaps prior to sale. If irrigation is frequently excessive, this will de-oxygenate the soil beneath and this leads to massive worm casting and the deposition of weed seeds from the soil beneath on the surface of the sowing mulch.

Rogueing-out weeds is very valuable, especially if this is combined with the establishment of the sown vegetation by the end of the first growing season and where this canopy has just about closed over. If at this point the landscape is almost free of major weeds, then typically, the future is very promising as the balance in the community is very much in favour of the sown as opposed to invading species. Where weeds dominate, the long term prognosis is generally poor, although in meadow vegetation types that are cut as hay in summer, this process is very helpful in checking many common weeds. Rogueing-out weeds is generally best left to about June/July in the northern hemisphere, as by then the difference in the vigour of the weeds and sown species is obvious in most cases. Where staff have limited ability to differentiate between weed and ornamental plant seedlings, then a useful process is to 'arm' these staff with 'thumbnail' photographic images (laminated on an A4 card, which individuals hang around their necks on a lanyard whilst weeding) that can be used as a quick identification guide (Hitchmough, unpublished). This greatly helps identification and reduces the risk of weeding-out the sown species. Sowing some of the seeds in small labelled pots helps maintenance staff become familiar with the seedlings too, a vital part of the training process.

With hay meadow-type vegetation, mowing off at a height of 75 mm can be used to reduce weed competition in the first year, and is particularly useful with annual weeds. With prairie vegetation types that are less tolerant of cutting during the growing season this process is potentially less beneficial as regrowth of the prairie species post-cutting is often slower. Prairie-like vegetation typically requires more input during the first year to establish successfully.

Conclusions

- Grasslands and forb-rich meadow communities generally predominate where abiotic or biotic factors inhibit the establishment of taller, woody vegetation. Such factors include temperature and moisture extremes, but also fire, grazing pressure or mowing (hay-making).
- These types of meadow are found as native plant communities in many parts of the world, and often contain a high proportion of very attractive species that can be used to support biodiversity and provide exciting visual experiences for people living in urban areas.
- In urban areas a move away from close-mown turf grass to meadow-like communities has a number of advantages in terms of energy use, labour costs and urban biodiversity. Developing an iconic, species-rich and colourful 'traditional meadow' with consistent performance year-on-year can be challenging, however. A useful compromise is to introduce spring-flowering geophytes, or limit the floral contribution to the more vigorous forbs such as *Taraxacum* (dandelion) and *Ranunculus* (buttercup) spp. through altered management of the sward, e.g. timing and frequency of cutting.
- Due to eutrophication increasing soil productivity, in most urban grasslands, forb density and diversity is low and very slow to develop even with sensitive management. Most desirable species are often locally extinct and hence provide no seed source. Similarly, the chances of the more colourful and floriferous forb species regenerating from the seed bank are low.
- Flowering forbs can be introduced to grass-dominated swards, but chances of successful establishment are strongly determined by the vigour of the grass species present (e.g. as determined by nutrient/moisture levels), the size of transplant (larger plants being more resilient than plugs or seed for example), shade and mollusc tolerance of the introduced species, and the persistence/patience of the management team to introduce plants over a number of years to try and build up a viable population.
- In many cases a more realistic strategy is to create meadows by sowing from scratch, as a means of accelerating the developmental process. Although this requires a larger input of capital resources at the outset, the use of low productivity substrates such as crushed building materials or subsoil, changes the habitat sufficiently to enable a much wider range of species to persist and thrive.
- The key to developing a community of colourful flowering forbs is to choose species 'ecologically fit' for the site. This includes using species adapted to similar climates, soil fertility and moisture regimes, and in many situations also species that possess some tolerance to mollusc grazing pressure and are able to compete with grasses and other taller growing forbs for light.
- A combination of severe spring moisture stress plus competition with weeds from the soil seed banks often causes poor performance in many sown perennial meadows.

References

Cobham, R. (1990) *Amenity Landscape Management: a Resources Handbook*. Spon Press Ltd, London.

Davies, A., Dunnett, N.P. and Kendle, T. (1999) The importance of transplant size and gap width in the botanical enrichment of species-poor grassland in Britain. *Restoration Ecology* 7, 271–280.

Grime, J.P., Hodgson, J. and Hunt, R. (1996) *Comparative Plant Ecology: a Functional Approach to Common British Species*. Chapman and Hall, London.

Herbert-Bormann, F., Balmori, D. and Gebale, G.T. (2001) *Redesigning the American Lawn: a Search for Environmental Harmony*. Yale University Press, Connecticut.

Hitchmough, J.D. (2000) Establishment of cultivated herbaceous perennials in purpose sown native wildflower meadows in south west Scotland. *Landscape and Urban Planning* 51, 37–51.

Hitchmough, J.D. (2004) Naturalistic herbaceous vegetation for urban landscapes. In: Dunnett, N. and Hitchmough, J.D. (eds) *The Dynamic Landscape, Design, Ecology and Management of Naturalistic Urban Planting*. Taylor and Francis, London.

Hitchmough, J.D. (2009) Diversification of grassland in urban greenspace with planted, nursery-grown forbs. *Journal of Landscape Architecture* 4, 16–27.

Hitchmough, J.D. and Wagner, M. (2011) Slug grazing effects on seedling and adult life stages of North American Prairie plants used in designed urban landscapes. *Urban Ecosystems* 14, 279–302.

Hitchmough, J.D., de la Fleur, M. and Findlay, C. (2004) Establishing North American Prairie vegetation in urban parks in northern England: 1. Effect of sowing season, sowing rate and soil type. *Landscape and Urban Planning* 66, 75–90.

Hitchmough, J.D., Paraskevopoulou, A. and Dunnett, N. (2008) Influence of grass suppression and sowing rate on the establishment and persistence of forb dominated urban meadows. *Urban Ecosystems* 11, 33–44.

Humbert, J.Y., Ghazoul, J., Richner, N. and Walter, T. (2012) Uncut grass refuges mitigate the impact of mechanical meadow harvesting on orthopterans. *Biological Conservation* 152, 96–101.

Jiang, F.W., Dieter Jeschke, W.D., Hartung, W. and Cameron, D.C. (2008) Does legume nitrogen fixation underpin host quality for the hemiparasitic plant *Rhinanthus minor*? *Journal of Experimental Botany* 59, 917–925.

Junge, X., Jacot, K.A., Bosshard, A. and Lindemann-Matthies, P. (2009) Swiss people's attitudes towards field margins for biodiversity conservation. *Journal of Nature Conservation* 17, 150–159.

Lindemann-Matthies, P. and Bose, E. (2007) Species richness, structural diversity and species composition in meadows created by visitors of a botanical garden in Switzerland. *Landscape and Urban Planning* 79, 298–307.

Lloyd, C. (2004) *Meadows*. Cassell, London.

Marrs, R.H. (1985) Techniques for reducing soil fertility for nature conservation purposes: a review in relation to research at Roper's Heath, Suffolk, UK. *Biological Conservation* 34, 307–332.

Murray, P.J., Hopkins A., Johnson, R.H. and Bunn, S. (1996) Feeding preferences of the grey field slug (*Deroceras reticulatum*) for dicotyledonous species of permanent grassland. *Annals of Applied Biology* 128, 74–75.

O'Reagain, P.J. (1989) Leaf silicification in grasses: a review. *Journal of the Grassland Society of South Africa* 6, 37–42.

Riginos, C. (2009) Grass competition suppresses savanna tree growth across multiple demographic stages. *Ecology* 90, 335–340.

Smith, L.S., Broyles, M.E.J., Larzleer, H.K. and Fellowes, M.D.E. (2015) Adding ecological value to the urban lawnscape. Insect abundance and diversity in grass-free lawns. *Biodiversity and Conservation* 24, 47–52.

Townsend-Small, A. and Czimczik C.I. (2010) Correction to 'Carbon sequestration and greenhouse gas emissions in urban turf'. *Geophysical Research Letters* 37, L06707, doi:10.1029/2010GL042735.

Wells, T.C.E. (1980) Management options for lowland grassland. In: Rorison, I.H. and Hunt, R. (eds) *Amenity Grasslands: an Ecological Perspective*. Wiley, Chichester, UK.

8 Bedding and Annual Flowering Plants

Ross W.F. Cameron

> **Key Questions**
> - What is a 'bedding plant'?
> - How are they used in the landscape?
> - What biomes/environmental conditions have promoted the evolution of colourful flowering annuals?
> - Why have bedding plants a poor reputation in terms of environmental sustainability?
> - What is meant by 'plug technology'?
> - Why is F_1 seed so important to the bedding plant industry?
> - What attributes should growing media provide during intensive production of bedding plants?
> - How do temperatures and light regimes during production affect plant quality at retail?
> - What are the key stresses bedding plants are exposed to during the transport and retail stages?
> - How does water availability affect bedding plant performance after planting in the outdoor landscape?
> - What management techniques are involved in establishing flowering annuals through direct sowing into the soil?

8.1 Introduction

Bedding plants are annuals, biennials or short-lived perennials (often treated as annuals) that traditionally were 'bedded out' in flower beds in parks and other civil or municipal areas from the 1820s onwards. These formal bedding schemes were at their zenith during the Victorian era, and grew in popularity in the UK when the glass tax was repealed in 1845 (glasshouses being required to germinate and grow-on the tender 'half-hardy' annuals). 'Formal' bedding relates to plants being arranged in specific colour groups and designs, often using straight lines and geometrical patterns, rather than more ad hoc (informal) bedding plant arrangements. Before the tax on glass was relaxed, the usage of half-hardy flowering annuals tended to be restricted to gardens of large estate houses, whose owners could afford to grow-on plants under heated glasshouse conditions prior to bedding-out these colourful plants into the garden. After the repealing of the glass tax, the subsequent widespread use of glass and the introduction of newer, larger glasshouses meant that the germination and early development of bedding plants could be more easily justified.

The majority of bedding plant species and derived genotypes are utilized for their flamboyant flower colours and ability to produce flowers for much of the active growing season (repeat flowering), although some are popular due to foliage characteristics too, such as *Plectranthus scutellarioides* (formerly *Coleus blumei*). The free-flowering characteristics of bedding and other ornamental annual plants relates to key ecological adaptations associated with rapid flower formation and seed setting in their natural environment before adverse conditions ensue. For example, a number of bedding plant species are derived from desert ephemerals, which only have a limited period to complete their life cycles, after significant rainfall events have replenished soil moisture levels. Indeed, most annual flowering ornamentals originate from arid biomes where there are distinct rainy seasons either for discrete periods in winter, or alternatively in summer. Species adapted to climates with winter rains include *Eschscholzia, Nemophila, Clarkia, Limnanthes* (California); *Calceolaria, Salpiglossis, Nicotiana* (Chile and other parts of South America); *Gazania, Diascia, Lobelia, Dorotheanthus* (southern Africa);

Calendula, Lathyrus, Antirrhinum, Papaver (Mediterranean basin). Others evolved in biomes with summer rains, namely *Cosmos, Salvia, Zinnia, Tagetes* (Mexico and Central America) and *Impatiens, Primula, Dianthus* (China and central Asia).

Although, for today's gardening public, bedding schemes usually connote vibrant flower displays, foliage was important historically, in that 'carpet' bedding schemes developed with the aim of providing a close cover of foliage that formed a backdrop for the flowering plants. Low-growing species such as *Sedum, Echeveria, Alternanthera, Senecio* and *Sempervivum* being ideal for such situations. Carpet bedding arose in popularity during the 1850s and early 1900s showing itself to be highly adaptable to commemorative or civic planting, incorporating town names, shields and other designs.

Bedding plants are still used in civic town-centre and park displays (Fig. 8.1), as well as along roadsides (Fig. 8.2) and business parks, but are also utilized extensively in the private garden. Indeed, many garden centres in the northern hemisphere rely on the late spring sales of bedding plants to ensure they remain in profit for the rest of the year. Sales are very weather dependent and fair/sunny weather over Mothering Sunday, Easter, May Day Bank Holiday weekends and other country-specific holidays are often critical in determining if a garden centre or nursery is to have a successful season. The contribution of bedding plants to the horticultural sector is significant and the wholesale value of annual bedding plants in the USA is in the region of US$1.4–2.0 billion per annum (Anon., 2013).

8.2 Commercial Production of Bedding Plants

Bedding plants embrace a large range of species and cultivars, but the most popular include *Tagetes patula* (French marigold) and *Calendula officinalis* (marigold), *Begonia semperflorens* (begonia), *Pelargonium × hortorum* (geranium), *Impatiens walleriana* (busy Lizzie), *Impatiens hawkeri* (New Guinea impatiens), *Viola × wittrockiana* (pansy and viola), *Petunia × hybrida* (petunia) and *Salvia splendens* (salvia). In the USA, *Petunia* represents the greatest value of any single crop, although more pots of both *Viola* and *Pelargonium × hortorum* cuttings are sold (Tables 8.1 and 8.2). Traditionally, bedding plants were defined as hardy or half-hardy, the former being able to be sown from seed out of doors, compared to the latter that required protected environments (glasshouses or polythene tunnels) during the propagation and early growing-on phases. These half-hardy species were planted-out once threat from late spring frosts had receded.

Each year seed houses release new varieties of these 'old favourites' to maintain market interest, improve robustness and flowering attributes or introduce

Fig. 8.1. Bedding plants are still commonly used in formal designs within city parks.

novel strains (new, unusual colours or floral morphologies). Increasingly though, new species are being introduced and novel cultivars raised that fall outwith these old classifications of hardy/half-hardy. These are tender perennials that are propagated each year with the notion of providing summer flower colour, but with plants not being retained through to the following year. In the UK and elsewhere the prime functions of these plants are to fill containers, pots and hanging baskets around the house or on garden patios. Many derive from warm climatic regions such as southern Africa (*Osteospermum*, *Diascia*, *Gerbera*) or Central (*Zinnia*, *Dahlia*) or South (*Calibrachoa*) America and therefore do well on the hot summer microclimates of hard-surfaced patios. This has led to the additional term 'patio plants' being used to describe plants with these traits. Another distinction from the traditional bedding plant is that many of these species are perpetuated, not by sowing seed, but by being propagated vegetatively from cuttings e.g. the Surfinia range of petunias or *Fuchsia* hybrids.

Extending the season of interest has been a prime goal for bedding plant producers, and bedding species are not solely excluded to summer displays. Indeed, bedding is defined by the season of interest, with spring bedding exploiting biennials (sown one year to flower the next), or hardy, short-lived perennials that flower early in the growing season (February–May in northern latitudes). Biennials include *Erysimum* (biennial forms also classified as *Cherianthus* – wallflowers), *Bellis perennis* (daisy) and *Myosotis* (forget-me-not), and the short-lived perennials typified by *Viola* (pansy) and *Primula* (polyanthus). These plants are frequently used in combination in formal schemes with spring-flowering bulbs, including *Tulipa* (tulip), *Hyacinthus* (hyacinth), *Narcissus* (daffodil) or *Muscari*. Bedding schemes were transformed by the introduction of the winter-flowering pansy in 1979. Winter flowering pansy

Fig. 8.2. Bedding plants are often the first choice of plant subject for containers in public spaces or along roadsides.

Table 8.1. Numbers of trays, hanging baskets and pots of different bedding plants sold as wholesale products in 2013 in the USA. (Source: Anon., 2014.)

		Numbers (million)		
Scientific name	Common name	Trays	Hanging baskets	Pots
Petunia × hybrida	petunia	6.78	4.92	28.11
Pelargonium × hortorum (cuttings)	geranium	0.34	3.13	34.93
Pelargonium × hortorum (seed)	geranium	0.47	0.38	17.35
Viola wittrockiana	pansy and viola	6.76	1.11	30.43
Impatiens walleriana	busy Lizzie	5.52	2.19	15.09
Begonia semperflorens	begonia	3.94	2.04	21.89
Impatiens hawkeri	New Guinea impatiens	0.28	2.25	16.92
Tagetes and *Calendula*	marigold	3.43	0.12	10.66
Other		17.8	14.65	186.57

Table 8.2. Wholesale value of trays, hanging baskets and pots of different bedding plants sold in 2013 in the USA. (Source: Anon., 2014.)

Scientific name	Common name	Value (US$ million)			
		Trays	Hanging baskets	Pots	Total
Petunia × hybrida	petunia	53.97	30.17	45.83	129.97
Pelargonium × hortorum (cuttings)	geranium	3.98	23.01	86.18	113.17
Pelargonium × hortorum (seed)	geranium	3.47	2.62	18.21	24.3
Viola wittrockiana	pansy and viola	58.4	6.16	34.7	99.26
Impatiens walleriana	busy Lizzie	44.5	11.6	17.85	73.95
Begonia semperflorens	begonia	31.27	12.82	27.82	71.91
Impatiens hawkeri	New Guinea impatiens	2.79	16.43	33.75	52.97
Tagetes and *Calendula*	marigold	29.34	0.53	11.43	41.3
Other		148.6	110.03	354.53	613.16

(and later the smaller-flowered viola version) were bred to produce flowers in the short days of winter, possess enhanced tolerance to cold, rain and wind, and to retain a compact habit when weather conditions improved. In reality, their most abundant floral displays are associated with the more mild periods in autumn and spring, when flower production is supported by (relatively) good irradiance levels and warmer temperatures. Such traits are shared by other autumn-/spring-flowering species such as *Cyclamen* and *Primula × polyantha*). Summer remains the paramount season for bedding plants, however, with the greatest variety of genotypes being available for flowering from May to September in the northern hemisphere.

Seed companies often produce cultivars of many bedding plant species within a series (or strain). These usually have a similar genetic background, although cultivars within the series will have different flower colours or markings. Selecting a variety of coloured cultivars from the same series allows the grower to provide the market with variety whilst keeping cultural requirements uniform. Much effort is put into developing new series and cultivars within a series, and due to the demand for novelty an individual seed house may introduce as many as 80–100 new selections each year.

Much of this breeding effort has been centred on developing more flamboyant strains, with emphasis on double or multiple flower forms. To increase the impact of flower colour, traits have been preferentially selected to increase petal size, or alter the hues within the petal. This has often resulted in the elimination or diminished size of other parts of the inflorescence, including the nectaries and pollen sacs. To this extent certain genotypes, and bedding plants in general, have been criticized for reducing their potential to supply invertebrates (as well as pollen/nectar feeding bats and birds) with 'high-energy' food sources. Varieties of *Dahlia*, for example, with 'open flowers' allowing easy access to nectaries/pollen were found to be visited by pollenating insects 20 times more frequently than forms bred for their double or multiple petals (Garbuzov and Ratnieks, 2014). Concerns have also been expressed by the use of genetic modification (GMO technology) for ornamental plant varieties, where the introduction of new genes promises unusual colour 'breaks', e.g. a blue flowered *Dianthus* (carnation) or increased drought tolerance, but where these modifications may enhance risk due to factors such as pollen-mediated gene flow, 'weediness' characteristics, enhanced vigour and resilience with implications for invasion of natural areas, as well as direct harm to non-target organisms (e.g. invertebrates that may feed off the modified plants). In reality, the introduction of genetically modified ornamental plants has been limited to date, largely by the regulatory costs of introducing the new genotypes. This is due to ornamentals having a much smaller market value than food crops and the fact they are often traded internationally, i.e. the market does not justify the registration and trialling costs required for analysis and risk assessment when the sales potential is limited, and there is a requirement to meet the regulations of multiple countries.

Although there has been a trend in recent years for bedding plants to be sold in individual pots, reflecting the marketing of more exclusive cultivars and larger individual plants that can command a greater price, the traditional bedding plant is sold in a tray or 'flat'; these often being composed of

6, 8 or 12 plants within a tray. The original wooden seed trays that seedlings were pricked out into had no compartments, but modern plastic and polystyrene trays are divided into a number of cells, with the volume of individual cells dictating the size of plant required at the retail stage. Most bedding plant sales rely on some of the plants in the tray being in flower (in colour) to attract the customer and help guarantee a sale. In contrast, etiolated plants with numerous dead/dying flowers, chlorotic leaves and dry growing media indicates the tray has exceeded its 'shelf-life' and should be removed from retail or sold at discount. The objective for grower and retailers is to ensure most trays of plants are sold before they reach this stage. Some bedding species are now sold 'in the plug' (e.g. 45 × 30 mm), i.e. small plants not in flower, with the notion the customer will grow them on for themselves before they are planted in the final location.

8.3 Propagation and Production Factors

Despite the trend for cuttings to maintain clonal stock, the majority of bedding plants are still propagated from seed. Spring/summer bedding species are sown from late December into March (northern hemisphere), although batches of the same cultivar may be sown sequentially in a scheduled production to allow these cultivars to be marketed over an extended period. In large-scale production nurseries, seed is sown into cellular plug trays via a mechanized seed planter. Plug trays come in different sizes (number of cells × volume of individual cell), and may range from 50 to 800 individual cells. The use of plug cells, which encourages the development of robust, coherent 'rootballs', helps the young seedlings transplant into their final trays without excessive root disturbance (transplant shock). The small volume of the plugs, however, requires the growing media to be very fine grade so it flows into the spaces easily. Fine-milled and sieved coir or peat is commonly used for this purpose, as are smaller grades of perlite and vermiculite. As with seed distribution in plug trays, transplanting the young plant from the plug tray to the cell pack or final tray is done mechanically. Compared to traditional seed sowing in open trays and pricking-out by hand, plug technology has the advantage of:

- Faster transfer of young plants and reduced labour inputs – each plug being a discrete unit that can be handled by machinery.
- Accelerates production times as no transplant shock (check to growth) occurs.
- Regulates growth and keeps seedling size uniform as each seed has the same space/substrate volume, which acts to hold back more vigorous seedlings and encourages the slower developing ones to catch up.
- Optimizes handling and space utilization in glasshouses/polytunnels.
- Reduces the risk of root diseases spreading, as each plug is isolated.
- Allows large numbers of plug plants to be shipped to other nurseries (e.g. some nurseries will buy in plugs rather than grow from seed themselves).

The disadvantages are largely due to capital costs, with mechanized handlers of seed and seedlings being expensive.

Seed

Reliable and consistent germination is key to ensuring plants are produced in predictable schedules. Sowing date is calculated based on the marketing period for any given crop, but the length of production schedules can vary depending on time of retail. Early spring season batches, for example, take longer to produce, due to predominantly cooler growing temperatures or lower irradiance levels, compared to those seed batches sown later under warmer/brighter conditions (Fig. 8.3). Sowing dates, therefore, are carefully calculated by factoring-in, not only the time to retail the crop, but also the environmental factors that are likely to affect growth rates and flower initiation. Consistency in this respect, within a bedding plant series is increased by the use of F_1 seed.

F_1 seed is derived from two parental strains that each have been in-bred (self-pollinated) to develop a consistent phenotype (pure line). These parental pure-lines may have a specific desirable trait, such as unusual flower colour, or good branching habit. The F_1 is obtained by crossing these two pure lines to produce seeds that inherit both positive traits. The F_1 generation retains the consistency of the parental lines, with the selected seed promoting a uniform, high-quality bedding plant crop. In addition, as the seeds come from two very distinct parental lines, they possess hybrid vigour, i.e. the progeny tend to be stronger, more vigorous plants than either of the parental in-bred lines. In this way it is feasible to

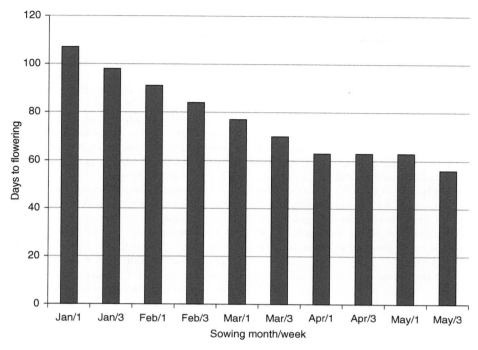

Fig. 8.3. The effect of sowing date on time to flowering based on a *Petunia × hybrida* crop grown in the northern hemisphere under protected (glasshouse) conditions.

express both the outstanding qualities of the parental plants and new desirable characteristics from the hybrid itself. In addition to qualities including vigour, true-ness to type, frequent and consistent flower production and a high level of uniformity which hybrid plants enjoy, other characteristics such as precocity to flowering (earliness), disease resistance and good plant form have been incorporated into most F_1 hybrids. F_1 seed tends to be more expensive than other seed types, not only due to the fact that each year new crosses need to be made from the two parental lines (ensuring there is no cross-contamination from any other pollen sources), but also to the fact that the parental pure lines themselves may take many years to develop. Perhaps eight or nine generations of self-pollinated plants are required before a pure line is created.

F_2 bedding plant hybrids also exist, resulting from the self- or cross-pollination of an F_1 plant. These retain some of the desirable traits of the original F_1 and can be produced more cheaply as no intervention in the pollination is required.

Heavier or larger seeds are desirable in that there is a correlation between size and viability. Larger seeds tend to have higher germination rates, germinate more rapidly than smaller ones and promote stronger early development of the seedling. This is because seed size relates to the resources the seed receives during its development on the parent plant. Thus, well-managed parental stock (kept under high irradiance, with consistent supplies of water and nutrients, and maintained free of pests and pathogens) is a requirement to ensure seed quality is optimized.

To guarantee that only the best quality seeds are actually sown, a number of techniques are employed by seed houses. These include 'refining' the seed, where they are cleaned and physically separated by size, shape, weight or density. Mechanical techniques have been developed to remove unusual or misshapen seeds from any given batch of seed. Seed may be 'primed', i.e. soaked in a solution with a high osmotic potential so that the seeds embibe enough water to increase fresh weight, but then further developmental processes are arrested due to the inability to embibe any further water. In essence, a number of pre-germination metabolic processes are activated, but emergence of the radical does not occur. The seeds are then removed from the solution, dried, packaged and labelled with a 'sow-by date'. Nevertheless, they are 'primed' to germinate once in contact with soil and exposed to the moisture in the substrate.

The advantages of primed seed are that germination is more uniform across the batch and seed can tolerate a wider range of environmental conditions at the germination stage. Another technique is pelletization, where seeds are covered with a thick coating to increase their volume and thus facilitate sowing by a mechanical seeder. Pellets are composed of a coating material held together by a binding substance, and include various starches, cellulose, clay and vinyl polymers in their construction. The binder substance needs to be strong enough to avoid the pellet being broken or chipped when mechanically handled, yet pliable enough to allow the pellet to split when the seed enlarges and the radicle/hypocotyl emerge.

Bedding plant propagation is divided into four stages:

- Stage 1: radicle emerges from the seed coat.
- Stage 2: stem and hypocotyl (seedling leaves) emerge.
- Stage 3: first true leaves emerge and develop.
- Stage 4: seedlings become established in the plug and are ready for transplanting, shipping (moving on to other nurseries or garden centres) or holding.

Plug technology – including the requirement for capital investment in mechanized seed sowers, roller benches and carefully controlled environmental germination rooms – means that some nurseries now specialize in plug production alone, selling-on plants to 'finishing' nurseries, rather than encompassing the whole production process. Conversely, other nurseries will solely buy in plugs, transplant these on to larger module sizes and become more specialized on the husbandry required to control growth and meet the specific demands for finished plants in the garden centre or landscape markets.

Germination

Bedding plants are largely derived from annual, biennial or short-lived perennial species, many of which have developed evolutionary strategies to germinate under conditions that will increase the chances of establishment and survival. These may vary though depending on the specific ecophysiology of the species. As such, some bedding plant species require a range of environmental parameters to facilitate germination, whereas others may have a different set of optimal criteria. This is best demonstrated with the requirement for light. Germination is promoted in species such as *Amaranthus*, *Nicotiana*, *Lobelia*, *Cosmos* and *Antirrhinum* when light is provided, whereas light inhibits germination in *Cyclamen*, *Nigella*, *Tropaeolum* and *Phlox*. The wavelength of the irradiance is the fundamental component in many cases, with the ratio of red to far-red light influencing germination through the action of phytochrome.

Plant origin also impacts on temperature requirements. Species from tropical and subtropical regions require germination temperatures >10°C (below which physiological chilling injury of the seed/seedling may occur). Other species can tolerate temperatures as low as 5°C. Some 'cool-requiring' species require temperatures <25°C, with germination being inhibited at higher temperatures (e.g. *Delphinium*, *Primula* and *Viola*). During germination, media moisture content should be sufficient to allow the seed to absorb water, whilst retaining oxygen within the medium. The media needs to be consistently moist as temporary drying out shortly after germination is a common source of seedling failure.

Media

The relatively small volumes of growing media (substrate) accessible to individual plants in plug or cell trays requires that the medium must have high performance criteria. The medium needs to hold water between consecutive irrigations, maintain aeration through an effective matrix of pores, retain sufficient nutrients and make these available to developing roots, whilst buffering against excess ion release by providing a high cation exchange capacity (CEC). Cation exchange capacity regulates the form and proportion of chemical ions that bind to the particles of the medium, compared to those found 'freely available' in the substrate solution. The former are generally unavailable to plant roots, whereas the latter are readily absorbed by the roots. Removal of ions by roots from the solution, however, results in the further release of ions from the particles themselves due to osmotic gradients. The ability to absorb and release ions in this dynamic process, however, varies with the composition of the medium. The advantage of materials such as peat that have high CEC, is that they can hold large quantities of nutrient ions, but only make these available to roots, as the roots themselves deplete their immediate supply. In this way effective CEC helps avoid the build-up of high ionic concentrations in the solution that would directly damage roots. Ideally, the media itself should be relatively inert, minimizing the potential release of any injurious chemical elements, or altering pH or ionic concentrations radically. Consistent performance is also a requirement and this can be challenging when the medium itself is not of a uniform composition between different batches or suppliers. With media such as peat or green compost, there is variation in the source material and hence key performance criteria may differ between suppliers or even at different times of the year. In addition, media should be delivered and stored in such a manner as to minimize risk due to weed seed contamination or inoculation with plant pathogens.

Most bedding plant producers in the UK and elsewhere still rely on peat as the main component of the growing media, although coir and amending mixes with bark are becoming more common as concerns are raised over the environmental implications associated with peat exploitation. In England and Wales, UK, 312,000 m^3 of peat were used by the bedding plant sector alone during 2005 (Waller, 2006), with 87% of production reliant on peat. (This equated to 29% of the professional market use of peat.)

Peat is the primary growing media based on its high water holding capacity (some peats can hold 6–8 times their own weight in water) and good aeration. It has a low bulk density making it popular within horticulture for providing volume but not weight. It decomposes slowly and can release low concentrations of nitrogen (N) (being composed of 0.6–1.4% N). Sphagnum peat lowers pH, but this is amended by additions of calcium carbonate or magnesium carbonate, whilst ions associated with fertilizer supplements also influence its pH. One disadvantage of peat is the difficulty in re-wetting the substrate once it becomes excessively dry (i.e. its hydrophobic nature), and so surfactants are added to the media to aid water absorption where there is a risk of this occurring.

Fine barks are also used in bedding plant mixes. Barks of softwood tree species are often composted before use to degrade phytotoxic organic compounds and break down larger fragments. The composted bark is then sieved to provide different fractions sizes, with 3–10 mm diameter particles being used in 'growing-on' mixes.

Vermiculite and perlite are used on occasions within the bedding plant industry. Vermiculite has good CEC, water holding and aeration properties. In addition to being used as the sole medium in plug trays, it can be used as a light covering over seeds to aid germination. Vermiculite can compress (slump) over time after heavy watering and lose some of its beneficial properties. Perlite has greater stability and is used in peat mixes to further improve drainage and aeration.

Nutrition

Fertility is not an essential factor during germination as most seeds have enough stored nutrients. Fertilization in plug production normally starts as liquid feeding at production stage 2 with dilute concentrations (e.g. 25–50 ppm N) being applied at first, followed by increasing concentrations (100–330 ppm N) from stage 3. Actual rates of fertilizing will vary depending on species, growth stage, frequency of application, leaching rates as well as environmental conditions such as temperature and irradiance. The ratio of the macronutrients N:P$_2$O$_5$:K$_2$O and proportion of micronutrients applied alters with growth stage or time of year. (Note N:P$_2$O$_5$:K$_2$O is the convention for describing the relative proportions of nitrogen, phosphate and potassium fertilizers, respectively, but does not mean they are necessarily in these precise chemical forms.) Even the ratio of ammonium (NH$_4^+$) versus nitrate (NO$_3^-$) nitrogen may be important for some genotypes, and background pH of the medium also affects nutrient absorption by the plant. Young bedding plants of many species grow taller and have deeper green foliage after feeding with NH$_4^+$ compared to NO$_3^-$, although sub-factors such as pH and phosphate (P$_2$O$_5$) levels also influence the extent of responses. Ammonium nitrogen (NH$_4^+$) tends to acidify the medium whereas additions of nitrate (NO$_3^-$) raise the pH, making it more alkali. As the exchange of ions across the root membrane also alters the ratio of H$^+$ and OH$^-$ ions present, plants too directly influence the pH of the medium. Root activity of *Begonia*, *Celosia*, *Dianthus* and *Viola* reduce pH, whereas *Tagetes*, *Vinca* and *Zinnia* tend to raise pH. Paradoxically, on occasion these changes can induce a nutrient deficiency or toxicity in the plant.

During production stages 3 and 4 it is not uncommon for growers to fertilize the crop with liquid feeds based on 15-15-15 or 20-10-20, N:P$_2$O$_5$:K$_2$O, (the numbers relate to the percentage of each compound in the fertilizer) perhaps with supplemented additions of key formulas that include calcium or magnesium. At these later stages, some growers will provide nutrition through the use of controlled release fertilizer granules incorporated into the media, rather than apply via liquid feeds. These granules are coated in plastic and release nutrients slowly into the medium as the granule weathers, the process being temperature dependent. This form of fertilization can help growers differentiate between water and nutritional needs, optimize fertilizer use and reduce nutrient leaching. Small granule sizes help ensure each cell receives its due allocation of granules. The granules help extend nutritional feeding through to the retail and even garden stage too (e.g. 2–3 month formulations). The drawbacks are that growers can no longer regulate growth via alterations in the nutrients made available to the crop, and that as nutrient release is temperature dependent, mis-matches can occur between greatest release periods and crop need. For example, excessively warm periods early on during production release more nutrients than the crop actually requires at that particular stage. Growers will monitor crop schedules carefully with respect to the desired retail period, and adjust fertilization (along with irrigation, temperature or even irradiance) if crop development needs to be delayed or accelerated.

Temperature

Temperature can be controlled in glasshouses and polytunnels by providing a heat source, such as steam heat pipes, or cooling the environment through ventilation, e.g. ridge or side vents in a glasshouse, or roll-up sides in a polytunnel. In general, crops respond to increased temperature by increasing rates of vegetative growth and flowering. Generally, temperatures tend to be somewhat higher for stage 1, lowered for stages 2 and 3, and reduced further during stage 4. Prior to shipping and retail stages, temperatures may be dropped even further. Indeed, many crops are 'finished' outdoors during the warmer spring/summer months where although there is less control over temperature, the generally cooler temperatures and wind movement (thigmomorphogenesis) help plants harden-off (acclimate) before shipping and retail.

Optimum temperature ranges within each stage vary between species, but when housing mixed crops growers will seek compromise temperatures that suit the needs of all and will avoid excessively high or low temperatures that can impair metabolic development. Reducing the duration of individual production schedules is a key driver for growers so as to increase the number of crops that can be grown within a given season. Providing the optimum temperature accelerates time to flowering, with many bedding plant species responding to temperatures between 25 and 30°C (Fig. 8.4).

Temperature affects the vegetative habit of bedding plants, with high temperature (>25°C) encouraging internode extension whereas lower temperatures (10–14°C) promote short internode sections resulting in very compact plants. Day temperatures have a stronger role in inducing these characteristics than night temperatures. Research with *Petunia* showed that plant height increased by 25 mm when night temperatures were increased from 10 to 30°C, whereas the same temperature changes when applied during the day enhanced height by 114 mm (Kaczperski *et al.*, 1991).

Production schedules that 'force' crops and encourage larger plants and rapid flowering rates, however, do not necessarily enhance the number of flowers produced (excessively high temperatures can even abort flowers initiated). A study involving two cultivars of *Petunia*, 'Easy Wave Coral Reef' and 'Wave Purple' showed that the rate of flower development increased and that the date of first flowering was earlier when daily mean temperature rose from 14 to 26°C (Blanchard *et al.*, 2011). Time of first flowering decreased from 51 to 22 days for *P.* 'Easy Wave Coral Reef' and from 62 to 30 days for *P.* 'Wave Purple' (all plants being grown under a light integral of 12 mol m^{-2} d^{-1}; the light integral is derived from different intensities of irradiation and photoperiods). In contrast, high temperatures reduced the number of flowers present at the onset of the flowering period, especially at lower light integrals. For example, flower number at 23°C and 4 mol m^{-2} d^{-1} was reduced to one-third of those of other plants grown at 14°C with 18 mol m^{-2} d^{-1}. Lower temperatures and greater light integrals also encouraged shorter, more compact plants (Fig. 8.5).

Light

Bedding plant development is regulated by three components of light – irradiance (intensity of light), photoperiod (daylength) and wavelength or spectrum (light quality). For those growing in high latitudes, natural irradiance levels early on in the

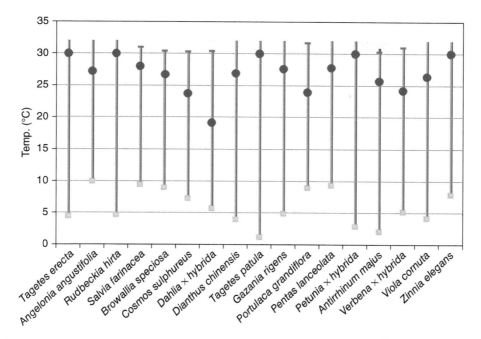

Fig. 8.4. Temperatures that optimize flowering rates in a range of bedding species (●), with known temperature minima (■) and maxima (—) where flowering can still occur. (Modified from Blanchard and Runkle, 2011.)

Fig. 8.5. Lower growing temperatures and high levels of irradiance (left) help keep a *Petunia* × *hybrida* crop more compact.

production schedule can be insufficient – e.g. winter light levels in northern European countries being only 20% of those of mid-summer. Reduced transmission rates due to dirty glass/polythene sheeting and associated infrastructure (frames, stanchions, heating units, fans etc.) of the production environment make the problems around insufficient natural irradiance even worse. Under such conditions, supplementary 'growth' lighting is provided to increase photosynthetic capacity of the crop, by augmenting natural light. Electric lights designed to emit light between 400 and 700 nm wavelength (photosynthetically active radiation – PAR) are used and traditionally have involved high pressure sodium, mercury, metal halide or fluorescent lamps. Increasingly, light emitting diodes (LEDs) are being researched as alternatives due to their lower power consumption and the fact that they can provide more specific photosynthetically appropriate wavelengths such as in the red and blue spectra. Despite the additional costs, supplementary lighting can be vital for some crops due to the positive effects on plant development. Providing supplementary lighting during early phases of growth is particularly important. By supplementing natural light with artificial sources, benefits accrue such as accelerated flower development, higher plant quality through a more compact growth habit (i.e. induced by shorter internode sections) and greater side-branch numbers, as well as an enhanced total biomass. Supplemental light reduces production time across a range of species. In *Cyclamen persicum* 'Metis Scarlet Red', for example, exposing plants to increasing daily light integrals reduced time to flowering to 75 days (17.3 mol m^{-2} d^{-1}) compared to 133 days (1.4 mol m^{-2} d^{-1}). Increasing daily light integrals from 1.4 to 11.5 mol m^{-2} d^{-1} increased flower numbers from 0 to 15 per plant, and average leaf number from 9 to 28 (Oh *et al.*, 2009).

Photoperiod regulates flowering in a number of bedding species, but manipulation of photoperiod has been under-utilized for bedding plant crops (in contrast to flowering pot-plants and cut flowers) when grown to meet their normal marketing and retail period. Many spring-flowering species are sensitive to photoperiod and flowering time can be brought forward by extending the natural photoperiods. Flowering was hastened by 10 days in *Osteospermum* 'Pink Whirls' when plants were maintained under 16 h compared to 8 h photoperiods (Pearson *et al.*, 1995). In *Petunia* 'Express Blush Pink', flower development rate increases linearly with lengthening photoperiod up to 14.4 h d^{-1}, after which there is no further effect (Adams *et al.*, 1998).

As most 'long-day' plants are actually 'short-night' plants, providing a night break illumination can be used as an effective means to induce flowering without the energy requirement of extending daylength for a number of hours at the end of each day. Researchers are actively investigating ways to do this whilst minimizing the electricity costs to growers. For example, Blanchard and Runkle (2010) advocated the use of high pressure sodium lamps with oscillating aluminium reflectors that provide an intermittent beam of light over the growing area. As long as plants were within distance to receive irradiance levels >2.4 µmol m^{-2} s^{-1} (about 7–10 m from the lamp in their study) flowering potential was not compromised. They estimated that such a system could save 83% of the costs compared to banks of less efficient incandescent lamps. This is one area where technological advances with light-emitting diodes may improve the economic factors further.

Light spectrum (quality) also controls the flowering response through gene activation, with red/far-red light as well as other wavelengths such as blue impacting on some species (Hori *et al.*, 2011). The use of plastic films or filters that alter light wavelengths, for example by reducing the ratio of far-red light, can improve crop quality (e.g. *Viola, Petunia* and *Dianthus*) through the inhibition of excessive shoot elongation (stretching or etiolation) and increasing the branching habit. As outlined above, the assessment of the daily light integral (DLI) can prove useful to bedding plant growers as it correlates with the amount of energy plants receive on a daily basis; it is defined as the amount of photosynthetic light that 1 m^2 receives during one day, so is dependent on irradiance, wavelength and photoperiod.

Irrigation

Rapid growth in bedding plants is achieved by maintaining a moisture content within the growing medium that provides both enough water to avoid any check in growth and good levels of aeration to the roots. Small plug/pot volumes mean that irrigation tends to be regular (two or more times a day in warm periods) and consistent. For example, in some nurseries 'ebb and flow' systems are used to ensure all plugs receive full saturation. Ebb and flow involves flooding of the bench that plant trays are placed on so that the media within cell plugs is inundated with water. The benches are then drained after a number of hours, leaving the cells at container capacity (maximum water holding capacity, whilst allowing large pores to drain and refill with air).

Uniformity of irrigation is vital if crop uniformity is to be maintained, as inconsistencies in watering lead to plants of variable size, thereby reducing the crop quality at the retail stage. The aim for the grower is to achieve a crop where each tray of plants is identical to the next in terms of stature and coverage. This does not necessarily imply though that irrigation should be kept optimum, as this encourages excessive shoot growth (optimum growth not being the same as optimum quality). Reducing the amount of irrigation can improve crop quality by fostering a more compact plant habit, and potentially improve shelf-life by pre-acclimating the plants to subsequent drought events (for example by increasing cell osmotic potential, which allows cells to resist desiccation during subsequent water shortages). Growers often 'dry-down' their crop during the final stages of production to both control excessive shoot growth, but also as a pre-conditioning treatment (hardening) prior to shipping the plants off the nursery. This should not be confused with sending the plants out with dry growing media, however; and final irrigations on the nursery should be thorough to ensure water availability is kept at a maximum during the shipping period.

Many growers are experienced at 'drying-down' the crop though visual assessments, but van Iersel et al. (2010) advocated that soil moisture sensors provide more precise control. In studies where the volumetric moisture content of the medium was held at different levels ranging from 0.05 to 0.40 $m^3\ m^{-3}$ water content, crop responses to water deficits were observed in *Petunia*. Drought stress reduced growth at water availability thresholds <0.25 $m^3\ m^{-3}$ and size was strongly correlated with the total amount of water applied. Plants on low water contents though showed high levels of physiological adaptation (altering their physiology through reductions in leaf water potential, osmotic potential and tolerating some loss of turgor pressure), yet plants in all treatments survived, irrespective of the severity of drought. Van Iersel et al. concluded that precise irrigation control can reduce water consumption whilst regulating shoot growth and can actually result in enhanced crop quality. De Graaf-van der Zande (1990), however, also investigating *Petunia* as well as *Verbena*, suggested that despite effective growth control, consistent low water availability overall resulted in loss of crop quality at the retail stage. In contrast though, once plants were planted outdoors in their final flowering position, regrowth and longevity of plant performance were excellent. Varying the moisture content during production (i.e. cycles of wetting and drying), however, was shown to improve quality at retail, but interestingly had no lasting advantage on development after planting out. The degree to which different species can be dried down or held on low irrigation regimes varies with species (perhaps again relating back to the parent species ecophysiology and climatic preferences), with adaptation capacity dependent on the level of stomatal control and osmotic adjustment (Nemali and van Iersel, 2008).

Growth regulation – chemical and management tools

As indicated above, excessive lengthening of the internodal sections of the stem (etiolation) reduces the marketable quality of a bedding plant, with retailers requiring a uniform crop of compact plants. Factors such as extremely high day temperatures, low irradiance levels or a low red to far-red light ratio promote etiolation. To overcome these effects, growers apply chemical growth regulators that inhibit internode elongation. Most of these chemicals have anti-gibberellin properties (gibberellins being the plant hormone group most closely associated with stem elongation).

Growth regulators such as paclobutrazol and chlormequat chloride have a good reputation for controlling excessive growth and maintaining a compact habit in many annual flowering species. A number also have the advantage of encouraging precocious flower formation, with opportunities to market the crop earlier than usual, e.g. flurprimidol on *Pelargonium × hortorum*, although responses are not universal across all species. Chemical growth regulators, however, are expensive and may have adverse environmental or human health impacts, or lead to long-term or permanent stunting of the crop plants (e.g. failure to grow after planting out in the landscape). Indeed, the advantages of controlling growth and maintaining a compact habit during retail do not necessarily translate into improved quality once planted out in the landscape (Fig. 8.6); for example, the excessive use of chemical growth regulators used to keep plants compact during production can subsequently result in poor growth and stunting once

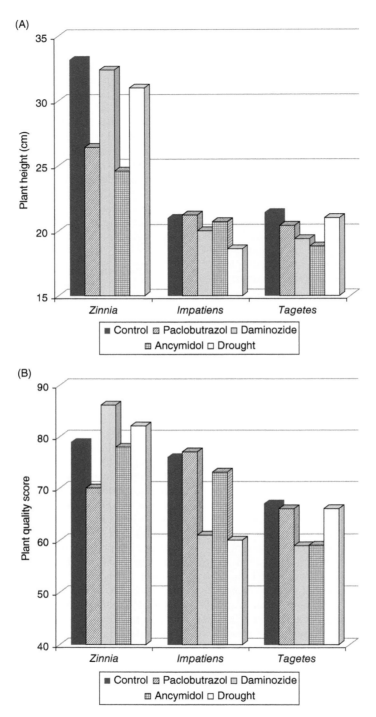

Fig. 8.6. The effect of plant growth regulator application (paclobutrazol, daminozide, ancymidol) or reduced irrigation regimes ('drought') on (A) plant height and (B) perceived overall quality, in *Zinnia*, *Impatiens* and *Tagetes*. (Modified from Latimer, 1991.)

placed in a garden setting. Similarly, suboptimal root growth in the tray due to growth regulators or excess fertilizer, may impair establishment once planted out. Due to the increasing costs of registering agrochemicals too, unless the market demand is sufficiently high for such chemicals (for example they have applications relevant to agriculture as well as horticulture) they may simply be withdrawn, as the costs of trialling and registration outweigh any potential profit from sales for the parent agrochemical company.

Environmental manipulation is also exploited to control plant extension growth. This is either through restricting water availability at certain key stages (see section on irrigation, above), or manipulating temperature within the production glasshouses. By altering temperature regimes between day and night periods (the 'difference in temperature' or DIF) more compact plant specimens are encouraged. In general, a warm night and cold day results in more compact, smaller plants. A 17/23°C day/night regime (i.e. negative DIF of 6°C) for example, decreased plant height in *Begonia, Calceolaria, Impatiens, Pelargonium, Petunia* and *Salvia* compared to constant temperature regimes (Mortensen and Moe, 1992). An alternative strategy to DIF is DROP – a short-term temperature reduction either shortly before or shortly after sunrise (or an alternative 'lighting phase'). The effectiveness of DROP seems to correlate with the differential in temperatures the plant experiences (larger differentials or longer durations of lower temperature promoting more compact habit; Langton *et al.*, 1992; Langton and Cockshull, 1997). These techniques presuppose that temperature can easily be controlled in this manner and that it is cost effective to do so.

Regulating irrigation to reduce water availability (and/or increase the electrical conductivity (EC) of the growing media) results in a compact growth habit, by inhibiting cell elongation. This needs to be implemented carefully and requires equipment that can uniformly distribute the irrigation water. Precise irrigation control accommodates homogeneous crops and avoids injury due to excessive periods of moisture stress. Mechanical brushing of crops (thigmomorphogenesis) is an alternative mechanism to inhibit internode extension. This form of mechanical stress is thought to elicit ethylene synthesis in the shoot tip and inhibit growth through the action of this hormone. In *Viola*, 10–20 brushes a day, 5 days a week gave a growth reduction of 25–30% without an adverse effect on crop quality

(Garner and Langton, 1997), but similar approaches were relatively ineffective in *Pelargonium* and *Petunia*.

Most of the factors outlined above help crops retain quality during subsequent shipping and retail (see below). These 'pre-conditioning' or 'toning' treatments include reducing watering during production, but ensuring a final thorough watering before transport itself. Growing the crop at cooler temperatures (2°C less), brushing the crop and applying a single chemical growth regulator were also promising techniques for maintaining quality during transportation.

8.4 Sustainable Production

The predominant use of polypropylene and polystyrene trays and pots to hold bedding plants has led to criticism relating to excessive waste associated with production. Recycling of these materials takes place in some countries, but large amounts still end up in landfill sites or other waste streams. This has acted as a catalyst for research into biodegradable containers that intrinsically have a lower carbon footprint and degrade rapidly after use (although retaining their integrity during production remains an important consideration). Some products are now composed of biodegradable materials such as paper, cardboard, wood, or indeed growing media held in place via gels and polymers – for example pots comprising peat or coir fibre. A number of these pots allow the plant to be placed directly in the ground without removing it from the container; the roots breaking through the organic material to reach the parent soil. Although such approaches are considered more sustainable, concerns arise as to how they impact on production (growers citing problems due to higher costs, problems with storing and handling, premature degradation, lack of consistency/reliability in the product, greater volumes of irrigation water required and potential impacts on crop quality). Nevertheless, some materials have proven useful. Kuehny *et al.* (2011) evaluated a range of biodegradable pot types with crops of *Pelargonium, Catharanthus* and *Impatiens* and they concluded that marketable plants for both the retail and landscape markets were achievable. Specimens held within biocontainers and directly planted in the landscape generally performed well, matching the performance of plants grown in plastic containers and planted out in the conventional manner. They advised, however, that growers may need to alter production protocols to attain the optimum performance from such new pot types.

McCabe *et al.* (2014) postulated that pots made of natural fibres could be strengthened by the inclusion of certain polymers, thereby improving their reputation with commercial growers. They compared the effectiveness of containers made from wood, coir and paper with and without coatings of four biopolymers (tung oil, polyamide, polylactic acid and polyurethane; the latter derived from castor oil rather than petroleum-based hydrocarbons). All the coatings, bar the tung oil, increased the strength and durability of containers and reduced the requirement for frequent irrigation. Plants grown in such containers were larger, healthier and of a better quality than plants grown in uncoated or tung oil-coated containers. Indeed, plants grown in coated paper or coir pots were comparable to plants grown in conventional plastic containers. In addition, using biopolymers slowed the degradation of fibre pots when exposed to soil, but critically did not halt it. Overall, the authors recommended the use of paper pots coated with polyurethane, as this water solvent biopolymer was cost effective, simple and easy to apply and resulted in plant development akin to that of conventional pots. Although biodegradable pots may be challenging for growers, Yue *et al.* (2010) indicated they are more popular with consumers compared, for example, to recycled pots.

Although consumers favoured biodegradable pots, Yue *et al.* (2010, 2011) found that there was little appeal for ornamental plants to be grown within organic production certification schemes, as is common with edible crops. Presumably, this is because the consumer is connoting some health benefit with organic food production (still a controversial point scientifically), and this is irrelevant if the plants are not to be consumed. In contrast, work by Burnett and Stack (2009) suggested that buyers of ornamental plants would pay a premium of 10% for bedding plants grown to organic standards. Perhaps more of a motivating factor for gardeners buying bedding plants is where they are grown, with some evidence for preference in buying from local producers, both to reduce the environmental footprint associated with long-distance transport as well as a means to support local businesses and employment.

Motivations for growers to adopt sustainable practices are not always easy to translate into practice, especially if such practices cannot provide a clear economic justification. Factors affecting the adoption of more sustainable practices by growers were divided into five areas: environmental regulation, value to the customer, growers' attitudes toward sustainability, age, and operation size. Approximately 65% of ornamental plant growers surveyed thought sustainable production was important to the environment, and indeed 63% directly employed some sustainable practices in their operations. Positive attitudes, however, did not predict adoption behaviours. Sustainable approaches were often restricted in practice, through both concerns about ease of implementation and the perceived risks involved. Education was seen as a means to help growers overcome these concerns (Hall *et al.*, 2009).

8.5 Transport and Retail Stages

Growers of bedding plants spend much time and investment in ensuring that uniform, high-quality crops leave their nurseries and can thus ensure a high market value. Unfortunately, much, if not all, of this effort can be lost through poor transportation of the crops or inadequate storage and management at the point of retail. Bedding plants are particularly vulnerable in this respect, more so than many other ornamental crop types, due to the limited capacity to hold moisture (and nutrient) reserves in the small volumes of media held within the tray and plug systems. In contrast to most nurseries, there may not be the facilities or staff knowledge/time to ensure that crops are watered effectively (or at all) during transport and retail stages, and growers need to try and account for these factors when attempting to enhance crop 'shelf-life'. Practices for handling and maintaining bedding plant crops can vary radically depending on the retail outlet, market profile and expected financial return from the crop. At one end of this spectrum bedding plant trays may be left on trolleys at the point of retail (i.e. having to cope with significant shading), inadequately or infrequently watered, and exposed to damaging low or high temperatures or strong wind – the sort of conditions that might be experienced on a supermarket forecourt. In contrast, at the other end of the spectrum the crops may be laid out on specially designed retail-benches, with automated irrigation and good levels of ventilation and irradiance.

As bedding plants are a perishable product the logistics of moving them from nursery to retail area requires careful consideration and planning. Some nurseries use their own vehicles to bring plants to market, but larger business enterprises will use haulage contractors, with specialized 'cool-chain' lorry trailers. Plants may undergo more than one journey, with plants being brought to a central hub before being reorganized and dispatched to different garden centres across a region. As such the 'product' can be in transit for in excess of 48 hours (Fig. 8.7).

During these periods the crop will be without light, may be exposed to variations in temperature and humidity, and have limited ventilation. Crops will be closely spaced, being packed onto Danish trolleys (trolleys with 3–4 shelves to allow ease of movement for large numbers of plants at any one time), and with such confinement can encourage the build-up of ethylene gas, as well as plant pathogens such as *Botrytis cinerea*. In addition, it is unlikely that the crop will be watered during transit and will be reliant on the last watering at the nursery to remain hydrated.

In practice, inappropriate temperatures, exposure to ethylene and poor water management are likely to be the key factors determining bedding plant viability/quality during transit and retail. Armitage (1993, 1994) divided species into two groups based on temperature optima: 10–13°C, which includes *Calendula officinalis*, *Viola* × *wittrockiana* and *Petunia* × *hybrida*; and 15–17°C, which includes *Begonia semperflorens*, *Impatiens walleriana* and *Celosia plumosa*. Notably, these groupings appear to align closely with differences in chilling tolerance between species. Other studies suggest that the temperatures during transportation can be cooler still (Nell and Reid, 2001).

One practical problem may be the mismatches between temperature regimes set for the lorry trailer, and what the crop actually experiences. Høyer (1997) showed that compared with the thermostat set point of 16°C, mean temperatures actually recorded during different deliveries varied between 11.3°C and 21.6°C with up to 3°C differences within the body of a truck at any one time. Excessively high temperatures cause reduced shelf-life and loss of quality, especially if maintained over a prolonged period. Controlled experiments designed to simulate transport conditions indicated that storage in the warmth (22°C) for 24–48 h reduced quality in *Viola*, *Petunia*, *Cyclamen* and *Impatiens* (Cameron *et al.*, 2008). In addition to elongated shoots and subsequent loss of uniformity and habit, plants often had a characteristic yellow coloration to the new shoot tips or leaves compared to cooler storage

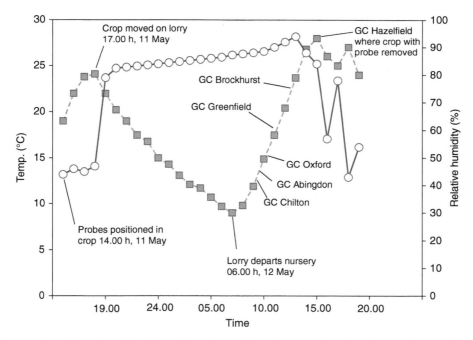

Fig 8.7. Temperature (■) and humidity (○) profiles of a bedding plant crop placed in a lorry trailer and distributed to a number of garden centres (GC) in the UK. (Source: Cameron *et al.*, 2008.)

for the same period. Overall, shorter-term storage at lower temperatures (10–12°C) was optimal for most species in these studies. Even so, loss of quality could be evident at these temperatures after extended storage, exhibited, for example, after 48 h in *Cyclamen*, *Primula* and in some cases *Petunia*.

Sensitivity to ethylene is also genotype dependent (Smith, 2001), as in the following examples:

- Very sensitive: *Impatiens walleriana*, *Pelargonium × hortorum*, *Salvia splendens* and *Antirrhinum majus*.
- Moderately sensitive: *Begonia semperflorens*, *Solenostemon scutellarioides* and *Petunia × hybrida*.
- Insensitive: *Calendula officinalis*, *Lobularia maritima*, *Tagetes* spp. and *Zinnia elegans*.

Higher transportation temperatures and poor ventilation increase the risk of ethylene injury. Ethylene concentrations as low as 0.01–5 ppm (0.01–5 µl l^{-1}) may be injurious to some species, but degree of damage is often a function of exposure time; thus emphasizing the need for good ventilation during transport and storage. Whilst monitoring the environmental conditions plants were exposed to during transportation Cameron *et al.* (2008) showed that relative humidity can reach 100% across a range of temperature regimes. Certain species demonstrated sensitivity to high humidity, e.g. *Petunia* and *Impatiens*, with flowers being particularly susceptible to injury and distortion (Fig. 8.8).

In their studies of bedding plant production and transport, Armitage (1986, 1993) and Cameron *et al.* (2008) highlighted the key issues that strongly influenced post-production quality (Box 8.1). Lack of sufficient/accurate watering during post-production stages and hence loss of quality through drought has been a concern for many bedding plant growers, especially since there is limited volume of media associated with most pack and other bedding plant containers. Cameron *et al.* (2008) found examples of commercial crops arriving at their destinations at both ends of the watering spectrum; some trays were characterized by dry growing media (the medium contracting away from the sides of the module and plants beginning to wilt) and others where foliage and flowers were drenched in water. In subsequent controlled studies *Impatiens* showed no effects of leaf wetting but there were slight negative influences on plant habit with *Petunia* and *Primula*. Flowers were more prone to injury than foliage, with *Petunia*, *Cyclamen* and *Viola* displaying flower damage when stored wet. Ideally, growers should aim to irrigate crops thoroughly 1–2 h before loading, allowing foliage to dry before being placed on lorry trailers.

The use of wetting gels and surfactants has been investigated with respect to ensuring that adequate moisture is retained in the peat-based media during transit, and to help re-wetting at the retail stage, once trays have been left excessively dry. Million *et al.* (2001) working with *Chrysanthemum*

Fig. 8.8. High relative humidity associated with flower elongation and smaller petal size in *Petunia × hydrida*. From left to right: polytunnel control; 22°C >90% relative humidity for 48 h; 12°C >80% relative humidity for 48 h.

Box 8.1. Factors to consider to avoid loss of bedding plant quality during transportation from nursery to consumer (Armitage, 1986, 1993; Cameron et al., 2008).

For the grower:

- Avoid too small a module/cell size with little capacity to store moisture.
- Avoid media that tend to become hydrophobic and are difficult to re-wet. Adding 10% loam or even hydrogels, may help in this respect.
- Late seasonal sowings tend to experience higher production temperatures, and these plants can struggle to retain quality during production.
- Chemical growth regulators may help 'hold' crops at an appropriate size and stop rapid shoot extension ('bolting') when the plants experience warm conditions/low irradiance in transit.
- Adopt acclimation 'toning' practices that 'harden' plants before moving off the nursery. When flower buds become visible in the crop consider:
  - reducing watering frequency by 50%;
  - reducing fertility concentration by 50%;
  - reducing temperature by 3–5°C; and
  - brushing the crop to keep plants compact before retailing.
- 'Clean' the crop at dispatch stage – removing any dead leaves and flowers.
- Water the crop in dispatch, ideally 1–2 h before loading, as this helps to keep the media wet, but any surface moisture on leaves/flowers has time to dry off.
- Use electrical/battery operated machinery in confined spaces as these are less likely to emit ethylene or other hydrocarbons that do the crop damage.
- If contracting out transport of the crop ensure a reliable haulage company is employed, including the use of appropriate temperature controlled wagons.

For the haulage company:

- Ensure crops are stored at suitable temperatures, e.g. 10–12°C.
- If a non-refrigerated lorry is used then travel in evening/early morning when coolest.
- Avoid rough handling of the crop or trolleys, this breaks stems, leaves and flowers and promotes ethylene release.
- Engines emit large amounts of ethylene – turn diesel and petrol engines off when loading the crop, or at least avoid the crop being placed near the vehicles exhausts.
- Try to avoid the driver exceeding permitted driving hours during the transit – a perishable crop of bedding plants will soon deteriorate if drivers additionally need to schedule in 6–7 hours of sleep.

For the retailer:

- Store/site trolleys with trays carefully if there is no room on the retail shelves themselves.
- Retailing the plants on Danish trolleys and suchlike is acceptable practice to help avoid double handling of the crop, but ensure the plants are inspected and, if needed, watered on arrival. Also remove polythene wraps to avoid temperature/ethylene build up.
- Car parks are useful for impulse purchases but are an adverse environment for most plants due to the open nature of the site (wind, direct sunlight and exposure to vehicle emissions).
- Ensure staff, including seasonal summer staff, are trained on basic plant handling and in particular, watering procedures.

For the customer:

- Do not select plants that look wilted or over-stretched.
- Place plants in a shaded car boot, not exposed to direct sunlight, after purchasing. (Ideally, make the garden centre the last destination on any shopping trip.)
- Plant out in the garden soon after purchase (at least within a day or two of purchasing).
- Remove any dead flower heads or broken branches.
- Ensure they are well-watered after planting – if the media in the tray is dry, soak the trays beforehand.

and *Petunia* concluded that surfactants needed to be added as a drench immediately prior to shipping the crop, to obtain any effect. Antitranspirants applied to the foliage of bedding species has been attempted, but with only limited success in extending shelf-life (Gehring and Lewis, 1980). Snider et al. (2003), however, claimed that spraying *Impatiens* with lysophosphatidylethanolamine (LPE), a naturally occurring lipid, increased flower retention after drought stress. The use of buffered phosphate fertilizers applied during production have also been cited as eliciting tolerance to drought, through their action of extending root proliferation within the module and inducing more compact shoot growth (a consequence of which is reduced leaf area and lower transpiration rates (Borch *et al.*, 2003)).

Abscisic acid (ABA) is a naturally occurring growth regulating 'hormone' found within plants and is often known as the 'stress hormone' as it tends to increase during stress events, and thus regulates the plant's responses to adverse factors. Abscisic acid mediates drought stress responses by closing the leaf stomata and reducing water loss. Trials have attempted to use exogenously applied ABA and some of its derived compounds to increase drought tolerance whilst plants are in transit and through the retail phase. Applying ABA drenches (500 mg l^{-1} ABA) delayed the onset of wilting and extended shelf-life by 2.4 d (*Viola*), 2.7 d (*Impatiens* and *Salvia*), 3.0 d (*Tagetes*), 3.7 d (*Pelargonium*) and 4.3 d (*Petunia*) (Waterland *et al.*, 2010a). Unfortunately, ABA application can lead to leaf chlorosis and even abscission, which obviously reduces the overall quality at retail. Working with *Viola*, Waterland *et al.* (2010b) showed that both spray (250 or 500 mg l^{-1}) and drench (125 or 250 mg l^{-1}) applications of ABA induced leaf yellowing. Stage of development may be a critical factor and young plants at the plug stage and finished plants (11 cm diameter pots) with one to two open flowers have been compared for susceptibility to chlorosis and the effectiveness of the sprays to induce drought tolerance. Both plugs and finished *Viola* developed leaf chlorosis after ABA applications, but symptoms were generally more severe in finished plants. The individual application of benzyladenine (BA), gibberellic acid (GA$_{4+7}$), or the ethylene perception inhibitor 1-methylcyclopropene before ABA application had no effect on the development of the ABA-induced leaf chlorosis. Pre-treatment, however, with *mixed* applications of benzyladenine (BA) and gibberellic acid (GA$_{4+7}$) were more promising in preventing leaf chlorosis. Kim and van Iersel (2011) working on *Salvia splendens* 'Bonfire Red' suggested that 250–500 mg l^{-1} of ABA provided both effective stomatal closure and a minimized risk of leaf abscission (Fig. 8.9). Such applications remain questionable in terms of economic benefits, but further development may mean more cost-effective treatments can be implemented to prolong drought tolerance.

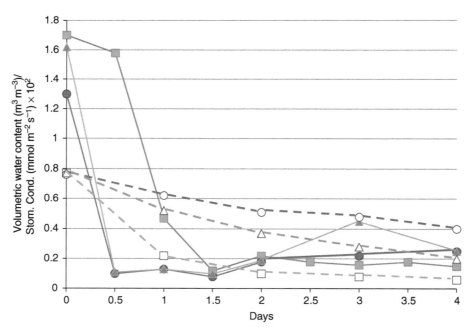

Fig. 8.9. The effect of abscisic acid (ABA) drenches on volumetric substrate water content (m^3 m^{-3}) and stomatal conductance (mmol m^{-2} s^{-1}) over the first few days of drought imposition in *Salvia splendens* 'Bonfire Red'. Water content: □, control; △, 500 mg l^{-1} ABA; ○, 2000 mg l^{-1} ABA. Stomatal conductance: ■, control; ▲, 500 mg l^{-1} ABA; ●, 2000 mg l^{-1} ABA. (Source: Kim and van Iersel, 2011.)

8.6 Establishment in the Landscape

Most bedding plants excel in full light or semi-shade, whereas heavily shaded areas promote etiolation and limit the number of flowers induced. Prior to planting outdoors, soil should be cultivated to provide a good 'crumb structure' with any resident weeds being removed. The cultivation of the soil will encourage weed seed germination – species such as *Senecio vulgaris* (groundsel) and *Stellaria media* (chickweed) in northern Europe – and cultivation sometime prior to the planting of the bedding plants can be used positively to promote early weed seed germination, at which point these can be hoed-off or sprayed with a contact herbicide. Even so, some hoeing of the ground may be required to keep weeds in check until the bedding plants cover the soil surface.

As with any other plant groupings, bedding and other annual flowering species vary in their adaptations and preference for different soil and climatic conditions (Table 8.3, Fig 8.10).

Irrigation

Bedding plants have a high requirement for water until they become established, and irrigation should take place at planting, and on a frequent basis thereafter. Hanging baskets, pots and small containers may need irrigation daily and in some cases twice daily throughout the growing season; automated drip systems with timer control proving useful under such circumstances (Fig. 8.11). One downside of automated irrigation systems is that once set up, they are infrequently recalibrated and can result in over-watering of bedding plant schemes. Automated systems based on timers have been estimated to use 60% more water than comparable systems regulated by tensiometer assessments of soil moisture. The latter control systems though have also been linked to smaller plants and some reduction in floral display (Scheiber and Beeson, 2006).

Rather than relying on tap water (with its requirements for purity and health criteria) irrigation water of a lower quality is appropriate for bedding plants. In temperate regions, this 'frees up' supplies of clean water for drinking and bathing, a point that becomes paramount during drought periods, where water is prioritized for domestic supplies. In drier regions mains water may not be used in the landscape at all, with supplies dependent on boreholes, ex-industrial or recycled water, natural water features or from sources known to have a higher EC than is compatible for drinking purposes. The question arises, however, how much can the quality of water be compromised before affecting plant development or soil ecology? Bedding plants have shown variable tolerances to water with high EC. In studies by Niu *et al.* (2010), all plants survived exposure up to 7.4 dS m^{-1} (decisiemens per metre – a measure of salinity) with the exception of *Capsicum* 'Purple Flash', but growth was reduced in the faster growing cultivars (Fig. 8.12). In conclusion, these researchers consider that bedding plants may be irrigated with saline water at salinity of 4.0 dS m^{-1} or less provided that well-drained soils or substrates are used to prevent excessive salt accumulation in the root zone.

In comparisons of water quality from various sources, *Viola* and *Antirrhinum* grown in the landscape performed well when irrigated with recycled water or water runoff from nursery beds (Arnold *et al.*, 2003). *Viola* growth during the cool season in Texas, USA, increased steadily during the winter months before increasing rapidly in April prior to suffering dieback and then declining under hotter conditions. Differences in growth and flowering were small between recycled, runoff and tap water treatments, but the extra addition of salt reduced plant growth increments particularly in later winter and early spring when more frequent irrigation was required. Growth in *Antirrhinum* tended to be greatest early on with the use of tap water, with those irrigated with nursery runoff water performing better than with either wetland recycled or elevated salt water. Differences between runoff and tap water-irrigated plants, however, were not evident in later spring measurements.

Bedding plants have a reputation for high water use in landscape settings, a trait that means many local authorities have questioned their future widespread use due to climate change and potential restrictions on irrigation water. In a similar manner financial restrictions in local authority budgets can mean that the relatively intense management associated with bedding plants (annual cultivation in the glasshouse), but also high labour requirements related to watering in containers, tubs and hanging baskets, results in their replacement with less intensely managed species. This has spurred on research to assess more closely what the water requirements of bedding plant species are, and how performance would be compromised if more restrictive irrigation regimes were employed.

Blanusa *et al.* (2009) investigated the competitive effects for water in a hanging basket scenario by vary-

Table 8.3. Common bedding plant species and soil/temperature preferences.

Scientific name	Common name	Horticultural traits
Dry or free-draining soil/warm temperatures		
Calendula officinalis	pot marigold	Single or double daisy-like flowers in orange, gold, cream or yellow hues. Long flowering season. Used as a 'companion plant' to help avoid aphid infestation.
Celosia	cockscomb	*C. argentea* var. *plumosa* and *C. spicata* have upright 'feathery' flower plumes, whereas *C. cristata* has some varieties with crested plume, similar to a cockscomb.
Dorotheanthus bellidiformis syn. *Mesembryanthemum criniflorum*	Livingstone daisy/ice plant	Iridescent, star-shaped, daisy-like blooms in shades of orange, yellow, white, cream, pink and crimson. Low growing with narrow succulent leaves.
Eschscholzia californica	Californian poppy	Single row of petals of yellow or orange in true species, but hybrids now also have reds and pinks.
Erysimum cheiri	wallflower	Adapted to grown out of walls and other dry stony environments.
Gazania spp.		Hot conditions and well-drained soils.
Iberis umbellata	candytuft	Flat inflorescence in shades of white through to deep pink – easy to cultivate.
Linaria spp.	toadfax	Similar, but smaller flowers to *Antirrhinum*, with strains such as 'Northern Lights' and 'Fairy Bouquet' popular.
Nemesia spp.		Adapted to warm, sandy soils, and found growing naturally in sand flats in southern Africa. Wide range of flower colours, including white and red bicolour form called 'Danish Flag'.
Oenothera spp.	evening primrose	Most varieties available are upright 200–800 mm tall with normally yellow flowers, although white and mauve also available.
Portulaca grandiflora	moss rose	A trailing habit with single or double miniature rose-like flowers.
Zinnia spp.		Single, double or multi-petal forms in a range of pinks, reds, yellows and orange. Rich, free-draining soil.
Moderately rich, moisture retentive, but free-draining soil – warm temperatures		
Alcea rosea	hollyhock	A biennial. The quintessential cottage garden plant – large spires of single or double flowers in a wide range of colours.
Amaranthus spp.	love lies bleeding	Long drooping panicles of flowers often red or yellow. Likes warm conditions, moist but nutrient poor soils.
Antirrhinum majus	snapdragon	Common name derived from the way the base of the dorsal petal can be squeezed to 'open' the ventral petal, like a dragon (or rabbit's) mouth. Does best on light, but fertile soils.
Centaurea cyanus	cornflower	Wild form has bright, deep blue flowers, but cultivated ones include red, pink and white too. Does better in a cool rather than a hot climate.
Clarkia amoena	godetia	Open, fan-shaped flowers of various hues of pink, red, salmon and white.
Consolida spp.	larkspur	Member of Ranunculaceae, similar to the perennial *Delphinium*.
Coreopsis spp.	tickseed	Prefers a consistently moist, if not wet soil. Flowers dominated by red and yellow hues. Good nectar source for pollinating insects.
Cosmos spp.		Open or slightly cup-shaped flowers, often on long stems. Tall plant useful in open, 'airy' spaces.
Dianthus spp.	pinks, carnations, sweet William	Flowers dominated by pinks, white and red. Strong scent in some cultivars. Most prefer high pH soils with lime.
Diascia spp.		Short-lived perennial now an archetypal patio plant, used in container and hanging baskets.
Helianthus annuus	sunflower	Popular due to the large size of plant, large leaves and grand flower heads in yellow, ochre and brown.
Ipomoea tricolor	morning glory	Climbing vine with various colours of white, purple or blue flowers. *I.* 'Heavenly Blue' has luminous sky-blue flowers. Prone to red spider mite, but high humidity and temperatures help keep this pest at bay.

Continued

Table 8.3. Continued.

Scientific name	Common name	Horticultural traits
Lathyrus odoratus	sweet pea	Annual climbing plant, used for cutting flowers for interior décor and as colourful screen outside. Nutrient-rich, free-draining soil optimizes growth and continued production of flowers.
Linum spp.	flax	Rapid growing with flower colours in blue, red or white.
Lobelia erinus		One of the oldest bedding plants, having been grown in Europe since 1752. Both compact and trailing forms, with small flowers in blue, red or white. Often used in hanging baskets.
Matthiola incana	stock	Flowers in white, mauves and pink. Dwarf varieties have been bred to stop lodging (flower heads bending over).
Nigella damascena	love-in-a-mist	Habit of self-seeding and returning year on year.
Papaver spp.	poppy	Shades of red, pinks and white. *P. commutatum* is the ladybird poppy with black spots at the base of the petals. *P somniferum* is the opium poppy, with glaucous blue/grey foliage and flowers in red, white mauve and purple.
Pelargonium × *hortorum*	geranium	Common in red, pink, white and mauve flowering forms. Scarlet varieties very characteristic of window boxes in southern Europe and the Mediterranean region.
Petunia × *hybrida*		Tolerates moderately dry soils, but flower damage evident after periods of wind and rain. Highly popular bedding plant, but not the easiest to produce via seed due to *Pythium* and *Phytophthora* diseases. Wide range of flower colours, including coloured centred, veined and striped petal forms. Surfinas are a smaller flowered group that are useful in hanging baskets due to their trailing habit.
Tagetes spp.	African and French marigold	Despite the common names for *T. erecta* (African marigold) and *T. patula* (French marigold) both are actually native of Mexico and central America. Typified by bright yellow and orange flower heads.
Tropaeolum	nasturtium	Vigorous trailing plant with flowers in red, yellow and orange. Excessively rich soils result in rampant plants with fewer flowers.
Salvia splendens		Usually a bright scarlet and popular in formal bedding schemes, now other variants available.
Verbena × *hybrida*		Upright forms are usually 200–300 mm high, although trailing forms can have longer stems. Wide range of colours.
Viola × *wittrockiana*	pansy/viola	Pansy flowers tend to be larger than viola, although both forms can be clear colours or with a black blotch at the petal base. Perform best in fertile, moist soil.
Moist, free-draining soil with some shade		
Begonia semperflorens		Best in part-shade in hot climates. Predominantly pink, red and white flowers.
Impatiens × *hybrida*	New Guinea impatiens	Needs shade in hot climates, good for containers and attractive foliage.
Impatiens walleriana	busy Lizzie	Tolerates shade and fleshy leaves prone to drought-induced wilting if inadequate moisture supply. Multiple hues of pinks, reds and white, with some striped petal forms.
Limnanthes douglasii	poached egg plant	White flowers with a yellow centre reminiscent of a poached egg.
Lunaria annua	honesty	White or purple flowers with large papery seed-heads.
Myosotis sylvatica	forget-me-not	Prefers some shade and moist soil in warm climates, but good spring bedding species in the more temperate regions. Pale blue flowers, but also now white and pink forms.
Primula spp.	primula and polyanthus	Excellent spring-flowering bedding plants. Variations in flower colour, size and stem length. Avoid excessively dry soils.
Wet or moist soils – cool temperatures		
Mimulus	monkey musk	*M. luteus* and derived hybrids require damp soil, although other species are xerophytes adapted to arid soils and dry woodlands.

Fig. 8.10. Careful selection of species is important with respect to climate and other environmental considerations. Here *Nemesia*, a genus that performs well under warm conditions, is being used to good effect in Brisbane, Australia.

Fig. 8.11. Drip irrigation pipes being used to help establish recently planted-out bedding.

ing numbers of individual plants of two bedding species (*Petunia* 'Hurrah White' and *Impatiens* 'Cajun Violet') within the basket and recorded responses to reduced irrigation volumes. To optimize performance *Petunia* required approximately 30% more water than *Impatiens*. Under a reduced irrigation regime in which plants only received 25% of the full watering, flower number, plant height and flower size in *Petunia*

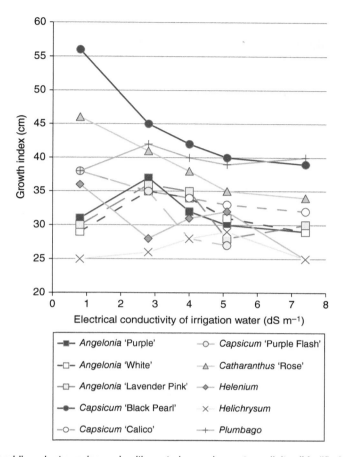

Fig. 8.12. Variation in bedding plant species and cultivars to increasing water salinity. (Modified from Niu et al., 2010.)

were reduced by 50%, 33%, and 13%, respectively. Despite there being significant reductions in key ornamental plant parameters, these were not proportional to the reduced irrigation. Indeed, visually, there was often little loss of ornamental properties in *Petunia* under the lower irrigation regime; the plants appeared compact and retained relatively good flower cover in proportion to their biomass. For *Impatiens*, however, the growing of single plants at 25% watering was plausible, but the addition of a *Petunia* plant at this regime (i.e. low watering plus competition) was detrimental to plant quality in the *Impatiens* (i.e. flower numbers were reduced by 75% compared with a well-watered single specimen). Therefore, for *Impatiens* to have acceptable quality in a mixed container with *Petunia*, near-optimal irrigation was required. Rather than eliminating bedding plant displays outright, however, the data presented suggested that acceptable bedding displays could be still achieved even when water consumption was reduced by 67% below optimum.

As outlined above, the use of hydrogels (organic polymers) has been evaluated to determine if these can help store water in the growing media and release moisture to plant roots when required. Adding such gels to the soil itself is more expensive, but has been attempted. Adding hydrophilic polymers to a sandy soil was shown to better support growth and flowering in *Petunia* (relatively drought sensitive) under dry conditions, compared to control plants (Boatright *et al.*, 1997). No significant differences were noted for species with greater drought tolerance though (*Tagetes* and *Catharanthus*). Under wetter conditions the advantage of polymer additions were less clear, and may have been detrimental to growth and flowering.

Nutrition

On nutrient-poor soils, fertilizer addition can help plant performance in the landscape, aiding establishment and promoting longer periods of growth and flowering. Although often recommended, the precise volumes of fertilizer to add is based on limited research. Comparisons using different rates of nitrogen applied to nutrient-poor subsoil (Shurberg et al., 2012) showed that less nitrogen is needed than may commonly be assumed. These studies involving *Zinnia, Catharanthus, Melampodium, Dianthus, Viola* and *Antirrhinum* used nitrogen fertilizer applied in the landscape at 0, 96, 192, 288 and 576 kg ha^{-1}. Regression analysis indicated that all species required nitrogen inputs at annual rates exceeding 400 kg ha^{-1} to achieve maximum size, shoot biomass, and photosynthetic activity. Plants of acceptable quality, however, were produced at 200–300 kg ha^{-1} nitrogen. Shurberg et al. (2012) believe that adopting these lower rates will provide acceptable aesthetic performance whilst reducing fertilizer costs and the potential for nutrient losses in runoff or leachate. Where the addition of inorganic fertilizers is not appropriate due to concerns linked to environmental sustainability, then organic alternatives can be sought or the environmental horticulturalist may consider species better suited to nutrient-poor soils, e.g. leguminous plants that can fix their own nitrogen, such as *Trifolium* (clover) or *Lathyrus* and *Pisum* (peas), or those that can still flower relatively well under such conditions (e.g. *Calendula, Dianthus* and *Tropaeolum*).

Concerns over peat extraction and limited supplies of bark-based growing media have led to research investigating the potential of wood-based growing media. These notoriously immobilize nitrogen through the activation of microbial activity, resulting in potential nitrogen deficiencies in the plants. These concerns have extended to the exterior landscape where wood amendments may interfere with nitrogen availability. To some extent these issues can be surmounted by the addition of higher than normal rates of nitrogen fertilizer, thus providing sufficient nitrogen for microorganism activity and plant development. Wright et al. (2009) compared how bedding plants (*Begonia, Solenostemen, Impatiens, Tagetes, Petunia, Plectranthus, Salvia* and *Catharanthus*) grown in either composted pine bark media or composted pine wood media developed once planted out into a landscape setting. As long as some fertilizer (48.6 kg ha^{-1} nitrogen) was added to the landscape, soil plant development was comparable irrespective of the initial growing media used in the growing container.

8.7 Annual Flower Beds from 'Direct Sowing'

The colourful, exuberant displays of annual flowering plants within formal and informal 'beds' is not solely dependent on intensively-produced nursery plants, sold on in pots and trays. Many individual gardeners, local authorities and other landowners produce flowering displays by sowing seed directly into well-prepared soil. These usually rely on so-called 'hardy' annuals, although low temperature tolerance is not the only factor that defines these plants. *Tropaeolum* (nasturtium), for example, is not frost tolerant, but is sufficiently vigorous and robust to grow from a seed planted in late spring to become a 2-m-long flowering 'vine' within 6 weeks. There are overlaps with many of these species in terms of traits and management approaches to that of extensive displays of arable flowering annuals (cornfield mixes). Nevertheless, such plants have traditionally been used in a different context, particularly with where they have been sown – i.e. in garden settings to fill vacant spots between shrubs and herbaceous plants, as well as in designated flower beds of their own.

Hardy annuals can be sown from seed in autumn (depending on cold tolerance) to give early displays the following summer. Indeed, a first sowing in autumn (e.g. September in the northern hemisphere, when soil temperatures are still conducive to good germination rates), followed by a subsequent sowing in spring, provide an opportunity to maximize the display throughout the summer. Autumn sown seed often overwinter as seedlings, well able to capitalize on increasing temperatures and photoperiods in spring to maximize photosynthesis.

Seed beds are prepared by ensuring the soil is free of weeds and cultivated to provide a fine tilth (good soil crumb structure and even surface without large clods). Direct sowing into this cultivated soil involves either a broadcast technique (scattering seed over the soil surface) or drilling (sowing in rows). Designating areas for the different genotypes to be sown may be implemented by marking out with sand or a similar material, and this helps plan for variations in flower colour, height and habit. The aim with broadcasting is to sow the seed evenly and mixing the seed well with sand beforehand improves the distribution. A moderate level of soil fertility tends to suit many hardy annuals. Excess nutrients tend to promote aggressive weed growth, but very impoverished soils will not allow these species to attain the rapid growth rates and quickness to flowering that is their characteristic trademark. Heavy, damp 'cold' soils may inhibit germination, whereas a light, but fertile free-draining soil is optimum for many genotypes. As annual flowering plants cover many genotypes and even different ecological niches (cornfield annuals which exploit the light and rapid release of nutrients that corresponds to the annual cultivation of fields, as well as ephemerals evolved from arid areas where their short life cycles are adapted to the temporary periods of rain), some of these points cannot be generalized; not least with plants such as *Lathyrus odoratous* (sweet pea), which despite being a nitrogen 'fixer' actually responds well to deep organic, nutrient-rich soils.

Sowing in drills has the advantage of knowing where the seedlings will emerge from and hence easing management tasks such as hoeing out competing weeds. It also allows larger seed to be sown at a pre-set distance, ensuring the young plants have adequate space to develop, without crowding each other and competing for water and light. Once sown, seeds are often covered with a thin layer of soil or sand, e.g. 2–5 mm. Although some species may be reliant on light to improve germination, shallow covering often still provides enough irradiance, but has the advantage of disguising the presence of the new seed from birds and small mammals. The use of direct-sown annuals, compared to glasshouse-produced tray/pot-grown stock, is seen by some as a more sustainable way to provide seasonal colour in the landscape. The fact that these plants are not dependent on peat-based media, elated temperatures and irradiance for early season production, relatively high nutrient, growth regulator and water inputs, or on the energy involved in transporting them from producer to consumer, are all seen as advantageous from a sustainability perspective.

8.8 Cornfield Annuals and Annual 'Meadows'

The term 'annual meadow' is a bit of a misnomer, as technically meadows are usually defined as perennial grassland communities, which in practice are more often than not cut for hay (although some meadows such as alpine meadows would not necessarily be cut). The term is used for annual plant communities though, where horticulturally speaking there is a riot of colour induced by flowering plants arranged in an informal manner. Much of the practice and philosophy around annual meadows relate to the flowering annual plants that were traditionally associated with arable crops, before the advent of chemical herbicides largely removed their presence. The most 'common' colourful annual 'cornfield' plants in Western Europe are *Papaver rhoeas* (field poppy), *Glebionis segetum* formally *Chrysanthemum segetum* (corn marigold), *Anthemis austriaca* (corn chamomile), *Centaurea cyanus* (cornflower) and *Agrostemma githago* (corn cockle), but a number of these are anything but common now in the rural landscape. These species are often sown together in ornamental displays to recreate the impression of cornfields of yesteryear. One – the field poppy – is also used on its own within public space plantings, as it has a 'revered' place in Western culture as the symbol of death, blood and self-sacrifice associated with World War I. The arable fields of Belgium and northern France 'turning red' with poppies after the soil had been disturbed by munition shells and mortars. These traditional arable field annuals have been augmented by other species to further embolden and extend the season of interest of annual flowering communities (Table 8.4, Fig. 8.13).

Most of these annual species are adapted to germinating and growing quickly; setting flower in as little as 6 weeks after germination. They do best in a weed-free, light, moisture retentive soil, with some degree of fertility. In the past, a limited flowering period has been one of the drawbacks, but the use of successive sowings or increasing the range of species that flower at different times can help extend the interest. As implied, the flower display is only a single 'annual' event, and although some viable seed may remain in the soil from one year to the next, it is normal practice to resow seed again the following winter/spring.

8.9 Pests and Pathogens

As bedding plants themselves represent a wide range of genotypes, this is reflected in an extensive range of pests and pathogens. Nevertheless, certain pests and pathogens are encountered on a frequent basis when growing bedding species, although they may vary between those encountered under protection and those outdoors (depending on climate).

During early phases of production particularly, inappropriate or irregular watering, temperature and humidity control leads to fungal diseases commonly

Table 8.4. Range of species with colourful flowers now commonly used in annual flower communities ('annual meadows').

Species	Common name	Predominant colour
Agrostemma githago	corncockle	purple
Ammi majus	bishop's flower	white
Atriplex hortensis	orache	red
Centaurea cyanus	cornflower	blue
Coreopsis tinctoria	tickseed	yellow/red
Cosmos bipinnatus	cosmos	pink/white/red
Delphinium ajacis	larkspur	blue
Dimorphotheca sinuata	African daisy	orange/yellow
Echium vulgare	viper's bugloss	blue
Eschscholzia californica	Californian poppy	orange
Glebionis segetum	corn marigold	yellow
Iberis umbellata	candytuft	pink
Linaria maroccana	fairy toadflax	purple/white/pink
Linum grandiflorum 'Rubrum'	flax	red
Linum usitatissimum	flax	blue
Malcolmia maritima	Virginia stock	purple/ white /pink
Matricaria inodora	scentless mayweed	white
Nemophila insignis	baby blue eyes	blue
Nigella damascena	love-in-a-mist	blue
Papaver rhoeas	field and 'Shirley' poppies	red/pink
Papaver somniferum	opium poppy	purple/red/pink
Rudbeckia hirta	black-eyed Susan	yellow
Salvia horminum	salvia	blue/purple

Fig. 8.13. Annual flowering plants are increasingly being used in informal 'meadow'-type landscapes, in this case *Eschscholzia californica*, the Californian poppy. (Image courtesy of Mahsa Mohajer.)

described as damping-off (*Pythium*, *Phytophthora* and *Rhizoctonia* spp.), grey mould (*Botrytis cinerea*), root rot (*Pythium*, *Phytophthora* and *Thielaviopsis* spp.), and crown rot (*Sclerotinia* spp.), with some crops also susceptible to rusts (e.g. *Puccinia* spp.) and powdery mildews (*Oidium*, *Plasmopara* and *Peronospora* spp.). Environmental conditions that stress the young plants through excessively high or low temperature, overwatering or poor drainage, oscillations between too wet and too dry a growing media, or prolonged periods when moisture droplets are present on the leaf surfaces, predispose plants to these pathogens. Similarly, excessive nutrient levels or pH that cause damage to root tissues through high ion concentrations cause surface wounding and weakening of the plant, thus allowing pathogens to infect more readily.

Common pests encountered across a range of bedding plant species include aphids (e.g. *Aphis gossypii* and *Myzus persicae*), thrips (e.g. *Frankliniella occidentalis* and *Heliothrips haemorrhoidalis*), whitefly (e.g. *Trialeurodes vaporariorum*), fungus gnat (*Bradysia* spp.) and red spider mite (*Tetranychus urticae*). The use of integrated pest management (IPM) and biocontrol agents is now mainstream in controlling such pests in protected environments. Integrated pest management is a change in philosophy in how pests (and pathogens) are dealt with. Rather than waiting for a large infestation of a pest to occur and reacting by applying a heavy dose of pesticides, the approach is one of proactive management. For example, growers develop a site-specific management strategy that includes careful assessment of pest problems through vigilance and regular monitoring of the crop. In this way potential problems can be identified and dealt with quickly before they become a significant threat to the crop.

Intervention is on a needs only basis, and is determined by compatible, effective management tactics, i.e. cultural, physical, mechanical, chemical and biological. As crop monitoring is an essential component to success, reliable and experienced personnel 'crop walkers' are used to identify potential problems. This may not be restricted to inspection of the plants themselves (close inspection of foliage, flowers, and root systems) but may also involve regular monitoring of the glasshouse environment (presence of weeds, debris and spilled growing media), growing media within the pots/trays, irrigation and nutrient regimes, water quality and locating water puddles and other environmental parameters (aided via IT technology as required). For example, computer programs can record temperature, humidity or (if outside) precipitation levels and give predictions of pest/pathogen outbreaks before they occur. Successful IPM programmes involve evaluation of the processes, modifying management strategies as needs dictate. Management activities will include regular cleaning/disinfection of benches and equipment, the use of insect traps to monitor population levels (in IPM certain levels of pest/pathogen presence can be tolerated and action only activated once populations reach a particular threshold), or the sending off of leaf tissue or growing media samples to an external laboratory to identify possible pathogens. IPM strategies should be timely and pest specific, because a missed diagnosis can delay implementation of the proper set of controls.

These pests and pathogens also cause problems outdoors, yet here other problems become more prevalent. Certain bedding plant species are very susceptible to damage by molluscs (slugs and snails) and protection may involve either chemical or physical means (manual removal, use of mulches, physical barriers such as copper bands, or biological control via nematodes). Susceptibility often depends on growth stage (seedlings being most prone to 'fatal' browsing), or plant organs. Species used as bedding specimens for which there is some evidence of tolerance or unpalatability with slugs and snails include: *Ageratum*, *Alyssum*, *Anchusa*, *Antirrhinum* (some), *Aquilegia*, *Arabis*, *Argyranthemums* (some), *Aubrietia*, *Begonia*, *Bidens*, *Calendula*, *Cheiranthus*, *Coleus*, *Cosmos* (some), *Dianthus*, *Diascia*, *Eschscholzia*, *Fuchsia*, *Gaillardia*, *Gazania*, *Geranium*, *Gypsolphila*, *Iberis*, *Impatiens*, *Lantana*, *Leucanthemum*, *Myosotis*, *Nasturtium*, *Nemesia* (some), *Oenothera*, *Origanum*, *Oxalis*, *Papaver*, *Pelargonium*, *Phlox*, *Polygonum*, *Portulaca*, *Pulmonaria*, *Rudbeckia* (some), *Saxifraga*, *Scabiosa*, *Sedum*, *Sempervivum*, *Verbascum*, *Verbena* and *Veronica*.

Many bedding plants are protected from molluscs using slug pellets (composed of metaldehyde or methiocarb); such chemicals though are non-selective and can harm non-target species such as domestic dogs and cats, wild birds, mammals or reptiles that either eat the pellets directly or ingest the poisoned molluscs (methiocarb being the more persistent of the two). There are also claims that high use of these pesticides can contaminate water courses, although this may relate more to their use in extensive agricultural crops, rather than in gardens. Nevertheless, alternatives are sought and ones currently being used include aluminium sulfate and copper sulfate; these work by direct contact with the mollusc and efficacy may be limited to the smaller slug and snail species. Various forms of traps and chemically treated bands are also used to deter slugs from moving towards the susceptible plants. Biocontrol products rely strongly on nematodes, e.g. *Phasmarhabditis hermaphrodita*, which are applied to the planting area to increase the parasitic nematode population and increase likelihood of slugs and snails being infected.

Not all nematodes predate on molluscs, however, and some groups are herbivorous, and can infect and damage root systems or occur in the leaves and other foliar parts of the plant. Again, species susceptibility is determined by both the host plant genotype and nematode species (Fig. 8.14).

Conclusions

- Bedding plants are annuals, biennials or short-lived perennials, used in the landscape for their flamboyant floral and foliage displays. Their name is derived from the tradition of growing these plants in pots and bedding them out in flower borders (beds). Formal bedding schemes relate to regular geometrical planting patterns, whereas informal bedding is a more natural style of planting.
- Bedding plants are still popular, although a greater proportion are now used in the domestic garden, with gardeners reliant on a bedding plant production industry that sells trays and pots of plants through garden centres and other retail outlets. This is augmented with gardeners producing their own bedding plants from seed, or directly sowing annuals into the soil to provide colourful displays during spring and summer.
- The intensive production of bedding plants year on year is under scrutiny from an environmentally sustainable perspective, due to reliance on peat growing media, heat and light energy to establish seedlings/encourage early flowering, heavy use of water and nutrients, and waste associated with packaging. The industry is responding by investigating alternative media and lower input production systems, as well as developing containers that are biodegradable. Allied to this there is growing interest in promoting genotypes that are attractive to wildlife so that both aesthetic and ecological goals can be attained simultaneously.
- Many bedding plant genotypes are derived from F_1 seed. These provide plants that are relatively 'true to type' and may have hybrid vigour (enhanced qualities compared to the parent lines). F_1 seeds are generated by crossing two inbred parental lines to reduce the variability within the genotype. Often F_1 seed provides benefits such

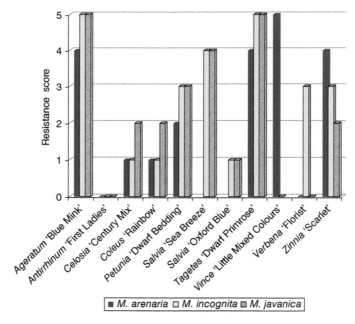

Fig. 8.14. Variability in bedding plant species/cultivar resistance to root knot nematodes – *Meloidogyne* spp. (Modified from Krueger and McSorley, 2010.)

- as more uniform germination, larger flowers and more vigorous growth.
- During the professional production of bedding plants, crop quality is enhanced by careful manipulation of temperature, irradiance, irrigation supply, nutrition and the use of chemical plant growth regulators. This is demanded by the market requirements to produce uniform, highly quality flowering plants at the point of sale.
- Transporting plants from the nursery to the garden centre is a challenge, as the perishable nature of the crop predisposes plants to risks such as prolonged lack of irradiance, lack of water, temperature extremes and exposure to potentially high humidity and ethylene concentrations. A number of research avenues are being explored to help improve crop resilience during shipping of the crop and to improve shelf-life on the retail bench.
- As attention focuses on the more sustainable production and management of bedding plants, it becomes more critical that plant selection reflects the landscapes the plants will find themselves grown in. Matching plant tolerance more closely to the environmental conditions of the landscape will help reduce reliance on additional water, nutrients and pesticides.

References

Adams, S.R., Hadley, P. and Pearson, S. (1998) The effects of temperature, photoperiod, and photosynthetic photon flux on the time to flowering of *Petunia* 'Express Blush Pink'. *Journal of the American Society for Horticultural Science* 123, 577–580.

Anon. (2013) Floriculture Data 2013. United States Department of Agriculture, National Agricultural Statistics Service. Available at: www.nass.usda.govCharts_and_Maps/Floriculture_Crops (accessed 14 December 2014).

Anon. (2014) United States Department of Agriculture, National Agricultural Statistics Service. Floriculture Crops, 2013Summary.www.nass.usda.gov/Statistics_by_State/New_Jersey/Publications/Floriculture_Statistics/FlorCrop-06-19-2014.pdf. (accessed 14 December 2014).

Armitage, A.M. (1986) Influence of production practices on post-production life of bedding plants. *Acta Horticulturae* 181, 269–277.

Armitage, A.M. (1993) *Bedding Plants: Prolonging Shelf Performance: Postproduction Care and Handling*. Ball Publishing, Chicago, Illinois.

Armitage, A.M. (1994) *Ornamental Bedding Plants*. CAB International, Wallingford, UK.

Arnold, M.A., Lesikar, B.J., McDonald, G.V., Bryan, D.L. and Gross, A. (2003) Irrigating landscape bedding plants and cut flowers with recycled nursery runoff and constructed wetland treated water. *Journal of Environmental Horticulture* 21, 89–98.

Blanchard, M.G. and Runkle, E.S. (2010) Intermittent light from a rotating high-pressure sodium lamp promotes flowering of long-day plants. *HortScience* 45, 236–241.

Blanchard, M.G. and Runkle, E.S. (2011) Quantifying the thermal flowering rates of eighteen species of annual bedding plants. *Scientia Horticulturae* 128, 30–37.

Blanchard, M.G., Runkle, E.S. and Fisher, P.R. (2011) Modeling plant morphology and development of petunia in response to temperature and photosynthetic daily light integral. *Scientia Horticulturae* 129, 313–320.

Blanusa, T., Vysini, E. and Cameron, R.W. F. (2009) Growth and flowering of *Petunia* and *Impatiens*: effects of competition and reduced water content within a container. *HortScience* 44, 1302–1307.

Boatright, J.L., Balint, D.E., Mackay, W.A. and Zajicek, J.M. (1997) Incorporation of a hydrophilic polymer into annual landscape beds. *Journal of Environmental Horticulture* 15, 37–40.

Borch, K., Miller, C., Brown, K.M. and Lynch, J.P. (2003) Improved drought tolerance in marigold by manipulation of root growth with buffered phosphorus nutrition. *Hortscience* 38, 212–216.

Burnett, S.E. and Stack, L.B. (2009) Survey of the research needs of the potential organic ornamental bedding plant industry in Maine. *HortTechnology* 19, 743–747.

Cameron, R.W.F., Wagstaffe, J., Brough, W. and Davis, A. (2008) Bedding plants: benchmarking current transport and distribution practices with the aim of identifying factors that determine plant quality during transit and methods to maintain quality. Agriculture and Horticulture Development Board, PC 234.

de Graaf-van der Zande, M.T. (1990) Watering strategies in bedding plant culture: effect on plant growth and keeping quality. *Acta Horticulturae* 272, 191–196.

Garbuzov, M. and Ratnieks, F.L. (2014) Quantifying variation among garden plants in attractiveness to bees and other flower-visiting insects. *Functional Ecology* 28, 364–374.

Garner, L.C. and Langton, F.A. (1997) Brushing pansy (*Viola tricolor* L.) transplants: a flexible, effective method for controlling plant size. *Scientia Horticulturae* 70, 187–195.

Gehring, T.M. and Lewis, A.J. (1980) Extending the shelf life of bedding plants. III. Antitranspirants. *Florists' Review* 165, 65.

Hall, T.J., Dennis, J.H., Lopez, R.G. and Marshall, M.I. (2009) Factors affecting growers' willingness to adopt sustainable floriculture practices. *HortScience*, 44, 1346–1351.

Hori, Y., Nishidate, K., Nishiyama, M., Kanahama, K. and Kanayama, Y. (2011) Flowering and expression of flowering-related genes under long-day conditions with light-emitting diodes. *Planta* 234, 321–330.

Høyer, L. (1997) Investigations of product temperature management during transport of pot plants in controlled temperature trucks. *Gartenbauwissenschaft* 62, 50–55.

Kaczperski, M.P., Carlson, W.H. and Karlsson, M.G. (1991) Growth and development of *Petunia* × hybrids as a function of temperature and irradiance. *Journal of the American Society for Horticultural Science* 116, 232–237.

Kim, J. and van Iersel, M.W. (2011) Abscisic acid drenches can reduce water use and extend shelf life of *Salvia splendens*. *Scientia Horticulturae* 127, 420–423.

Krueger, R. and McSorley, R. (2010) Susceptibility of flowers and bedding plants to root-knot nematodes. Document ENY-061, Entomology and Nematology Department, Florida Cooperative Extension Service, Institute of Food and Agricultural Sciences, University of Florida. http://edis.ifas.ufl.edu/in850 (accessed 14 October 2014).

Kuehny, J.S., Taylor, M. and Evans, M.R. (2011) Greenhouse and landscape performance of bedding plants in biocontainers. *HortTechnology* 21, 155–161.

Langton, F.A. and Cockshull, K.E. (1997) A re-appraisal of DIF extension growth responses. *Acta Horticulturae* 435, 57–64.

Langton, F.A., Cockshull, K.E., Cave, C.R.J. and Hemming, E.J. (1992) Temperature regimes to control plant stature: current UK R&D. *Acta Horticulturae* 327, 49–60.

Latimer, J.G. (1991) Growth retardants affect landscape performance of zinnia, impatiens, and marigold. *HortScience* 26, 557–560.

McCabe, K.G., Schrader, J.A., Madbouly, S., Grewell, D. and Graves, W.R. (2014) Evaluation of biopolymer-coated fiber containers for container-grown plants. *HortTechnology* 24, 439–448.

Million, J.B., Barrett, J.E., Nell, T.A. and Clark, D.G. (2001) Late season applications of media surfactant improve water retention and time to wilt during post-production. *Acta Horticulturae* 543, 235–244.

Mortensen, L.M. and Moe, R. (1992) Effects of various day and night temperature treatments on the morphogenesis and growth of some greenhouse and bedding plant species. In: *II European Workshop on Thermo-and Photomorphogenesis in Plants* 327, 77–86.

Nell, T.A. and Reid, M.S. (2001) *Flower & Plant Care: the 21st Century Approach*. Society of American Florists, Alexandria, Virginia.

Nemali, K.S. and van Iersel, M.W. (2008) Physiological responses to different substrate water contents: screening for high water-use efficiency in bedding plants. *Journal of the American Society for Horticultural Science* 133, 333–340.

Niu, G., Rodriguez, D.S. and Starman, T. (2010) Response of bedding plants to saline water irrigation. *HortScience* 45, 628–636.

Oh, W., Cheon, I.H., Kim, K.S. and Runkle, E.S. (2009) Photosynthetic daily light integral influences flowering time and crop characteristics of *Cyclamen persicum*. *HortScience* 44, 341–344.

Pearson, S., Parker, A., Hadley, P. and Kitchener, H.M. (1995) The effect of photoperiod and temperature on reproductive development of Cape Daisy (*Osteospermum jucundum* 'Pink Whirls'). *Scientia Horticulturae* 62, 225–235.

Scheiber, S.M. and Beeson, R.C. (2006) *Petunia* growth and maintenance in the landscape as influenced by alternative irrigation strategies. *HortScience* 41, 235–238.

Shurberg, G., Shober, A.L., Wiese, C., Denny, G., Knox, G.W., Moore, K.A. and Giurcanu, M.C. (2012) Response of landscape-grown warm- and cool-season annuals to nitrogen fertilization at five rates. *HortTechnology* 22, 368–375.

Smith, T. (2001) *Caring for Plants in the Retail Setting*. University of Massachusetts Extension service leaflet. www.umass.edu/umext/floriculture/fact_sheets/business_management/retail.html (accessed 23 November 2014).

Snider, A., Palta, J.P. and Peoples, T. (2003) Use of lysophosphatidylethanolamine (LPE), a natural lipid, to enhance opening and retention of flowers on bedding plants experiencing water stress during retail sales. *Acta Horticulturae* 628, 849–853.

van Iersel, M.W., Dove, S., Kang, J.G. and Burnett, S.E. (2010) Growth and water use of petunia as affected by substrate water content and daily light integral. *HortScience* 45, 277–282.

Waller, P. (2006) A review of peat usage and alternatives for commercial plant production in the UK. Horticultural Development Council, CP 41.

Waterland, N.L., Campbell, C.A., Finer, J.J. and Jones, M.L. (2010a) Abscisic acid application enhances drought stress tolerance in bedding plants. *HortScience* 45, 409–413.

Waterland, N.L., Finer, J.J. and Jones, M.L. (2010b) Benzyladenine and gibberellic acid application prevents abscisic acid-induced leaf chlorosis in pansy and viola. *HortScience* 45, 925–933.

Wright, R.D., Jackson, B.E., Barnes, M.C. and Browder, J.F. (2009) The landscape performance of annual bedding plants grown in pine tree substrate. *HortTechnology* 19, 78–82.

Yue, C., Hall, C.R., Behe, B.K., Campbell, B.L., Dennis, J.H. and Lopez, R.G. (2010) Are consumers willing to pay more for biodegradable containers than for plastic ones? Evidence from hypothetical conjoint analysis and nonhypothetical experimental auctions. *Journal of Agricultural and Applied Economics* 42, 757–772.

Yue, C., Dennis, J.H., Behe, B.K., Hall, C.R., Campbell, B.L. and Lopez, R.G. (2011) Investigating consumer preference for organic, local, or sustainable plants. *HortScience* 46, 610–615.

9 Lawns and Sports Turf

Ross W.F. Cameron

> **Key Questions**
> - What are the functional, recreational and aesthetic uses of turf grass?
> - What is the difference between a 'cool-season' and a 'warm-season' turf grass?
> - Why does the length of the grass (height of cut) vary with different sports?
> - What is 'thatch' and 'mat'?
> - List the requirements placed on turf grass when used in an elite sport venue, such as a golf green within an international championship course.
> - How does design and management of the golf green help achieve these requirements?
> - In what ways can these management practices be made more sustainable in terms of resources and energy used, and what are the likely impacts on the wider environment?
> - Where and when are more sustainable approaches to turf management particularly warranted?
> - How could a garden lawn be managed to improve its value as a wildlife habitat?

9.1 Introduction

Turf grass covers a range of sward types, largely defined by the height of the grass and the species composition. These include the highly managed 'elite' sports turfs which epitomize golf, bowling and tennis as well as the close-mown 'fine' lawns of private residences and heritage gardens. But utilitarian public-park swards and green space, as well as 'rougher' less-manicured turf used on roadside verges and banks, are also 'managed' areas, albeit perhaps less intensively so. Once the sward is allowed to grow above a certain height, and broadleaved plant species become a significant component of the community, then these grass areas tend to be referred to as meadows (see Chapter 7). There is also commonality with pastures and 'rangeland' where the primary purpose of the sward is to provide grazing for livestock. It is perhaps the intensively managed short grass swards, popular due to their aesthetics and consistency as sports surfaces that represent the most challenged form of green space as far as sustainable environmental management is concerned.

The role of turf grass within horticulture tends to be divided up into three categories: functional, recreational and aesthetic. Functional aspects include providing a walkable surface where there is an intermediate level or low frequency of pedestrian traffic. Grassed areas are employed as temporary car parks, camping sites or for hosting fêtes, fairs and other non-permanent activities. Aircraft runways that accommodate light aircraft can be composed of turf, especially in rural areas where landings may be infrequent. Grass verges provide an accessible and relatively safe zone at the side of roads or buildings. Such roadside verges may help trap dust and de-activate vehicle emissions and other pollutants. Grass perimeters around airports are thought to prolong aircraft engine life due to reducing the incidence of dust. Turf surfaces act as a catchment and filter-out waterborne pollutants, including heavy metals, and both the activities of the plants themselves and the associated rhizosphere microflora can deactivate organic pollutants, such as pesticides and light hydrocarbons. A mantle of turf aids soil to resist wind and water erosion and helps to stabilize steep slopes or waterside embankments. Soil covered with turf (*Festuca arundinacea*, tall fescue) was shown to reduce soil erosion (10–62 kg ha^{-1}) compared to bare soil alone (223 kg ha^{-1}) during a 30 min intensive rainstorm (Beard and Green, 1994). Damaged, compacted or contaminated soils are improved though the action of grass, with root activity promoting fissures and pores that aid drainage and aeration. Organic matter released from root mucilage and leaf clippings encourages the formation of a favourable, 'crumb-like', soil structure. A cover of turf grass promotes the activity of earthworms and other invertebrates. Their burrowing habits, in turn, improve the drainage capacity of the sward, and by converting leaf litter into organic humus they enhance soil structure and fertility. Grass alters its surrounding microclimate, with lawns and meadows functioning as natural cooling/humidifying agents. Last but not least, functional benefits include the direct economic impact, with the growth and retail of turf and grass lawn seed being significant industries, augmenting the labour and employment centred around turf management.

Perhaps it is the recreational aspects associated with turf that are most in the public eye. Despite the advent of synthetic sports-playing surfaces, grass is the most frequent surface used in competitive sports (with sports turf management being a discrete professional discipline in its own right). Grass remains popular for sport due to its ability to provide a cushioning surface that helps minimize risk of injury to the participants. It also facilitates an ideal (or unique) playing surface for some sports. For example, with tennis the speed of the ball off the playing surface is accelerated when compared to alternatives such as clay (crushed brick or stone) or synthetic carpets. In golf, grass is the only feasible playing surface in temperate climates, due to the scale of the area involved.

Grass has extensive use in non-formalized sport, through its predominance in parks, children's play areas and domestic gardens. Here it also provides the forum for relaxation and social interaction through a wide spectrum of activities, spanning events from school fêtes to family picnics. The third category – that of the role of turf in landscape aesthetics – is considered in the following section.

9.2 Role of Turf in the Landscape

Within the landscape, lawns provide open space and opportunities for freedom of direction and movement. They act as an aesthetic foil to other key features including commonly: buildings, trees, floral beds, pathways and water features. Indeed, humans seem to have a preference for landscapes where one-third of the area is composed of vertical structure and two-thirds is devoted to open space, and grass is invariably the preferred choice for the horizontal, open space plane. Without the dominant role of grass many famous gardens and landscapes would feel much more claustrophobic or have a cluttered feel to them. They would lose their grand effect, namely the scale of the landscape that extensive lawns provide – as typified by Lancelot 'Capability' Brown through the 'English landscape style'. Similarly, many iconic landscape views and vistas would not exist, if the open ground that the grass cover affords were to be lost.

In the USA, turf grass covers a substantial amount of land, encompassing approximately 20 million ha and representing a real-estate value of US$40 billion. This is largely due to the 80% (85 million) of households that participate in outdoor lawn or garden activities. To put this into context, amenity turf grass is estimated to cover an area three times larger than any other irrigated crop type in the USA (Milesi et al. 2005). It is partly due to this extensive nature of turf coverage that has led to turf itself becoming a 'battlefield' between those wishing to promote the benefits of intensively managed green space, and those that argue that the management of such space is non-sustainable and contributes to the global depletion of natural resources, including oil, gas, minerals, aggregates and clean water. Indeed, for some, 'the lawn' has become a metaphor for society's lack of environmental awareness, and our ability to exploit dwindling or increasingly costly resources, for largely aesthetic reasons. These conflicts are perhaps best illustrated by the numerous studies showing the advantages of turf grass in terms of its carbon sequestration potential (Raciti et al., 2011) and the disadvantages with respect to carbon dioxide (CO_2) and other greenhouse gas emissions, associated with its traditional management (Townsend-Small and Czimczik, 2010a, b) or establishment in inappropriate locations (Selhorst and Lal, 2013). In a review of life cycle analyses for different ground-cover cover types in Georgia, USA, Smetana and Crittenden (2014) suggested traditional lawn turf (intensively managed and comprising non-native species) performed poorly compared to planting with native prairie grasses (Fig. 9.1). Other studies suggest the wider environmental, social and health aspects need to be considered too; for example some of the arguments for and against lawns and intensively managed sports turf are summarized in Table 9.1.

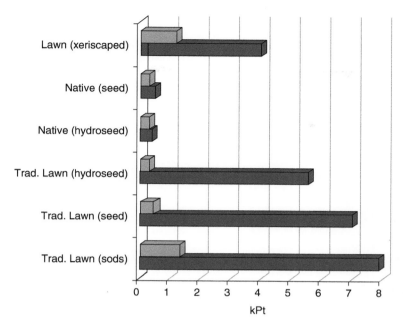

Fig. 9.1. The environmental impact (life cycle analysis in kilopoints[a]; kPt) modelled for a 'traditional' lawn species (*Cynodon dactylon*) grown from turf sods, seed, hydroseed or managed as a xeriscape, compared to native prairie grass species (four species grown as a mix) grown from seed or hydroseed. Data was based on grasses grown in Georgia, USA, and represents the environmental impact after installation (pale bars) and after 10 years maintenance (dark bars). (Modified from Smetana and Crittenden, 2014.)

[a]Kilopoint (kPt) is equal to the impact of one European citizen on environment (integrated summary of multiple impact categories) during one year. Impact factors include carcinogens, respiration of organics, respiration of inorganics, climate change, radiation, ozone layer, ecotoxicity, acidification/eutrophication, land use, mineral and fossil fuel consumption.

9.3 Grass Genotypes – Physiology and Traits

So why is the grass sward such a predominant part of the horticultural landscape? The key is in the nature of grass itself, in that the growing point (meristem) is located at the base of the stem, allowing stems and associated leaves to regrow readily after being cut or grazed. This ability to regenerate means that a low stature sward can be maintained perpetually, allowing the lawn to be a walkable (or playable) surface. Turf grasses spread either by developing daughter plants close to the base of the mother plant – in a process known as tillering – or through initiating lateral stems which can grow above ground (stolons) or below ground (rhizomes). The first 'lawns' would have developed out of natural grassland and meadows, where grazing resulted in the maintenance of a low, easily-traversed sward.

Abiotic factors such as moisture-deficient soil or exposure to wind can result in the sward being naturally short in stature too. For example, early golf courses in Scotland, UK, were reputably developed from coastal sand dune systems, where the combination of free-draining sandy soil, wind, salt exposure and grazing by sheep (and/or rabbits) resulted in a fine, short turf; short turf that was ideal to allow a small ball to roll and bounce. A short, close-knit turf has greater resilience to wear and tear than turf composed of taller stems of grass; another factor that favours short turf being promoted for paths and playing surfaces. Grass species are the predominant constituents of formal lawns, although it should be noted that other 'green mantles' exist, composed of species such as *Matricaria chamomilla* or *Chamaemelum nobile* (chamomile), *Hypnum* or *Thuidium* spp. (moss) and *Trifolium* spp. (clover).

Turf grasses are divided into 'cool-season' (Table 9.2) and 'warm-season' (Table 9.3) grasses depending on the climate. The cool-season grasses have evolved from alpine, acidic moorland or coastal strand conditions. They tend to be fine-leaved and form close-knit swards when close-mown. Free-draining conditions,

Table 9.1. The physical, social and environmental advantages and disadvantages associated with turf landscapes and lawns.

Advantages	Disadvantages
Opportunities for sport – both formal and informal. Active sport seen by many policy makers as a means to combat sedentary lifestyles and reduce risks of obesity-related diseases, including stroke, late-onset diabetes and coronary heart disease.	Extraction and transport of mineral deposits for construction; top-dressing and fertilizer. Use of oil and natural gas in agrochemicals and fertilizer synthesis, and in producing energy for maintenance equipment.
Sequestration of atmospheric carbon. Carbon storage is thought to be comparable to agroecosystems, but not quite as good as forest systems. Carbon sequestration per year estimated as 0.52 t C ha^{-1} with permanent grassland over sandy loam soil.	Oxidation of carbon in soil disturbance. Release of nitrous oxide N_2O (greenhouse gas) from volatilization of nitrogen fertilizers.
Rainfall capture and infiltration and attenuation of water flow into water courses. Aids floodplain function acting as a capture point for flood waters.	Potable water used for irrigation. Potential source of pollution to ground and surface water.
Contribution to urban cooling and heat island mitigation.	Source of noise pollution, due to routine maintenance activities.
Psychological benefits of green swards.	Allergens, including allergic rhinitis (hay fever).
Biodiversity – value depends on what it is being compared to as an alternative; biodiversity in a natural turf system is greater than in tarmacadam or synthetic turf alternatives. Grassland habitat of various forms. E.g. certain golf courses have designated sites of special ecological interest.	Pesticide use, especially if not targeted to pest species, impact on biodiversity and pollutant source.
Location for children's play.	Location for dog fouling. Potential to cause toxocariasis (infection of the roundworm parasite *Toxocara canis*). Especially so in areas where dog walking and children's play overlap.

e.g. sandy subsoil and the avoidance of organic matter, such as thatch (dead grass leaf-blades), help to keep pathogen pressure at bay. Many species are derived from European ecotones and have been introduced to North America and elsewhere. Most cool-season grasses are C_3 plants, i.e. they fix carbon into the three carbon molecule, 3-phosphoglycerate. In contrast, warm-season grasses predominate in tropical/semi-tropical regions and grow at temperatures >10°C, with optimum growth at temperatures 25–35°C. They are active during spring and summer, but often go dormant in cooler months, with a tendency to turn brown. Many warm-season grasses are quite drought tolerant, and can survive temperatures >45°C. The warm-season grasses are C_4 plants, i.e. carbon is fixed into the four carbon molecule, oxaloacetate. The biochemical pathways associated with the transport and fixation of CO_2 are more efficient than in C_3 plants, so C_4 plants have both a lower requirement for nitrogen and their stomata are not required to open for so long prior to meeting their carbon requirements. This latter aspect means they are likely to lose less water through the stomata, and hence, can be more efficient in terms of their water utilization.

These tables outline the common species of grass used, but within each species there are numerous cultivars, specifically bred and selected for particular traits, e.g. improved pathogen tolerance, more compact habit or greater capacity to stay green when under stress. The development of new cultivars is a prolonged, costly and sophisticated business. Plant breeding (crossing, selecting and conducting trials) typically takes a decade or more, with further investment required in the production of seed, stock-holding, manufacturing and marketing before varieties can be released into the trade. Different breeders (seed houses) have their own range of cultivars, covering the spectrum of requirements for

Table 9.2. Commonly used cool-season turf-grass species.

Species	Common name	Habit and characteristics	Uses
Agrostis canina	velvet bent	Stoloniferous and forms a very dense turf. Relatively good drought tolerance, but more thatch production compared to other *Agrostis* spp.	Fine golf and bowling greens, where drought may be a problem.
A. capillaris	brown top bent	Rhizomatous and stoloniferous growth pattern, but with short sections between plants. Occurs on poor acid soils – under such conditions can be an aggressive colonizer. Tolerates cool, damp conditions, but has poor wear tolerance.	Fine-leaved, tufted habit makes it ideal for fine lawns. A cool temperate species, it is used in western Europe, New Zealand and north-west North America.
A. castellana	Highland bent	Short rhizomes and very fine seed. Cooler climate areas with dense growth pattern. Poor wear tolerance.	Fine lawns and provides good winter colour.
A. stolonifera	creeping bent	Leafy stolons. Forms a tight-knit sward that tolerates close mowing.	Putting greens/bowling greens.
Festuca arundinacea	tall fescue	Tuft forming. Coarse turf grass, but generally not as good as *Lolium perenne* due to slower establishment and coarser leaf texture. Good in dry, low fertility soils.	Useful in rough areas where there is requirement for taller erect turf, such as airfields.
F. ovina	sheep's fescue	Tuft forming. Very hardy – adapted to moorland conditions. Tolerates close mowing.	General lawn use plus finer lawns and greens.
F. filiformis	fine-leaved sheep's fescue	Tuft forming. Finer leaved than *F. ovina*. Relatively poor wear tolerance. Does not tolerate heavy damp soils.	Good for ornamental lawns especially on well-drained soil. Due to poor wear tolerance only for lawns with little traffic. Some cultivars also suited for low maintenance, environmentally sensitive areas, where parent species can be found.
F. rubra subsp. *commutata*	Chewing's fescue	Tuft forming. Quick to establish but poor competitor. Moderate tolerance to wear. Tolerance to shade, cold and drought. Cultivars vary but some have greater tolerance to pathogens compared to *F. rubra* subsp. *litoralis*. Prone to forming thatch on acid soils. Colours up well in spring.	Does best on well-drained conditions. General lawn use plus finer lawns and greens.
F. rubra subsp. *rubra*	strong creeping red fescue	Rhizomatous with wide range of natural habitats and hence tolerant of salt and drought. Deep rooting but slow to recover from injury. Intolerant of wet or heavy clay soils. Less tolerant of close mowing compared to finer-leaved fescues.	General lawns, without excessive requirement for close mowing.
F. rubra subsp. *litoralis*	slender creeping red fescue	Rhizomatous with finer leaves than *F. rubra* subsp. *rubra*. Good binding capacity in the sward matrix. Not particularly disease tolerant.	Good on extreme environments. Close mowing tolerance, so fine lawns and greens.
F. longifolia	hard fescue	Tuft forming. Drought, shade and heat tolerant – with some capacity to stay green over summer.	Fine turf, especially in infertile soils and hot dry conditions.

Continued

Table 9.2. Continued.

Species	Common name	Habit and characteristics	Uses
Lolium perenne	perennial rye	Tufted perennial with broad leaves. Common, robust species. Quick to establish and withstands wear. Strong grower that requires nutrient-rich soil.	Sports turf due to hard wearing character. Species does not tolerate close mowing, although modern cultivars more tolerant of this. Some cultivars have 'stay green' genes to maintain green hue when under stress.
Poa pratensis	smooth-stalked meadow/ Kentucky blue	Rhizomatous growth. Can be slow to germinate and establish, but once it does, has good wear tolerance. Robust with some drought tolerance.	Used with *L. perenne* to provide good wear resistance in e.g. soccer and rugby pitches. Also used in general grass areas – banks – golf fairways.

Table 9.3. Commonly used warm-season turf-grass species.

Species	Common name	Habit and characteristics	Uses
Cynodon dactylon	Bermuda	Leaves are borne on stems with long internodes giving a branched appearance. Close mowing is required to avoid branching and promote a more finely textured sward. Species requires full sun and is drought tolerant. An aggressive species that can become invasive where it is not desired. Dies back if exposed to frost.	Sports turfs in tropics and for general robust lawns. Aggressive nature means will recolonize bare areas quickly.
Eremochloa ophiuroides	centipede	Originates from Southeast Asia. Slow-growing species that grows prostrate to the ground and produces medium textured, pale-coloured sward. Over-fertilization can cause die-back. Turns brown when dormant. Poor salt, wear, and cold tolerance, and prone to nematode damage.	Prefers acidic, infertile soils and has low maintenance requirements. General lawns and low maintenance rough areas.
Paspalum notatum	Bahia	Native of South America – now common in south-east USA. High drought tolerance, doing well on sandy, infertile lawns. Invasive and may become a weed in other species lawns. Its own open habit means other species can invade. Good pest and pathogen tolerance.	Rather coarse textured leaf. Used for erosion control alongside roads and banks. Robust habit means also used for low maintenance lawn open areas and wildlife habitat.
Stenotaphrum secundatum	St. Augustine	Water-use efficient with few pest problems. Broad leaf blades with creeping growth habit that nevertheless forms a dense prostrate turf that is generally weed free. Some drought tolerance, with good shade and salt tolerance, but tends to produce thatch.	Lawns and general use, but not wear tolerant enough for sports pitches. Shade tolerance means it can be used around buildings and near trees.

Continued

Table 9.3. Continued.

Species	Common name	Habit and characteristics	Uses
Zoysia japonica, *Z. matrella* and *Z. tenuifolia*	zoysia	Spreads by stolons and rhizomes, providing a very dense turf, with stiff leaf blades. Exceptional wear tolerance for a warm-season grass with both good drought tolerance and moderate shade tolerance. Better cold tolerance than other warm-season species, but turns brown after frost and does this earlier than other species. Generally low water and nutritional requirements.	Lawn grasses and due to wear tolerance used on golf courses, parks and athletic fields. Adaptable to a range of soil types and pH.

different amenity functions, soil type, climate, pest/pathogen pressure and stress tolerance.

The seed houses will promote mixtures of grass species/cultivars to provide a 'blend' of genotypes for a specific function. For example, a grass seed mixture for a golf course fairway might be composed of 35% *Festuca rubra* subsp. *rubra* (strong-creeping red fescue), 25% *F. rubra* subsp. *litoralis* (slender-creeping red fescue), 25% *F. rubra* subsp. *commutata* (chewing's fescue), 10% *Poa pratensis* (Kentucky bluegrass) and 5% *Agrostis castellana* (Highland bent). This blend allows each species/cultivar to exploit a distinct 'ecological niche' within the turf matrix, and ideally the blend encompasses a robust community of genotypes that provides a resilient playing surface. Sowing a mix also ensures that if conditions are not conducive to some of the species incorporated within the blend, then the complementary species will still ensure that a sward develops. In reality, swards rarely grow in the exact proportions to the percentage of seed applied, and indeed over a period of time a significant component of the turf can be attributed to 'weed' grass species such as *Poa annua*, which naturally colonizes (particularly damaged) turf (Table 9.4). *P. annua*, however, is often considered undesirable in sports turf, due to its shallow rooting/poor drought tolerance, preponderance to turn yellow and its characteristic of forming unsightly flower/seed heads that can interfere with playing capabilities of the turf, e.g. altering the natural roll of a golf ball.

9.4 Grass Genotypes for More Sustainable Management Practices

The drive for more sustainable approaches to turf management has activated interest in identifying new genotypes of grass that provide acceptable performance whilst reducing inputs and lowering the environmental impact of turf management. It has been long recognized that 'low input' grasses exist such as the warm-season Bahiagrass (*Paspalum notatum*) and cool-season hard fescue (*Festuca longifolia*), but to date, these species have had only limited acceptance due to a lack of performance capability. Attention has now turned to selecting and breeding new cultivars that can provide a good compromise between desired performance criteria and sustainable management regimes. Development of seashore paspalum (*Paspalum vaginatum*) during the 1990s resulted in cultivars (e.g. Salam, SeaIsle 1, SeaIsle 2000, and Seadwarf) with finer textured leaves and greater tolerance to low mowing. These were considered to be improved sufficiently to meet the standard requirement for golf course turf. Moreover, they had increased pest tolerance, reduced water requirements, an ability to tolerate poor quality water (e.g. saline/brackish), and lower and more efficient nutrient retention compared to traditional warm-season golf course turf (Duncan and Carrow, 2000). In Iran, selections of native *Lolium* with good stress tolerance characteristics have proved promising compared to current commercial varieties. Similarly, in the USA native genotypes are being examined for greater stress tolerance, in an attempt to select for, or breed in, traits that reduce

Table 9.4. Transition in grass species composition on soccer pitch over a 10-year period.

Species	Sowing (%)	After 10 years (%)
Lolium perenne	25	0
Poa annua	0	24
Poa trivialis	75	2

energy, chemical or water inputs. These include *Buchloe dactyloides* (buffalograss), *Bouteloua gracilis* (blue grama), *B. curtipendula* (sideoats grama), *Distichlis spicata* (saltgrass), *Agropyron cristatum* (crested wheatgrass), *Koeleria cristata* (prairie junegrass) and *Deschampsia cespitosa* (tufted hairgrass) (Johnson, 2000). Dwarf selections of *Cynodon dactylon* (Bermuda grass) have shown not only a compact growth habit but a reduction in mowing frequency and 25% less requirement for nitrogen fertilizer.

It should be noted too that climate change may alter the geographical distribution and adaptive capacity of turf grass species with, for example, warm-season grasses being used in more northerly/southerly latitudes than is currently the case. As they are C_4 grasses, the warm-season species require up to 50% less water to fix CO_2 and have substantial lower evapotranspiration rates than their cool-season counterparts. Increasing the adaptive range of warm-season grasses, therefore, has been viewed as a strategy for water conservation. For example, breeding *Cynodon dactylon* varieties that have greater cold tolerance could provide alternatives to the cool-season grasses in many locations, thereby replacing the water inefficient species with genotypes that have lower water-use requirements (Taliaferro, 2000).

9.5 Cultural Procedures

Grass sward from seed

For extensive areas of lawn and sports fields, the generation of the sward is usually accomplished by sowing seed. This has the advantage over the other principal way of establishing a sward (i.e. by laying turf sods) through lower costs, more choice of grass species within the blend, and potentially allied to this, better adaptation of species to the site. Turf establishment on inaccessible locations, or where the landscape is convoluted by difficult angles or slopes is easier too, via seed. The amount of seed required is calculated by weight; but the number of seeds present per unit of weight can vary widely between species. For example, 1 g equates to 700 seeds of *Lolium perenne*; 1000 seeds of *Festuca rubra* subsp. *commutata*; 1500 seeds of *Poa pratensis*; or indeed 14,000 seeds of *Agrostis castellana*. Rate of sowing depends on a balance between the need to cover the ground rapidly and the costs involved. Typical rates can vary between 4 g m^{-2} and 40 g m^{-2}. Species that subsequently self-propagate via stolons and rhizomes require lower rates of seeding compared to those that have a tuft-forming habit.

Species, cultivar, germination rates and company/brand can all influence price, with 1 kg of seed retailing typically at between £3.50 and £40.00 in the UK (US$5.60–64.20). Time of sowing depends on local climatic conditions. In the UK, the seed of cool-season grass species has been traditionally sown in early autumn or mid-/late spring, when temperatures are high enough, but there is also sufficient moisture held within the soil. In the northern hemisphere there are many points in favour of a September sowing. The ground retains warmth after the summer, whilst early autumnal rains increase soil moisture levels. High levels of germination in September will provide sufficient time for new roots to develop before the onset of frost, and competition from weeds will be minimal during winter. Strong establishment at this stage aids tolerance to drought in the following summer.

Where irrigation is an option, then sowing can take place at moderate temperatures throughout the year. Even within the cool-season grasses there is some variation in their temperature tolerances for germination, with *Festuca* tolerating cooler conditions (8–18°C) compared to *Agrostis* (12–21°C), *Lolium* (7–25°C) and *Poa pratensis* (15–25°C). Some specialist seed mixes are designed to germinate at temperatures as low as 3.5°C (e.g. certain *Lolium* cultivars). For genera such as *Festuca* though, temperatures >22°C at night can inhibit germination. Successful establishment of the seedlings is reliant on avoiding drought or cold stress shortly after sowing. Again, depending on species, germination rates should be in the region of 70–95%.

Seed of warm-season genotypes is invariably sown in late spring/early summer. These species cope better with warm summer temperatures and the advantage of early establishment is that it helps to resist weed pressure from *Poa annua* during the subsequent cool autumn period.

Seed bed preparation

If grass is to be seeded on soil (in sports turf it is often sown on to a sand, peat or other free-draining medium) then residual weeds need to be removed through the application of herbicides, followed on by a fallow period and cultivation. The non-selective glyphosate (N-(phosphonomethyl)glycine) is often the preferred herbicide for clearing up existing weeds and unwanted grass species. After weeds have died (usually about 10 days after application) these can be raked off. The land is then left fallow and reapplications of herbicide employed as required, after which there is cultivation (tillage), though the extent of this will depend on the area and future function of the turf. The purpose of tillage is to create a fine seedbed that facilitates seed germination and establishment, whilst also contouring the ground to provide a smooth, even surface. Hard clods of earth and stones are removed at this stage. On larger scale sites the soil may be ploughed, disked and harrowed to improve the soil structure. On smaller sites the soil may be rotovated and then consolidated or even hand dug and settled with a rake. Irrespective of the precise methods, the objectives are the same; namely to allow recontouring or aligning of the soil surface, incorporate any residual organic matter, break up any clods of soil and prepare the soil for seed germination through the development of a fine, crumb structure.

Many lawns are still constructed with a 2% gradient to help encourage surface drainage. For sport pitches this usually means the centre of the playing field is the highest point with a slight gradient to the edges. For domestic lawns, gradients are aligned to help avoid surface runoff being directed towards the dwelling property. A lawn surface should be true – i.e. no sharp changes of gradient or undulations as these encourage the turf to be 'scalped' (bare patches) due to the action of the passing blades of the lawn mower.

For elite sports grounds and indeed most golf course greens an artificial soil surface is constructed to ensure enhanced play performance. This is largely achieved through optimizing drainage, and reducing the incidence of turf disease. A fundamental element is maximizing the drainage whilst maintaining sufficient moisture in the growing medium to support strong grass growth. The rationale behind these artificially constructed playing surfaces is to ensure play can continue during periods of heavy rainfall, not that such surfaces are necessarily easier to maintain than traditional turf. Indeed, maintenance can often be challenging.

Golf greens represent the leading edge of these 'manufactured' soils for turf development. The United States Golf Association (USGA) recommends golf greens to be constructed by providing drainage channels and a number of layers of free-draining media above the parent soil (or subsoil, if the topsoil has been removed) (Anon., 2014). The soil is contoured to fit the requirements of the pitch or green, and then compacted to form a stable surface. A series of subsurface drains are embedded within the parent soil ensuring that any main drains are placed in line with the greatest falls in height. This encourages water to flow to the lowest point on the site, and exit from there. The main drains are connected to a series of lateral drains (held in trenches >150 mm wide × 200 mm deep) and usually spaced at regular intervals to maximize water capture. On regular, geometric sport facilities such as soccer or rugby pitches these drains will be laid out in a formal lattice or herringbone pattern, whereas in irregular areas such as golf greens they may act as a series of 'branches' leading to the main drains. Lateral drains should be orientated to aid water movement towards natural, water-collecting depressions in the landscape. Although clay pipes were the traditional land drain in many countries, plastic perforated drains (>100 mm diameter) tend to predominate now. Drains should be angled to allow a gradient of at least 0.5% to encourage water to flow, and placed so that the perforation holes are at the base. Once the pipes are laid, the trenches should be back-filled with gravel to a depth >25 mm.

For the playing surface itself, a layer of gravel (blanket layer) is placed over the parent soil (sublayer), again to a depth of at least 100 mm. Above this is the root zone layer where the grass roots will proliferate. The blanket layer is composed of pre-washed crushed stone or pea gravel (6–12 mm diameter) that is likely to resist weathering. (This is in contrast to 'softer' sandstone and limestone aggregates that degrade into smaller particle sizes over time). Much controversy exists as to the relationship between the blanket layer and the root zone layer above it, in that inappropriate aggregate sizes in both, allow fine sand particles to migrate downwards from the root zone and infiltrate the gravel, thereby reducing the pore space for water to drain. In addition, a sharp contrast in particle sizes between the fine sands in the root zone and the gravel in the blanket layer contributes to a 'perched' water table – where drainage from the root zone is impeded by a lack of conductivity and capillary action between the two media types. To avoid this, gravel selected for the blanket layer should conform to a standard where the particle grades are: 0% >12 mm diameter; 10% max. <2 mm diameter and 5% max. <1 mm diameter.

In addition, the smallest 15% of the gravel particles in the blanket should compare in size ('bridge') with 15% of the root zone's largest particles. This helps to form pores that inhibit further migration of particles from the root zone into the gravel.

Another factor that aids successful bridging is that the diameter that best represents the smallest 15% of gravel particles (D15) must be ≤8 times the diameter that represents 85% of the smallest root zone particles (D85). For example: if 15% of the blanket particles are smaller than 4 mm, then 85% of the root zone particles need to be 0.5 mm or smaller. Similarly, to maintain good permeability of water across the root zone/blanket interface the D15 value for the blanket's gravel needs to be ≥5 times the D15 value for the root zone.

Where these criteria are not met the USGA recommends an intermediate (blinding) layer to be placed between the gravel and root zone (Fig. 9.2). This requires the blanket gravel to have: 65% of particles at 6–9 mm diameter; ≤10% of particles >12 mm diameter; and ≤10% of particles <2 mm diameter. This also requires the blinding layer to have 90% of particles 1–4 mm diameter.

As can be seen for these computations, careful grading of the gravel and root zone particles is essential to ensure compatibility!

These days, the root zone itself is invariably composed entirely of sand, although older recommendations suggested 20% of either soil or peat in the mix. In reality, degradation of the dead grass (thatch) means that the organic component can increase in the top strata of the root zone mix, without any intentional additions. As with the gravel there are recommendations with respect to composition of the root zone (Fig. 9.3).

The root zone mix needs to represent a media with a total pore space in the region of 35–55%, of which the air-filled porosity is 5–30%, and capillary (potential water-filled) porosity 15–25%. Hydraulic conductivity when saturated with water should equate to 150 mm h^{-1}. The root zone media is frequently blended by the company supplying the substrates or direct from a media manufacturer whose product meets the desired specification. It is good practice, however, that the required specifications are validated by a soils science laboratory or similar professional body. Thoroughly mixed root zone material is placed over the gravel or blinding layer, then consolidated to provide a uniform depth of 300 mm. Care is required to avoid intermixing of the root zone with the basal gravel.

Seeding and base fertilizer application

Fertilizer requirements can be assessed from soil nutrient and pH samples taken from the site. For sand-based root zones, base fertilizer is normally a requirement. Nitrogen is mobile and may not be applied at this stage, but potassium and phosphorus

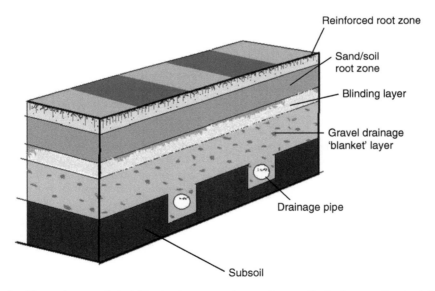

Fig. 9.2. Structural layers in a sports turf. The top layers may be sand or sand/soil mix, sometimes reinforced with synthetic fibres. A barrier or blinding layer is introduced to stop the smaller sand/slit particles migrating into the gravel and blocking drainage pores.

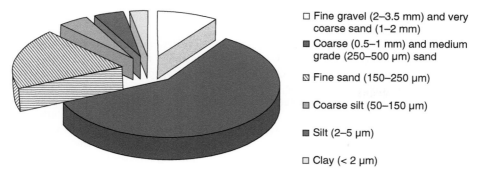

Fig 9.3. Typical particle types and proportions recommended for the root zone substrate mix used in the construction of golf greens. In reality organic matter also accumulates over time in this zone.

are incorporated prior to sowing (to a depth of 120–180 mm). Lime (Ca(OH)$_2$) may also be added at this early stage if the parent soil is excessively acidic. Nitrogen is usually applied at the time of sowing, or shortly after. Nitrogen forms a relatively small proportion of the fertilizer during early establishment phases, e.g. 6% N; 9% P$_2$O$_5$; 10% K$_2$O at 35–70 g m^{-2}. The top soil or root zone is consolidated by rolling and then raked to provide a fine tilth. Seeds are applied and the area rolled again to ensure good contact between seed and soil, thus allowing the developing seedling sufficient access to moisture.

Ideally, seeds should be placed approximately 6 mm below the soil surface; if they rest on the surface they are prone to being dislodged by the runoff of excess surface water. Placed too deep and the seedling's carbohydrate and protein resources may be over-taxed as it grows through the soil to reach the light. Uniform distribution of seed is important to ensure an even cover to the new sward, and avoid bare patches that will be prone to weed invasion. Mechanized seeders are available and cover a variety of scales of planting. These range from hand-held seeders, through to tractor mounted hoppers and distributors. Cultipack seeders work on the principle of spreading seed evenly along a line from a hopper or series of hoppers, but also at the correct planting depth using a ridged roller. This leaves characteristic parallel lines of grass seedlings along 'mini-furrows', but the grasses soon spread to colonize the inter-furrow spaces.

Where there is a requirement to sow grass on inaccessible sites, or where it is desirable to keep machinery off the soil to avoid compaction, then hydro-seeding can be employed. Hydro-seeding is the process in which grass seed is pumped onto the landscape in a stream of aqueous solution. The solution may contain fertilizer, mucilage gels to aid adhesion and even mulch materials (e.g. peat or vermiculite) as well as water.

Establishing a sward with turf

Lawns and sports pitches can be established by laying down sections of pre-grown turf (known as sods). These can be translocated from elsewhere on a given site, for example from one area of a golf course to another. Most frequently, however, they are grown on specialized turf farms. These turf farms are frequently located in regions with easily cultivated sandy or peat-based soils and the turf grown in conventional fields. Alternatively, some businesses now provide the sods using soil-less systems to cultivate the turf. The turf sections are lifted with the minimum amount of substrate (usually ≤10 mm) to keep weight down, but with a sufficient shoot/root matrix to provide integral strength to the turf sod (allowing for handling, transport and laying without sections breaking). Rolls of turf (typically 2.0 × 0.5 m) should be delivered from the farm, placed *in situ* and irrigated; all within 24–48 h. The turf needs to be transported damp to minimize root death, and stored during transit to avoid overheating. Heat stress is exacerbated by factors, including:

- high transport and storage temperatures;
- poor air movement through the roll or stack;
- high nitrogen ratios in the leaf tissues;
- the presence of pathogens; and
- transportation of turf with an excessively long leaf-blade length (height of cut) or where seed heads have formed prior to lifting.

Lawn establishment via turf grass is more expensive compared to seed (turf sods cost £1.50–7.50 m^{-2} (US$ 2.50–12.50 m^{-2})), but has the advantage of providing an instant visual effect. Additionally, if sods are well-laid and managed carefully, especially through appropriate irrigation scheduling, they can offer a fully playable surface within a few weeks of laying. Under optimum growing conditions new roots from the turf will start to integrate effectively with the parent soil on the site within a few days.

As with seed, successful establishment of the turf requires similar cultivation and preparation of the parent soil, ensuring the turf is laid on a fine weed-free tilth. The parent soil should be uniformly moist (but not excessively wet) at the time of laying. Rolls of turf can be laid by hand or by unrolling from bars fixed to a tractor. Ideally, the sections of turf should be set out in a staggered manner to avoid four corners meeting at any one point, as these corners may shrink and 'turn up' if desiccated. Topsoil or other substrates can be placed in any gaps between strips of turf, again to help avoid excessive drying. The turfs themselves need to be consolidated by light tamping to ensure effective contact between the root mat and the parent soil, and to remove any air gaps or ruffles. On steep slopes the turf may need to be staked in place to resist movement and slippage by gravity, with stakes being removed once roots have integrated with the soil beneath. Turf rolls placed across rather than down a slope will avoid rivulets of water forming channels between the rolls during periods of heavy rainfall.

Newly-laid turf should be irrigated (ideally within 30 min of laying) and then watered relatively frequently, e.g. daily in warm weather, over the subsequent 2–3 weeks. Once roots have penetrated into the underlying soil, irrigation can be applied less frequently, but in larger, single volumes. Frequency of irrigation during the establishment phase will depend on temperature and wind speed, as well as location. Southern aspects and more urbanized areas tend to have higher evapotranspirational demand than areas that are north-facing or have more surrounding greenery.

9.6 Lawn Maintenance Practices

Maintenance activities and time commitments are dictated by the use of the sward, size of area and the machinery available. Nevertheless, maintenance is dominated by mowing, irrigation nutrition management and controlling pests, weeds and pathogens.

Mowing

Mowing determines the playing quality of the turf, and its ability to tolerate wear. For a typical, intensively managed sward, mowing will represent 60% of costs. Mowing is required to keep the sward neat, true, vigorous and in a suitable condition for its purpose. It is also a very effective means of eliminating weed species that are intolerant of frequent close cutting. Mowing is most frequent for those playing surfaces that require a very close-knit sward such as golf greens. Indeed, golf greens may be mown daily during the growing season to ensure the height of the sward remains between 4 and 5 mm, with its height being increased to 6 mm in winter. A very low cut height of 3 mm may be achieved for short occasions but, if prolonged, this reduces the photosynthetic capacity of the grass with corresponding reductions in shoot density, rhizome numbers, root number and length. Such close-mown sward is more susceptible to drought stress and pathogens, and has a lower wear tolerance. Similarly, excessively long grass swards (e.g. >50 mm) have very poor wear tolerance, and even a single passage of a pedestrian can result in lodging (bending) of the stems. The severity of cut influences the physiology of the grass plant, and there is a balance to be struck between the functionality and predisposition to stress. For example, there is a direct relationship between height of cut and the depth of the root system. So close-mown swards have shallower roots systems (Fig. 9.4) and hence require greater frequencies of irrigation and fertilization to avoid water stress or nutrient imbalances.

The height of cut is strongly influenced by the desired playing quality of different types of sports turf. The mowing regime on golf greens is designed to influence the speed of the ball across the grass – the more challenging championship courses aiming for very 'fast' greens. The speed of a ball across a green is measured via a stimpmeter – a rod with a groove that allows a demonstration ball to roll onto the green from a consistent height and angle. For tournaments, the roll distance determines the speed of the green, e.g. fast, >2.75 m; moderate, 2.60–2.75 m; slow, <2.60 m. Green 'speed' is affected by the uniformity, smoothness, firmness and the resilience of the sward. Resilience (hardness or bounce) affects the capacity of the turf to absorb the shock of a golf ball hitting it and to 'hold' the ball on the green.

In sports such as tennis and cricket, the bounce of the ball was traditionally aided by modifying the soil

Fig. 9.4. Rooting depth is proportional to length of the grass shoots and leaves. Close mowing encourages a more compact, tightly knit sward (left), but results in a shallower root system that is more prone to drought stress, compared to a sward that has a higher height of cut (right).

constituents too, for example by providing a harder surface through the use of clay, and then using a roller to compact this soil surface. The higher bulk density and poorer drainage characteristics of clay topsoil, means more wear on the turf itself. To accommodate this, cricket squares are designed to incorporate a number of 'pitches': the area between the wickets in which the ball is bounced (and batted) and where there is the highest player footfall. Rather than renovate a pitch between consecutive matches, a new pitch is selected and the old pitch left unused for the grass to re-establish (Fig. 9.5). In other sports the roll or the bounce of the ball may be less important in comparison to the qualities the turf provides in terms of player safety and integral strength. Rugby pitches, for example, will have a longer sward length to provide greater cushioning when players tackle one another, and the longer leaf blades support a deeper root system that resists damage during scrumming.

Frequency and height of mowing relate to a compromise where there is a need to maintain an effective playing surface, improve wear tolerance, avoid abiotic and biotic stress but also to ensure that management, particularly labour, costs are not excessive. The height of cut associated with different sports reflects the requirements of each and the varying demands for wear tolerance (Table 9.5). Sports where there is a high footfall such as soccer, and where there are shearing and compression forces on the turf, have a high component of wear resistant *Lolium perenne* incorporated, maintained at a medium height of cut to aid resilience. Such pitches may also have artificial fibres (e.g. polypropylene) sown into the turf to improve physical robustness and help retain the playing quality should the grass die-out in localized patches. For many intermediate lawns and sports turfs, mowing is implemented when the grass is 1.5 times the required height, so a lawn that ideally should be at a height of 4 mm would be cut when at 6 mm. Exceptions to this rule include the avoidance of cutting the sward when excessively wet, during frost episodes or indeed when unduly dry, i.e. during

Fig. 9.5. A cricket ground is composed of a central square, an 'in-field' and an 'out-field' area. The topsoil within the central square has a relatively high clay content to allow the ball to bounce well. The square is divided into a number of 'pitches'. Due to the heavy wear and tear on the sward, a single pitch is not used continuously, but rather a new pitch is selected for each match, thus ensuring that the old pitch has time to recover. A new cricket pitch is being prepared before a match, with the old worn pitch (left of mower) being left for the grass to re-grow.

Table 9.5. Typical heights of cut for different sports.

Turf	Height of cut (height after cutting, mm)
Golf green	4–6
Bowling green/cricket square	4–6
Fine lawns	5–6
Golf tee	7–8
Golf fairway	10–12
Ornamental lawn	10–12
Soccer	12–25
Baseball	12–20
Rugby	25–50

drought periods when the plants may enter a quiescent state. Even within periods of fine weather, removing moisture such as morning dew from the leaves (switching) prior to cutting has advantages. Not only is the effectiveness of the mowing improved, it limits problems due to clippings adhering to and clogging the machinery.

At the onset of the growing season it is good practice to keep the height of cut relatively high during the first few mowing sessions. This tends to avoid excessive stress on the turf; it being wise to provide two or more successive cuts over a period of a few days, lowering the height of the blades on each occasion. This allows some capacity for the recently cut plants to reassimilate photosynthates, and for the sward to re-impose some tolerance to mowing stress, rather than the plants suffering a significant loss of leaf mass on a single, first occasion.

Mowing of less-intensively used swards

Where criteria for sport or other intensive forms of use are less relevant, for example turf grass around business parks, then considerably more sustainable approaches to mowing are feasible. Reductions in mowing frequencies are achievable with consequent savings in energy and labour inputs. Reducing the frequency of cut though means mowing at a higher sward height. This is acceptable for strong growing, erect grasses genera such as *Lolium* (perennial ryegrass) and *Paspalum* (Bahiagrass), many species still being able to be cut readily at even 100 mm height. Naturally shorter growing species, on the other hand, which may be out-competed by weeds

(*Zoysia*) or have unattractive features (*Cynodon*; Bermuda grass) are best cut at a lower height, e.g. 50 mm maximum.

Grass mowing machinery

The function of the sward influences not only the frequency of mowing, but also the type of mowing instrument used. There are five basic models of 'mower' to cut grass: the cylinder (reel) mower, rotary mower, flail mower, reciprocating mower and the strimmer.

The cylinder mower comprises a stationary bed knife, a cylinder with rotating blades and one or more rollers. These can be 'driven' by one of the mower's side-wheels, which is attached to the cylinder via gears or chains. Cylinder mowers are ideal for small intricate areas and where there is a requirement for a low mowing height. They have a sheering action on the grass leaves with the rotating blades working in tandem with the fixed bed knife to give a clean cut. Mowing quality is a function of the sharpness of the cylinder blades and proper adjustment of the bed knife in line with these. Cleanness of cut is influenced not only by the sharpness of the blades, but also by the number of blades on the cylinder (more giving a finer, even cut), the rotational velocity of the cylinder and the speed of forward motion of the mower. The most effective mowing is accomplished when the mowing height (the height the blades are set above the ground) is equivalent to the distance between successive clips of the cylinder blades, i.e. the distance the mower travels between one cut and the next. Where this is not in proportion, the cut height is not uniform and a wave pattern can be left on the grass (marcelling). The clean cut associated with cylinder mowers results in this type of mower being standard for golf and bowling greens and the more aesthetically demanding fine ornamental lawns. The rollers too are used to extenuate the aesthetic aspects in that they leave the grass bending in the direction of cut, and therefore can be used to form the characteristic stripe patterns frequently found on lawns and sports fields. The intricate cutting mechanism and the preponderance of the blades to notch when they hit a hard object means they are not so suited for longer grass and rougher areas.

Rotary mowers work on the principle of a horizontal blade rotating at high speed, and this blade fractures the grass rather than cutting it. These are energy-efficient and robust machines, ideal for typical utility lawns, parkland and rough areas of grass along roadsides and other verges. Rotary mowers can either be placed on a frame with wheels (Fig. 9.6) or suspended on a cushion of air, in a similar fashion to a hovercraft. The rotating blade has some tolerance of impact with hard objects, but can scalp the grass (cut into the substrate) on uneven topographies.

Flail mowers work on a similar principle to rotary ones, but have the blades attached by pivots to the central horizontal arm, rather than the arm itself being the blade. These free-swinging blades are designed to deflect away from any hard foreign object they hit, thereby reducing the risk of projectiles exiting from the skirting below the mower. The action of the blades and skirting means that the grass is finely chopped into a mulch, rather than being left as discrete stalks of grass. Maintenance is centred around ensuring the blades remain sharp

Fig. 9.6. A gang of three rotary mowers (two forward, one aft) on a 'ride-on' machine.

and that there is free movement of the blade on the horizontal arm.

The reciprocating blade mower is less frequently encountered compared to other types, but works on a similar principle to a hedge trimmer, i.e. there are two reciprocated saw-like blades that work in unison to clip the grass blades. The machines are useful for activities such as meadow cutting, as they can deal with long grass and provide a uniform height of cut. The disadvantage is that foreign objects, sticks, stones, plastic debris etc., can get trapped between the rows of blades.

The strimmer has a single shaft with either a high speed rotating blade or nylon cord at the base, which fractures the grass stems. The advantage of using nylon cord is that the cord extends as it breaks off against hard objects, thus avoiding the need to stop the machine. Strimmers are primarily used for small areas of long or rough grass, and are beneficial in sites that are uneven or where grass needs to be cut close to walls and other objects. Care should be taken, however, when operators are working close to trees and other desirable plant types as these can quickly strip a young tree of its bark.

Mowers are powered by a variety of means, including pedestrian-driven push mowers, four-stroke or two-stroke petrol engines, electrically powered machines and even solar powered instruments. They can be mounted on or be integral to small tractors ('sit on' mowers). Where large areas of turf need to be mown, 'gangs' or groups of mowers can be attached together and pulled by tractor, or as was previously done and is being promoted again by some landscape managers, by horse. Smaller areas such as domestic lawns can be mown 'automatically', by radio or computer controlled systems. Increasingly, the form and power source of the mowers is not just determined by the function of the turf, labour available or direct cost, but also the running and maintenance costs, and indeed the environmental footprint associated with the mower.

Alternative and low energy mowers and other turf machinery

Lawn mowing and other turf maintenance activities are thought to contribute significantly to engine emissions in the USA, e.g. approximately 1% of total motor engine petrol consumption per annum. An estimated 2.7–5.5 billion litres of petrol and up to 500 million litres of diesel are used to mow and trim lawns each year. Two-stroke mowers also consume oil in their fuel, and most mowers consume engine oil in their crankcases. Mowing of public land or sports facilities contributes 35% of this total fuel use, with domestic maintenance of garden lawns comprising the remaining 65% (Anon., 2011). Professional lawn maintenance teams can mow for up to 7 h per day, and use 4000–9000 litres of fuel each year, depending on land use, length of growing season and climate. As such, alternative and more efficient forms of fuel are considered appropriate for lawn maintenance activities. These include powering mowers with alternative hydrocarbon fuel sources, e.g. biodiesel, compressed natural gas, propane (liquefied petroleum gas) or electricity/battery powered machines or solar powered mowers. The alternative hydrocarbons still produce CO_2, but may burn more cleanly, reducing the proportion of other contaminants released to the atmosphere, or may be from more sustainable sources such as recycled fats and vegetable oils. Electricity is more sustainable when the power stations that generate it rely on renewable sources such as wind, water or solar energy.

Grass clippings

Clippings from mowing can either be collected by the mower catch-bin or left to be reincorporated into the sward as organic matter. The pros and cons of this depend on the volume of clipping, the impact on the playability of the surface and the nutritional regimes employed. Clippings are a source of plant nutrients, for example 3–5% of dry matter is nitrogen, and so where incorporation is feasible then there are opportunities to reduce the volumes of additional fertilizer required. Clippings are blamed for the build-up of thatch – dead organic matter at the base of the plant crowns, increased incidence of plant pathogens and promoting a soft, spongy turf. In reality, these claims are probably overstated, certainly in situations where active soil macro- and microflora degrade the leaf clippings quickly, and where the clippings are not left in thick mats on the surface of the turf. Regular mowing means that the volume of biomass left after any individual mowing is relatively limited anyway. The removal of clippings to landfill sites, or even to centrally located composting facilities, results in additional transport costs and an enhanced carbon footprint. It also results in a net loss of nitrogen from the lawn system, with a subsequent demand to replace this (usually through artificial fertilizers).

Hence, mechanisms to deal with the clippings *in situ* are likely to grow as holistic, sustainable management approaches become more widely adopted.

Thatch and mat

The space between the growing leaf blades of the sward and the parent soil is often filled with dead organic matter either in a partially or fully decomposed state. The top stratum is known as 'thatch', and the underlying layer as 'mat'. The mat layer is effectively the interface between thatch and soil, hence it may be composed of living roots and stolons in addition to organic residues and soil particles. The 'soil' particles themselves may be a consequence of topdressing activity, where sand is brushed into the sward to even-out the surface. Excessive accumulation of thatch (>13 mm) is considered undesirable as it impedes water infiltration and creates localized dry spots on the turf which are difficult to re-wet (hydrophobic spots). Similarly, the penetration of fungicides and liquid fertilizers are impeded. The fact that roots ramify through the open-textured thatch too, also means these roots are predisposed to greater environmental stress, e.g. drying and temperature extremes. Thatch may increase susceptibility of the grass to pests and pathogens as well as affect playing qualities of sports turf – e.g. the 'spongy' nature may result in poor ball bounce or roll.

Thatch accumulation is influenced by a range of physical, chemical and biological factors. Excessive build-up is inhibited by the presence of a functional microbial population, the activity of macrofauna (primarily earthworms), good aeration and a medium or slightly alkali pH (6.7–8.5). Conversely, excessive thatch can result from compacted, poorly aerated and more acidic soil. Thatch accumulation is traditionally mitigated through cultural practice, such as topdressing (adding sand or other 'open' aggregates) and/or physical cultivation techniques, such as verticutting (also known as vertical mowing, scarification or dethatching) or core removal (aerification). As microbial degradation of the organic material is a key component, some products that promote microbial activation or even introduce new microorganism populations are commercially marketed for this purpose.

The effects of lawn management on thatch development was studied by McCarty *et al.* (2007) over a 2-year period using swards composed of creeping bentgrass (*Agrostis stolonifera* var. *palustris*). In untreated swards, the organic matter content increased by 32%. In contrast, in those swards where core removal (tining) was practised in conjunction with verticutting procedures to remove the thatch/improve aeration, organic matter declined by 19%. Water infiltration rates into the sward increased with both verticutting (by 40–65%) and core removal (by 127–168%). Adding artificial populations of microorganisms though was not found to enhance organic breakdown rates. The authors concluded that sand topdressing alone was insufficient for managing thatch/mat levels in an established creeping bentgrass lawn, and that coring plus verticutting provided significant additional advantages.

Verticutting and similar activities are often carried out twice in the annual maintenance cycle: once at the start of the growing season and again at the end (i.e. in northern climates March/April and September/October). Care should be taken however, on those surfaces where the soil integrity is important, e.g. verticutting and coring should be avoided on a cricket square where it is actually desirable to have a compacted soil (combined with a high clay content) to enhance the bounce of the ball.

Aeration and drainage

Heavy traffic and wear on turf causes soil compaction. Compacted soils reduce oxygen availability to the roots and increase bulk density, providing greater physical resistance to root development. They may also impede water infiltration and drainage. Puddling on the soil surface frequently indicates a compacted soil. Different types of activity affect the degree of soil compaction and other stresses on the turf. Sheer volume of footfall traffic can cause compaction on sports turf, with certain locations on the pitch experiencing more compaction than others, e.g. the diamond pattern on soccer pitches (Fig. 9.7). Most compaction occurs in the top 30 mm of the soil profile, although movement of heavy machinery over a sward induces compaction to a greater depth.

Over localized areas, compaction is relieved by tining with a garden fork, although specialized machines cater for larger areas. As well as providing a macro puncture hole in the soil surface, the tining action also creates smaller fissures and micropores down the soil profile. Solid tines punch a hole into the soil surface, whereas hollow tines physically remove a plug (core) of soil. The advantage of the

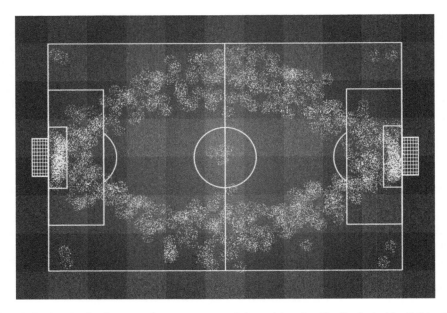

Fig. 9.7. The relative levels of soil compaction on a soccer pitch as determined by the typical footfall in different locations. This leads to a characteristic diamond pattern, with high wear and tear of the turf around active points such as the goal mouths.

latter is that there is less likelihood of localized compaction adjacent to the hole. Tining machines either comprise a roller with a series of solid, slit or hollow tines attached, or alternatively may utilize a vertical action, where soil plugs are removed at close spacing, but with minimal disruption to the turf surface. Normally tines are 75–100 mm long; however, longer 150–200 mm tines are useful to deal with the aforementioned machinery-induced compaction. Cores removed can be 6–18 mm in diameter.

At the top end of the machinery spectrum some instruments use vibrating tines pushed into the soil that tilt slightly before removal – this tends to lift the turf surface as well as create airways beneath it. Hard, dry soils are not conducive to tining and most aeration/decompaction takes place in autumn, winter, or early spring when the soil is moist, but not waterlogged or frozen. Tining is also best avoided during dry periods as the activity increases the sward surface area by as much as 120%, hence increasing the evaporative demand on the grass and leaving it much more prone to desiccation. Topdressing (filling the holes with sand or similar substrate) and or irrigating immediately after tining reduces this risk. Topdressing has the added advantage of filling the cores with sand, rather than encouraging soil particles to migrate back into the voids once the cores themselves break down.

Turf reinforcement

Turf reinforcement is used to improve wear resistance, particularly in sports turf, but systems also exist to facilitate vehicle movement over turf, or for parking. Heavily walked grassed paths may use polypropylene or steel wire reinforcement mesh to protect the turf and the soil structure from pedestrian traffic. In sports turf, sand root-zone reinforcement can be accomplished with thin polypropylene fibres or mesh to improve the structural strength of substrate by binding the sand particles more effectively. Other systems, e.g. TSII (Motz Group) or Desso GrassMaster systems, reinforce the turf directly. These integrate living grasses with synthetic fibres either through the use of pre-formed mats that can be placed on the substrate or, as in the case of the Desso system, where synthetic turf fibres 200 mm long are injected (or 'sewn') by machine into the natural turf root zone. The advantages of these systems are that if the natural turf wears down through excessive trafficking then the synthetic materials remain, providing some functionality and, as they are usually dyed green, they have some aesthetic attributes too.

9.7 Shade

Shade tolerance in amenity turf has always been a consideration when attempting to establish grass under heavy shade in the proximity of buildings or trees. It has become particularly relevant over the last decade or so where sports stadia have become larger and spectator stands, or even enclosed roofs, block out a proportion of the solar irradiance. Some turf species have greater shade tolerance than others, notably *Festuca arundinacea, Deschampsia caespitosa* and *Poa supina,* but the adaptations that plants use to cope with shade tend to work against the characteristics of quality sports turf (e.g. stretched 'etiolated' leaf/shoot growth). Hence for the most part, solutions rely on technical approaches, including: the use of mobile lighting rigs, chemical growth inhibitors to encourage a more compact plant habit, 'moveable pitches' where turf is growing in full light then translocated in sections to the shaded area of the stadium, frequent re-turfing or using turf-stabilizing materials to help maintain the playing quality. Many of these approaches adopted, of course, are not particularly sustainable in terms of resource management. Some stadia are designed to take account of these factors. For example, certain new stadia have translucent roofing panels on stands to permit the transmission of photosynthetically active radiation to the pitch. Similarly, the Emirates Stadium in London, UK, was designed to avoid complete enclosure of the stand sides, thus allowing irradiance to penetrate in above the seating areas. A number of strategies have involved growing 'alternative' pitches in full light in parallel to that in the stadium; they are then moved in as pallet-sections of pre-grown turf. In Phoenix, Arizona (USA), the entire pitch at the Arizona Cardinals American Football Stadium is grown and maintained outside the stadium and moved in under the stand via a complex roller system.

Stadia present other problems to the turf too. Other environmental problems include a lack of air movement, which can encourage fungal pathogens. There may be pressure to use the stadia for alternative sources of income. The Madejski Stadium in Reading, UK, hosts both soccer and rugby union on alternative weekends – sports that impose different and compounding pressures on the sward. Alternatively, stadia may be used for music concerts, boxing matches and even stockcar racing. Often the turf is covered for protection under such circumstances, but nevertheless the exclusion of light, albeit only temporarily, can weaken the grass.

9.8 Nutrient Management

Fertilization is one of the more controversial elements of turf management with respect to its environmental impact (see Chapter 2). As with other plant species, turf grasses have a requirement for macronutrients – mainly nitrogen, phosphorus and potassium, but also calcium, magnesium, sulfur – with essential micronutrients, including iron, manganese, zinc, copper, boron, molybdenum and chlorine. Much of the plant growth is determined by the ratio of carbon (derived from photosynthesis) to nitrogen; indeed this is a key factor that influences much of the dynamics of 'turf ecology' in general.

Optimum fertilizer management for turf grass is a subjective topic. Many garden lawns and amenity areas are, in effect, self-sustaining; they rely on no artificial fertilizer at all, with the required nitrogen being supplemented by atmospheric fixing of nitrogen, bacterial fixation or from 'recycled' organic matter degradation. Similarly, phosphate and potassium are sourced initially from the soil, and again recovered from natural processes where grass clippings are retained and broken down *in situ*. In 'high quality' sports turf, however, fertilizer may be applied regularly to ensure grass growth is strong and resilient to the wear imposed on it by the playing activities (or in ornamental lawns regular close mowing). Even here, however, seemingly high quality lawns may subsequently succumb to disease, often attributed to a high nitrogen regime.

Nutrition management then is essentially a response to the demands placed on turf when utilized as a sports surface or where great emphasis is placed on the aesthetics. This is particularly so where traffic levels are high, where the play characteristics set specific criteria (such as ball roll speed etc.), and/or where the environmental conditions are suboptimal for turf growth.

The level and frequency of nutrient inputs therefore reflects the function and management of the sward. Frequent mowing of a sward grown on a sand substrate and where clippings are removed continuously will require much greater fertilizer input than those situations where the turf is on an organic or clay soil (which can store nutrient ions through greater cation exchange capacity) and where the clippings are retained and reincorporated within the turf. It should be noted that under certain circumstances fertilizer applications can exceed the requirements of the turf, and that this not only poses potential for nutrient leaching and runoff, but also faster shoot growth rates may occur than is actually desirable; hence more frequent mowing and higher labour costs.

Nitrogen, phosphate and potassium

Nitrogen is normally considered the instrumental element in determining turf nutrition regimes (it comprises 3–5% of plant dry weight). It is essential for healthy growth, being a component of chlorophyll, plant enzymes and DNA. Excessive application, however, is often blamed for:

- inappropriate C:N ratios;
- excessive shoot growth;
- poor root development;
- increased susceptibility to certain pathogens; and
- reduced tolerance to abiotic stress, including cold, heat, drought and pedestrian or vehicle trafficking.

Nitrogen is absorbed as nitrate (NO_3^-), nitrite (NO_2^-) and/or ammonium (NH_4^+) ions. Nitrogen fertilizers can be either fast or slow releasing, and can be absorbed conventionally through the roots, but also as liquid feeds with some uptake via the foliage. Fast-releasing forms which are readily available for absorption include ammonium nitrate (NH_4NO_3), ammonium sulfate (($NH_4)_2SO_4$) potassium nitrate (KNO_3) and urea ($CO(NH_2)_2$). These fast-releasing, highly available forms of nitrogen tend to be characterized by high solubility, rapid growth responses in the turf and availability at a range of temperatures, but have potential to cause phytotoxic responses (leaf scorching) and to be excessively leached through the soil or lost through volatilization of gaseous ammonia. Slower-release forms include urea formaldehyde ($C_3H_8N_2O_3$), isobutylidene diurea ($C_6H_{14}N_4O_2$), sulfur-coated urea (controlled-release granules) as well as organic materials such as activated sewage sludge (Milorganite), turkey litter, and soybean-based extracts. Although nitrogen can still be lost from the system through leaching, denitrification and volatilization of slow release forms, this is less of a problem compared to the faster-releasing compounds. The fate of nitrogen after fertilizer application is complex. Ammonium and nitrate ions may be immediately available to the plants, or may require some further degradation of the source material. Plants may compete with microorganisms for the available nitrogen, and a proportion can be 'locked-up' by microbial dynamics. This may become available again at a later stage or alternatively be lost as nitrogen (N_2) and nitrous oxide (N_2O) gases and by denitrifying bacteria.

Phosphorus plays an essential role in plant cell division and growth: it is a constituent of cells' 'power source', adenosine triphosphate (ATP) and is a vital component of DNA and RNA. It is recycled and better conserved within plant tissues than nitrogen. Similarly, it is relatively immobile in the soil and less readily leached. Although present in the soil/substrate, phosphate is not always readily available to the plant; it is often converting to insoluble forms such as aluminium phosphate ($AlPO_4$) or iron phosphate ($FePO_4$), where the solubility may be dependent on soil pH. Due to the dynamic equilibrium associated with these compounds in the soil matrix, however, as plant roots absorb phosphate ions and the concentration in the soil solution decreases, more are in turn released from the compounds, again becoming available to the plant in due course. To this extent, apart from new swards established on sand or other substrates with inherently low phosphate levels, phosphate is rarely a limiting factor to turf grass development. Typical phosphorus-containing fertilizers include superphosphate and triple superphosphate (the active component in both being $Ca(H_2PO_4)_2$). These are formed from phosphate ores treated with sulfuric or phosphoric acid. The long-term sustainable use of these ores is now questioned and alternative sources of phosphate may be required in future.

Potassium normally constitutes about 2.5% of the dry weight of turf grass, and is used to regulate cell osmotic relations and water uptake. Potassium is included as a fertilizer as it reputedly increases tolerance to wear and other stresses, although in reality this may be more to do with developing an appropriate nitrogen:potassium (N:K) ratio. Stress tolerance may relate more to avoiding an imbalance of nutrients, rather than individual ions specifically having a 'magic bullet' direct role in stress alleviation.

The ratio of the principal elements can be varied depending on the time of fertilizer application and whether the aim is to address specific needs or deficiencies. In 'complete fertilizers' (ones that contain nitrogen, phosphate and potassium) the amount of nitrogen applied is often 2–3 times greater than potassium, with the phosphate ratio dependent on the soil type. So a 10:1:5 $N:P_2O_5:K_2O$ may be applied in soils with high residual phosphate levels, whereas on a sand-based substrate a ratio of 2:1:1 may be more appropriate. Applications are best split over the year to even out availability and avoid excessive leaching, so a typical programme for a fescue lawn in the north-east USA (Anon., 2003) where the clippings are returned may follow:

early June: (2:1:1); 50 kg ha^{-1} N, 25 kg ha^{-1} P_2O_5, 25 kg ha^{-1} K_2O;

early September: (2:1:1); 50 kg ha^{-1} N, 25 kg ha^{-1} P_2O_5, 25 kg ha^{-1} K_2O;

mid-November: (4:3:0); 50 kg ha^{-1} N, 37 kg ha^{-1} P_2O_5.

Late autumn fertilizer application has been cited as improving the coverage and visual quality of turf, particularly in cold climates, although responses have not been uniform across species, e.g. *Festuca rubra* (red fescue) and *Agrostis stolonifera* (creeping bent) showed better responses compared to *A. canina* (velvet bent) and *Poa annua* (annual meadow grass) (Kvalbein and Aramid, 2012).

Leaching and runoff

As legislation increases in an attempt to tackle issues linked to leaching and runoff, alternative sources of fertilizer application or improved irrigation management are being investigated. However, most sources of nitrogen leaching and runoff (although not exclusively so) relate to nitrogen application at higher than recommended rates, excess rainfall after fertilization, and fertilization at a time when turf is not actively growing; it being argued that properly maintained lawns provide a relatively effective system for absorbing nutrients.

Using more precisely controlled irrigation and fertigation systems (liquid fertilizer applied in the irrigation) was deemed promising by Snyder *et al.* (1984) in reducing nitrogen leaching. Comparisons were conducted in the management of *Cynodon dactylon* × *C. transvaalensis* (hybrid Bermuda grass) grown on a free-draining sand substrate. Irrigation was applied daily either as a set amount or through a more precise tensiometer-controlled system; the latter also being used for fertigation. This liquid feeding was also compared to the use of granular ammonium nitrate or sulfur-coated urea pellets.

A combination of tensiometer-controlled irrigation and nutrients being applied via fertigation or urea pellets resulted in the lowest levels of nitrogen leaching (<1–6% of total nitrogen) while maintaining acceptable turf grass quality. In contrast, ammonium nitrate granules in combination with more conventional daily irrigation produced the greatest nitrogen losses (22–56% of total nitrogen applied).

Even the grass species selected may influence nutrient leaching. Trenholm *et al.* (2012) found that *Stenotaphrum secundatum* 'Floratam' (St Augustine grass) was more successful than *Zoysia japonica* 'Empire' (zoysiagrass) in reducing the amount of leaching (Fig. 9.8). Increasing frequency of irrigation on *S.* 'Floratam' across a range of different nitrogen fertilizer rates (49–490 kg ha^{-1}) had little effect on the amount of nitrate leached from the sward, whereas similar regimes with *Z.* 'Empire' corresponded with more leaching as the frequency of irrigation increased (Fig. 9.9). Subsequently, lower rates of nitrogen fertilizer were recommended for *Zoysia* spp. not only to reduce leaching, but also to help avoid disease and improve the coverage of the turf. Timing was important; for both genotypes lawn quality was improved by applications of nitrogen in early summer (Fig. 9.9).

Soil conditions and differences in root morphology and/or architecture play an important role in nutrient uptake capacity. Evaluations on *Poa pratensis* (Kentucky bluegrass) cultivars showed that nitrate losses averaged 4.7% of applied nitrogen in non-compacted soil across a series of trials, whereas compaction of the subsurface soil layers increased losses to 8.9%. As such, interventions such as tining and verticutting, which improve water infiltration rates to the soil, can reduce runoff potential.

Phosphorus, despite being less mobile than nitrogen in the soil column, contributes to runoff pollution too, and subsequent eutrophication of water courses. This is especially so where the extractable phosphorus levels are high, such as in a recently applied liquid fertilizer, and where there is limited opportunity for the ions to bind to soil particles, e.g. pure sand zones, as the sand particles tend to 'fix' phosphorus less effectively than clay particles. Unlike nitrogen, where the pollutant source can dissipate relatively easily with time, phosphate pollution is more problematic due to the persistent nature of this element. Mismanagement of fertilizer application on turf grass is an important component of this. Column leachate studies have shown that phosphorus levels from intensively managed

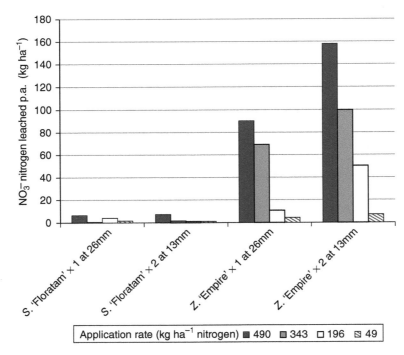

Fig. 9.8. The effect of fertilizer application rate (490, 343, 196 or 49 kg ha^{-1} nitrogen) on annual nitrogen leaching on lawns composed of *Stenotaphrum secundatum* 'Floratam' (St Augustine grass) or *Zoysia japonica* 'Empire' (zoysiagrass), when irrigation is applied once per week at 26 mm depth or twice per week at 13 mm depth. (Modified from Trenholm *et al.*, 2012.)

turf can be as high as 1–13 mg l^{-1}, well in excess of designated standards for surface water (0.3 mg l^{-1}). Tissue phosphate levels have been cited as being optimal at 3.4–3.9 g kg^{-1} in grasses (e.g. *Agrostis stolonifera*) and fertilizer applications often exceed that which is needed. Controlled release forms of phosphate tend to provide less leachate than soluble fertilizers, especially when applied at low rates (e.g. 5 kg ha^{-1}) and irrigated at low frequency. King *et al.* (2012) evaluated conventional phosphate fertilizer regimes for golf courses, with a lower input regime (reduced rate, low dose applications, and organic formulations). Management practices that involved lower application rates and the use of organic phosphorus formulations resulted in substantial reduction in the dissolved, reactive and total soil phosphorus concentrations. Due to the perceived problems of high phosphate leaching in sand-based substrates, it has been recommend that a two-pronged approach is adopted: turf managers need to explore the options available through organic formulations and lower dose applications, and fertilizer manufacturers should develop and make available commercial fertilizer blends with reduced or zero amounts of phosphate incorporated.

9.9 Irrigation

Increasing demand for fresh water from domestic and industrial sources has decreased the allocation of water for irrigation of non-food crops, especially in warmer/drier climatic regions. A lack of irrigation water has been a limiting factor to the development of new amenity turf areas in such locations, and existing features have had to ration and manage water resources more effectively, including evaluating alternative sources of water (saline water or recycled waste water). In parts of south-west USA 75% of natural water supplies may be considered unfit for human consumption due to moderate or severe degrees of salinity and a proportion of this could be used for irrigation. Likewise, recycled water accounts for irrigation in 37% of all golf courses in the south-west USA (Throssell *et al.*, 2009). In addition to alternative water sources, developing strategies to maintain acceptable turf

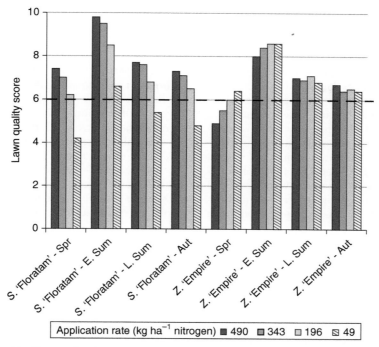

Fig. 9.9. The effect of fertilizer application rate (490, 343, 196 or 49 kg ha^{-1} nitrogen) on lawn quality (6 being deemed acceptable lawn quality) of *Stenotaphrum secundatum* 'Floratam' (St Augustine grass) or *Zoysia japonica* 'Empire' (zoysiagrass), when fertilizer applied either in spring (Spr), early summer (E. Sum), late summer (L. Sum) or autumn (Aut). (Modified from Trenholm *et al.*, 2012.)

quality with considerably less water use has been a primary goal of turf grass researchers and managers.

Irrigation is not just required on turf grass when there is inadequate precipitation. It may also be called upon when plant evapotranspiration exceeds the available supply in the substrate (i.e. substrate moisture holding capacity is not great), or where moisture movement upwards through capillary action is insufficient to meet the demands of the transpiring leaf. Hence, the characteristics of the root zone, particularly depth-of-rooting, influences how much moisture is available to the leaf canopy. Shallow-rooted turf requires more frequent, but less intensive irrigation than deep-rooted turf. In a somewhat circular relationship, root growth is very much determined by soil moisture, and conversely irrigation practices profoundly affect root growth and survival. There is a form of irrigation known as 'syringing'; this is used to offset temporary wilting, and is largely targeted at keeping the foliage hydrated or cool, rather than penetrating the substrate. It is employed to help the turf avoid the excessive heat of midday, or to be used on a newly established piece of turf where root penetration into the substrate is limited. As such, these irrigation schedules tend to apply only limited volumes of water at any one time. Not all irrigation practices just relate to moisture requirements or leaf cooling either. Irrigation is used to help better distribute fertilizers down the soil profile and avoid localized high concentrations of nutrients sitting on the foliage, potentially causing shoot and leaf damage (scorch).

Irrigation control and scheduling

The principle behind effective irrigation control is that irrigation should be applied whilst the substrate matric potential is high enough that the soil can still supply water to meet plant demand; in essence avoiding excessive moisture deficits and the potential of incurring hydrophobic zones in the sward. The aim is to avoid true drought stress to the grass, whilst still allowing moisture levels to decrease between consecutive irrigation events. Irrigation traditionally was managed by time clocks or the ground manager's perceptions of irrigation needs. Using feedback systems, however, that monitor soil moisture status or use algorithms based on climatic data (evapotranspiration and rain-based sensors) can evaluate moisture status accurately. Soil moisture sensors, evapotranspiration controllers, and/or rain sensors improve irrigation effectiveness and are known as 'smart controllers'. These are designed to irrigate the turf based on measured or estimated water depletion so that water is added to meet plant water needs while minimizing losses through runoff. Soil moisture sensors can be based on electrical capacity or impedance signals initiated by a probe (frequency domain) and where the signal alters based on the soil moisture content. Alternatively, neutron probe and tensiometer-based systems can be utilized.

The advantage of sensor-based systems is that they control irrigation supply more effectively, reduce the volume of water wasted as well as maintain acceptable turf quality. Compared to standard timer-based control clocks programmed on the basis of historical water use, irrigation controlled by soil moisture sensors has shown savings of anywhere between 0 and 74% based on location, sward composition and other management regimes. Rain-based sensor systems reduce water use over conventional controls by 7–45%, depending on local circumstances. Turf quality in soil sensor systems has generally been found to be acceptable for the most part, but dry weather and low threshold settings (set points) have resulted in cases of marginally acceptable to unacceptable turf quality (McCready et al., 2009; Cardenas-Lailhacar et al., 2010). Grabow et al. (2012) compared soil moisture sensors with an evapotranspiration-based controller. They found that a soil sensor that was linked to two moisture thresholds, one to initiate irrigation (at 21% soil moisture content) and one to switch it off (set at 30% soil moisture content), was the most effective system. This provided the best compromise between saving water (24%) and maintaining good quality turf. Soil sensors that had a single set point, but could switch off the irrigation without a full 30 min irrigation cycle being completed saved more water (39%) but also risked poor quality turf during dry, warm periods. In contrast, treatments based on evapotranspiration controllers resulted in good quality turf but applied 11% more water, on average, than the timer-based system. This may have been due to an overestimation of the reference evapotranspiration value that was used to schedule the irrigation.

Adaptation to reduced irrigation schedules has been noted in certain turf grass genotypes, and may actually improve quality. Reduced irrigation in warm-season *Cynodon dactylon* × *Cynodon transvaalensis* (hybrid Bermuda grass) and *Paspalum vaginatum* (seashore paspalum) cultivars indicated that 75–83% of standard reference evapotranspiration (ET) rates maximized turf grass quality across all the turf grasses tested. Acceptable quality was achievable in *Paspalum* at 75–80% of standard ET, whereas irrigation could be lowered to 66–75% of standard ET in *Cynodon* whilst still maintaining acceptable quality. *Cynodon* 'Midiron' was found to be particularly tolerant of the lower irrigation regimes but one downside was that deficit irrigation delayed growth activation in the spring ('greening-up'), e.g. delaying it by 6 weeks under a 60% of standard ET rate (Bañuelos et al., 2011).

Frequency as well as volume of irrigation can alter turf dynamics too. Less frequent irrigation has been recommended to improve turf quality through enhanced shoot density and encouraging deeper rooting, thus mitigating drought impacts (Jordan et al., 2003). This appears to be true for both cool-season and warm-season grasses. Where irrigation is controlled automatically, applications tend to take place during the night or early morning thus reducing losses from direct evaporation but also avoiding the periods where the sward is in use, or when maintenance activities are taking place. To avoid excessive surface runoff, the irrigation may be delivered in a series of cycles, thus allowing time for the water to infiltrate fully into the turf and substrate between consecutives applications.

Over-application of irrigation is encouraged by poor distribution of water across and within the turf. Hydrophobic patches (dry spots) resistant to wetting are a frequent occurrence in turf, where thatch builds up. The problem is induced by organic acids from the degrading thatch that cover the sand particles rendering them difficult to re-wet upon drying. These dry spots encourage grounds managers to prolong irrigation in an attempt to ensure effective irrigation coverage. This means that non-problematic areas receive considerably more water than required and that water is wasted. The use of surfactants or wetting agents can overcome this problem, by reducing the surface tension of the dry spots and aiding the penetration of water. Park et al. (2005), for example, claim that surfactants improve *Cynodon dactylon* (Bermuda grass) performance whilst reducing the requirement for irrigation overall. During the drier seasons, surfactants reduced irrigation volumes by 50% and were associated with increased photosynthetic rates, stronger shoot growth, higher soil moisture levels and reduced the symptoms of hydrophobicity.

Water quality and salinity stress

Grass species vary in their tolerance to saline soils and water sources. *Agrostis stolonifera* (creeping bentgrass), *Cynodon dactylon* (Bermuda grass), *Zoysia* spp. and *Stenotaphrum secundatum* (St Augustine grass) tend to have greatest tolerance, with *Festuca arundinacea* (tall fescue) and *Lolium perenne* (perennial ryegrass) intermediate, and *Poa pratensis* (Kentucky bluegrass), *Agrostis capillaris* (browntop bentgrass) and certain forms of *Festuca rubra* (red fescue) least tolerant. While the use of saline water has become an increasingly accepted strategy for conserving potable water, the detrimental effects to plants and the underlying soil ecology and physical structure resulting from salt accumulation has become apparent. The level of salinity determines potential use for turf grass situations. Electrical conductivity values of 0.75–2.25 dSm^{-1} should be restricted to non-sensitive species and used in situations where the soil drainage characteristics encourage ion leaching, 0.25–0.75 dSm^{-1} is used where there is some leaching potential and <0.25 dSm^{-1} presents water supplies that are deemed low risk. In addition to problems induced by high sodium and chloride ion concentrations, poor quality water can also be defined by high boron and bicarbonate ion levels that likewise cause phytotoxicity effects.

Salinity stress is particularly problematic when grass seeds are germinating and during the early establishment phases. Studies have evaluated the type of irrigation application (overhead sprinkler versus subsurface capillary) with water of either saline (electrical conductivity (EC) = 2.8 dS m^{-1}) or potable (EC = 0.6 dS m^{-1}) quality (Schiavon *et al.*, 2013). Results with *Poa pratensis* and *Festuca arundinacea* showed that seedling density was greater on overhead sprinkler-irrigated plots than on subirrigated plots. Higher EC levels were measured in the subirrigated plots, with highest sodium concentration (26.3 mmol$_c$ l^{-1}, i.e. millimoles of charge per litre) correlating with the saline water source. Despite this, subirrigation was shown to be a feasible way of establishing cool-season turf grass; but subsequent development was delayed compared to sprinkler irrigation.

In similar approaches, Sevostianova *et al.* (2011a) used established plots of seven different cool-season grass species, however, in this case both sprinkler and subsurface irrigation with saline water in semi-arid conditions was deemed to reduce turf quality to below an acceptable level, with the exception of *Festuca arundinacea* (Fig. 9.10). In contrast, evaluations of warm-season grasses generally showed greater tolerances to saline irrigation, although poor quality was evident with seasonal effects, i.e. during winter periods when cultivars were exposed to both saline stress and suboptimal temperatures (Sevostianova *et al.*, 2011b). Poor temperature tolerance may also explain some of the results for the cool-season cultivars, e.g. low scores for *Puccinellia distans* irrespective of salinity levels (Fig. 9.10). These studies were conducted in New Mexico, USA, where typical temperatures in this case may have been supra-optimal for *P. distans*.

In addition to naturally saline water sources, reclaimed water from other industrial processes or from agricultural land or practises (e.g. cleaning fruit and vegetables) can be used as an irrigation source. Reclaimed water that receives treatment and a high level of disinfection is used for turf irrigation and not deemed a threat to public health. However, the quality of reclaimed water, as with that of saline sources, differs from that of drinking quality water or rainfall. Depending on the quantity and timing of nutrients supplied by reclaimed water, the need for additional fertilizers could be significantly reduced. So, applying fertilizers at recommended rates without accounting for the nutrients in reclaimed water runs the risk of applying more nutrients than is necessary, exceeding levels that can be used by the plant or retained by the soil; the knock-on effects of which are excess nutrients running-off or leaching to groundwater and degrading water quality. Where nutrient management can be controlled though, there are advantages to using reclaimed water. The turf can help filter out pollutants before the water re-enters groundwater aquifers or river systems. Indeed, turf grasses are actively utilized by regulatory agencies as a tertiary treatment of effluent or reclaimed water. Snyder and Cisar (2000) indicated that golf courses could tolerate reclaimed water with up to 10 mg l^{-1} nitrogen, without the sources of nitrogen leaching into aquifers.

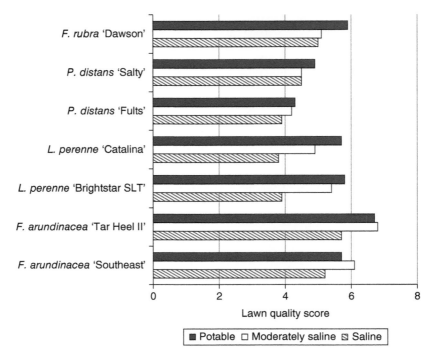

Fig. 9.10. The effect of water quality (potable, 0.6 dS m^{-1}; moderately saline, 2.0 dS m^{-1}; saline, 3.5 dS m^{-1}) on turf quality scores across a range of cool-season turf cultivars: *Festuca arundinacea* 'Southeast', *F. arundinacea* 'Tar Heel II', *Lolium perenne* 'Brightstar SLT', *L. perenne* 'Catalina', *Puccinellia distans* 'Fults', *P. distans* 'Salty' and *Festuca rubra* 'Dawson'. (Modified from Sevostianova *et al.*, 2011a.)

Lawns and sustainable water use

If irrigation of sport and other large recreational facilities is a political topic in terms of environmental management, almost more so is the domestic lawn. Many home-owners link a neat, well-tended lawn to aspects of affluence and even social respectability, or see the lawn as a cultural link back to a colonial past – where grand houses were offset by verdant lawns. Hurd *et al.* (2006) suggested that for western USA, educational background, regional culture as well as water costs were significant determinants of landscape choice and allied irrigation requirements. A number of studies report that volumes of water used in the landscape relate to the proportion of public to private land, and to income level within different residential groups; more affluent areas using more water. It is perhaps no surprise then that even in relatively dry regions such as central USA, surveys reported that 45–60% of home-owners thought it was moderately/very important their lawns looked green all the time.

Despite potable water being a precious resource, 61–63% did not know how much water their lawns required and 71–77% did not know how much water they applied to their lawns when they irrigated. Of those surveyed, however, 65–77% ranked water conservation at the same level of importance as maintaining a green lawn (Bremer *et al.*, 2013). The lack of understanding about turf grass water requirements and how much is actually being applied are significant issues relating to excessive water use. Over-irrigation is a common problem in domestic landscapes, especially in the transition periods between summer and autumn when evapotranspiration demand decreases quickly but irrigation systems schedules are not re-set accordingly (Salvador *et al.*, 2011).

To compound these problems the use of simple irrigation controllers based on time clocks has actually resulted in increased domestic water use compared to manual control of irrigation. This is due to the fact that many owners do not re-set the time clocks or find the instructions to do so too complicated. For example, Mayer *et al.* (1999) reported that 47% more water was used via automated irrigation compared to manual control due to 'the set-and-forget' mentality. As these systems frequently overwater, they exacerbate the problems associated with runoff and leaching, as well as actually running the risk of reducing the quality of the sward. In essence, the use of automated irrigation controllers appears to save home-owners time, not water. In contrast, the more sophisticated irrigation control models based on rainfall or evapotranspiration estimation do tend to save water use even in a domestic situation, e.g. 11–75% savings compared to manual operations.

Landscape irrigation in the USA had been cited as contributing approximately 9–48% of total municipal water use (St Hilaire *et al.*, 2008), whereas in parts of Spain it contributes 38% in spring to 69% in summer (46% of annual total water use) (Domene and Saurí, 2006). The Environmental Protection Agency in the USA reported that a typical lawn consumes 37,854 l of irrigation water beyond rainwater per year (Vickers, 2001).

There are opportunities to save water through better design too. In city-centre streetscapes, lawns are often incorporated in almost a tokenistic manner, with relatively small areas of turf being used alongside roads and pavements, as short strips or squares in formal civic gardens, or as centres to traffic roundabouts. Such limited areas of turf are 'high maintenance', both due to the demands placed on them from an aesthetic perspective and from the extreme environmental factors they may be exposed to (high evaporative demands due to urban heat island effects, air pollution, traffic wear and compaction, road-salt stress etc.). Irrigating such small and irregular-shaped swards via sprinklers can be very inefficient, due to the volume of water that misses the target area. In these circumstances larger areas of grass or other forms of green infrastructure may represent 'better return' for the resources, time and effort required to maintain them to a sufficiently high standard.

Attitudes to turf and water conservation

Due to its relative rarity in natural landscapes, paradoxically, turf is valued highly in countries with a Mediterranean climate, despite its high water-demanding characteristics. Price of water though does seem to be a moderating factor on the domestic use of water for irrigation, at least in some cultures. Inefficient irrigation practices in Spain corresponded to situations where high home-owner income was combined with low water charges; where charges increased water was used more responsibly. In those situations where water was cheap, aesthetics of landscapes and the right to apply irrigation were considered more important by the residents than saving water; this despite Spain, in its entirety, being considered as a drought-prone country (Salvador *et al.*, 2011). Research by Hurd *et al.* (2006), however, indicates that education programmes and 'moral persuasion' in the drier regions of the USA can have a positive influence toward water-conservation within the landscape. Studies in New Mexico showed home-owners were conscious of the issues around water conservation and were prepared to help regulate use of the State's water resources. There was support to limit traditional turf grasses (e.g. *Poa pratensis,* Kentucky bluegrass) and to increase the areas planted to native, natural, and water-conserving landscapes. For example, 92% favoured limiting turf grass to <25% of the area around public buildings. A significant minority, however, (40%) were not content with their current (dry, xerophytic) landscapes, suggesting that impediments to adopting more coherent water conservation measures remain. There is also evidence that those living in drier climates still aspire to having green and luxuriant landscapes around them. Martin *et al.* (2008) indicated that many residents still prefer to have a high proportion of irrigated lawn or shrubs borders, compared to adopting xerophytic vegetation per se (Fig. 9.11). This is despite public campaigns for reductions of landscape water use, and the fact that many residents stated that desert landscapes were aesthetically pleasing (Yabiku *et al.*, 2008).

The legacy of where home-owners had previously lived (i.e. usually the 'lush' greener landscapes of the eastern USA or Europe) and their preferences for landscape design type were significant predictors of front garden landscapes. Residents preferred green spaces with many textures and colours to open bare landscapes. Women and long-term residents particularly were more averse to dry landscapes. Interestingly, stronger environmental attitudes did not lead to greater preference for xeric desert landscapes per se, but did lead to compromises on the amount of turf grass preferred in lush landscapes. Overall, residents living in this arid landscape placed water conservation third on their

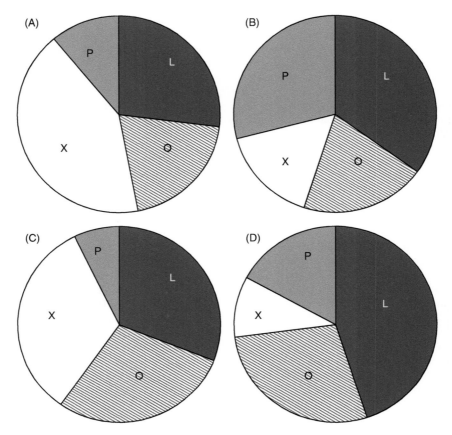

Fig. 9.11. The landscape composition of existing (A) front gardens, (B) back gardens, and the stated preference for (C) front gardens, (D) back gardens, in arid landscape of Phoenix, Arizona (USA). L, 'lawn' (turf grass and overhead irrigation); O, 'oasis' (turf grass and overhead irrigation mixed with landscape trees and shrubs, drip irrigation and decomposing granite mulch); X, xeric (desert-adapted trees and shrubs, drip irrigation and decomposing granite mulch); P, patio or other use. (Modified from Martin, 2008.)

priority list after requirements for low maintenance and aesthetically pleasing gardens.

In future, water use for turf is likely to be determined by both the cost of the resource and legislation. In San Diego, California (USA), local authorities restrict irrigation to <12 min per week, and in other parts of California there is the incentive of US$150 rebate for every 50 m² of natural turf replaced with synthetic turf as local governments attempt to reduce water consumption. In Victoria and New South Wales, Australia, authorities restrict the quantity and timing of irrigation on all sports surfaces, resulting in the temporary closure of some facilities to focus irrigation onto a limited number of surfaces (James, 2011).

In Spain, new golf course developments must be self-sufficient for irrigation water, and these are encouraged to link up with housing schemes in their locality where waste water from the housing can be collated and recycled for irrigation purposes.

9.10 Pesticide Use and Integrated Pest Management

Pesticides are still regularly used in turf management, but the intensity is again strongly dependent on the use of the sward – being higher on golf courses and other intensively-managed turf. Attitudes to pesticide use and actual practice vary strongly based on regional policies. In Quebec, Canada, the government environmental agency enacted a 'Pesticide Management Plan' in 2003 (Cisar, 2004) to reduce non-target impacts of agrichemicals, notably concerns over human and animal health. This resulted in bans on pesticide use on government-owned land, withdrawal of pesticide use within 3 m of any water body and prohibition of pesticide use on non-golf playing fields where these were used by children under the age of 14. Golf courses in the province were required to submit a detailed pesticide-use report and provide a 3-year plan for their reduction of pesticides. In another Canadian province, Ontario, the domestic use of lawn pesticides has been banned on a community-by-community basis (Cisar, 2004). Increasingly, the use of chemical pesticides, especially of compounds that may have some persistence in the ecosystem, will be hard to justify for non-food crops and amenity horticultural situations.

Strategies being adopted by professional land managers to reduce pesticide use include:

- Complying with local/national regulations on handling, storage and use of pesticides.
- Implementing cultural control programmes that minimize the likelihood of pathogen/pest problems arising. This includes understanding 'cause and effect' between management and potential problems, i.e. avoiding creating those environmental conditions that favour pathogens, invertebrate pests or weeds. Such control programmes have included integrated pest management (IPM), biocontrol, and best management practices (BMP) that have evolved to reduce pesticide application and improve efficacy with tangible impacts.
- Setting tolerance levels and implementing a regular monitoring programme to highlight problems earlier in their development, and deal with them in a more timely manner.
- Seeking expert advice where appropriate to properly identify problems so that the most integrated approach to managing them can be achieved.

Despite these approaches there may be incidences where pesticide application is considered necessary. Here too, though, mitigating actions include reducing the toxicity of the chemicals used through decreasing concentrations or the range of active ingredients in each agrichemical. Similarly, developing compounds with more specific modes of action that reduce the impact on non-target species is more commonplace. For example, mole crickets (*Scapteriscus* spp.) which damage *Cynodon dactylon* (Bermuda grass) were traditionally controlled in the USA by broad spectrum organophosphate (OP) and carbamate pesticides at relatively high rates and frequencies. In contrast, fipronil (that is, (RS)-5-amino-1-(2,6-dichloro-4-(trifluoromethyl)phenyl)-4-(trifluoromethylsulfinyl)-1H-pyrazole-3-carbonitrile) is now used and is applied precisely to soil via slit injection systems, at only a 0.1% active ingredient rate, once every 6 months (Cisar, 2004).

The use of 'natural' products approved by 'organic-growing' organizations is increasingly popular. Attempts to control the invasive weed *Urochloa subquadripara* (tropical signalgrass) have been undertaken through the use of bioherbicides, consisting of a cocktail of fungal species that attack this species, but are benign to other turf grasses (Chandramohan *et al.*, 2003). Other non-synthetic materials, including alfalfa pellets, corn gluten, cornmeal, garlic extract, molasses, liquid seaweed and vinegar, have been used as biopesticides to control weeds, insects, and plant pathogens or promote suppressive soils through stimulation of 'competitive' microbial populations (Cisar, 2004). Developing soil microbial populations that are either aggressive or competitive to the pathogenic organism, or exploiting species of mycorrhizal fungi that can afford better protection to the turf grass, are likely to be key strategies to improve cultivar resistance to disease (Duller *et al.*, 2010). Biocontrol agents are being developed to control nematodes that damage grass roots. These bionematicides include bacteria such as *Pasteuria* spp., *Purpureocillium lilacinus*, *Bacillus firmus* and *Myrothecium verrucaria* (Wilson and Jackson, 2013).

Success with biological control agents in a laboratory situation, however, does not always translate into effective application in the field, and further evaluations are necessary for some biocontrol strategies. For example, McGraw and Koppenhöfer (2008) investigated the positive use of parasitic nematodes as biocontrol for the *Listronotus maculicollis* (annual bluegrass weevil). Over a 3-year study, combinations of nematode species (*Steinernema carpocapsae*, *S. feltiae* and *Heterorhabditis bacteriophora*), application rate and timing produced strong numerical, yet few statistically significant, reductions. *H. bacteriophora* reduced first generation late instars by between 69% and 94% in at least one field trial (instar being the phase of larval development). At low instar (larvae) densities, *S. feltiae* provided a high level of control (94%) but gave inadequate control for higher densities (between 24% and 50% suppression). Time of application was also important, with introductions timed to coincide with larvae entering the soil (fourth instars) being optimal (McGraw and Koppenhöfer, 2008).

Where biocontrol and IPM options work, there are tangible reductions in pesticide use. (IPM takes a 'holistic' approach to pest management by understanding how predation rates and life cycles are influenced by environmental influences and wider management strategies, and not just the efficacy of pesticides.) Grant and Rossi (2004) reported that golf greens managed via IPM received 28–46% fewer pesticide applications than those managed by conventional means. Due to the nature of the processes involved, such as complex biological and environmental interactions, results do vary with location, predominant grass species and year; when *Agrostis canina* (velvet bentgrass) was managed by IPM it outperformed *A. palustris* (annual bluegrass/creeping bentgrass). It has been observed that control via IPM is more challenging during wet years, especially when rain periods coincide with warmer summer months.

In addition to management strategies to reduce chemical pesticide use, better gene selection for abiotic resistance in the breeding process is paramount. This may include the use of genetically modified cultivars of turf grass, where legislation allows (Cisar, 2004). Breeding disease-resistant varieties of turf grass can improve tolerance in the short/medium term, but pests and pathogens will continue to adapt too. Some of the longer term solutions may involve a change in mind-set of the turf users. The development of more sustainable husbandry techniques will need to go hand-in-hand with better education of the turf user (or in the case of sport, additionally, the spectator). For sports turf, this will require further exploration of the relationship between aesthetics and performance (in terms of ball/surface and player/surface interaction) (James, 2011).

9.11 Less Intensive Management

Although turf culture is one of the more challenging areas for environmental and sustainable management practices to be adopted, especially in relation to sports turf, advances are being made. New concepts and approaches are challenging the need for heavy chemical use, many local authorities and other landowners are questioning their requirement to mow extensive areas of infrequently used grass on a regular basis, and some home-owners are now more tolerant of non-grass species becoming present in their garden lawns. Not all solutions, however, are deemed to be held within the true philosophy of sustainability, for example the use of artificial turf to avoid mowing and pesticide applications (but with strong negative connotations for biodiversity); or even the application of growth-retarding chemical compounds (plant growth regulators) to help reduce the carbon footprint of mowing. On a more positive note, however, attention is being attributed to changes in attitude and philosophy, for example the adoption of genotypes better adapted to local conditions and changing attitudes to water, pesticides, fertilizer use and indeed more fundamentally what a lawn should look like and be used for.

Genotype selection for 'low input' systems

It is argued that as restrictions on water use, fertilization, and pesticide applications continue to increase, then turf managers will need to re-evaluate the choice of species used for top-quality lawns. Using golf fairways in the northern USA as the context for their research, Watkins *et al.* (2011) examined which grass species performed best when managed under carefully controlled, low-input regimes. Seventeen grass genotypes were evaluated for their performance in relation to differential traffic levels (zero, three, or six passes per week using a drum-type traffic simulator) and two mowing heights (19 and 25 mm). *Festuca ovina* ssp. *ovina* (sheep fescue), *F. rubra* ssp. *commutata* (Chewing's fescue), *Agrostis capillaris* (browntop bentgrass), and *A. canina* (velvet bentgrass) proved to be the most promising and were highlighted as candidates for further study for situations *in situ*. Genetic studies also help aesthetics and attitudes. The Institute of Biological, Environmental and Rural Sciences in Aberystwyth, Wales (UK), discovered a naturally occurring 'stay-green' gene in *F. pratensis* (meadow fescue), which through conventional cross-breeding has been introduced to *Lolium* sports turf cultivars. This allows the turf to remain green for longer when exposed to drought stress or low irradiance conditions during midwinter, thus making these cultivars more acceptable for premier sports stadia.

Even the concept of grass-only swards is now being challenged. Utilizing nitrogen-fixing clover, especially the small-leaved white clover, *Trifolium repens* form *microphyllum*, has received attention for its contribution to 'low input' turf communities. Not only does this species provide nitrogen to the soil matrix when there is root turnover but the small leaves improve sward colour in the winter. How these incorporated species affect play quality though, is not yet fully evaluated.

Growth regulation of the sward

One way to limit energy and labour costs is to reduce the frequency of mowing. Plant growth regulators and reduced irrigation have been evaluated for this purpose. Mefluidide (N-(2,4-Dimethyl-5-(((trifluoromethyl)sulfonyl)amino)phenyl) acetamide) has shown promise in this regard, it outperforming ethephon (2-Chloroethylphosphonic acid) and trinexapac-ethyl (4-(Cyclopropyl-a-hydroxymethylene)-3,5-dioxo-cyclohexanecarboxylic acid ethylester) in suppressing grass growth on *Agrostis stolonifera* lawns (McMahon and Hunter, 2010). Although mefluidide decreased grass growth it produced lower visual ratings. It was noted that all the plant growth regulators increased phosphorus levels in leaf tissue but had no impact on other nutrients or root zone pH. A number of additional advantages have been linked to growth regulator use; these include:

- the need for less irrigation;
- a better close-knit turf with fewer bare patches;
- more intense green colour (some products); and
- a reduced likelihood of seed heads forming on the fine-lawn species.

The extensive use of chemicals, however, to regulate grass growth is controversial, mostly from a cost perspective but also over some environmental concerns. Mefluidide, for example, is no longer registered for use in the landscape within the European Union, due to costs of trialling and registration and limited market application.

Artificial turf

An alternative to irrigated lawns that is now being seriously considered on its environmental benefits (rather than just playing quality) is artificial turf. (Artificial turf can be the preferred surface for certain sports, such as hockey where there is a high requirement to ensure a consistent role of the ball and to avoid it 'embedding' in the turf.) Certain households in California, USA, have adopted artificial turf to help save water. Some golf courses have used artificial turf for the greens, but there are now golf courses that are entirely composed of synthetic grass (e.g. the Echo Basin Golf Course, Colorado, USA). Although such courses are likely to reduce their environmental impact via lower energy, water and chemical use, the wider use of artificial surfaces can hardly be recommended for their contribution to animal and plant habitat; at least not when advocated on the larger scale.

Areas where less intensive management is warranted

Despite all the emphasis on high quality turf grass and the requirements for intricate management regimes, not all areas of turf need to be managed intensively. Many local governments are realizing the costs of frequent mowing regimes, and are considering land use change, or even just a relaxation in the intensity of current management activities. In countries such as the UK, Australia and the USA, roadside verges are frequently managed as 'rough grassland', with perhaps only a single annual cut to stop succession to woody plant species taking place. Similarly, there may be no requirement for pesticides, fertilizer or irrigation, and the mixed plant communities that evolve can be resilient to invasive weed invasion. Management of such sites is sometimes referred to as integrated vegetation management (IVM) where the aims are to reduce financial costs, reduce resource use and encourage wildlife. This holistic approach requires an understanding of the life cycle, seasonal cycle, species composition and population dynamics of grassland vegetation (Navie *et al.*, 2010). When managed well, this approach can reap tangible benefits to local wildlife, with the development of more biodiverse plant communities (Akbar *et al.*, 2009) and a more heterogeneous physical structure to the sward. Integrated vegetation management has been utilized recently to restore prairie plant communities as well as encourage populations of specific fauna taxa, including ants, bees, butterflies, diurnal moths, beetles and some birds and small mammal species.

This change of attitude has also been reflected in the sphere of the domestic lawn too, in some countries at least. Whereas traditionally the inclusion of broadleaved dicots in fine lawns was seen as an anathema, and that such 'weeds' were an indication of poor management (or gross negligence!), greater tolerance is now forthcoming for the inclusion of dicot species (Fig. 9.12). To survive, many of these dicot species either possess some tolerance to low mowing akin to grasses, or have a prostrate plant habit that allows them to avoid the mower blades. Whether the presence of these species within the sward is intentional or not, surprising numbers of genotypes can be found in domestic lawns. Surveys have shown 159 species in UK lawns (Thompson *et al.*, 2004), 83 in German lawns (Müller, 1990) and 139 in the lawns of Christchurch, New Zealand, alone (Horne *et al.*, 2005). Although the inclusion of certain dicots can disrupt the uniformity of the lawn, others that are more effective at retaining their colour (e.g. *Trifolium* spp., clovers) during stress may contribute to retaining the 'greenness' of the sward. Similarly, these and related 'nitrogen-fixing' species contribute to the nitrogen balance of the turf, thereby reducing the reliance on artificial forms of nitrogen. As the term 'weeds' used above implies, some species of dicots are readily integrated into the grass community, but the likelihood of any specific species becoming prominent will be dependent on the key ecological factors associated with the lawn, namely location, soil type and nutrient status, moisture availability, irradiance intensity, length of grass cutting and other disturbance regimes. Typical dicots found or easily incorporated into non-intensively managed European cool-season lawns are shown in Table 9.6. By using the appropriate types of dicots, lawns can stay green for longer during drought periods. Some 'sustainable' mixed species lawns require only 50% of the conventional volume of irrigation, with mowing frequencies also being reduced to once every 2–3 weeks (Cook and VanDerZanden, 2011).

Grasslands for conservation and biodiversity are covered in more details under Chapter 7.

Future directions

Developing truly sustainable systems of construction and management for turf grass remains challenging. Priority needs to be given to the notion that not all grassed areas require the same approach. Much of the public green space can be managed in a more benign, less energy- and resource-intensive manner, with reductions in frequency of mowing and adopting systems that allow for better 'cycling' of carbon and nitrogen, such as the re-incorporation of mower clippings. Similarly, mind-sets need to change with respect to the private lawn, where nutrient inputs, fossil fuel energy consumption and water are often considerably more excessive than they need be. Greater consideration also needs to be given to the

Fig. 9.12. Low-growing dicot species such as *Ranunculus repens* (creeping buttercup) can survive in less intensively managed lawns. Although still seen as a weed species by some gardeners, this and other lawn flowering species have an increasingly important conservation role to play in providing nectar and pollen sources for insects.

Table 9.6. Herbaceous perennial dicots (forbs) that can be effectively integrated with lawns of 30–60 mm height.

Species	Common Name	Characteristics
Achillea millefolium	yarrow	Drought tolerant, pale cream flowers.
Ajuga reptans	bugle	Moist, well-drained soils. Blue flowers. Can be invasive in North America.
Anthyllis vulneraria	kidney vetch	Dry soils of high pH, yellow, sometimes red/orange, flowers.
Bellis perennis	common daisy	Moist soils, although has some drought tolerance. White flowers sometimes flushed pink.
Campanula rotundifolia	harebell	Adaptable to wide pH, acid and chalkland grass communities. Distinctive pale blue, bell-shaped flower.
Cardamine pratense	cuckoo flower	Moist/wet lawns, pale mauve flowers in spring.
Galium verum	lady's bedstraw	Very drought tolerant. Compatible with lawn grasses – acid yellow flowers.
Chamaemelum nobile	chamomile	Pale green leaves, white flowers – some tolerance to drought. Also used as a monoculture sward – chamomile lawn.
Glechoma hederacea	ground ivy	Moist soils and tolerates shade. Broadleaved, but blends well with lawn grasses.
Helianthemum nummularium	common rock-rose	Dry grassland, will not tolerate excessive low mowing – sulfur yellow flowers.
Hieracium brunneocroceum	orange hawkbit or fox and cubs	Medium/moist soils. Basal rosette of green, oval, hairy leaves and tall stems topped with orange, daisy-like flowers.
Hieracium pilosella	mouse-ear hawkweed	Light soils, basal oval leaves, with yellow flowers. Can be invasive in non-native countries.
Hypochaeris radicata	cat's ear	Tolerant of a wide range of soil conditions. Daisy-like yellow flowers.
Lotus corniculatus	bird's-foot trefoil	Sandy, free-draining soils. Tolerates a degree of trampling and mowing, yellow and red pea-like flowers.
Lotus pedunculatus	greater bird's-foot trefoil	Moist soils and wet pastures. Similar, but slightly larger than *L. corniculatus*.
Leontodon hispidus	rough hawkbit	Dry soils with neutral or high pH. Rosette-forming perennial with ability to regenerate after close mowing. Yellow flowers.
Medicago lupulina	black medick	Short-lived perennial that dies out if sward gets too long. Does best when there is a degree of soil disturbance. Small, yellow 'ball-like' floret head.
Oxalis corniculata	yellow woodsorrel	A number of prostrate, spreading *Oxalis* species can survive in lawn turf, the yellow-flowered *O. corniculata* often being associated with drier or disturbed soil.
Plantago lanceolata	ribwort plantain	Rosette-forming perennial with leafless, silky, hairy flower stems, Commonly found in pasture land and rough lawns, especially on high pH soil. Brown/grey flower panicles appear on stalk above the grass.
Plantago media	hoary plantain	A more refined plant than *P. lanceolata*, *P. media* grows in moist lawns. Produces a slender stalk with pink-white flowers.
Potentilla erecta	tormentil	Neutral to low pH soils. Wet and moist soils, and often associated with grasses of acidic heathlands. Small, open cup-shaped, yellow flowers.
Primula veris	cowslip	Neutral to high pH soils. Medium to dry lawns, with yellow clusters of flowers appearing on a stalk held above the sward.
Primula vulgaris	primrose	Moist and heavier soils than *P. veris*. Pale yellow flowers amongst the foliage.
Prunella vulgaris	self-heal	Fairly drought tolerant, neutral to high pH soils. Purple conical flowers at mowing height.
Ranunculus repens	creeping buttercup	Moist to wet lawns. Creeping habit, with bright yellow, buttercup flowers displayed just above the foliage.

Continued

Table 9.6. Continued.

Species	Common Name	Characteristics
Taraxacum officinale	dandelion	Moist and heavy clay soils. Persistent deep fleshy tap-root, allows for quick regeneration. Robust leaves and yellow daisy flower, with conspicuous globe seed head.
Trifolium repens	white clover	Low growing and tolerant of mowing – prefers clay soil, but tolerant of others. Can 'fix' atmospheric nitrogen. Leaves stay green under stress and white flowers are nectar source for insects.
Trifolium fragiferum	strawberry clover	Tolerant of wet and saline soils. Pink fruiting head resembles a strawberry.
Veronica filiformis	slender speedwell	Grows from prostrate rhizomes, below the height of cut. Prefers moist soils. Attractive blue or mauve flowers.
Veronica chamaedrys	germander speedwell	More refined than *V. filiformis*, with a deeper blue flower.
Viola odorata	sweet violet	Shady or moist lawns. Found in woodland edges, banks and hedgerows but can invade turf swards. Blue-purple or white flowers.
Viola riviniana	dog violet	Wide range of soils, except very low pH or very wet. Foliage can be conspicuous within the lawn. Blue or purple coloured pansy-like flower. No scent to the flower.

siting of lawns in the first instance, the component species and the composition of the physical attributes, as well as the type and intensity of the maintenance, to improve the environmental performance of turf grass. Elite sports turf will always be the most demanding aspect of turf management, but even here utilizing slower release fertilizers, recycled or low-grade irrigation water, sustainable and locally sourced aggregates, IPM programmes and developing grass genotypes bred for their stress tolerance and low input requirements (rather than playing quality and aesthetic attributes) should go some way to improving the environment credentials of the sector.

Conclusions

- Turf grass provides a range of functional uses: a playing surface for formal and informal sport; aesthetic value and psychological benefits; land stabilization and avoidance of soil erosion; a foil to buildings and other infrastructure; a safe haven adjoining roadways; low-use pathways and a walkable surface; habitat; microclimate modification; and the ability to trap aerial and waterborne pollutants.
- The ability of grass plants to grow from an active meristem at the base of the plant, and essentially to continuously 'renew' themselves, allows them to play a unique role in the landscape, not least being resilient against considerable wear and tear. This also means though that maintenance and management issues can be significant under certain circumstances and the criteria imposed on the grass sward.
- 'Cool' season grasses associated with turf in mild temperate climates tend to be C_3 plants, whereas their tropical/subtropical counterparts 'warm' season grasses are usually C_4.
- A range of grass species have been selected for turf culture and lawn use, based on their ecological adaptations to factors such as tolerance to drought, acidity, low nutrients and physical abrasion. Many thousands of genotypes, however, have been bred from these species to hone these attributes further.
- Sport turf surfaces and lawns can be established either from seed or laying down sections of turf (sods) pre-grown elsewhere.
- Sport places the most demand on turf, and the interaction between the soil (substrate) and the grass genotypes used is critical in ensuring the playing quality of the turf is maximized for any given sport. Often this requires the turf species to grow in exceedingly sharp-draining substrates, and be dependent on artificial sources of nutrition (fertilizers), additional water (irrigation), chemical additives (e.g. pesticides and growth regulators) and environmental manipulation (mowing, tining etc.) to establish and thrive. The sustainability of some of the approaches is open to question, and

- research agendas are aimed at reducing resource inputs and negative environmental impacts.
- Mowing accounts for most of the labour and fiscal costs associated with maintaining elite sports turf.
- Grass swards can be self-sustaining in terms of nutrient balance, especially if nutrients are recycled via grass clippings; however, for most intensively managed sports swards, appropriate management of nutrients, especially through the use of externally applied fertilizers, is the key to the functional success of the turf.
- Similarly, despite the resilience of grass species (many of which are drought adapted) irrigation is the other key factor that largely determines playing function and the quality of the turf.
- Although many sport turfs require high levels of inputs, other grass swards are often over-managed and the high energy and resource inputs are not justified.
- More sustainable approaches to turf management are available, and include physical, biological and psychological factors. Changing attitudes to the aesthetics of turf and the functions required of it can significantly reduce the resources required to maintain turf. In addition, enhanced management approaches can reduce the requirements for high volumes/high quality water, energy and artificial chemicals that are currently used in many situations where turf grass is grown. Reducing such inputs are likely to have positive benefits in terms of mitigating effects due to waterborne pollution, carbon-based energy use and the overuse of potable water.

References

Akbar, K.F., Hale, W.H. and Headley, A.D. (2009) Floristic composition and environmental determinants of roadside vegetation in north England. *Polish Journal of Ecology* 57, 73–88.

Anon. (2003) Turfgrass fertilization: a basic guide for professional turfgrass managers. Available at: plantscience.psu.edu/research/centers/turf/extension/factsheets/turfgrass-fertilization-professional (accessed 7 November 2014).

Anon. (2011) Clean cities guide to alternative fuel commercial lawn equipment. US Department of Energy. Available at: www.nrel.gov/docs/fy12osti/52423.pdf (accessed 14 November 2014).

Anon. (2014) USGA Recommendations for a Method of Putting Green Construction. Report by the United States Golf Association Green Section Staff. 11pp. Available at: www.usga.org/content/dam/usga/images/course-care/2004%20USGA%20Recommendations%20For%20a%20Method%20of%20Putting%20Green%20Cons.pdf (accessed 2 February 2016).

Bañuelos, J.B., Walworth, J.L., Brown, P.W. and Kopec, D.M. (2011) Deficit irrigation of seashore paspalum and Bermudagrass. *Agronomy Journal* 103, 1567–1577.

Beard, J.B. and Green, R.L. (1994) The role of turfgrasses in environmental protection and their benefits to humans. *Journal of Environmental Quality* 23, 452–460.

Bremer, D.J., Keeley, S.J., Jager, A.L. and Fry, J.D. (2013) Lawn-watering perceptions and behaviours of residential homeowners in three Kansas (USA) cities: implications for water quantity and quality. *International Turfgrass Society Research Journal* 12, 23–29.

Cardenas-Lailhacar, B., Dukes, M.D. and Miller, G.L. (2010) Sensor-based automation of irrigation on Bermudagrass during dry weather conditions. *Journal of Irrigation and Drainage Engineering* 136, 184–193.

Chandramohan, S., Stiles, C. and Charudattan, R. (2003) Bioherbicide. *Florida Green*, Autumn 2003, 48–52.

Cisar, J.L. (2004) Managing turf sustainably. In: *Proceedings of the 4th International Crop Science Congress*, 26. Brisbane, Australia, September 2004.

Cook, T.W. and VanDerZanden, A.M. (2011) *Sustainable Landscape Management: Design, Construction, and Maintenance*. John Wiley & Sons, Hoboken, New Jersey.

Domene, E. and Saurí, D. (2006) Urbanisation and water consumption: influencing factors in the metropolitan region of Barcelona. *Urban Studies* 43, 1605–1623.

Duncan, R.R. and Carrow, R.N. (2000) *Seashore paspalum: the environmental turfgrass*. John Wiley & Sons, Hoboken, New Jersey.

Duller, S., Thorogood, D. and Bonos, S.A. (2010) Breeding objectives in amenity grasses. In: Boller, B. (ed.) *Fodder Crops and Amenity Grasses*. Springer, New York, pp. 137–160.

Grabow, G.L., Ghali, I.E., Huffman, R.L., Miller, G.L., Bowman, D. and Vasanth, A. (2012) Water application efficiency and adequacy of ET-based and soil moisture–based irrigation controllers for turfgrass irrigation. *Journal of Irrigation and Drainage Engineering* 139, 113–123.

Grant, J.A. and Rossi, F.S. (2004) Evaluation of reduced chemical management systems for putting green turf. *USGA Turfgrass and Environmental Research Online* 3, 1–13.

Horne, B., Stewart, G.H., Meurk, C.D., Ignatieva, M. and Braddick, T. (2005) The origin and weed status of plants in Christchurch lawns. *Journal of the Canterbury Botanical Society* 39, 5–12.

Hurd, B.H., Hilaire, R.S. and White, J.M. (2006) Residential landscapes, homeowner attitudes, and water-wise choices in New Mexico. *HortTechnology* 16, 241–246.

James, I.T. (2011) Advancing natural turf to meet tomorrow's challenges. *Proceedings of the Institution of*

Mechanical Engineers, Part P: Journal of Sports Engineering and Technology, 1754337111400789.

Johnson, P.G. (2000) An overview of North American native grasses adapted to meet the demand for low-maintenance turf. *Diversity* 16, 40–41.

Jordan, J.E., White, R.H., Vietor, D.M., Hale, T.C., Thomas, J.C. and Engelke, M.C. (2003) Effect of irrigation frequency on turf quality, shoot density, and root length density of five bentgrass cultivars. *Crop Science* 43, 282–287.

King, K.W., Balogh, J.C., Agrawal, S.G., Tritabaugh, C.J. and Ryan, J.A. (2012) Phosphorus concentration and loading reductions following changes in fertilizer application and formulation on managed turf. *Journal of Environmental Monitoring* 14, 2929–2938.

Kvalbein, A. and Aamlid, T.S. (2012) Impact of mowing height and late autumn fertilization on winter survival and spring performance of golf greens in the Nordic countries. *Acta Agriculturae Scandinavica, Section B – Soil & Plant Science* 62, 122–129.

Martin, C.A. (2008) Landscape sustainability in a Sonoran Desert city. *Cities and the Environment* 1, Article 5.

Mayer, P.W., DeOreo, W.B., Opitz, E.M., Kiefer, J.C., Davis, W.Y., Dziegielewski, B. and Nelson, J.O. (1999) *Residential End Uses of Water*. AWWA Research Foundation and American Water Works Association, Denver, Colorado.

McCarty, L.B., Gregg, M.F. and Toler, J.E. (2007) Thatch and mat management in an established creeping bentgrass golf green. *Agronomy Journal*, 99, 1530–1537.

McCready, M.S., Dukes, M.D. and Miller, G.L. (2009) Water conservation potential of smart irrigation controllers on St Augustine grass. *Agricultural Water Management* 96, 1623–1632.

McGraw, B.A. and Koppenhöfer, A.M. (2008) Evaluation of two endemic and five commercial entomopathogenic nematode species (Rhabditida: Heterorhabditidae and Steinernematidae) against annual bluegrass weevil (Coleoptera: Curculionidae) larvae and adults. *Biological Control* 46, 467–475.

McMahon, G. and Hunter, A. (2010) Determination of the effects of plant growth regulators on *Agrostis stolonifera* and *Poa annua*. In: *XXVIII International Horticultural Congress on Science and Horticulture for People (IHC2010): International Symposium* 937, 161–168.

Milesi, C., Running, S.W., Elvidge, C.D., Dietz, J.B., Tuttle, B.T. and Nemani, R.R. (2005) Mapping and modeling the biogeochemical cycling of turf grasses in the United States. *Environmental Management* 36, 426–438.

Müller, N. (1990) Lawns in German cities. A phytosociological comparison. In: Sukopp H., Hejný S. and Kowarik, I. (eds) *Urban Ecology: Plants and Plant Communities in Urban Environments* 209–222. SPB Academic Publishing, The Hague.

Navie, S.C., Hampton, S.J., Bloor, N. and Zydenbos, S.M. (2010) Integrated management of mown vegetation in eastern Australia. In: *17th Australasian Weeds Conference. New Frontiers in New Zealand: Together we can Beat the Weeds*. Christchurch, New Zealand, 26–30.

Park, D.M., Cisar, J.L., McDermitt, D.K., Williams, K.E., Haydu, J.J. and Miller, W.P. (2005) Using red and infrared reflectance and visual observation to monitor turf quality and water stress in surfactant-treated Bermudagrass under reduced irrigation. *International Turfgrass Society Research Journal* 10, 115–120.

Raciti, S.M., Groffman, P.M., Jenkins, J.C., Pouyat, R.V., Fahey, T.J., Pickett, S.T. and Cadenasso, M.L. (2011) Accumulation of carbon and nitrogen in residential soils with different land-use histories. *Ecosystems* 14, 287–297.

Salvador, R., Bautista-Capetillo, C. and Playán, E. (2011) Irrigation performance in private urban landscapes: a study case in Zaragoza (Spain). *Landscape and Urban Planning* 100, 302–311.

Schiavon, M., Leinauer, B., Serena, M., Sallenave, R. and Maier, B. (2013) Establishing tall fescue and Kentucky bluegrass using subsurface irrigation and saline water. *Agronomy Journal* 105, 183–190.

Selhorst, A. and Lal, R. (2013) Net carbon sequestration potential and emissions from home lawn turfgrasses of the United States. *Environmental Management* 51, 198–208.

Sevostianova, E., Leinauer, B., Sallenave, R., Karcher, D. and Maier, B. (2011a) Soil salinity and quality of sprinkler and drip irrigated cool-season turfgrasses. *Agronomy Journal* 103, 1503–1513.

Sevostianova, E., Leinauer, B., Sallenave, R., Karcher, D. and Maier, B. (2011b) Soil salinity and quality of sprinkler and drip irrigated warm-season turfgrasses. *Agronomy Journal* 103, 1773–1784.

Smetana, S.M. and Crittenden, J.C. (2014) Sustainable plants in urban parks: a life cycle analysis of traditional and alternative lawns in Georgia, USA. *Landscape and Urban Planning* 122, 140–151.

Snyder, G.H., Augustin, B.J. and Davidson, J.M. (1984) Moisture sensor-controlled irrigation for reducing N leaching in Bermudagrass turf. *Agronomy Journal* 76, 964–969.

Snyder, G.H. and Cisar, J.L. (2000) Monitoring vadose-zone soil water for reducing nitrogen leaching on golf courses. In: Clark J.M. and Kenna M.P (eds) *Fate and Management of Turfgrass Chemicals* 743. American Chemical Society, Washington, DC, pp. 243–254

St Hilaire, R., Arnold, M.A., Wilkerson, D.C., Devitt, D.A., Hurd, B.H., Lesikar, B.J. and Zoldoske, D.F. (2008) Efficient water use in residential urban landscapes. *HortScience* 43, 2081–2092.

Taliaferro, C. (2000) Bermudagrass has made great strides – and its diversity has barely been tapped. *Diversity* 16, 23–24.

Thompson, K., Hodgson, J.G., Smith, R.M., Warren, P.H. and Gaston, K.J. (2004) Urban domestic gardens (III): Composition and diversity of lawn floras. *Journal of Vegetation Science* 15, 373–378.

Throssell, C.S., Lyman, G.T., Johnson, M.E., Stacey, G.A. and Brown, C.D. (2009) Golf course environmental profile measures water use, source, cost, quality, management and conservation strategies. *Applied Turfgrass Science* 6, doi:10.1094/ATS-2009-0129-01-RS.

Townsend-Small, A. and Czimczik, C.I. (2010a) Carbon sequestration and greenhouse gas emissions in urban turf. *Geophysical Research Letters* 37, L02707.

Townsend-Small, A. and Czimczik, C.I. (2010b) Correction to 'Carbon sequestration and greenhouse gas emissions in urban turf'. *Geophysical Research Letters* 37, L06707.

Trenholm, L.E., Unruh, J.B. and Sartain, J.B. (2012) Nitrate leaching and turf quality in established 'Floratam' St Augustine grass and 'Empire' zoysiagrass. *Journal of Environmental Quality* 41, 793–799.

Vickers, A. (2001) *Handbook of Water Use and Conservation*. WaterPlow Press, Amherst, Massachusetts.

Watkins, E., Fei, S., Gardner, D., Stier, J., Bughrara, S., Li, D. and Diesburg, K. (2011) Low-input turfgrass species for the north central United States. *Applied Turfgrass Science* 8, doi:10.1094/ATS-2011-0126-02-RS.

Wilson, M.J. and Jackson, T.A. (2013) Progress in the commercialisation of bionematicides. *BioControl* 58, 715–722.

Yabiku, S.D., Casagrande D.G. and Farley-Metzger, E. (2008) Preferences for landscape choice in a Southwestern desert city. *Environment and Behavior* 40, 382–400.

10 New Green Space Interventions – Green Walls, Green Roofs and Rain Gardens

ROSS W.F. CAMERON

> **Key Questions**
> - What are the three different green wall systems and their distinguishing features?
> - How does the green wall system selected affect the plant types that might be used within it?
> - In what sort of locations would green walls be used?
> - What are the different types of green roofs that might be placed on a building?
> - What does the depth of substrate on a green roof influence?
> - What factors might be considered when increasing the environmental value of a green roof?
> - For what purposes are rain gardens useful?
> - How does their design help in their functionality?
> - What traits should plants used within rain gardens possess?

10.1 Introduction

As global urbanization increases, there is a desire to seek mechanisms that allow nature to integrate into the built world in some form or another. In reality, this is often at a small scale as pressure on, and costs of, urban land rise. Over the last 20 years there has been growing interest in green roofs and roof gardens, green walls, city allotments and urban farms, urban nature reserves and green spaces used as part of an integrated approach to sustainable water management; so-called sustainable urban drainage systems (SUDS). The latter include features such as swales, rain gardens and reed beds. Some of these features are not necessarily new phenomenon (e.g. the Hanging Gardens of Babylon – now thought to have possibly been located at Nineveh in 700 BCE) but their prominence is rising as humans attempt to grapple with a range of environmental and social issues aligned with population growth, urbanization, non-sustainable consumerism and climate change. These and other factors are drivers for change too with respect to people's attitudes to green space and how they value those spaces. The concept of the sky garden – green oases placed on the sides or tops of multilevel tower blocks – has arisen over the last 10 years. At the other extreme, guerrilla gardening has taken hold in a number of cities as citizens 're-utilize' unused urban open space and attempt to make their immediate neighbourhood more vibrant and engaging through the 'spontaneous' planting of flowers, fruit and vegetables.

This chapter briefly explores three of these 'rediscovered' phenomena (green walls, green roofs and rain gardens) and outlines their roles, functions, design and management.

10.2 Green Walls

Green walls tend to be divided into different categories (Köhler, 2008). 'Green façades' are where plant-root balls are placed either in the ground or in pots and the shoots grown up the side of a building or other vertical structure such as a metal frame. These green façades usually comprise perennial climbing plants (vines), annual climbing species or wall-shrubs. Climbing species can either fix themselves to walls through morphological features such as leaf tendrils, adhesion pads or aerial roots, or can be trained up a trellis or other framework against the wall. Some green façade systems have plants grown in troughs either at the top of the wall or at intermediate levels, and the plants trail down the wall.

'Living walls', in contrast, support plants that either root into the wall or have cells of substrate embedded in/on the wall. These cells or compartments are often supplied with water and nutrients through artificial irrigation/fertigation systems.

A third system used for specific problematic sites is the 'retaining living wall'. These are part of an engineered solution to steep slopes, where vegetation is used to stabilize the soil and prevent erosion. Retaining living walls are designed to provide structural strength which resists the lateral forces on the slope that would normally result in soil particles moving. Some systems can perform on slopes up to 88° and many have capacity for variable slope angles as flat as 45°. Most systems are modular in nature and combine inert physical elements with plant material, e.g. geo-textile bags in conjunction with interlocking units, metal, concrete, plastic cellular confinement mats or woven plant mats. The eventual aim, however, is to provide a green mantle when mature, where the underlying structural elements are no longer visible.

A fourth designation is also used – that of 'biowalls'. These are similar to living walls but tend to be designed to improve indoor air quality and humidity; they can be composed of microorganisms or populations of primitive plants (e.g. Bryophyta) as alternatives to higher plant communities.

Green walls are a component of modern urban green infrastructure and contribute to a range of ecosystem services, including: habitat provision for urban biodiversity (Francis and Lorimer, 2011); intercepting precipitation and reducing runoff rates; providing thermal insulation in winter (Cameron *et al.*, 2015); screening out aerial particulate matter and improving air quality (Currie and Bass, 2008); attenuating noise (Veisten *et al.*, 2012); contributing to psychological well-being and improving the aesthetics of the cityscape; potentially reducing urban air temperatures, so helping to mitigate urban heat island effects; and lowering surface temperatures of buildings thereby reducing the reliance on mechanized air conditioning (Cameron *et al.*, 2014). (See Chapters 2 and 3 for more extensive reviews of these subjects.)

Green façades

Although the term has not always been used historically, green walls have a long pedigree when the use of climbing plants or vines against a building wall are taken into consideration (Fig. 10.1). In Europe, *Hedera helix* (English ivy) is a common woodland plant that will happily adhere itself to stone and brickwork as well as tree bark. As it self-seeds itself into cracks and apertures, this species would have had to be actively removed from walls

Fig 10.1. Flowering climbing plants or vines, such as this *Clematis montana*, historically have been used to decorate dwelling places.

if undesired, rather than needing any active encouragement to grow. Old ruined buildings today are often covered with *Hedera*, and it is a point of contention as to whether these ruins are being further damaged by the plant, or indeed, whether the stems of the plant are actually keeping the ruin intact! Despite the arguments that rage around *Hedera* and whether it damages masonry and increases dampness around walls, it is still a common enough veneer to houses today – a point reinforced in that there are numerous cultivars of various leaf forms and colours available from any garden centre or nursery. This includes cultivars of *Hedera helix*, but also of other *Hedera* species. Due to its evergreen nature, close dense foliage and relatively rapid growth rate, *Hedera* is a common species used by the professional green wall sector. To offset concerns, it is often grown on a trellis or framework that fixes to the wall, rather than encouraged to grow up the wall itself. *Hedera* is not the only species with tendril or rootlets that may explore weaknesses in the masonry. Other species include *Ipomoea* (morning glory), *Parthenocissus quinquefolia* (Virginia creeper), *P. tricuspidata* (Boston ivy), *Calystegia sepium* (hedge bindweed) and *Hydrangea anomala petiolaris* (climbing hydrangea).

Other 'traditional' wall climbing plants used on housing, or other domestic structures (e.g. to screen unsightly garages, sheds and outhouses) have been selected either due to their interesting form and colour of foliage, or because they have attractive flowers (Table 10.1). Over and above the aesthetic criteria, fruit trees, fruiting vines and vegetables have been well utilized against walls, due to the favourable microclimate associated with these locations. In the northern hemisphere walls with southern aspects are advantageous as heat traps, and hence have been used in more temperate climates to help early season shoot development, to protect blossom from frost and to allow fruit to ripen successfully. As such, the growing of fruit such as apricots, peaches, nectarines, almonds, figs, grapes and kiwi are feasible at latitudes much further north than would be the case if they were grown as free-standing bushes/trees. Although, it should be noted that these genotypes are, on occasion, still protected by additional lean-to glasshouse structures covering the wall to further increase temperatures.

Green façades can be attached to existing walls or built as free-standing structures. There are various forms on the market, but most use either a trellis system or a cable system attached to the walls of the building. Trellis systems tend to come as modular units that can be built up the building or used in parallel along the base of a building. Cable systems involve the insertion of spacers, rods or hooks at regular intervals up the building and these are connected via a series of stainless steel cables mounted vertically or horizontally. The cables are held under tension to keep their rigidity. They provide the physical structure the plants twine around, or are tied in to (where accessible). Other smaller-scale systems are used to provide greenery, privacy, screens or instantaneous 'living fences' around car parks, patios and walkways. Some are even used to provide permanent or temporary shade to glazed façades.

Wall trellis/cable systems need to be installed professionally, as the weight of the system needs to be carefully calculated. This is not only with respect to the physical aspects of the wall itself, for example being drilled into brick, façade tiles etc. and the forces put on these materials, but also the potential weight of the plants in future. This is particularly so when the effects of strong winds and snow loads need to be accounted for. As the materials involved are steel or other metals, then design and engineering needs to allow for thermal expansion and contraction.

Costs vary with the complexity and scale of the system employed, but typical installation costs would be £165–220 m^2 (US$250–330). Once plants are established, most of the maintenance involved is in trimming-off excess shoots that could coil themselves around drain pipes or clog gutters with their leaf litter. This might be done once a year. Wind damaged shoots need to be cut out too on an occasional basis, and some species will be more susceptible than others to cold winds. Certain genotypes may, in addition, be sensitive to de-icing salts used on roads and should not be planted in locations where there could be runoff or spray from roadways. Whilst attending to the plants it is also prudent for maintenance teams to inspect fasteners, cables, anchors and existing building materials for any loss of integrity at the same time.

Living walls

Living walls take their inspiration from those plants in nature that can survive on stone or brick walls, by seeding into cracks in the wall and deriving enough moisture and nutrients from the apertures within the wall. Such plants are normally adapted for

Table 10.1. Examples of genotypes commonly grown as green façades. C, climbing habit; S, free-standing, T, often trained against wall to maximize benefits.

Species	Common name	Attributes
Foliage		
Celastrus scandens	American bittersweet	C – distinctive leaves that turn golden yellow in autumn.
Garrya elliptica 'James Roof'	silk tassel bush	S – grey-green leaves, and long male flower 'tassels'.
Hedera colchica	Persian ivy	C – evergreen climber with some forms variegated gold and green, e.g. *H.* 'Sulphur Heart'.
Hedera helix	English ivy	C – a wide variety of cultivars, including variations based on leaf size and shape.
Hedera hibernica	Irish ivy	C – similar to *H. helix* and adapted to mild maritime climates.
Parthenocissus henryana	Chinese Virginia creeper	C – vigorous, large, deciduous climber, with dark green leaves turning red in autumn. Produces dark blue berries.
Parthenocissus quinquefolia	Virginia creeper	C – famous for its leaves that turn flame red and orange in autumn.
Parthenocissus tricuspidata	Boston ivy	C – ovate or three-lobed leaves that colour-up purple and crimson in autumn.
Flowers		
Apios americana	American groundnut	C – reddish brown flowers, but also edible tubers.
Akebia quinata:	chocolate vine	C – semi-evergreen, scented brownish purple flowers in early spring, followed by fleshy purple fruits.
Bignonia capreolata	cross vine	C – related to *Catalpa* spp. (Indian bean trees). Long, tubular, red and yellow flowers which can have a coffee-like fragrance.
Campsis radicans	trumpet vine	C – red or yellow tubular, trumpet-like flowers.
Ceanothus spp.	Californian lilac	S – used as a wall shrub in temperate areas, although often derived from arid zones in south-west USA. Flowers usually various shades of blue or purple.
Clematis spp.	clematis	C – wide range of cultivars from 10 m high *C. montana* types to new dwarf 'patio' cultivars. Flower colours across red, pink, blue, purple, white and even yellow ranges.
Coronilla valentina		C – blue green leaves with yellow pea-like flowers.
Hydrangea anomala subsp. *petiolaris*		C – a self-clinging, deciduous species. It possesses 'heads' of white lace-cap flowers in summer as well as attractive purplish bark and heart-shaped leaves.
Jasminum spp.	jasmine	C – various species and cultivars, including the yellow-flowered winter jasmine – *J. nudiflorum*. *J. officinale* and cultivars have strong scent. Some species have a semi-evergreen habit depending on climatic conditions.
Lonicera spp.	honeysuckle	C – includes *L. periclymenum*, *L. japonica*, *L. henryi*, *L. sempervirens* and *L.* × *tellmanniana*. Various colours of yellow, white, red and orange, often with combinations of more than one. Also famous for their sweet fragrance in the evening, designed to attract moths and other nocturnal pollinators.
Passiflora spp.	passion flower	C – includes *P. incarnate* and *P. caerulea*. Usually blue or purple flowers, but white and red colours (*P. alata*) exist too.
Rosa	rose	T – includes the rambling (very vigorous, single flowering period) roses and climbing roses, of less stature, but often repeat flowering habit. Wide range of colours.
Solanum crispum	potato vine	C – cultivars tend to have purple, blue or white flowers with an orange centre.
Trachelospermum spp.		C – *Jasminum*-like plants with white flowers.
Wisteria		C – cultivars derived from *W. sinensis* and *W. floribunda*. Panicles of purple, white or pale pink flowers depending on cultivar.

Continued

Table 10.1. Continued.

Species	Common name	Attributes
Fruit		
Citrus spp.		T – a wide range of *Citrus* genotypes can be trained against walls, including lemon, lime, orange and grapefruit in Mediterranean-type climates (or under glass in temperate regions).
Ficus carica	fig	T – large green lobed leaves and green-purple 'fruit'. Fruiting is enhanced by restricting the root growth, so can be grown in a pot beside the wall. In *Ficus* the 'fig fruit' is actually the stem of an inflorescence, with the flowers inside, these being pollinated in nature by the wasps within the family Agaonidae.
Malus	apple	T – apples can be trained against a wall through techniques such as espalier and cordon. Depending on the vigour of the rootstock and the degree of pruning trees can ultimately reach 5–6 m in height.
Pyrus	pear	T – culinary and dessert pears – can be trained in similar way to *Malus*.
Prunus spp.		T – this includes the plums, cherries, apricots, peaches etc., many of which benefit in terms of ripening from the warmer microclimate around a wall.
Vitis		C – grape vines can rapidly clothe a wall in foliage and with careful pruning also yield acceptable quality fruit.

rock screes or arid stony soils (lithosols) and include species such as *Cymbalaria muralis* (ivy-leaved toadflax), *Phlox subulata* (Fig. 10.2), *Aubrieta* × *cultorum* and *Centranthus ruber* (valerian). They may take their inspiration from these plants but most living walls have water supplied to the plants through drip irrigation seep hose or hydroponic systems using rockwool or some other material that allows effective capillary water movement. This enables a much wider range of species to be grown, including for example *Carex* species (sedges), that would normally be associated with quite wet locations.

Living wall systems vary in their construction and design, with commercial companies frequently designing and patenting their own system. The market also varies depending on context and situation. At one end of the spectrum there are DIY kits for the domestic market that can be fixed together and put on a house or garden wall, and are often no more elaborate than a range of planting troughs and modules linked together to form a single unit. At the other extreme there are very elaborate bespoke systems that are custom-fitted to a wall, and may cost the equivalent of £2–3 million. This is particularly so when the wall is an integral part of a prestigious new iconic building or forms a city-centre landmark piece (Fig. 10.3). Broadly speaking, however, two approaches tend to dominate the designs utilized. These are hydroponic systems and substrate-based systems (Anon., 2014).

Hydroponic-based living walls can be constructed from either modular containers or large panels. To avoid risk of dampness to the building (and to provide an extra layer of thermal insulation) hydroponic systems are usually discrete from the wall, with an air gap being left between the building and the irrigated wall. The wall is attached to the building by way of brackets or a stand-alone framework. The back of the living wall is composed of a waterproof sheet or lining, and immediately in front of that is the inert medium that conducts the water and anchors the plants. This comprises materials such as rock or stone wool (basaltic rock or industrial slag that is heated and spun into fibres), horticultural foams or felt fabrics. These materials conduct and supply water to the plants, but also drain freely allowing oxygen to pass through to the plant root-systems. As most hydroponic media are inert, they allow effective control of nutrient concentrations and any structural decay is a prolonged process.

Fig 10.2. Plants adapted to alpine conditions with frequent exposure to cold and wind, such as the *Phlox subulata* cultivar depicted here, are usually considered most appropriate for living walls. This cultivar also requires good drainage; living walls can be at either the wet or dry end of the spectrum depending on the system adopted, substrate type and irrigation frequency.

Fig. 10.3. The living wall at the Caixa Forum, Madrid, Spain, designed by Patrick Blanc. (Image courtesy of Richard Bisgrove.)

New Green Space Interventions

Where slabs of rockwool, or similar material are used, it is the nature of such material that the base holds more water or is wetter for longer, due to the action of gravity, than areas near the top. Designers take advantage of this and often more drought-adapted species are placed at the top (e.g. *Helianthemum nummularium, Salvia officinalis, Silene schafta*) and those more tolerant of prolonged wetting, or which require more consistent and higher volumes of water, are located nearer the base (e.g. *Asplenium scolopendrium, Hosta sieboldiana, Primula vulgaris*). Nutrition in hydroponic systems needs to be carefully managed to avoid excessive or inadequate ions in the solution, and this is controlled via fertigation systems that dose the irrigation at intermediate intervals with nutrients, when ionic concentrations drop below certain thresholds.

Living walls with substrate-based systems tend to use discrete cells, modules or troughs. These are orientated on an independent, structurally secure metal rack or framework, anchored directly to the wall or placed on a grid structure adjacent to the building wall. The advantage of these systems is that individual cells or modules can be removed if plants fail or become overcrowded. Irrigation in module systems may be via pipes linking the modules or from drip lines run along the tops of the modules, and an individual dripper supplies water to each unit. Substrates vary in composition and are still being researched for optimum performance, but as with any successful medium the notion is to maximize water-holding capacity, whilst minimizing anaerobic conditions, i.e. keeping them well-drained and aerated. Substrate systems by nature have a greater buffering capacity than hydroponic systems; for example if a pipe should fail there usually is some moisture reserve within the module to keep plants alive for a few days. The downsides, however, include nutrients becoming exhausted over time, and conversely the possible concentration of excessive salts if the nutrition is not carefully managed or if there are additional ions in the water supply (calcium, sodium etc.).

Excessive water dripping off a building, due to over-irrigation or poorly positioned drippers, can be a problem, particularly so during cold conditions or where periods of frost occur, as the resultant 'ice-slide' on the pavement below poses something of a hazard to pedestrians. Excessive water shedding off the building may also prove problematic to any plantings at the base of the wall or indeed to the building fabric itself. As such, drip trays are used frequently to capture excess irrigation water from the growing medium as well as water droplets that drip off foliage. Often this water is recycled and pumped back to the top of the building for re-use. The capacity of the drip trays and holding tanks should be sufficient to hold an entire irrigation cycle's water volume.

The relatively restricted root volume available within a living wall, plus exposure to high irradiance (if in full sun) and wind, has resulted in a predominance of low-growing plant forms being favoured for living walls. Although there are different rationales for living walls, aesthetics tends to play a significant component in many walls systems, and as most walls have a high public profile the 'need to look good' is often at the forefront of a client's mind. This is somewhat in contrast to the underlying philosophies that have driven the green roof movement – where early principles were based around ecological agendas, such as replicating rare and unusual habitats associated with urban brownfield sites.

The fact that living walls are viewed from the streetscape, however, has dictated that plantings should be interesting, attractive, uniform in cover and easily maintained. Although many plants die back to a rootstock or become leafless in winter, large brown patches on the sides of walls are not always appreciated by the clients who have commissioned the wall. As such, an extra dimension to the plant selection is the inclusion of evergreen species, or species with foliage that remains attractive on dying, e.g. grass and *Carex* spp. Other plants may be included because they:

- have a prostrate growth habit and cover the wall fabric and structural parts;
- provide a few weeks of colour and contrasting form through their flowers;
- have attractive evergreen foliage or foliage of an unusual colour; and
- help provide a specific ecosystem service, e.g. supplying nectar or pollen for invertebrates or birds such as hummingbirds or honeyeaters.

These factors determine species choice and as a consequence alpine plants (Table 10.2), small woody plants including evergreen species (Table 10.3), small herbaceous perennials (Table 10.4) and grasses and ferns (Table 10.5) tend to dominate

Table 10.2. Examples of alpine plant species used in green walls within temperate climates and their attributes.

Groups/species (cultivars)	Attributes/prominent colour
General attributes/traits	
Dwarf-growing with usually either prostrate or hummock-forming growth habits. Some are silver-leafed or have numerous leaf hairs (hirsute) to avoid moisture loss and protect from UV light. Most alpines flower in spring. Alpines are useful for green walls due to their tolerance to drought and exposed conditions, low temperatures but also high wind speeds.	
Armeria maritima (thrift)	Flowers pink, clump forming.
Aubrieta × *cultorum*	Flowers purple, blue or magenta.
Dianthus spp.	Flowers pink, white or red. Leaves deep green or grey.
Herniaria glabra (green carpet)	Leaves small, deep green. This plant is known simply as 'green carpet' due to its habit of forming a uniform sward of mid-green.
Phlox subulata, *P. douglasii* (alpine phlox)	Flowers blue, purple, pink, pink/white striped. Leaves fine-leaved texture, mat forming.
Pulsatilla vulgaris (pasque flower)	Flowers blue, white or wine red.
Saxifraga spp.	Flowers white, pink and yellow. Some hummock/clump-forming; others prostrate 'mats' of foliage.
Sedum acre (mossy stonecrop)	Flowers sulfur yellow – very low growing/drought tolerant.
Sedum album	Flowers white. Foliage persists throughout winter, but turns red to bronze as temperatures become cooler. *S. album* 'Coral Carpet' is a selected form that colours up cherry-red in summer.
Sedum ewersii	Flowers pink. Leaves blue-green.
Sedum hispanicum var. *minus*	Flowers pink or white. *S. hispanicum* var. *minus* 'Blue Carpet' has grey/blue foliage which turns deep blue to purple in winter.
Sedum hybridum 'Immergrunchen'	Flowers yellow.
Sedum kamtschaticum	Flowers yellow. *S. kamtschaticum* var. *floriferum* 'Weihenstephaner Gold' has orange tips to the flowers giving an overall gold appearance.
Sedum reflexum	Flowers yellow. Branches look like miniature conifer trees. Forms with deep ice-blue foliage contrast well with the yellow flowers.
Sedum sexangulare	Flowers yellow. Leaves mid-green in summer turning orange to red in autumn and eventually brown. It derives its name from the leaf arrangement in rows of six.
Sedum spurium	Flowers red, white and pink. Leaves rosettes of green or bronze green, leaves turning red and orange in some cultivars. *S. spurium* 'Summer Glory' has deep pink/red flowers in mid-summer; contrasting well with deep green foliage.
Sempervivum montanum	Flowers purple/red. Rosette leaves with flowering stems held well above the canopy. A very variable species with numerous forms.

living walls in temperate climates. Despite the wide range of plants available it needs to be remembered that walls by-and-large are fairly inhospitable environments, with wind and excessive temperatures in summer placing a strain on a plant's ability to regulate water distribution and hence leaf temperature. Similarly, unlike natural soil, substrates in containers and hydroponic mats have a relatively small volume, meaning that thermal buffering capacity is low; hence roots can be exposed to frost that would not normally be a problem with plants in natural soil at ground level. Compared to stems and leaves, roots are much more susceptible to injury from sub-zero temperatures. As outlined above, for many urban wall systems aesthetics are important as they are in the public eye. Complementing functionality with an attractive colour scheme or range of leaf forms is therefore a consideration (Fig. 10.4).

In the tropics (and indoors in buildings of temperate climates) epiphytes such as bromeliads and orchids are more important in wall systems, and augment evergreen vines and shrubs.

Table 10.3. Examples of shrubs and dwarf sub-shrubs used in green walls within temperate climates and their attributes.

Species/cultivars	Attributes/prominent colour
General attributes/traits	
Woody plants used on walls tend to be those that are prostrate in habit or the dwarf forms of larger genotypes. Many are from Mediterranean, semi-arid climates or from heathlands, with sclerophyllous or hirsute leaves. Other types used though are shrubby climbers or ground-cover plants, often adapted to the shade of a forest floor environment. Flowering, foliage and even fruiting characteristics can be important for selection on wall systems.	
Ceanothus repens	Flowers blue. Leaves deep green. Evergreen or semi-evergreen in exposed locations.
Centranthus ruber (valerian)	Flowers red, pink or white flowers. Drought adapted – can grow in stone and rubble walls without irrigation.
Cerastium tomentosum (snow-in-summer)	Flowers white. Leaves grey. Vigorous-growing ground cover, for high light locations.
Convolvulus cneorum	Flowers white. Leaves grey.
Cotoneaster dammeri	Flowers white. Leaves green. One of the more prostrate cotoneasters spreading to 2 m, flowers in early summer, followed on by bright red berries.
Cistus spp. (sunrose)	Flowers white, pink. Various hybrids including *Cistus* × *hybridus* (*C. populifolius* × *C. salviifolius*) and *C. purpureus* (*C. ladanifer* × *C. creticus*) types – even hybrids formed with *Halimium* spp., i.e. × *Halimiocistus*, which can bring in yellow flowers. Not all genotypes are cold tolerant.
Erica spp. (heath)	Flowers purple, pink, white. Leaves green, bronze. Includes *E. carnea* hybrids such as 'Myretoun Ruby' and 'Springwood White' which have a degree of lime tolerance. Careful selection of cultivars can allow for year-round flowering.
Euonymus fortunei	Leaves green or variegated with silver or gold.
Hedera spp. (ivy)	Leaves green or variegated with silver and gold. Trailing or climbing habit over the wall.
Hebe spp.	Flowers white, pink, purple, blue. Leaves various forms and colours in blue/green but also variegated forms some with pink edges to leaves, e.g. *H.* 'Magicolors'. Wind tolerant where temperature not excessively low.
Helianthemum spp. (rockrose)	Flowers white, red, pink, orange, yellow. Leaves mid-grey.
Hyssopus officinalis (hyssop)	Flowers spikes of whorled, tubular blue flowers. Leaves aromatic, linear leaves. Popular with pollinators.
Iberis sempervirens (perennial candytuft)	Flowers white. Leaves mid-green. Short stature bush.
Lavandula spp. (lavender)	Flowers purple, mauve, white. Leaves green/grey. Both *L. angustifolia* (English lavender) and *L. stoechas* (French lavender) do well on walls. Good for attracting pollinating insects when in bloom.
Lithodora diffusa	Flowers electric blue. Leaves mat-like mid-green. *L.* 'Heavenly Blue' is common and reliable.
Origanum spp. (marjoram)	Flowers pink. Leaves aromatic grey/green.
Pachysandra terminalis (Japanese spurge)	Flowers small white flowers. Leaves pale green, serrated. *P.* 'Green Carpet' particularly low-growing variety – shade tolerant.
Potentilla fruticosa	Flowers orange, yellow, white, pink, red. Leaves green.
Rosmarinus officinalis Prostratus Group	Flowers blue. Leaves mid-green. Prostrate forms of common rosemary.
Salvia officinalis (sage)	Flowers purple. Leaves blue/green or silver or purple hues.
Thymus spp. (thyme)	Flowers purple. Leaves green bronze, gold.
Vinca minor (periwinkle)	Flowers blue/purple. Leaves green. Trailing habit and shade tolerant.

Due to the inaccessible nature of the plantings on wall systems, maintenance can be one of the more costly activities. Physically locating personnel close to the wall may require the provision of a cherry picker, genie lift or even scaffolding. This will only be feasible if there is space on the ground for the equipment to be moved into. Some larger green wall structures are maintained through the use of special gantries that are held by cables from the roof. Safety is also a para-

Table 10.4. Examples of herbaceous perennials and geophytes used in green walls within temperate climates and their attributes.

Groups/species (cultivars)	Attributes/prominent colour
General attributes/traits	
Herbaceous or semi-herbaceous plants that tend to keep a low-growing habit, again selected for their tolerance to exposed conditions and ability to spread across the wall. Aesthetic characteristics such as interesting leaf colour, form or high impact or prolonged flowering capabilities also important.	
Ajuga reptans	Flowers blue. Leaves green/bronze. Shade tolerant. *A.* 'Black Scallop' has dark red/purple leaves.
Allium schoenoprasum (chives)	Flowers purple, blue. Leaves green strap/needle-like.
Artemisia spp. (tarragon)	Flowers small clusters of yellow flowers. Leaves green or blue/grey. Aromatic.
Bergenia spp.	Flowers white, red, pink. Leaves green with large bold rounded form.
Chamaemelum nobile (chamomile)	Flowers white. Leaves small, fern-like, scented. *C.* 'Treneague' is a very low-growing, non-flowering selection that can be used to represent a uniform green plane.
Erigeron spp.	Flowers white, daisy-like. *E. karvinskianus* gives a pleasing multi-tone effect with pink and yellow in flowers too.
Euphorbia amygdaloides 'Purpurea' (wood spurge)	Flowers acid yellow/green. Leaves deep purple.
Fragaria vesca (alpine strawberry)	Flowers white. Leaves mid-green. Also enhanced by small red fruit in summer.
Helleborus spp.	Flowers white blooms in winter. Shade-adapted species that dies back in summer.
Hemerocallis spp. (day lily)	Flowers wide range of colours, with individual blooms only lasting one day. Leaves strap-like. Dwarf forms most suitable for walls.
Heuchera spp.	Flowers loose panicles of red, pink or white flowers. Leaves very wide range of colour tones, including lime-green, deep-red, almost black, yellow, copper and tawny.
Hosta spp.	Flowers purple, white. Leaves green, variegated, gold. Large bold leaves.
Liriope muscari	Flowers blue spikes. Leaves green, strap-like. *L.* 'Monroe White' has white flowers.
Mentha spp. (mint)	Flowers white or mauve. Leaves mid-green through to olive grey/green. A wide range of species – most having strong aromatic scent. Responds to trimming back in summer to re-shoot from base.
Omphalodes verna (blue-eyed Mary)	Flowers bright blue with white centre similar to forget-me-nots. Leaves green.
Petroselinum crispum (parsley)	Flowers small umbels of white. Leaves finely divided and bright green. When used *en masse* leaves resemble miniature forest.
Pratia pedunculata	Flowers blue. Leaves fine-textured green. Can be very vigorous when conditions suit.
Primula vulgaris (primrose)	Flowers yellow and blooms present early in the growing season. Leaves mid-green with rough texture.
Silene schafta (autumn catchfly)	Flowers purplish-pink. Leaves mid-green, lance-shaped. Mat-forming semi-evergreen perennial.
Sisyrinchium striatum	Flowers pale yellow in slender spires. Leaves grey-green strap-like.
Stachys byzantina (lamb's ears)	Flowers purple. Leaves soft textured and light grey.
Tellima grandiflora	Flowers spires of pale yellow/green flowers. Leaves emerald-green, scalloped. Shade tolerant.
Teucrium chamaedrys	Flowers purple. Leaves medium green, similar appearance to *Nepeta* and *Salvia*.
Tiarella spp.	Flowers panicles of small star-shaped flowers in white or pale pink. Leaves green or purple-centred green. Shade tolerant and even flowers well in shade.
Viola spp.	Flowers blue, purple. Leaves mid-green. Perennial forms used rather than the bedding plant types, including cultivated forms such as *V.* 'Martin', *V.* 'Irish Molly', *V.* 'Columbine' and *V. odorata* 'Queen Charlotte'.

Table 10.5. Examples of ferns, grasses and grass-like plants used in green walls within temperate climates and their attributes.

Groups/species (cultivars)	Plant type/attributes
General attributes/traits	
Ferns often do well in shaded areas, or locations with high humidity. Grasses can vary in their tolerances, but many prefer free-draining situations. These plants are often used in green walls for their ability to 'lift' or 'lighten' the planting composition, due to their fine texture and movement in the wind.	
Acorus gramineus 'Variegatus' (sweetflag)	Grass-like. Variegated leaves of green and cream – prefers the wetter locations in walls.
Asplenium trichomanes (maidenhair spleenwort)	Fern. Hardy, evergreen fern native to NW Europe well-suited to planting in a dry or shaded wall. Forms a rosette of dark-stemmed, pinnate fronds with small, rounded or oblong segments.
Blechnum spicant (deer fern)	Fern. Hardy, shade-loving and adapted to cool moist locations. Large pinnate fronds give a bold image. Needs to be in a medium where the pH remains low, i.e. requires acidic irrigation water.
Carex albula syn. *comans* 'Frosted Curls'	Grass-like. Pale silver green foliage – does best in regions with mild winters.
Carex oshimensis 'Evergold'	Grass-like. Characterized by dark green leaves with a central strip of creamy yellow. Forms low evergreen hummocks.
Cyrtomium fortunei (Japanaese holly fern)	Fern. Requires rich moist, but well-drained medium. Semi-evergreen with arching fronds of a mid-green colour.
Deschampsia flexuosa 'Goldtau'	Grass. Forms a compact tuft of foliage, with spires of flowering shoots with pale cream flower heads, followed by brown seeds.
Dryopteris affinis (golden shield fern)	Fern. Bright yellow-green when unfolding, followed by fronds coloured rich green. Semi-evergreen habit.
Festuca glauca	Grass. Prefers full sun, and free-draining conditions. Improved versions have strong consistent blue colour foliage, e.g. *F. glauca* 'Intense Blue'.
Luzula nivea, L. sylvatica (woodrush)	Grass-like. Prefers moist conditions, and tolerates shade. Gold forms such as *L. sylvatica* 'Aurea' help lighten dark locations.
Phyllitis scolopendrium (hart's tongue fern)	Fern. Evergreen, non-pinnate leaf type of mid-green hue. Moist, cool shady conditions.
Polypodium vulgare	Fern. Requires shade, but otherwise copes well with a range of moisture conditions – adapted to the cool, moist climates of NW Europe. Long, leathery, dark green fronds with a slight sheen, and illustrates an attractive, lacy texture when viewed from a distance.
Polystichum polyblepharum (Japanese tassel fern)	Fern. Provides interest all year round. As the fronds unfold they turn a glossy deep green and are presented in a slightly recurved rosette. The fronds have a slight reflective sheen.
Uncinia rubra (red hook sedge)	Grass-like. Mahogany-red sedge grass, moist free-draining media in sun or part shade to exploit the foliage hue.

mount consideration, and personnel may need to be kitted out with a harness, descent and safety rope, rope protector, rope-grabbing tool, descent mechanism, lanyard and suction cups depending on the height and size of the wall, as well as the ubiquitous 'high-vis' jacket and hard hat. High buildings are not only prone to exposure to the wind and high wind speeds, but their geometry often encourages 'unexpected' gusts of wind, and workers need to be attached to the building in some way at all times. Hazards due to dropped equipment or loose modules also need to be considered, and road/pavement areas below the wall may need to be fenced off to protect the public.

Bio-walls

The term 'bio-walls' describes indoor living walls. The original objectives of bio-walls centred around the use of plants and their root-associated microbial populations to remove aerial pollutants from the interior space. Over time this definition has been blurred as more bio-walls have been constructed for both aesthetic and well-being objectives. Walls implemented

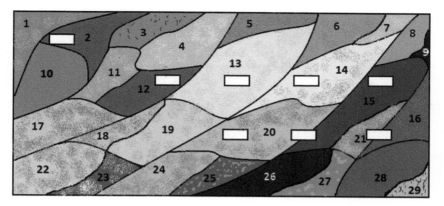

Fig. 10.4. Typical green wall plant community – plants chosen for aesthetic character and reliability within the wall system. In this system irrigation is applied by drip line, and water percolates through the system by gravity, thus plants at the top of the system are more drought-adapted than those at the bottom.
Key to species/cultivars: 1, *Nepeta* 'Six Hills Giant'; 2, *Ajuga reptans*; 3, *Salvia officinalis*; 4, *Stipa tenuissima*; 5, *Potentilla* 'Helen Jane'; 6, *Armeria maritima*; 7, *Phlox subulata* 'Emerald Blue'; 8. *Aubrieta* 'Violet Queen'; 9. *Aeonium* 'Zwartkop'; 10, *Pachysandra terminalis*; 11, *Rosmarinus officinalis*; 12, *Hebe* 'Autumn Glory'; 13, *Helianthemum* 'Wisley Pink'; 14, *Lavandula angustifolia*; 15, *Heuchera* 'Purple Palace'; 16, *Knautia macedonica*; 17, *Centranthus ruber*; 18, *Dianthus* 'Feuerhexe'; 19, *Stachys byzantina*; 20, *Dryopteris affinis*; 21, *Sedum spectabile*; 22, *Cistus* 'Sunset'; 23, *Tiarella cordifolia*; 24, *Euonymus fortunei* 'Emerald Gaiety'; 25, *Hemerocallis* 'Pink Damask'; 26, *Ophiopogon planiscapus* 'Nigrescens'; 27, *Vinca minor*; 28, *Polypodium vulgare*; 29, *Carex flacca*.

to 'clean' air are designed to have air passing over the plant/microbial communities and so act as natural filters. This is usually accomplished by a circulation fan, although passive air systems have been used too.

Bio-walls are used to reduce levels of indoor volatile organic compounds (VOCs) that are derived from materials and activities commonly found in offices (Table 10.6, see also Chapter 11). As the rhizosphere bacteria require carbon sources to develop small chain carbon atoms, the VOCs are utilized and catabolized to release energy, thus degrading their activity as pollutants. Although the rhizosphere has a high diversity of microbial species, it is believed that some of these microorganisms developed their ability to degrade VOCs based on environmental pressures or the lack of availability of other carbon sources. Although much is made about the plants in a bio-wall, in many ways their primary function is to act as hosts for the microorganisms, and to increase the surface area that these microbial communities can grow on. Increasing the microorganism's access and time of exposure to the VOCs is important. The VOCs are readily soluble and a film of moisture is important to improve contact, and slow rates of air passage as well as operating temperature of approximately 20°C are optimal for maximizing microbial activity and hence removing the VOCs from the air.

Although bio-walls are also sometimes quoted as being useful for providing oxygen and removing carbon dioxide, it is unlikely such factors are significant, not least because irradiance levels in interior environments are often too low to optimize photosynthesis in the plants contained within the wall.

Bio-walls are a novel way of introducing greenery to the interior environment and have the advantage of taking up less room than a conventional interior planted landscape. This, however, needs to be balanced with the costs of specific systems and ease of access to areas behind the wall (e.g. access to irrigation pipes, drippers etc.). Nevertheless, bio-walls are becoming a more common feature in lobbies, reception spaces and other communal areas within buildings, with the psychological health benefits of viewing plants being one of the main motivators for the inclusion of such features (see Chapter 3).

10.3 Green Roofs

Green roofs are vegetated or semi-vegetated landscapes located on top of a building or other form of built structure. They vary in complexity based largely on the depth and characteristics of the substrate placed on the roof, as this determines the amount of water and nutrients available to the plant

Table 10.6. Guidelines for maximum exposure limits to common indoor contaminants as derived from various US and Canadian sources.

Contaminant	Maximum exposure limit	Exposure time
Benzene	1 ppm	Short term (8 h)
Carbon dioxide	1000 ppm	
Carbon monoxide	9 ppm	Long term (24 h)
	25 ppm	Short term (1 h)
Formaldehyde	40 ppb	Long term (8 h)
	100 ppb	Short term (1 h)
Hydrogen sulfide	0.01 ppm	
Microbial contaminants, such as viruses, fungi, mould, bacteria, nematodes	Low as possible	
Naphthalene	1.9 ppb	Long term (24 h)
Nitrogen dioxide	11 ppm	Long term (24h)
	90 ppm	Short term (1 h)
Ozone	20 ppb	Long term (8 h)
Particulate matter ($PM_{2.5}$)	Low as possible (<65 µg m^{-3} of air)	
Sulfur dioxide	5 ppm	Short term (8 h)
Toluene	0.6 ppm	Long term (24 h)
	4.0 ppm	Short term (8 h)
Trichloroethylene	100 ppm	Short term (8 h)

Sources: Health Canada - www.hc-sc.gc.ca/ewh-semt/air/in/res-in/index-eng.php; Indoor Air Quality Handbook - www.tsi.com/uploadedFiles/_Site_Root/Products/Literature/Handbooks/IAQ_Handbook_2011_US_2980187-web.pdf

communities that develop. Their *raison d'être* too varies significantly from city to city and building to building. Most green roofs are formed from a series of layers placed horizontally above a building's roof and these layers comprise a waterproof sheet to protect the roof, a root barrier to stop roots penetrating the roof zone itself, a mat or insulating layer to further protect the roof, a drainage layer and a filter sheet to stop particle migration to the lower zones (Anon., 2014). Above this is the substrate to support the plant roots and finally the foliar canopy of the plants. Some systems will contain modular structures that are clipped together and may contain wells to hold water and provide structural integrity to the green roof. The angle of the roof will determine the construction principles as well as physical matters, such as the volume of moisture held within the system. More traditional designs may have none of the above! Thatched and turf roofs, used historically, would have plants growing out of them, especially in a moist climate. These 'roofs' would have simply been constructed by placing sods of turf over branches and these set upon the roof timbers. Better structural integrity and protection to the living quarters are provided these days, but the concept of using the native vegetation as a roof covering has not been lost (Fig. 10.5).

The rationale for green roofs is multifaceted. They are promoted because of their environmental benefits (see Chapters 2 and 4). This includes providing wildlife habitat on space that would otherwise be fairly inhospitable to most species. They improve the energy efficiency (Castleton et al., 2010) and noise mitigation of buildings, reduce the thermal flux on a building, contribute to urban cooling, have potential to absorb certain aerial pollutants, extend the life of the physical roof material as well as help detain rainfall and reduce subsequent runoff (Figs 10.6 and 10.7). They also prove popular due to being architectural features in their own right and adding to the distinctiveness of the building. Depending on the green roof design, they may also be social space for the residents or workers of the building as well as a much needed space to grow food in a congested urban world. In light of the fact that roofs can represent up to 32% of the horizontal surface area of urban conurbations, there is tremendous potential to mitigate many of the environmental problems associated with urbanism through a wider adoption of green roof technology (Oberndorfer et al., 2007).

Green roof typology

The extent to which vegetation grows on a green roof depends on substrate, type, depth and climate (Nagase and Dunnett, 2010). These points frequently

Fig. 10.5. Green roofs using native plants have been used by a number of environmental organizations to help blend their buildings into the surrounding landscape. The use of living 'turf' roofs such as this in the Western Highlands of Scotland, UK, is reminiscent of the traditional roofing material of past times.

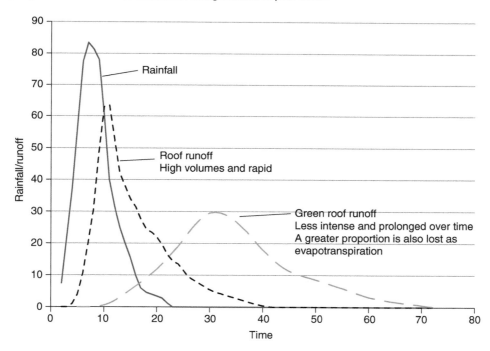

Fig. 10.6. Model of rainfall and water runoff dynamics on conventional and green roof systems.

define the function and type of roof employed (Table 10.7; Fig 10.8). Costs tend to increase with the more complex systems and the depth of substrate required (largely due to weight considerations – see below), but in theory, large areas of a city's horizontal surface area could be converted to green using extensive or semi-extensive vegetated roof systems aligned with contemporary building design. This could be the case

New Green Space Interventions

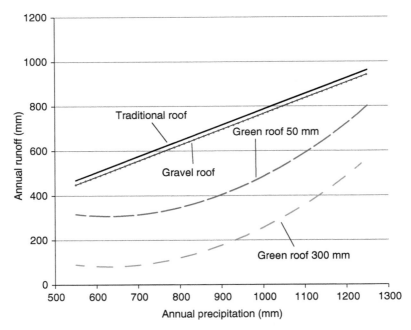

Fig. 10.7. Relationships between the annual runoff and annual rainfall for various roof types, including shallow substrate (50 mm, extensive) and deeper (300 mm semi-extensive) green roofs. (Modified from Mentens et al., 2006.)

for cities within temperate climates and adequate rainfall, but remains challenging for those in drier Mediterranean or semi-tropical climates, where rainfall may be more intermittent (Williams et al., 2010).

Weights and load bearings

Deeper substrates correlate with greater physical mass, more water-holding capacity and larger plants; factors that all add to the weight on a roof. It is absolutely paramount that a build structure can tolerate the weight placed on it by the addition of a green roof system. This is where the environmental horticulturalist needs to bring in a structural engineer to seek the right guidance on approaches to individual buildings. For new buildings, the potential weight load can be calculated and the building engineered sufficiently to meet these demands, although additional costs may be significant. For retrofitting an existing building, an independent assessment will need to be carried out to determine whether the installation will meet the existing structural capacity inherent in the building, or whether this will need to be modified. Load-bearing characteristics are divided into three categories: (i) The 'deadload', namely the weight of all the built elements and all components associated with the roof assembly, including plants, substrate and water at full capacity in the soil and also on the plants. This needs to take account of the fact that trees and shrubs may put on considerable biomass over the lifespan of the building. (ii) The 'liveload' takes account of any human activities on the roof. This may be weight of people, e.g. during social gatherings, but also maintenance or other mobile equipment that may be taken onto the roof from time to time. (iii) The 'transient load' is the weight conferred when the building may move in the wind or even during seismic activity.

In Victoria, Australia, the typical weight loadings for ornamental plants are: turf, 5.1 kg m^{-2}; succulents and low grasses, 10.2 kg m^{-2}; herbaceous perennials and sub-shrubs, 10.2–20.4 kg m^{-2}; larger shrubs, 30.6 kg m^{-2}; small trees (≤6 m), 40.8 kg m^{-2}; medium trees (≤10 m), 61.2 kg m^{-2}; large trees (≤15 m), 150 kg m^{-2} (Anon., 2014).

Many green roofs are placed on flat roofs, but pitched or sloped roofs are feasible too. Shallow roof gradients, up to a 10° pitch tend to use the same materials for a flat roof. At steeper pitches than this, however, additional support and infrastructure is

Table 10.7. Characteristics of different green roof types and the typical plant/animal communities they can host.

Green roof type	Characteristics	Typical plant/animal communities
Extensive lightweight	No added substrate per se, just pre-grown ≤25 mm medium attached. Limited biodiversity potential. Low water-holding capacity, some with irrigation installed.	Bryophyta (moss), *Sedum* spp. and other xerophytes.
Super lightweight	Consists of thin (12 mm) drainage board, a filter, fleece/water retention mat and pre-grown vegetated mat about 25 mm in thickness. Minimum building loadings required and can be used on some retro-fitted buildings. Can be drought prone and irrigation sometimes supplied.	*Sedum* and *Sepervivum* spp. grown on the vegetated mats.
Extensive	Substrate ≤100 mm. Not normally irrigated. Planted with plug plants into substrate or pre-grown vegetation mats. Limited water-holding capacity. Low maintenance. Can be undulated topography with deeper and shallower areas of substrate.	As above, but also flowering annuals, e.g. *Centaurea, Coreopsis, Linaria*; low-growing perennials, e.g. *Aubretia, Campanula, Delasperma, Dianthus, Sisyrinchium, Thymus*; geophytes, e.g. *Allium schoenoprasum, Iris pumila, Muscari azureum* and *Tulipa turkestanica*.
Semi-extensive	Substrate depth 100–200 mm. Many of the advantages of extensive roofs whilst allowing a wider range of species. Sometimes irrigated. Relatively good rainwater retention and detention capacity. Slightly higher maintenance, e.g. during plant establishment or if weed species become problematic.	'Widening range of grasses, annuals, geophytes and herbaceous forbs become feasible.'
Brown roof	Substrate depths vary, e.g. 100–200 mm, and sites are usually sculptured to provide a heterogeneous landform. Designed to replicate the ecological niche of terrestrial ex-industrial brownfield sites and utilize aggregates such as crushed brick and concrete rubble, sands and gravels. Important for certain invertebrate species and bird species that colonize such sites. Relatively good rainwater retention and detention capacity.	In NW Europe useful habitat for *Phoenicurus ochruros* (black redstart) and *Motacilla alba* (pied wagtail).
Biodiverse/wildlife (extensive)	Where depth and substrates allow, a range of natural/semi-natural vegetation communities can be replicated, e.g. chalk grassland, heathland, mesic meadow etc. Natural colonization often encouraged. Depending on habitat and species present, human access may be restricted or limited at times. Frequently include a pond and wetland areas. Relatively good rainwater retention and detention capacity.	Due to the ability to manipulate the substrates effectively, green roofs provide a good opportunity to alter conditions to suit particular species or communities. Flower-rich meadows have been created, including populations of terrestrial orchids. Even amphibian/reptile species have been found on these roofs, despite being many storeys up in some cases.
Intensive	Substrate depth ≥200 mm. Intensive maintenance required. Sometimes irrigated. Composed of woody or herbaceous ornamental plantings. Good rainwater retention/detention capacity. Usually good access for people provided.	Wide range of herbaceous perennials, smaller woody perennials, geophytes and flowering annuals.
Lawn roof	Laid down to turf and used for amenity functions or aesthetics. High maintenance. Good rainwater retention/detention capacity.	Species vary with climate and depth of substrate.
Roof food garden	Plants can be grown in soil (e.g. ≥200 mm) or in planters/containers. Soils need to be nutrient-rich organic media. Access to irrigation required. Usually useful social space.	Usually annual vegetables and vines such as *Pisum* (pea)/*Lycopersicon* (tomato), although full fruit trees have been used, e.g. the Reading International Solidarity Centre (RISC) garden, UK.

Continued

Table 10.7. Continued.

Green roof type	Characteristics	Typical plant/animal communities
Roof gardens	Replicating a ground-level garden at height. Buildings need to be designed to cope with the weight of deep substrates and large plants (including small trees) when soils and canopies hold their maximum amount of water. Some depth may exceed 500 mm. Where depths are not sufficient plants are grown in raised beds and planters too. Various styles, but often reflecting the most sought after penthouse locations, i.e. highly stylized. Usually very good rainwater retention/detention capacity.	Depending on site/climate, plants may need to have some tolerance to wind and high light exposure. Typical plants might include *Betula* spp. (birch, including coloured bark asiatic types), dwarf *Pinus* spp. (pine) and grasses such as *Stipa* and *Festuca* spp.

Fig. 10.8. Extensive and semi-extensive green roof systems have the greatest potential to turn large areas of the city over to a green mantle, albeit the number of plant species that can survive may be relatively small on such shallow soil systems. (Image courtesy of Green Roof Centre, University of Sheffield.)

usually required to ensure sheeting layers, substrates and plants all stay in position. More complex infrastructure increases direct costs and the steeper angles have implications for access, with knock-on effects for maintenance costs. Some roofs have sloping angles >45° – this does not preclude the use of green roofs, but the technology begins to resemble that of green walls. As with flat roofs, the best way of improving the structural integrity of the system is to have an effective integration of plant roots and substrate, and good foliage cover of the roof surface; plant roots and stem structure helping to bind the system together.

Roofs, whether flat or otherwise, can be dangerous places to carry out maintenance activities. Most roofs that are accessible require a balustrade or retaining fence structure to ensure personnel are safe. Some parts of roofs are stronger than others and restricted zones may need to be identified and cordoned off in some manner. There may also be instrumentation and building infrastructure 'plants' such as air conditioning units or fume vents, again where it is inappropriate for maintenance staff to be close to. Equipment and even water/plant material may be difficult to access on the roof and these factors need to be considered when designing and managing the green roof.

Substrate technology

Growing substrates for green roofs are typically more inorganic in nature that those used in most other scenarios covered by environmental horticulture. This partially reflects the history of green roofs where drought-adapted species (e.g. houseleeks – *Sempervivum* spp.) were encouraged to colonize shallow (or even no) substrates; or where the ambitions were to replicate the stony aggregate nature of many brownfield sites. There is also the assertion that coarse inorganic materials such as gravels and stones allow rapid drainage, thus not holding excessive weights of water on the roof, before filtering into the physical drains. Many green roof substrates do have an organic component, but frequently it is as low as 15–35%, at least for extensive systems. Organic materials used include peat, coir, composted bark and green waste compost. Inorganics may use a local quarried gravel, sand or lava rock (scoria) or 'recycled' materials such as crushed brick, crushed tiles, powdered fuel ash, pumice, chemically inert foams or furnace slag. Although, if some of these materials are technically deemed as a 'waste', their use may not comply with local or national regulations. The organic component is in the minor part of the ratio for a variety of practical reasons, including limited longevity before degradation, physical slumping (loss of oxygen and water-holding capacity) and possible hydrophobicity once it dries out (highly likely in a non-irrigated shallow-depth, extensive roof system). Once dry the organic component may be difficult to re-wet. Fear of fire on a roof also disinclines the use of flammable organics such as peat and bark.

Specialized lightweight materials may also be utilized where the weight on the roof is a concern. These include leca (expanded clay granules), rockwool, expanded shale as well as pumice and lava rock. Some of these materials, however, do not come from sustainable sources or involve high energy inputs to achieve their desired properties.

Specifications for substrates for extensive/semi-extensive roofs include the ability to drain freely, whilst holding enough moisture to sustain plant growth outwith heavy rainfall events. The ability to detain and retain precipitation water during storms is one of the advantages of green roofs, but the responsibility for this falls largely on the substrate, hence the need for these attributes to be 'fit-for-purpose'. Substrates need to retain their structural integrity and be stable and consistent in their performance over time. Not least, they need to support plant growth with adequate amounts of nutrition, moisture and air to encourage a strong root system (Fig. 10.9). This is especially important in shallow substrates, where the plant may need to regenerate from its root system or dormant crown after drought.

Fig. 10.9. Substrates can be designed to detain/retain rainwater, whilst providing enough moisture to support attractive vegetated landscapes, as in this design by Nigel Dunnett in the UK. (Image courtesy of Green Roof Centre, University of Sheffield.)

Specifying the substrate and its parameters are only part of the problem. Moving tonnes of substrate onto a roof is extremely challenging and is more than likely to overtax the domestic elevators in a building. The logistics of moving substrate and other materials to the roof needs to be considered early in the planning process. High rise roofs will require the hire and siting of large construction cranes; those lower to the ground may have the substrates pumped-up from street level via a powered blower. Either way space is required on the streets surrounding the building for the machinery and the storage of bulk bags holding the substrate. Substrates, especially the lighter types, can be blown up onto the roof using a compressor pump attached to a hose and the hose hoisted onto the roof. The blowing process, however, can redistribute the particle distribution; lighter particles being drawn up first, and some re-mixing may be required once the substrate is on the roof. Ensuring the substrate is moist beforehand will reduce this risk and also stop dust blowing around the worksite. It is also important to ensure the entire volume of substrate is not piled in the one location and pressurising the roof at that point. Wherever possible the substrate should be evenly distributed as it is transported onto the roof.

Irrigation

Not all green roof systems need irrigation infrastructure, although some sort of watering facility would be useful during the establishment phase of the vegetation, even of the most highly adapted xerophytic species. The requirement for irrigation is based on climatic factors, the site's own microclimate (degree of shade or exposure to prevailing winds), substrate depth and constitution, plant choice and anticipated final size of plants and wider functional requirements. Green roofs that are designed to keep the building's occupants cooler during summer months for example, may benefit significantly if water is applied to the plants and the substrates. In most cases, irrigation design is influenced by the nature of the water supply resource, for example, is it a potable water supply or can rainwater be harvested and stored? Important factors to consider will include quantities of water available and pressure of the supply. If some form of grey water is used, the quality of that source becomes a consideration, even if there is some dilution with potable water or rainwater. Plant selection should be strongly determined by the likely availability of the water supply. Having an inexpensive and environmentally sound, regular supply of water is key to widening the plant species selection available, optimizing growth and improving establishment rates, as well as improving the long-term success of the roof.

Irrigation, in general, follows the principles for other landscape typologies (e.g. see Chapters 5 and 9), although there are a few specific factors to consider. Preferred irrigation distribution systems will depend on the area to irrigate, volume and frequency of irrigation required, site characteristics (wind strength and direction), plant type (e.g. overhead spray lines and sprinklers may not be conducive to Mediterranean and other drought-adapted species with pubescent leaves, or plant species prone to leaf disease when the leaves are frequently saturated), cost and ease of maintenance. Sprinkler systems which would normally have acceptable distribution ranges at ground level may end up watering only half the roof area (and perhaps those pedestrians on the leeward side of the building at street level below) due to the high wind speed encountered on roof tops. Micro-sprays and pop-up sprinklers though, located close to the roof level, are acceptable on roofs if there is a protective parapet to shield them from the wind. Perforated drippers and seep hoses are useful, notably if utilized under low water pressure scenarios (as may be the case on a roof). Surface drip/seep systems are readily visible and accessible but subsurface systems may be more difficult to maintain and can be accidently dug up; this needs to be counterbalanced with the advantages of less water lost due to evaporation.

Unlike most other landscape situations, the use of water retention cells below the drainage layer can be capitalized upon. The excessive water that is stored here during high rainfall events can be re-utilized to supply the vegetation with water. In some systems/substrates this is directly from capillary action; however, this is usually poor in very stony aggregates and an additional capillary wick system can be exploited to raise the water to the levels of the roots.

Most extensive systems use *Sedum* spp. or other xerophytic plant species and drainage layer/substrate design needs to ensure these plants do not 'sit in water' after heavy rainfall events, but instead allow excess waterflow into the buildings' drainage pipes. Indeed, frequency of irrigation too should match the drainage and water-holding capacity of the substrate mix, and the requirements of the predominant plant species. Frequent irrigation of free-draining substrates may waste water, and timers should be set to allow some moisture deficit to occur before re-watering. Similarly, automated systems need to be able to accommodate changes in natural rainfall patterns and not overload the roof with water when already saturated (i.e. integration to a rainfall sensor is usually advisable). Although irrigation can help to keep the building cool, such as by maintaining plant evapotranspiration during warm periods of the day, it also needs to be recognized that water flowing through the substrate can itself transfer heat onto the building's roof surface.

How 'green' are green roofs?

It has been asked a number of times, do the benefits of green roofs actually outweigh the carbon equivalents used in their construction? Green roofs add to the infrastructure and construction costs of a building. This may involve the modification of columns and beams within the building and the roof tiles/slabs. In an attempt to reduce weight loadings lightweight plastic polymers are increasingly used in the construction of green roofs. These themselves, however, may have an environmental cost. Bianchini and Hewage (2012) reviewed the environmental benefits of green roofs by making a comparison between pollutants generated during their construction and their ability to absorb pollutants from the air, once fully functional. Pollutants from the polymers used included emissions of NO_2, SO_2, O_3 and particulate matter (PM_{10}). It was calculated that a green roof needs to be *in situ* for 13–32 years to offset the emissions associated with the manufacture of these polymers. The variation in the years calculated depended on the details of the roof construction and whether some or all the polymers used were recycled. For newly manufactured polymers the air pollutant rates were 1820 and 3960 kg ha^{-1} for extensive and intensive green roofs, respectively. This dropped to 680 and 1750 kg ha^{-1} when recycled materials were used, intensive green roofs having higher values due to the more extensive use of polymers in their construction. Green roofs, in comparison, may remove 70–85 kg ha^{-1} of aerial pollutants per annum, although this value is likely to be greater with intensive roofs where the plant biomass increases considerably over time.

In terms of carbon footprint (CO_2eq), the benefits of the green roof will again depend on the capacity for vegetative growth and its ability to be sequestered into the substrate and the woody fraction of the biomass, compared to the embodied carbon used in strengthening the building to accommodate the heavier, more intensive roofs. The precision of this relationship depends on many factors, not least where the materials used originate from and what their energy use has been for transportation.

Further research is required here, especially in relation to the cost/benefit scenarios of intensive roof systems. These account for a very small proportion of urban area at present, but potentially provide the greatest benefits, due to their potential for heterogeneous vegetation forms and landscape typologies, a wide range of species composition, deep, water-retentive substrates and large volumes of biomass that maximize ecosystem service delivery. Such roofs can incorporate highly stylized, 'outdoor rooms' and public space at one end of the spectrum through to ecologically rich, semi-natural habitats at the other. They range from the sky garden concept – highly designed 'aerial gardens' on the top and sides of 21st century skyscraper tower blocks, with the opportunity to provide luxurious open space and façades – to the urban forest concept, bringing woodland and other specified habitat types to city-centre roof tops. Issues relating to factors such as fire risk, water availability and quality, maintenance and access still need to be addressed, as well as the environmental implications of needing stronger support for green roofs. Nevertheless, if humans genuinely wish to 're-green' their highly urbanized environments and encourage closer contact with and appreciation of the natural world, then more ambitious and bold projects involving green roofs will be required, and remedies sought for any outstanding obstacles.

10.4 Rain Gardens

Rain gardens are a component of sustainable urban drainage systems which help regulate water flows across the urban environment and intercept sources of waterborne pollution. They are essentially a shallow depression in the landscape, with absorbent, yet free-draining soil (Fig. 10.10). They are important for detaining water and slowing the rate of water flowing into conventional drainage pipes and sewers, but also for retaining water (reducing

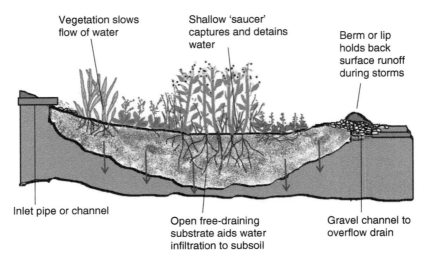

Fig. 10.10. Rain gardens are shallow water-capture basins, backfilled with free-draining substrate to detain surface runoff water and aid infiltration into the ground. Plants slow the rate of runoff, help trap particulate matter and remediate some pollutants. Through evapotranspiration they also help to dry out the soil, thereby recharging the capacity of the rain garden to hold further moisture after a subsequent rainfall event.

the volume of water entering drains, by losing a proportion to groundwater reserves and back to the atmosphere as evapotranspiration). The term swale often describes a grassy depression used to hold runoff water, whereas the term rain garden implies some degree of ornamental and functional planting scheme, although the two terms are often used interchangeably. Other terms used include bioretention strips (that retain water and dilute pollutants running off hard surfaces), stormwater planters (containers or planted raised beds which catch water from downpipes) and tree pits that are designed to capture and hold water temporarily, as well as to direct water to the tree to help improve its chances of survival. Rain gardens have been promoted most in the USA and Australia, where urban design has been closely engaged with managing stormwater events and also ensuring enough water is available to maintain urban green infrastructure.

Rain gardens are seen as an intervention – a place to temporarily trap and store excess water runoff. Many are still integrated with existing drainage systems, so once full, any excess water is carried into the conventional sewage system and does not overflow into adjacent roadways and houses. Although water and diffuse pollution management are the key *raison d'être* for rain gardens, they also provide a range of other ecosystem services, such as improved aesthetic value to an area, localized cooling, a haven for wildlife and recreational opportunities, including – where pollution is not a problem – opportunities for food growing.

Rain gardens are usually situated in a low depression where water would normally accumulate naturally or on a gentle slope in the landscape. The greater the gradient of the slope the more difficult it will be to site the rain garden, as remodelling the land to create a level perimeter becomes more challenging; slopes in excess of 1:8 are best avoided as these would require retaining structures and may look incongruous within the wider landscape. With the exception of stormwater planters, which are situated under a downpipe coming off a building's roof, rain gardens are usually situated some distance from a building (>3 m) to avoid any problems associated with overflow. Water coming off building roofs via downpipes can be led to the rain garden by additional pipes or channels cut out in the soil or even large sloping swales. These channels themselves are frequently brick, pebble or clay lined to stop soil eroding as water rivulets form (Fig. 10.11).

Water capture and infiltration

As it is desirable for a proportion of the water to percolate down through the soil, naturally free-draining soils, such as sands and chalks, are more appropriate for rain gardens than areas where the water table is

Fig. 10.11. Lining swales, channels and ditches that lead into rain gardens with stones or other hard aggregates stops soil erosion and keeps the watercourse from becoming turbid.

naturally high, or where heavy clay soils impede the movement of water downwards. Ideally, if the parent soil is permeable enough, then water should drain away at rates >50 mm h^{-1}. The media that is used above the parent soil is also important. Poorly draining media reduce infiltration and drainage rates even in those rain gardens that have a high surface area relative to their associated drainage area. Prolonged ponding of the area may not only impact on the drainage characteristics, but also erode public attitudes relating to unsightly algal growth and concerns about mosquito infestation. Often engineered media with a high sand content or gravel with only a limited proportion of smaller fine particles is advocated. Good *et al.* (2012) indicated that media composed of sand or sand/topsoil mixes demonstrated adequate hydraulic conductivity (sand: 800–805 mm h^{-1}, sand/topsoil: 290–302 mm h^{-1}). Slate gravel has been shown to have superior infiltration and conductivity rates than sand, but sometimes this can be too fast, and amendments with layers of organic matter such as green compost or pine bark achieve a good balance between conductivity and moisture retention (Riley *et al.*, 2014).

For soils that have poorer drainage capacity, then a larger rain garden may be required than in locations where water percolates through quickly. Irrespective of this, the width of rain gardens should be at least 3–5 m to provide enough vegetation to retain pollutants and help deactivate them (phytoremediation), as well as stopping water overflowing too quickly. Areas for rain gardens are dictated by space available, but Dussaillant *et al.* (2004) suggested that the space required to maximize the recharge of groundwater was 10–20% of the area of the contributing impervious surfaces.

Planting

Those plants used in rain gardens need to be able to tolerate inundation with water, but also periods where rainfall may be absent and the free-draining soil/substrate may retain little moisture. The depth of the media/storage zone affects the volume of moisture retained, but seems to have little impact on maximum duration of saturated conditions in the root zone, in practice. Perhaps this is due to the fact most subsoils are permeable and free-draining. Detailed evaluations of plant species for rain gardens are still in their infancy, although riverside species that are associated with temporary river systems (i.e. those that dry out in the dry season) may be a useful habitat for environmental horticulturalists to explore. Species such as *Eucalyptus camaldulensis* (river red gum) are thought to be able to tolerate both high levels of drought as well as periods of temporary inundation as they experience occasional and often drastic flooding following extreme rainfall events. From their studies in the USA, Turk *et al.* (2014) recommend that species such as *Betula nigra*, *B.* 'Duraheat', *Magnolia virginiana*, *M.* 'Sweet Thing', *Itea virginica*, *I.* 'Henry's Garnet', *Juncus effusus* 'Frenzy', *Panicum virgatum* 'Shenandoah', *Helianthus angustifolius*, *H.* 'First Light', and *Eupatorium purpureum* subsp. *maculatum* are useful for rain garden situations.

Maintenance activities centre around removing weeds until the desirable vegetation becomes established. As rain gardens have a higher probability of being moist compared to conventional gardens, predation pressure from slugs and snails can impair the establishment of young plants, so careful plant choice may be required (see Chapter 6). Annual cutting back of dead material from herbaceous plants will encourage space for new shoot growth in the following spring. Although the build-up of plant biomass over time will reduce water flow strength and rate, the development of an hydrophobic organic layer on the substrate surface may impede water infiltration. Removing organic matter on an annual basis and light cultivation of the soil surface can reduce the likelihood of this.

Pollutant control

The scale and composition of rain gardens influences their effectiveness as a pollution control mechanism. Early evaluations by Dietz and Clausen (2005) with a shallow system (25 mm depth) capturing water off a roof, suggested that there was limited capacity to ameliorate nitrite+nitrate nitrogen (NO_3-N), organic nitrogen or even total nitrogen levels, although action against ammonium nitrogen (NO_3-N) was more effective. Good *et al.* (2012) had more success with laboratory-based systems which investigated the amount of organic medium and sand in the topsoil. Systems with sand and a sand/topsoil mix demonstrated good heavy metal removal (with maximum removal rates being copper 83%, zinc 95% and lead 97%), although sand has not always been as effective at removing other contaminants such as nitrogen. Slate, on the other hand, has proven useful at removing nitrogen as well as phosphate under some circumstances (Turk *et al.*, 2014). Similarly, a two-phase experimental rain garden system using a gravel/soil mix was effective at removing the key nutrients as well as organic compounds such as herbicides. This system involved water moving from a saturated to an unsaturated zone in sequence, thus increasing overall retention time of runoff and improving bioremediation (Yang *et al.*, 2013). This new system was estimated to remove 91% of nitrate and 99% of phosphate in the runoff, whilst also removing large proportions of herbicide substances such as atrazine (90%), dicamba (92%), glyphosate (99%), and 2,4-D (90%). Fine particles washed off road and other urban surfaces can accumulate in rain gardens, and may interfere with pore structure and the ability of the system to deal with pollutants. Jenkins *et al.* (2010), however, suggested that although the proportion of fine particles was noted to increase in their long-term studies, there was no significant change in water infiltration potential. The increase in fines may be compensated by the actions of root systems as they explore the medium and open up further channels for water to move through.

Conclusions

- As demand for land increases within cities, the divisions between built and green infrastructure become blurred. Small-scale interventions such as green walls, roofs and sustainable urban drainage systems (SUDs) are used to provide contact with nature and deliver some degree of ecosystem service.
- Both green wall systems and green roofs vary in their construction with access and availability of water often dictating the type of vegetation that dominates in any given system.
- Green walls are usually divided into green façades, living walls (both with and without artificial irrigation systems), retaining living walls and bio-walls.
- Living walls vary in their complexity of design and associated production and maintenance costs. Broadly speaking modern systems break down into hydroponic systems and substrate-based systems.
- Green roofs have various designations too, but main types are extensive (shallow substrates, limited plant choice), semi-extensive (deeper profile substrates with greater plant choice) and intensive (allowing relatively large plants to be grown).
- The key consideration affecting the uptake of green roofs is their weight, with deeper substrate roofs requiring new buildings to be designed to cope with the additional weight loads, or retrofit buildings being re-engineered. As such, lighter extensive green roofs systems tend to be more popular.
- Much attention has been paid to the substrates used in green roofs, in an attempt to reduce rainfall runoff (detention and retention), insulate the building and protect the roof materials from weathering, as well as retain enough moisture and nutrients to allow plants to grow.
- Where natural water supplies are insufficient green roofs will employ irrigation systems, although the exposed conditions frequently encountered means that sprinklers distribute water at low heights, or that sub-irrigation drip or seep hose is popular. Water retention cells

below the substrate are also used commonly to help provide some reserve of water.
- Rain gardens are a component of SUDS and help detain excess runoff water and allow it to infiltrate into the ground, thereby reducing the risk of flash flooding during stormwater events. They also have a role in intercepting waterborne pollutants, although substrate composition can influence the effectiveness of this latter aspect.

References

Anon. (2014) A guide to green roofs, walls and facades in Melbourne and Victoria. Available at: www.growinggreenguide.org/wp-content/uploads/2014/02/growing_green_guide_ebook_130214.pdf (accessed 12 February 2015).

Bianchini, F. and Hewage, K. (2012) How 'green' are the green roofs? Lifecycle analysis of green roof materials. *Building and Environment* 48, 57–65.

Cameron, R.W.F., Taylor, J.E. and Emmett, M.R. (2014) What's 'cool' in the world of green façades? How plant choice influences the cooling properties of green walls. *Building and Environment* 73, 198–207.

Cameron, R.W.F., Taylor, J. and Emmett, M. (2015) A *Hedera* green façade – energy performance and saving under different maritime-temperate, winter weather conditions. *Building and Environment* 92, 111–121.

Castleton, H.F., Stovin, V., Beck, S.B.M. and Davison, J.B. (2010) Green roofs: building energy savings and the potential for retrofit. *Energy and Buildings* 42, 1582–1591.

Currie, B.A. and Bass, B. (2008) Estimates of air pollution mitigation with green plants and green roofs using the UFORE model. *Urban Ecosystems* 11, 409–422.

Dietz, M.E. and Clausen, J.C. (2005) A field evaluation of rain garden flow and pollutant treatment. *Water, Air and Soil Pollution* 167, 123–138.

Dussaillant, A.R., Wu, C.H. and Potter, K.W. (2004) Richards equation model of a rain garden. *Journal of Hydrologic Engineering* 9, 219–225.

Francis, R.A. and Lorimer, J. (2011) Urban reconciliation ecology: the potential of living roofs and walls. *Journal of Environmental Management* 92, 1429–1437.

Good, J.F., O'Sullivan, A.D. Wicke, D. and Cochrane T.A. (2012) Contaminant removal and hydraulic conductivity of laboratory rain garden systems for stormwater treatment. *Water Science & Technology* 65, 2154–2161.

Jenkins, J.K.G., Wadzuk, B.M. and Welker, A.L. (2010) Fines accumulation and distribution in a storm-water rain garden nine years post-construction. *Journal of Irrigation and Drainage Engineering* 136, 862–869.

Köhler, M. (2008) Green facades – a view back and some visions. *Urban Ecosystems* 11, 423–436.

Mentens, J., Raes, D. and Hermy, M. (2006) Green roofs as a tool for solving the rainwater runoff problem in the urbanized 21st century? *Landscape and Urban Planning* 77, 217–226.

Nagase, A. and Dunnett, N., (2010) Drought tolerance in different vegetation types for extensive green roofs: effects of watering and diversity. *Landscape and Urban Planning* 97, 318–327.

Oberndorfer, E., Lundholm, J., Bass, B., Coffman, R.R. Doshi, H., Dunnett, N., Gaffin, S., Köhler, M., Liu, K.K.Y. and Rowe, B. (2007) Green roofs as urban ecosystems: ecological structures, functions, and services. *BioScience* 57, 823–833.

Riley, E.D., Kraus, H.T. and Bilderback, T.E. (2014) Physical properties of varying rain garden filter bed substrates affect saturated hydraulic conductivity. *Acta Horticulturae* 1055, 485–489.

Turk, R.L., Kraus, H.T., Bilderback, T.E., Hunt, W.E. and Fonteno, W.C. (2014) Rain garden filter bed substrates affect stormwater nutrient remediation. *HortScience* 49, 645–652.

Veisten, K., Smyrnova, Y., Klæboe, R., Hornikx, M., Mosslemi, M. and Kang, J. (2012) Valuation of green walls and green roofs as soundscape measures: including monetised amenity values together with noise-attenuation values in a cost-benefit analysis of a green wall affecting courtyards. *International Journal of Environmental Research and Public Health* 9, 3770–3788.

Williams, N.S.G., Rayner, J.P. and Raynor, K.J. (2010) Green roofs for a wide brown land: opportunities and barriers for rooftop greening in Australia. *Urban Forestry & Urban Greening* 9, 245–251.

Yang, H., Dick, W.A., McCoy, E.L., Phelan, P.L. and Grewal, P.S. (2013) Field evaluation of a new biphasic rain garden for stormwater flow management and pollutant removal. *Ecological Engineering* 54, 22–31.

11 Interior Landscapes

Ross W.F. Cameron

> **Key Questions**
> - What sort of locations are house plants used in?
> - Why do many businesses invest in interior landscapes (plantscapes)?
> - How does the design and management of an interior planted landscape differ from an exterior one? What are the specific challenges?
> - How do interior plants interact with volatile airborne chemicals commonly found in modern buildings?
> - Not all interior plants come from the tropical rainforests; what other biomes contribute plant genotypes? How do their requirements and management vary?
> - What key properties make an ideal growing medium for an interior landscape?
> - What is meant by plant acclimation with respect to interior displays?
> - How does the implementation and management of interior landscapes work in practice?

11.1 Introduction

Interior landscapes (plantscapes) constitute the use of plants indoors or in semi-protected environments. These range from a single house plant placed on a window sill within a domestic dwelling, to extensive iconic biomes and atria that are significant tourist venues. The latter include the Gardens by the Bay, Singapore, or the Eden Project, UK. In Singapore, the protected structures are required to either keep plants cool as in the 'cloud forest biome' or grow them at lower humidities than would occur naturally outdoors, i.e. in the 'flower dome'. At the Eden Project almost the converse is true; the 'tropical biome' being designed to maintain plants within warm, high-humidity conditions or the 'Mediterranean biome' where the regimes are also warm, but relatively dry compared to outdoors. Such venues have been specifically designed to accommodate plant collections, but many other public plant displays are incorporated within existing buildings and structures, either exploiting light from nearby windows or being supplied with irradiance through artificial illumination.

Interior plants cover a range of contextual situations and are used in offices, schools, leisure centres, shopping malls, hotels, cafés and restaurants. Frequently they are utilized as part of a company's marketing strategy or for reinforcing a brand image or philosophy of an organization. Plantscapes are regularly utilized to help create a unique 'sense of place' and add to the visitors' experience of a given destination. Small trees and large shrubs with bold forms and strong textures are used to create an impression of grandeur, wealth, power or influence, such as in the foyer of a large commercial company or bank or within an iconic public space (Fig. 11.1). Robust and strongly textured species such as *Monstera deliciosa* (Swiss cheese plant), *Fatsia japonica* (Japanese aralia) and *Ficus elastica* (rubber fig) help promote feelings of strength, reliance and power.

In contrast, more intimate, relaxed or flowing planting styles as typified by species such as *Schefflera schizophylla* or *Ficus benjamina* (weeping fig) might be used to convey a sense of tranquillity or restfulness. Such a design style may be appropriate for a hospital entrance or in the offices of a charitable organization. Restaurants and hotels may exploit plants to reinforce their style of interior design, for example recreating an Edwardian ambience with palms and ferns that perhaps complements the wicker furnishings, e.g. *Howea forsteriana*

Fig. 11.1. Strong 'architectural' plants are often used to complement and emphasize the architecture of a building.

(Kentia palm). Alternatively, strongly architectural forms of plants are used to augment impressions of minimalist design and architecture, making good use of simple elegant shapes such as fastigiate *Sansevieria trifasciata* (mother-in-law's tongue) or bold spikes of *Dracaena draco* (dragon tree) and *Aechmea fasciata* (silver vase). Somewhat in contrast to the designed landscapes and gardens found outdoors, greater emphasis is placed on foliage indoors to provide form and texture (Fig. 11.2), with relatively little emphasis on flowers and colour (there are exceptions of course, such as the orchid display at Changi Airport, Singapore). Even when flowers are on display, bold and vibrant geometric styles are often still the in vogue choice for interior designers (Fig. 11.3).

11.2 Purpose and Function

There is evidence that plants have been grown indoors, or at least at the transition between outdoor spaces and building interiors, as early as 1000 BCE in China and perhaps parts of the Middle East. It was the development of the modern glasshouse and improved

Fig. 11.2. Foliage and indeed the whole plant form are important to create dramatic effects in interior landscapes.

technologies associated with protected growing of plants that largely resulted in the boom in popularity of 'house plants' and interior plant displays during the Victorian era, 1837–1901. This is a trend that still

Fig. 11.3. Species of plants are often selected for the dramatic form of their flowers in addition to any attributes purely based on colour.

exists today, with two distinct markets being associated with the use of plants in interior spaces. On the one hand, there is the domestic house plant market where plants are purchased by individuals and used within the home setting on window sills and within private conservatories; on the other hand, business offices, local authorities, retail arcades, leisure venues and other organizations will use large plant displays in specifically designed areas such as lobbies and atria to improve their interior décor. The role of interior plants in education and science should not be underestimated either. Northern Europeans, including the Victorians in the UK in the 18th and 19th centuries, went to elaborate lengths to design glasshouses within their botanic gardens to display the exotic tropical and semi-tropical species being discovered; a trend that has been repeated in recent years with new protected structures constructed in Wales, Cornwall and Surrey, UK, for conservation and educational purposes (Fig 11.4).

Most plants in a domestic household are used in small-scale displays, with plants held within individual pots and grown on a window sill or well-illuminated table. A number of these plant types are treated as disposable in that they are grown for a few months and then thrown away, for example *Euphorbia pulcherrima* (poinsettia) or *Rhododendron simsii* (pot azalea). Such plants are commonly 'forced' into flowering for instant appeal and used as presents on special occasions such as Christmas, Easter or Mothering Sunday. Skilled gardeners can encourage re-flowering (or in the case of *E. pulcherrima*, the recolouring of the young bracts), but most people find it easier to purchase another plant the following year.

In the case of plants used in large office complexes and shopping malls etc., the plants may not actually be sold per se, but rented from an interior landscaping company. Traditionally, the main driving force for interior plantscapes has been aesthetics. They complement the other criteria underpinning interior design, helping to contribute to a unique location or atmosphere and reflecting the tastes, attitudes and ethics of the building owner or occupier. With careful design, plantscapes can be used to influence environmental factors and improve the interior dwelling or working space by, for example, filtering out direct sunlight, reducing glare from reflective surfaces or acting as baffle against noise sources. In offices they are used to provide privacy and break up large open spaces into smaller 'rooms'. They provide focal points of interest or are carefully sited to visually block unsightly objects. They are also frequently exploited to help direct pedestrian traffic through a public space, channelling pedestrian flows through airports, railways stations, shopping malls etc. to avoid queues and congestion. Moreover,

Fig. 11.4. New glasshouse complexes are used to educate and inform the public, such as this one at the Royal Horticultural Society's garden at Wisley, Surrey, UK.

interior plantscapes facilitate opportunities to engage with living objects and promote a sense of nature – an increasingly important role as the average urban dweller now spends 90% of their time indoors. This latter aspect underpins the now widely cited benefits of including plants within interiors – most notably, of working environments.

Health and well-being aspects of interior plant displays

The psychological and physical benefits of exposure to plant-dominated environments is discussed in detail in Chapter 3, but key findings relating specifically to the interior of buildings are:

- enhanced positive mood and creativity (Larsen et al., 1998);
- the provision of a more attractive living or working environment (Park and Mattson, 2009);
- improved reaction time (Lohr et al., 1996);
- increased attention span and mental skills (Khan et al., 2005);
- lower levels of fatigue and fewer symptoms of physical discomfort (dry throat, coughing, dry skin) (Park and Mattson, 2008);
- lower systolic blood pressure (Lohr et al., 1996);
- lower electrodermal activity (Park et al., 2004);
- increased social interaction/engagement in patients with psychological health problems (Talbott et al., 1976);
- calmness, reduced anxiety and more rapid recovery from stress (Chang and Chen, 2005);
- higher tolerance to pain (Park and Mattson, 2009); and
- higher ZIPERS values (Zuckerman's Inventory of Personal Emotional Reactions Score) (Lohr and Pearson-Mims, 2000).

In a review of plants and their influence on human productivity, Bakker and van der Voordt (2010) assert that the most dramatic positive effects of plants include:

- an increase in shop sales by 12% (Wolf, 2002);
- an increase in personal response speed by 12% during simple recognition tests (Lohr et al., 1996);
- a reduction in symptomatic physical complaints by 23% in a study of 51 office employees (Fjeld et al., 1998); and
- a reduction in health complaints by 25% during a study of 48 employees of a hospital X-ray department (Fjeld, 2000).

Increasing the proportion of plants within an individual's viewpoint is also thought to enhance the positive effects, with greater reduction in anxiety and more positive responses reported

(Han, 2009). These results would allude to the fact that interior plants appear to have a significant role to play in locations such as hospitals, hospices and educational facilities as well as the working office.

Many interior landscapes are dominated by green foliage (rather than golden, variegated or bi-coloured plants), and this may be no surprise if the research carried out by Elsadek and Fujii (2014) holds true generically. These authors studied people's response to interior plant colour variation, with *Spathiphyllum wallisii* (green), *Cordyline terminalis* (green-red) and *Aglaonema pictum* (green-white) being utilized to assess visual stimulation. The species were selected to represent similar forms and habits, but different colours. Overall, most respondents preferred green plants over green-red or green-white plants (Fig 11.5). Using eye-tracking technology and monitoring brain stimulation it was evident that people exhibited different responses to the plant colours. The researchers concluded that there are practical implications to this research. Entirely green plants should be used predominantly where stress relief/ reducing anxiety is the over-riding criterion, e.g. in doctors' waiting rooms or hospitals, where the plantscape can help provide a relaxing environment. In contrast, the more striking green-red plants, which stimulated the motor area in the brain that controls muscles, may be better used in office environments to improve employees' productivity or in children's play areas to help stimulate creativity.

Modifying the interior aerial environment

Implications on physical health also relate to the way that interior plantscapes influence the indoor aerial environment. Plants are thought to contribute to improving air quality within buildings through modifications of humidity and an ability to remove or detoxify certain aerial pollutants. Atmospheric gases such as carbon monoxide (CO) and carbon dioxide (CO_2), sulfur oxides (SO_x), nitrous oxides (NO_x), ammonia (NH_3) and organic volatiles can diffuse indoors from the exterior environment. These are augmented by a cocktail of other aerial

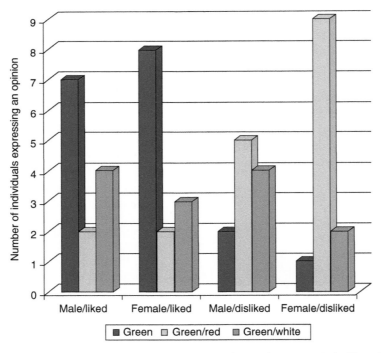

Fig. 11.5. Number of people expressing an opinion about plant colours when presented with an image of plants with either uniform green colour, or variegated with red or white patterns. (Modified from Elsadek and Fujii, 2014.)

pollutants generated indoors, which include particulate matter (dust) and volatile organic compounds (VOCs), of which 300 have been identified, e.g. formaldehyde, toluene, benzene and xylene. So-called 'sick building syndrome' has been attributed to VOCs and is a phenomena where occupants of the building suffer symptoms including headaches, nausea, dizziness, respiratory problems, dry throat and eyes, and complain of loss of concentration. These aspects have increased as buildings have become more airtight, with greater emphasis on reducing draughts and regulating temperature with air conditioning/central heating. VOCs are generated from office features such as furniture, printers and laminated surfaces over time, or can be released from products commonly used within the building such as detergents, paints, varnishes and polishes. In contrast to outdoor spaces, poor ventilation in buildings contributes to enhanced levels of common gases such as CO_2. Even this relatively benign gas has been associated with reduced workplace productivity (Seppänen and Fisk, 2006) and mental ability (Shaughnessy et al., 2006).

Plants remove particulate matter from the air by adsorbing it onto their leaf surfaces (Lohr and Pearson-Mimms, 2000). They also metabolize a range of aerial pollutant gases, either by direct action, or by hosting microbial organisms that themselves metabolize and break down the organic molecules. These may be located on the leaf surfaces as phylloplane bacteria and fungi, or in the growing media and around the roots, namely rhizosphere microorganisms. It is clear that plants absorb some of the gases and metabolize them directly – CO_2 of course being essential for photosynthesis. Tarran et al. (2007) found that by placing just three plants in an office environment CO_2 levels were reduced by 25% (and 10% in an air-conditioned building), but these plants were also effective at removing any of the more dangerous CO present (86–92% reduction). The ability to deal with small organic molecules, however, appears enhanced by the presence of bacteria and fungi (Wolverton and Wolverton, 1993; Tarran et al., 2007). By artificially enhancing the population of *Pseudomonas putida* bacteria on leaves of *Rhododendron indica*, De Kempeneer et al. (2004) showed that airborne toluene could be removed much more rapidly compared to leaves with only a background level of microorganisms. The role of the plant was essential, however, as the same effect was not apparent, when *P. putida* was placed on an artificial surface. Indeed, removal rates tend to be optimized when plants are growing well, i.e. with sufficient irradiance, at appropriate temperatures and humidities, and when not under water or nutrient stress. Plants and their allied microflora have been shown to aid the removal of SO_2 (Lee and Sim, 1999), NO_2 (Coward et al., 1996), mercury vapour (Bastos et al., 2004) as well as CO_2 and VOCs from interior environments (Wolverton and Wolverton, 1993; Dingle et al., 2000; Orwell et al., 2004; Tarran et al., 2007).

Which plants are best at this 'aerial scrubbing' is open to some scrutiny. When comparing relative removal rates by different plant species Wolverton and Wolverton (1993) ranked the order for formaldehyde ($\mu g\ h^{-1}$) as:

1. *Nephrolepis exaltata* 'Bostoniensis'
2. *Chrysanthemum morifolium*
3. *Dracaena deremensis* 'Janet Craigs'
4. *Hedera helix*
5. *Chamaedorea elegans*
6. *Cyclamen persicum*
7. *Sansevieria trifasciata*

For xylene, however, the order changed to:

1. *Chamaedorea elegans*
2. *Nephrolepis exaltata* 'Bostoniensis'
3. *Chrysanthemum morifolium*
4. *Cyclamen persicum*
5. *Sansevieria trifasciata*
6. *Dracaena deremensis* 'Janet Craigs'
7. *Hedera helix*

In similar approaches Sriprapat et al. (2014) evaluated 12 common house plants and demonstrated that *Sansevieria trifasciata*, *Kalanchoe blossfeldiana* and *Dracaena deremensis* were most effective at removing toluene, whereas *Chlorophytum comosum*, *Sansevieria ehrenbergii*, *S. hyacinthoides* and *Aglaonema commutatum* would be the species of choice for ethylbenzene removal. Some of the reactions are concentration dependent, however, with metabolism decreasing when background levels fall; thus plants will not necessarily eliminate the VOCs completely from a room. Removal rates are also under the influence of other factors, including size/volume of plant material present, air movement, irradiance, temperature, moisture status of the growing media and whether the plants or pots themselves release their own VOCs. Indeed, when Yang et al. (2009) assessed uptake rates accounting for different leaf areas, certain trends contradicted previous studies, although this approach was useful in identifying 'universally' effective species, e.g. *Hemigraphis alternata* and *Hedera helix* (Fig. 11.6).

As with exterior plants, there are counterclaims that certain interior plants may be detrimental from a human health perspective. As well as some species being toxic – e.g. *Dieffenbachia amoena* (dumbcane), *Solanum pseudocapsicum* (winter cherry), *Nerium oleander* (oleander) (Der Marderosian, 1976) – others are linked to contributing to human respiratory disorders, such as asthma, rhinitis and laryngitis; complaints have been associated with common species such as *Ficus benjamina* (weeping fig) (Axelsson *et al.*, 1985) and *Spathiphyllum wallisii* (peace lily) (Kanerva *et al.*, 1995), yet exposure levels may need to be high before symptoms become apparent. Indeed, most cases of illness or discomfort are associated with horticulturalists regularly maintaining the plants, rather than office employees, suggesting people need to have direct contact with the plants, or be working in close proximity to them for prolonged periods.

The capacity to sequester carbon through interior planting has been raised, although the benefits may have more to do with reducing CO_2 levels from offices and other rooms, rather than contributing significantly to global atmospheric CO_2 targets. Controlled environment studies have suggested that larger woody plant species/specimens are more effective at absorbing carbon than either smaller species/specimens or herbaceous subjects (Pennisi and van Iersel, 2012).

11.3 Interior Plant Requirements

The success of growing plants within buildings and other enclosed areas depends on a matrix of factors, including the levels of irradiance, temperature regimes, air quality and movement, and humidity. In addition, root systems need to be provided with appropriate levels of moisture, nutrients and oxygen, as well as a medium that provides sufficient anchorage. Tropical/semi-tropical plants housed in buildings located in temperate or cool climates will require heating, but

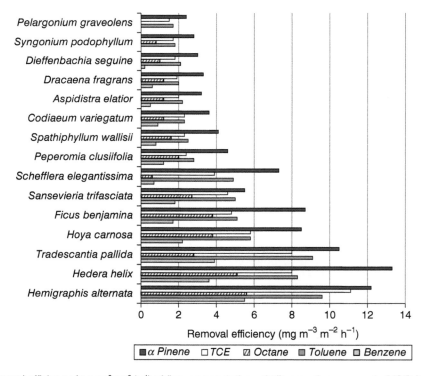

Fig. 11.6. Removal efficiency (mg m^{-3} m^{-2} h^{-1}) of five representative volatile organic compounds (VOCs) – benzene, toluene, octane, trichloroethylene (TCE) – when assessed on leaf area per species. (Modified from Yang *et al.*, 2009, and representing only a sample of the species tested.)

conversely other plants may require air conditioning or misting systems to keep them cool – this being a common problem for botanical glasshouses and atria in more tropical regions, due to the high amounts of solar energy these capture and retain.

Temperature

The choice of interior plant species will be determined by the predominant temperature in the building. For most buildings this will be set for human thermal comfort, e.g. in temperate regions this is considered to be 21°C during the day and 18°C at night. In locations specifically designed to cater for the plant collections, however, the temperature can be dictated by the plants' requirements, not necessarily that of the humans. Different plant collections will have different temperature requirements. It is not always safe to assume that plants require temperatures comparable to or higher than the human thermal comfort zone; some may prefer slightly cooler temperatures.

Interior plants can be broadly divided into four groups based on their natural distribution and biome:

- Cool temperate: day temperatures that vary with season from approximately 12 to 18°C with night temperatures of approximately 4–6°C.
- Warm temperate/cooler tropical (e.g. tropical montane regions): day temperatures that vary with season from approximately 21 to 26°C, with night temperatures of approximately 8–12°C.
- Moist tropical and subtropical: relatively stable temperatures across the season with day values of 28–32°C and night values of 14–16°C.
- Arid tropical (desert/semi-arid regions): high day temperatures up to 45°C, but tolerating 2–5°C at night (Fig. 11.7).

Diurnal variations are feasible in most office blocks and similarly used buildings, with heating systems being turned down at night. Most plants appreciate a cooler night temperature than day temperature. Recreating seasonal fluxes in temperature though may be harder to achieve, especially for cooler-adapted species, at least without inducing the wrath of the human workforce or occupants. Not all rooms in a building, however, are managed in the same way; entrance lobbies, hallways and some corridors may experience greater influence from exterior temperatures, hence making them more suitable for species that require some 'cool season' effect.

Temperature and irradiance interact, and excessively high temperatures are more detrimental when irradiance is suboptimal, as this induces relatively high rates of respiration but not photosynthesis,

Fig. 11.7. Arid tropical themes are popular in interior landscapes. Humidity levels are lower than for moist tropical plantings, and more conducive to typical humidities found within buildings. Care needs to be taken, however, with access and passage ways as many species have sharp thorns/leaf tips.

stressing the plant through a negative carbon balance. In such situations supplementary light will be required. Plants require high irradiance and consequently are placed beside windows, however, this is where central heating radiators are also frequently placed. Exposure to direct sources of heat such as this can scorch leaves, as well as cause localized drying of the air. Room temperature will also influence overall humidity levels and higher temperatures are likely to correspond to greater transpiration and the need for more frequent irrigation. Although most of the focus is on air temperature, root temperatures should not be forgotten. Planters that are located against exterior walls or sunk into the floor of the building below ground level may experience lower temperatures than the rest of the building, and plants may run the risk of root chilling injury.

Irradiance

Irradiance (light) is needed by interior plants for photosynthesis, photomorphogenesis (regulating plant development), synthesizing chlorophyll and other pigments, and regulating stomatal behaviour, as well as influencing factors such as leaf temperature, rate of transpiration and mineral uptake. In practical terms, irradiance is measured as the human eye perceives it, i.e. as lux (lumens m^{-2}), but plants have a preference for certain specific wavelengths of irradiance – and photosynthesis is highly dependent on red and blue spectral wavelengths. Irradiance as required by plants is therefore defined by the photosynthetically active region (PAR, i.e. 400–700 nm) and this is measured as µmol photons m^{-2} s^{-1}. Lux can be converted to µmol m^{-2} s^{-1}, but the conversion factors can vary depending on the light source, i.e. it will be different from daylight (multiply lux by 0.018) compared to say 'cool white' fluorescent tube lighting (multiply lux by 0.013).

It is thought that most plants will not survive below 400 lux, with shade-tolerant species coping with 400–750 lux, medium-light requiring species 750–1500 lux, and high-light species 1500–2500 lux. To put these values into context, on a clear sunny day outdoors irradiance can exceed 50,000 lux, whereas a typical office environment may only be 500 lux. It is often considered that house plants placed more than 1.5–2 m from a north-facing window or perhaps 3 m from a southern aspect one will begin to experience suboptimal irradiance. In buildings specifically designed to accommodate plantscapes, such as through the use of protected courtyards, conservatories or atria, energy for plant development and display is exploited through natural light. Protection from cold, wind and even excessively low humidity is provided by glass roofs/walls (or by other translucent materials), whilst still allowing the relevant irradiance wavelengths to pass through and be utilized by plants for photosynthesis and photomorphogenesis. In the design of such structures, however, care is required to ensure that features such as positional aspect, degree of shading (e.g. from adjacent walls or buildings), depth and width of glass (transmission) area are fully taken into consideration. Too much solar irradiance may lead to excessive temperature rises, and too much shade may lead to suboptimal irradiance for some species. For example, a deep narrow atrium with a small ground area, but surrounded by many storeys of built structure, may still not provide enough light to facilitate plant requirements. The planting design may itself also influence irradiance dynamics; tall plants with wide canopies add to shade problems for the lower strata of vegetation. Species included in the design for their flower displays may particularly require higher irradiance levels (Fig. 11.8). Similarly, greater irradiance intensity may be a prerequisite to ensure coloured or variegated foliage plants retain their characteristic hues (Fig. 11.9).

Although emphasis in plantscape design is often placed on maximizing irradiance (especially to ensure adequate irradiance levels in winter), in some regions of the world excessive irradiance can be equally problematic. For tropical 'forest floor' species adapted to mid or lower irradiances, direct summer sunlight and high temperatures (e.g. when combined with low humidity) can be damaging, and provision is required for venting of the glasshouse structures or providing additional shading. Shading can be accommodated by whitewashing glass, automated blinds, or the use of electrically controlled glass ('electrochromatic glass' or 'smart-glass') that changes its colour or transparency.

Where natural solar irradiance is absent or suboptimal, then artificial sources of light are required to provide either the plants' entire needs, or to supplement the natural irradiance. Incandescent, fluorescent, halogen, metal halide (high intensity discharge) lamps, or combinations of these, can be used to deliver the appropriate spectra and quantity of irradiance. Illumination from the lamps themselves may contribute to the interior ambience, but it should be noted that the spectrum of irradiance that is required for successful plant

Fig. 11.8. High irradiance is required for some species of flowering plant to flower well – and retain a compact growth form, such as with this *Pelargonium*.

Fig. 11.9. Attaining the best from variegated or coloured foliage also often relies on good levels of irradiance, in this case with *Solenostemon scutellarioides* (coleus).

development may be somewhat different to that perceived by the human eye, and appreciated by the human brain! Conversely, in most cases the illumination in interior spaces is dictated by the requirement for human use (appropriate brightness, spectrum and energy use characteristics) and there may need to be a compromise struck with that which is optimal for plant development. Such compromises need to be taken into account when specifying the type of light used in the interior space. Providing irradiance intensity is optimal, illumination durations in line with office working hours are usually sufficient to provide daily requirements of PAR (8–14 hours being ideal).

Traditionally, illumination sources used in commercial horticulture to support plant growth have been relatively inefficient in energy use. In recent years, more attention has been paid to low energy bulbs and lamps, with recent evaluations including the use of light emitting diodes (LED) to provide a photospectrum that promotes growth and development. To date the high relative costs of LED have delayed implementation of this technology. LEDs are a form of semi-conductor diode, with the wavelength (colour) of light generated dependent on the semi-conductor material used. This has the advantage that diodes can be manufactured where the irradiance quality matches directly the requirements of the photoreceptors of the plant – for photosynthesis this tends to be in the red and blue spectra. Where natural or broad spectrum artificial light is absent, however, a much broader range of spectra from LEDs are probably required to ensure that other morphological and plant development aspects are catered for. For example, introducing far-red light helps promote leaf elongation and increase biomass production. Even green light – a spectrum that is not normally absorbed in any great quantity (but rather reflected to give the green colour of plants) – may still be required to induce phenolic production and hence aid resistance to pests and pathogens. Despite these subtle requirements, LED technology may have a significant impact on interior displays in future as they can significantly reduce energy consumption used to irradiate plants. Compared to an incandescent bulb an LED light source may be 75% more efficient and last 25 times longer. Savings will increase further as technology improves. Blue LED light sources were only 11% energy efficient in the first models developed, but are now 49% efficient in terms of converting electrical energy to photon energy. Additional advantages in horticultural settings include the possibility of placing LEDs close to plants (including within the canopy) because they emit little heat and will not scorch leaves or be a danger to passing workers/members of the public. Unlike conventional bulbs they do not have fragile glass or elements that break easily and they can be integrated into digital control systems to allow variations in intensity and light quality if required, e.g. mimicking changes in light intensity and quality as would occur in the diurnal cycle in nature.

Air quality

The presence of plants may improve interior air quality, but contaminants within buildings can also be phytotoxic. Aerial pollutants and gaseous fumes can result in chlorosis and necrosis of foliage, epinastic growth, early bud abscission and even complete defoliation in severe cases (Ingels, 2009). Phytotoxic chemicals commonly used in offices include cleaning products, paints, varnishes and preservatives which may release ammonia, chlorine and volatile hydrocarbons. Chlorine used to disinfect bathing water can be a problem too in swimming pools and leisure centres, although these are offset by ensuring the plants are sufficiently far away from the pool to avoid direct splashing, and that there is forced air movement/ventilation to help reduce the background levels of chlorine in the air. In other circumstances, ethylene may induce epinasty (downward bending) and chlorosis in interior plants, due to concentrations building up in confined spaces with limited air circulation or replacement. Ethylene is synthesized from a variety of organic sources (including decaying plant material) but is also a by-product of hydrocarbon combustion. Even common household dust can be problematic – covering leaves, disrupting irradiance capture and blocking stomatal pores, thereby interfering with photosynthesis and transpiration. Dust is removed from the foliage by washing down the leaves periodically or, where feasible, cleaning them with a fine cloth.

Growing media

In contrast to most exterior plants in the wider landscape, many of which ultimately will grow in soil, interior plants are reliant for their entire lifespan on artificially constructed substrates held within containers, planters and planting beds. To avoid excessive weight on building floors, growing media need to be lightweight, whilst retaining the capacity to retain moisture. Peat or similarly lightweight organic materials are often used, but so too are inert materials, such as clay leca. Leca is an expanded clay pellet with a honeycomb microstructure that contributes to its light weight, but also enables it to hold sufficient volumes of air and water. Leca is also used in hydroculture; this involves the leca being placed in a planter which is then part-filled with water or a weak nutrient solution. Other inert material can be used as alternatives (e.g. perlite, rockwool, glass beads) and plants access the water through capillary action associated with these media. Although the media used in plantscapes is often lightweight, a sufficient bulk density (e.g. 0.15–0.75 g cm^{-3}) is still required to anchor the plants and avoid instability problems as they grow. Unlike natural soils, the substrates in interior plantscapes do not experience natural flushing through of nutrients with rainfall, and nutrient levels in the rhizosphere can build up over time. As such, care is required not to over-fertilize plants and to avoid root injury through excessive accumulation of ions. Liquid fertilizers are frequently favoured over base dressings of N, P_2O_5 and K_2O to help control nutrient application, with opportunities to flush through the media with excess water, or change old media for new, being sought. An additional requirement of the growing media is that it needs to be aesthetically pleasing, contributing to the overall design package. As such, ornamental mulches of cocoa shells, gravel or glass beads may be utilized to cover the substrate surface.

Irrigation

Irrigation systems vary depending on the medium used and the size of the planting area. Large plantscapes may possess their own automated irrigation system with drip lines being strategically placed to supply water to individual specimens. In large displays these are commonly linked to sophisticated control systems, automatically irrigating after set thresholds for light integrals, evapotranspiration demand or media moisture availability. Such systems still require occasional checking to ensure roots, detritus or precipitating calcium salts do not accumulate and block drainage areas or the application drippers. One disadvantage with discrete dripper type systems is that a moist 'cone' within the substrate develops, with limited lateral movement of moisture; this tends to encourage root growth around the moist zone at the expense of more extensive longitudinal root development. Hydroculture systems help provide a more uniform supply of water, and employ a subsurface reservoir, which irrigates the medium via capillary wicks or other mechanisms. The reservoir may have gauges to allow notification of when it requires refilling manually, or may be self-regulating with a feeder water pipe controlled by a ballcock to maintain water levels at a set height. With such hydroculture systems a certain depth of container may be required to ensure the capillary action of water rise is effective. For smaller scale plant displays and individual pot plants, then hand-watering by watering-can or hose may be employed, but this is labour intensive. It also runs the risk of water spillage on inappropriate places, including work stations and corridor/office floors.

11.4 Acclimatization to Interior Environments

Plants used in interior landscapes are conventionally cultivated in intensive production systems with the usual stipulations to maximize crop growth and quality, before being sold on. As many of the plant species used are derived from tropical/subtropical biomes, they tend to be produced either in outdoor nurseries in such climates, or within heated, well-lit glasshouse production systems in more temperate regions. Either way, plants may experience higher irradiance and humidity levels during production than they subsequently experience in their final planting location. With the advent of large corporate plantscapes in the 1970s it became evident that plants which were moved rapidly from nurseries with high irradiance to interior locations with lower irradiances, often experienced leaf chlorosis and abscission. This was due to a lack of acclimatization (acclimation) to the lower light irradiance within individual leaves of the plant. Changes in relative humidity can also exacerbate the stress on leaves. During the nursery stage, if plants are grown in glasshouses they may experience relative humidity levels of 85–90%, whereas office environments are closer to 45–50%, with some domestic homes <25%.

Today, interior plants are placed through an acclimation procedure where irradiance levels are reduced gradually by artificially shading the crop over the last 6–8 weeks of the production phase, and by progressively increasing the ventilation to reduce humidity levels. Irradiance intensity can be reduced by as much as 50% on 2–3 separate occasions over this period. Acclimation involves both physiological and anatomical changes in the plant leaf, e.g. chlorophyll being rearranged within the grana to provide greater exposure to the incidental radiation, altered nitrogen metabolism, reduced but more efficient photosynthetic capacity, the presence of smaller or less frequent stomata, the development of larger, but thinner leaves, and altered root to shoot ratios (Rodríguez-Calcerrada et al., 2008). Acclimation can be aided by less frequent irrigation and lower nutrient supply, in addition to reducing irradiance intensity and background humidity (Chen et al., 2005).

Once in the interior environment, inappropriate humidity or rapid constant air movement can still be problematic to the foliage. Draughts and direct placement close to open windows, heaters or air conditioning units should be avoided. Not only will these aspects affect temperature, but may decrease humidity or cause excessive movement of the foliage, resulting in necrotic lesions on leaves. Some species are more susceptible to low relative humidity and air movement than others. Whereas most broadleaved species tolerate the humidity typical of offices (i.e. 45–50%), bamboos, ferns and other fine-leaved plants may require 70% relative humidity once *in situ*. In contrast, species from arid climates, such as cacti, require lower values than those from the moist tropics. In large or high quality plantscapes, mist or fog nozzles can be specified in the design and automated to counteract low humidities. Systems tend to rely on mains water and compressed air is used to create the mist or fog. Spray nozzles should be orientated in the direction of the plant canopies to maximize the humidity effect but also to avoid accidental wetting of seating areas, paths or electrical equipment. As the system works by creating a vapour of moisture (and hence easily inhaled by humans), storage of water in tanks should be avoided to minimize risks due to *Legionella pneumophila* (Legionnaires' disease).

11.5 Pests and Pathogens

Interior plantscapes have led the way in the use of biocontrol agents to help control pests and pathogens. This has largely been out of necessity as the use of chemical pesticides in close proximity to humans (especially the general public, who may be unaware of the chemical application) and within enclosed spaces, has required that alternative solutions should be sought. Where chemicals have been used, they tend to be relatively benign products such as detergents or spot treatment with alcohols or oils. Minimizing stress through careful regulation of irradiance, humidity and irrigation is key to reducing the chances of infection and pathogen spread (Lockwood, 2000). For example, root pathogens such as *Pythium*, *Phytophthora* and *Rhizoctonia* can be more prevalent after plants have been overwatered or roots have been damaged by an excessive accumulation of fertilizer salts. Common plant pests are now regularly controlled by biocontrol means: *Tetranychus urticae* (glasshouse red spider mite) by *Phytoseiulus persimilis*, *Trialeurodes vaporariorum* (glasshouse whitefly) by *Encarsia formosa*, *Planococcus citri* (citrus mealybug) by the *Leptomastix dactylopii* wasp and *Coccus hesperidum* (soft brown scale insect) by *Metaphycus alberti* wasps (Table 11.1).

Where plants are heavily infested with pests or badly damaged by pathogens, they can be dug up and removed from the interior landscape and placed back in a glasshouse or quarantine area to be dealt with. In such locations the use of more virulent pesticides may be appropriate. Plants can then be restored to full health, before being placed back in the plantscape (assuming the original cause of the disease/pathogen in the interior landscape has been remedied in the meantime).

11.6 Managing the Interior Landscape

Most of the larger plantscapes are designed, installed and maintained by specialized landscape design or horticultural companies. These may provide a complete service, employing a wide range of personnel to cover the design and maintenance aspects. Some may have their own production nursery whereas others buy in plant specimens from specialist nurseries across the globe.

In terms of planting and maintenance a range of service levels may be offered. Plants may simply be sold to the client. Alternatively, the client may rent them off the interior landscape company, thus helping the client replace plants relatively easily if quality begins to deteriorate. Some companies too, offer design solutions, including developing a plantscape that fits with the existing or proposed interior, taking account of such infrastructure factors as site access, weight loading, colour schemes, design styles and suitability of locations for different plants – irradiance and photoperiods, draughts, seating arrangements etc. Once a plan is agreed the company will install and maintain the landscape, usually offering a service whereby landscape technicians will visit the planting on a regular basis to water, clean, prune, fertilize and inspect plants for pathogens, disease or stress effects. During such arrangements the plantscape companies may wish to insure themselves against any damage or other liabilities potentially caused by its personnel, when working within a client's building. The personnel need training, not only in the necessary horticultural skills, but also in ensuring a high customer service and being able to respond in a rapid and professional manner to the client's needs.

11.7 Environmental Sustainability

Although interior landscapes may demand high levels of energy, especially in terms of heating and irradiance (and paradoxically, in some locations, air conditioning and forced air-venting), they do bring a number of benefits to our living and working conditions. In many areas heat energy is already supplied by the requirements imposed on satisfying human thermal comfort (i.e. the central heating of occupied buildings), and indeed ironically the plant displays may be providing a respite from excessive internal heat through their localized cooling and humidifying effects. Some of the more luxuriant shopping malls will use plants and water in their

Table 11.1. Pests commonly found in interior landscapes and their biocontrol agents.

Pest	Symptoms	Biocontrol agents
Trialeurodes vaporariorum (glasshouse whitefly)	Sap-sucking white 'ghost-like' fly. Reduces plant vigour and multiplies rapidly. Transmits plant viruses, produces honeydew with corresponding sooty moulds.	Parasitic wasps: *Encarsia formosa* – one of the pioneer biocontrol agents – known to be able to parasitize 16 different species of whitefly around the globe. Each adult accounts for about 100 whitefly nymphs in its lifetime.
Planococcus and *Pseudococcus* spp. (mealybug)	Mealybugs cause leaf chlorosis through sap sucking activity, but plants also damaged by honeydew excretions and the sooty moulds that form on it.	Ladybird beetles: *Cryptolaemus montrouzieri* – Australian orange and black ladybird. Parasitic wasps: *Leptomastix dactylopii* – yellow parasitoid from Brazil. *Leptomastidea abnormis* – from southern Europe and copes with lower temperatures than other species. *Anagyrus pseudococci* – of Middle Eastern origin.
Myzus persicae (peach-potato aphid)		Ladybird beetles: *Adalia bipunctata* – two-spotted ladybird. Usually red with two black spots, although it can vary in pattern and colouring. Parasitic wasps: *Aphidius colemani* – will also control *Aphis gossypii* (melon, or cotton, aphid). *Aphidius ervi* – known to track aphid colonies by the volatile chemicals released from plants when under attack from aphids. At shorter distances the aroma of the honeydew allows the wasp to home-in on its host. *Aphelinus abdominalis* – a species that often walks rapidly rather than flies to its host. Once the female finds an aphid she touches it with her antennae, turns around, raises her wing tips and injects the ovipositor into the aphid. Parasitoid gall midges: *Aphidoletes aphidimyza* – tends to be active at night. Eggs are deposited in the aphid colonies and once hatched the larvae seek out an aphid, paralyse it and suck out the contents. Lacewings: *Chrysoperla carnea* – larvae prey on a wide range of species and are a good general pest-control agent. Known to attack several species of aphids, red spider mites, long-tailed mealybug, thrips, whiteflies, leafminers, small caterpillars, beetle larvae, as well as the eggs of leafhoppers and moths.
Tetranychus urticae (red, or two-spotted, spider mite)	Yellow spots followed by faint white webbing across the leaf.	Predatory mites: *Phytoseiulus persimilis* – fast-moving voracious predatory mite – confusingly a more obvious red orange colour than *Tetranychus*. *Neoseiulus californicus* – works better than *P. persimilis* at low humidity.

Continued

Table 11.1. Continued.

Pest	Symptoms	Biocontrol agents
Coccus hesperidum (brown soft scale insect) *Saissetia coffeae* (hemispherical scale insect) *Eucalymnatus tessellatus* (tessellated scale insect)	Brown soft scale is probably the most frequently encountered scale on plants indoors. Heavy infestations can encrust the stems and petioles of their host plant. They also settle on leaves, usually along midribs. Honey dew serves as a medium for the growth of sooty molds which in turn inhibit photosynthesis and make plants unsightly.	Parasitic wasps: *Metaphycus helvolus* – small black and yellow wasp – attracted to light sources, but will go back to plants at night to feed. *Metaphycus stanleyi* – for tessellated scale as well as some control of brown soft and hemispherical scale when it can access the scales before hardening. *Encyrtus infelix* – small black wasp with characteristic large head. Ladybird beetles: *Harmonia axyridis* (harlequin ladybird) – has become a concern in the natural environment of the UK due to its competitive nature against native ladybird species. *Chilocorus nigritus* – a small round entirely black ladybird species from India. Good at controlling armoured scale insects as well as soft scale. *Lindorus lophanthae* – lustrous black ladybird, with grey 'alligator-shaped' larvae. Will tackle mealybugs as well as scales.
Frankliniella occidentalis (western flower thrip) *Thrips tabaci* (onion, or tobacco, thrip)	White or transparent mottling on leaves, leaf abscission.	Predatory mites: *Amblyseius cucumeris* – need a temperature of ≥15°C. Hibernate in winter unless artificial irradiance is supplied. True bugs: *Orius insidiosus* – minute pirate bug, within the order Hemiptera – generalist predators consuming mites, aphids, and small caterpillars in addition to thrips.
Otiorhynchus sulcatus (vine weevil)	Characteristic semi-circular cut sections from leaf edges. Larvae damage roots of young plants.	Nematodes: *Heterorhabditis megidis* – requires temperatures above 12°C, so generally good for interior landscapes.

central café areas as a 'refuge' for shoppers and shop assistants, who have been exposed to high light and temperature levels within the shops themselves. Part of this experience may actually be psychological too – the greenery makes the location feel cooler and quieter than it actually is.

Alternative sources of energy to carbon-based fuels are appropriate for glasshouse complexes, and if these do not always completely replace conventional forms of energy they can at least augment them. Heat pump systems, for example, can reduce reliance on oil-based fuels by 30–40% in glasshouse systems used for commercial crop production. Also, solar energy within the complex can be utilized to heat solid interior objects such as walls and paths, or indeed water baths and pipes during daylight hours. These can be exploited as heat stores, allowing energy to be re-radiated out again and hence maintain high air temperatures during the night, without a requirement for additional heating. Although finding alternative sources of energy is important to some atria, this may be to do more with their 'overall message' and educational agendas. For example, the Eden Project in Cornwall, UK, is keen to reduce its carbon footprint as its whole ethos is about developing sustainable approaches to modern living and food production.

In reality, the actual acreage of protected structures used to maintain plant collections is very small within the wider urban context. As such it may be argued that the benefits such 'green' locations bring to urban living far outweigh their costs in terms of energy use. These benefits are not only those outlined above (physical, environmental, health and well-being, and engagement with nature), but also those stemming from these environments interacting with wider sectors of society to drive forward technology and innovation. Growing plants effectively within protected structures is a key component of space travel and exploration for example. Indeed, some of the advances in building/plantscape technology in recent decades have evolved out of research conducted by the North American Space Agency (NASA) as well as other space-related organizations and allied technological industries. This has included the development of interior 'biowalls' filtration systems to improve air quality and raise humidity within buildings or the recent evolution of low-energy light generation. The use of plantscapes in atria, biomes and conservatories, therefore, allow us to understand better plant requirements when grown under highly artificial conditions (and in turn contributes to our knowledge-base on aspects such as high-tech crop growing and future plant conservation strategies). Allied to this, plantscapes have been very much at the forefront of development of genuinely effective biocontrol strategies, and enabling complex plant communities to be sustained without reliance on pesticides. Equally importantly, such environments provide ready access to green space and help inform us about our relationship and reliance on nature. Most interior landscapes have a *raison d'être* based on aesthetics, but their value to society is much deeper than this.

Conclusions

- Interior plants are used in offices, schools, leisure centres, shopping malls, hotels, cafés and restaurants, as well as in the home. They are part of the interior décor and are utilized not only to improve the aesthetics of a room, but also to reinforce an image or style.
- Growing plants indoors can be challenging, and careful thought is required as to how appropriate irradiance, temperature and humidity are going to be provided. Similarly, infrastructure within the interior environment means that irrigation, nutrition and pest/pathogen control need to be astutely managed so as to avoid problems such as water leaks or the application of toxic chemicals.
- Plants in interior landscapes provide relaxing surroundings for the room users (employees, customers, home owners etc.), with evidence for their ability to reduce stress and promote positive feelings in the occupants of the room. They also physically contribute to a healthy environment by aiding the removal of aerial pollutants within a building. This includes particulate matter as well as volatile chemicals such as formaldehyde, toluene, benzene and xylene that are associated with modern buildings.
- This ability to 'scrub' aerial pollutants out of the building interior environment is largely, but not exclusively, due to the action of microorganisms found on the plant leaves and in the rhizosphere.
- Biocontrol of pests and pathogens is commonplace in interior landscapes, with the use of high toxicity pesticides restricted due to the largely public nature of the buildings. As such, this sector of horticulture has been at the forefront of implementing integrated pest management and pioneering the development and use of biocontrol agents.
- Although 'house plants' are privately purchased and owned, many business interiors and public spaces rely on the renting of plant material, and these are serviced by interior landscape companies. This enables the plants to be maintained and managed by professionals, who can address any problems readily.

References

Axelsson, G., Skedinger, M. and Zetterström, O. (1985) Allergy to weeping fig: a new occupational disease. *Allergy* 40, 461–464.

Bakker, I. and van der Voordt, T. (2010) The influence of plants on productivity: a critical assessment of research findings and test methods. *Facilities* 28, 416–439.

Bastos, W.R., de Freitas Fonseca, M., Pinto, F.N., de Freitas Rebelo, M., dos Santos, S.S., da Silveira, E.G. and Pfeiffer, W.C. (2004) Mercury persistence in indoor environments in the Amazon Region, Brazil. *Environmental Research* 96, 235–238.

Chang, C.-Y. and Chen, P-K. (2005) Human responses to window views and indoor plants in the workplace. *HortScience* 40, 1354–1359.

Chen, J., McConnell D.B., Henny R.J. and Norman, D.J. (2005) The foliage plant industry. In: Janick, J, (ed.) *Horticultural Reviews* 31. John Wiley & Sons, Hoboken, New Jersey, pp. 45–110.

Coward, M., Ross, D., Coward, S., Cayless, S. and Raw, G. (1996) *Pilot study to assess the impact of green plants on NO_2 levels in homes.* Building Research Establishment Note N154/96, Watford, UK.

De Kempeneer, L., Sercu, B, Vanbrabant, W. Van Langenhove, H. and Verstraete, W. (2004) Bioaugmentation of the phyllosphere for the removal of toluene from indoor air. *Applied Microbiology and Biotechnology* 64, 284–288.

Der Marderosian, A.H., Giller, F.B. and Roia, F.C. (1976) Phytochemical and toxicological screening of household ornamental plants potentially toxic to humans. I. *Journal of Toxicology and Environmental Health* 1, 939–953.

Dingle, P., Tapsell, P. and Hu, S. (2000) Reducing formaldehyde exposure in office environments using plants. *Bulletin of Environmental Contamination and Toxicology* 64, 302–308.

Elsadek, M. and Fujii, E. (2014) People's psychophysiological responses to plantscape colors stimuli: a pilot study. *International Journal of Psychology and Behavioral Sciences* 4, 70–78.

Fjeld, T. (2000) The effect of interior planting on health and discomfort among workers and school children. *HortTechnology* 10, 46–52.

Fjeld, T., Veiersted, B., Sandvik, L., Riise, G. and Levy, F. (1998) The effect of indoor foliage plants on health and discomfort symptoms among office workers. *Indoor and Built Environment* 7, 204–209.

Han, K.T. (2009) Influence of limitedly visible leafy indoor plants on the psychology, behavior, and health of students at a junior high school in Taiwan. *Environmental Behaviour* 41, 658–692.

Ingels, J.E. (2009) *Landscaping Principles and Practices*, 7th edn. Delmar, New York.

Kanerva, L., Mäkinen-Kiljunen, S., Kiistala, R. and Granlund, H. (1995) Occupational allergy caused by spathe flower (*Spathiphyllum wallisii*). *Allergy* 50, 174–178.

Khan, A.R., Younis, A., Riaz, A. and Abbas, M.M. (2005) Effect of interior plantscaping on indoor academic environment. *Journal of Agricultural Research* 43, 235–242.

Larsen, L., Adams, J., Deal, B., Kweon, B.-S. and Tyler, E. (1998) Plants in the workplace: the effects of plant density on productivity, attitudes, and perceptions. *Environmental Behaviour* 30, 261–281.

Lee, J.-H. and Sim, W.-K. (1999) Biological absorption of SO_2 by Korean native indoor species. In: Burchett, M.D. (ed.) *Towards a New Millennium in People-Plant Relationships*. Contributions from International People-Plant Symposium, Sydney, 101–108.

Lockwood, S.L. (2000) *Interior Planting: a Guide to Plantscapes in Work and Leisure Spaces*. Gower Publishing Ltd, Aldershot, UK.

Lohr, V.I. and Pearson-Mims, C.H. (2000) Physical discomfort may be reduced in the presence of interior plants. *HortTechnology* 10, 53–58.

Lohr, V.I., Pearson-Mims, C.H. and Goodwin, G.K. (1996) Interior plants may improve worker productivity and reduce stress in a windowless environment. *Journal of Environmental Horticulture* 14, 97–100.

Orwell, R.L., Wood, R.L., Tarran, J., Torpy, F. and Burchett, M.D. (2004) Removal of benzene by the indoor plant/substrate microcosm and implications for air quality. *Water, Air, Soil Pollution* 157, 193–207.

Park, S.-H. and Mattson, R.H. (2008) Effects of flowering and foliage plants in hospital rooms on patients recovering from abdominal surgery. *HortTechnology* 18, 563–568.

Park, S.-H. and Mattson, R.H. (2009) Therapeutic influences of plants in hospital rooms on surgical recovery. *HortScience* 44, 1–4.

Park, S.-H., Mattson, R.H. and Kim, E. (2004) Pain tolerance effects of ornamental plants in a simulated hospital patient room. *Acta Horticulturae* 639, 241–247.

Pennisi, S.V., and van Iersel, M.W. (2012) Quantification of carbon assimilation of plants in simulated and *in situ* interiorscapes. *HortScience* 47, 468–476.

Rodríguez-Calcerrada, J., Reich, P.B., Rosenqvist, E., Pardos, J.A., Cano, F.J. and Aranda, I. (2008) Leaf physiological versus morphological acclimation to high-light exposure at different stages of foliar development in oak. *Tree Physiology* 28, 761–771.

Seppänen, O.A. and Fisk, W. (2006) Some quantitative relations between indoor environmental quality and work performance or health. *Heat, Ventilation, Air-Conditioning and Refrigeration Research* 12, 957–973.

Shaughnessy, R.J., Haverinen-Shaughnessy, U., Nevalainen, A. and Moschandreas D. (2006) A preliminary study on the association between ventilation rates in classrooms and student performance. *Indoor Air* 16, 465–468.

Sriprapat, W., Suksabye, P., Areephak, S., Klantup, P., Waraha, A., Sawattan, A. and Thiravetyan, P. (2014) Uptake of toluene and ethylbenzene by plants: removal of volatile indoor air contaminants. *Ecotoxicology and Environmental Safety* 102 147–151.

Talbott, J.A., Stern, D., Ross, J. and Gillen, C. (1976) Flowering plants as a therapeutic/environmental agent in a psychiatric hospital. *HortScience* 11, 365–366.

Tarran, J., Torpy, F. and Burchett, M. (2007) Use of living pot-plants to cleanse indoor air – research review. In: *Proceedings of 6th International Conference on Indoor Air Quality, Ventilation & Energy Conservation – Sustainable Built Environment, Sendai, Japan*, 3, 249–256.

Wolf, K.L. (2002) Retail and urban nature: creating a consumer habitat. In: *Proceedings of International Plants for People Symposium, Floriade, Amsterdam*.

Wolverton, B.C. and Wolverton, J.D. (1993) Plants and soil microorganisms: removal of formaldehyde, xylene, and ammonia from the indoor environment. *Journal of the Mississippi Academy of Science* 38, 11–15.

Yang, D.S., Pennisi, S.V., Son, K.C. and Kays, S.J. (2009) Screening indoor plants for volatile organic pollutant removal efficiency. *HortScience* 44, 1377–1381.

Index

Page numbers in **bold** type refer to figures, tables and boxed text.

abscisic acid (ABA) applications 210, **210**
acclimation, interior plants 295–296
adaptability of urban species 82
aeration, grass swards 224, 239–240
age, of woody transplants 134
air conditioning (of buildings)
    causing external urban heat islands 11
    needs reduced by urban planting 12–13, 17
    required for indoor planting 291, 296
    use of plants around roof/wall units **31**, 276
air movement *see* wind
air quality
    of indoor (office) space 271, **272**, 288–290, 294
    positive/negative effects of urban vegetation 33–35, 279
alien species, invasive 36, 38, 91–92, 108
allotments
    biodiversity value 107–108
    social and health benefits 62
alpine meadows 183, **184**
alpine plants 123, 165
    for living walls **265**, 267
amenity horticulture 1
amphibians
    urban biodiversity 85, 87–88, 109
    value of garden ponds for 102–103, **103**
anecophytes 82
annuals
    direct sowing for flower beds 216–217
    germination 198
    'meadow' mixtures 217, **218**
    species for pollen/nectar for invertebrates **98**–99
    see also bedding plants
anoxia/hypoxia responses, roots 147, **148**
antisocial behaviour 55, 62, 63
antitranspirants 209
architectural plants 284, 285, **285**
arthropods, urban biodiversity 89, **90**, 116
artificial turf 250, 253
attention deficit disorder (ADD) 63
attention restoration theory (ART) 51–53, **52**, 61
attractiveness
    to invertebrates 168–169, **169**
    of landscapes, human preferences 53–54, **54**, 157
    maintenance, for living walls 266
    manipulation 174, 183

Bains report (1972, UK) 1–2
bare root planting stock 130, 135, 170
bedding plants
    commercial production 193–196, **194**, **195**, 198–206
    pests and pathogens 217, 219–220
    propagation from seed 196–198
    quality maintenance in transport/ retail 206–210, **208**, **209**
    species and establishment 211–216, **212–213**
    used in parks and gardens 192–193, **193**
bees, importance of parks and gardens 103
biennials
    bedding plants 194
    life cycle 157
    species for pollen/nectar for invertebrates **98**
bio-walls 261, 270–271, 299
biocontrol agents
    nematodes, parasitic 152, 220, 251–252
    psyllids, for Japanese knotweed control 92
    used against woody plant pests 152
    used for interior plantscapes 296, **297**–298
biodegradable plant pots 205–206
biodiversity
    definitions and scope 76, **76**
    garden surveys 83
    history of decline 175
    importance of habitat connectivity 77
    management style, effects 89, 91, 176, 180
    planning and legislation 4, 93
    public awareness 117, 177
    urban
        promotion and impacts 77, 80, 81–82, 116–117, 176
        species traits and categories 82–83
biofilms 32
biogenic volatile organic compounds (BVOCs) 34, **35**
bioherbicides/biopesticides 251
biophilia hypothesis 43–44, **44**, 70
birds
    behaviour adaptations 85
    biodiversity in cities 81, 83, 84–85, 87, 99
    green roofs as habitat 112, **115**, 115–116
    populations in urban parks 104–105, **106**
    road fatalities and amelioration 109

birdsong, human responses 49–50
blinding layer, sports turf substrates 232, **232**
block planting
    matrix approach, repeating small blocks 162
    standard style **161**, 161–162
botanic gardens, glasshouses 286, **287**, 291
broadcast sowing 188, 216–217
brownfield sites
    biodiversity value 109–111, 112
    carbon sequestration 23
    conversion to allotments 107–108
budding techniques **126**, 129
buildings
    energy efficiency and urban planting 12–18, **16**
    excess water flow management 266, 280
    high-rise, heat island effects 10
    interior air quality 288–290
    load-bearing calculation for green roofs 274
    see also air conditioning; interior plant displays
bulking carriers, for seed sowing 188, 216–217
burning-over (weed control) 172, **172**
butterflies, biodiversity in cities 81

canalization 25
canyon effects, urban
    temperature 11, 15
    wind channelling and air flow 18, 33–34
'Capability' Brown, Lancelot 101, 224
carbon sequestration 20–24, **21**, 176, **226**, 279
carpet bedding schemes 193
cation exchange capacity (CEC) 198–199
chemical growth regulators 130, 203–205, **204**, 253
child development
    benefits of green space 49, 62–63, **63**
    sedentary lifestyle and obesity 46, 48, 65
choice of plants *see* selection of plants
citizen science studies 83
climate change
    impacts
        incidence of pests and pathogens 38
        plant selection implications 133, 167, 230
        urban temperatures and human health 11
        weather events, severity and
            frequency 24, 29
    mitigation
        estimates and assessments 20–21
        options in garden design 29, **30–31**
        skills requirements 7–8
climbing plants
    air quality impacts 33
    choice of species for gardens **31**, **261**, 261–262
    cooling and insulation benefits 14, 17, **31**
    natural habitat 121
    nesting and food opportunities for birds 116
    pruning 143–144
    species for pollen/nectar for invertebrates 94–95
    types and urban uses 123, **123**

clonal (asexual) propagation 123
clump formation
    geophytes 159
    herbaceous perennials 158–159
collaboration, interdisciplinary 7–8
colour
    annual flowers, informal sowings 217, **218**
    in flower-rich meadows 175
    in matrix planting design 163
    provided by bedding plants 192, 194, 195
    psychological responses 50, 288, **288**
communities (ecological)
    creation from scratch (seed mixes) 183, 185–186
    grass-based 175–176, 183–185, **184**
    long-term development 173
community garden projects 62, 63, 107–108
community groups 154, 181
companion planting 106, 152, **212**
competitive asymmetry 180
conservation
    ecologists' attitude to horticulture 4, 7
    interactions with economic development 110–111
    priority species 99
    role of parks and gardens 103
    work, health benefits and risks 44–45, 56
container-grown plants
    biodegradable container materials 205–206
    garden/public space use of containers 194, **194**
    modules for living walls 264, 266
    nursery production 130
    pot sizes 170
    transplantation success 29, 134, 138
    used for introducing forbs to grassland 180–181
    watering 211, 214–215
contract growing, nursery material 170
controlled release fertilizer granules 200, 244
cool-season turf grasses 225–226, **227–228**, 247
coppicing 144
cornfield annuals 216, 217
corridors (habitat networks)
    importance for biodiversity 77, 87
    value of trees 99
cortisol, and stress levels 49, 57, 59
cricket ground management 234–235, **236**
crime 51, 63–64
crown management (tree canopy) 143
culverting 25
cuttings, for propagation
    bedding plants 194
    rooting success factors 127–128, **128**, 129
    techniques 124, **125**, 127
cylinder mowers 237

daily light integral (DLI) 200, 202
definitions
    biodiversity 76
    ecological concepts 78–79

open/green spaces and infrastructure 3
    plant life forms 121, 157–158
dementia, horticultural therapy 61
depression, restorative value of nature 49, 61
design
    formal, with bedding plants 192–193, **193**
    green infrastructure in cities 13, 19, 68, 249
    interior plantscapes 284–287, **285**, 296
    landscape 101, 148, 159–161, 224
    living walls 264, 266, **271**
    seed mixes 185–186
    sports grounds and stadia 224, 231, **236**, 241
    urban waterways 25
diet, healthy 64–65
direct sowing, annual flower beds 216–217
domestic violence 63
dormancy
    herbaceous plants/geophytes 157, 160
    seeds 124, 182
drainage
    impeded by soil compaction 239
    optimization for lawns and sports turf 231
    sustainable and integrated approaches 26, 27, 280
    traditional engineering solutions 24–25, 231
drills, for seed 217
drought tolerance
    effects of ABA drenches 210, **210**
    influence of mycorrhizae 138
    preconditioning of nursery stock 30, 134–136, **135**, 203
    trees and shrub species for xeriscapes 147–148, **149**
Dutch elm disease 149

ecology
    applied to public landscape design 160
    approaches and attitudes to horticulture 4–5
    principles/concepts in urban context 77, **78–79**, 181
ecosystem services
    benefits of urban vegetation 9–10, 25, **153**
    contributions of green walls 261
    disservices 34–35, 36, 38
    and environmental horticulture 2, 7
    functional aims in landscape design 160–161
    provided by invertebrates 88, 103
    types (Millennium Ecosystem Assessment) 9
Eden Project, UK 284, 298
electrical conductivity (EC) 205, 211, **215**, 247
emissions
    BVOCs emitted from plants 34, **35**
    carbon (dioxide), saved by planting 13, 20
    generated and absorbed by green roofs 279
    greenhouse gases, sources 22–23, 130, 172, 224
    from soil, mitigation methods 29
    from vehicles 33, 223, 238

energy efficiency
    of artificial lighting 294
    of buildings 12–18, **16**, 23, 298
environmental horticulture
    aims and objectives 75–76
    alternative terms 1–2
    current professional challenges 6–8
    definition and scope **2**, 43
    involvement of human agency 3–4
erosion
    river banks 25
    of soil, control measures 223–224, 261, **281**
establishment
    annual/bedding plants 211–216
    grass swards
        from seed 230–233, 247
        with turf 233–234
    herbaceous plants/geophytes 169–171
        in existing grassland 177–179, 180–183
    micro-habitats on green roofs 111–112
    seedlings 182, 186
    woody plants 29, **30**, 131–140
ethical issues
    choice of species for planting 5–6
    willingness to support conservation issues 249
ethylene
    causing transport/storage damage 207, 208
    effects on indoor plants 294
eutrophication 36, 108, 190, 243
evapotranspiration
    urban cooling effect 11–12
    variation between species 14, **15**, 17
    water supply dependence 16–17, 26, 245
evergreens
    herbaceous/geophyte planting design 163–164, 167
    rainfall interception 26, **27**
    used in living walls 266
    value for buildings insulation 14
evolution
    plant adaptations to climates 192–193
    psychological legacy 43–44

F_1 seed 196–197
fertigation systems 243, 266
fertilizer application
    base fertilizer for turf seeding 232–233
    in bedding plant production 199–200
    carbon footprint 22–23
    effect on turf quality 241–243, **245**
    input rates and costs 216
    over-use on managed grass swards 27–28, 35–36, 243–244, **244**
    water pollution from runoff 36
fitness
    ecological, under competitive pressure 181–182
    native and exotic species compared 5, 178–179

fitness (*continued*)
    physical, of humans 44–45, 48
    related to robustness, in horticulture 164–165, **165**
flail cutting
    diversification methods 177–179
    effects on grassland community 176–177
    machinery 237–238
    mowing regime 176
flooding
    accommodation in swales/rain gardens 26, 145
    avoidance strategies 24–25
    tolerance in woody plants 145–147, **146, 148**, 281
flowers, single/double 105, 169, 195
food security 65
forbs
    competitive types in grassland 178–179, **179, 185**
    decline in rough grassland 177
    definition 158
    establishment strategies 180–183
    European hay meadow species **180**
    suitable for integration in lawns 254, **255–256**
Forest School movement 62–63
forests
    'forest-bathing,' Japan, benefits 48
    global biomes **122**
    urban 152–154, **153**
        species composition 99

gang mowing 104, 176, **237**, 238
garden centres 130–131, **131**, 193
gardenesque design style 101
gardening
    design
        attitudes to lawns 249–250, **250**
        human preferences 44, **45**, 67–68, 70
        Japanese 67–68
        related to water use 28–33, **30–31**
        style trends 101, 159–160, 285–286
    health benefits 45, 48, 61
    health risks 56
    impacts of hard paving (for parking) 24
    introduced species and escapes 38, 92
    management for biodiversity 101–103, 105–109
    pejorative attitudes in ecology 4
    role in green infrastructure 8, 101
    roof gardens **276**, 277
    use of chemicals 35–36
gardens
    biodiversity data 83, 101
    conservation value 95, 102–103
    viability of mammal populations 84
genotypes
    F_1 and F_2 seed types 196–197
    new cultivar development 195, 226, 229, 252
    selection for low input systems 229–230, 246, 252–253

uniformity in clones, implications 123, 148
variability of introduced species 84
variation and ecological fitness 164–165
geophytes
    bulb selection and planting 170, 171
    establishment in grassland 177–178, **178**
    grown on green roofs 112
    incorporated in planting schemes 163–164
    palatability to slugs and snails 169, **169**
    perennation (bulbs, corms and tubers) 158
    population and clump expansion 159
    species for living walls **269**
    species for pollen/nectar for invertebrates **97**
germination, seeds 124, 182, 188, 198
glyphosate
    for competition-free planting gaps 181
    perennial weed control 140, 173
    toxicity 36
    used in site preparation 186, 230
GMO (genetic modification) technology 195, 252
golf courses
    grass species and characteristics 225, 229
    substrates for green construction 231–232, **233**
    use of chemicals 36, **37**, 244, 250–251
grafting techniques **126–127**, 128–129
grass
    clippings 238–239
    ground surface cooling effects 12
    growth habit and physiology 225–226
    management options
        alternative mowing regimes 176, 179–180, 234–237, **236**
        costs 7, 176, 179, 183
        intensity, in parks 104
        roadside verges 109, 253
    perennial weed species, control 173, 251
    rooting depth 234, **235**, 235, 245
    species and cultivars for turf 226, **227–229**, 229–230
    species for living walls **270**
    see also lawns; mowing machinery; turf
grass-based plant communities 175–176, 183–185, **184**, 223
'green' credentials
    claims, for ecosystem services 9, 33
    of green roofs 279
    indoor planting technologies 298–299
green façades **14**, 116, 260, 261–262, **263–264**
green infrastructure *see* infrastructure, green
green roofs
    benefits and potential 272, **276**, 279
        cooling and thermal insulation effects 14–16, **16**
        impacts on rainfall (stormwater) runoff 28, **273**, 274
        mitigation of aerial pollution 33–34
        noise attenuation potential 20
        wildlife biodiversity potential 111–116, **113, 114, 115**

construction and maintenance 274, 276, 278
  government promotion, Europe 111–112
  irrigation 112, 278–279
  limitations, compared with ground 116
  substrate materials and properties 276–277
  traditional 272, **273**
  types and characteristics 271–274, **275–276**
  weight loading on buildings 274
green spaces
  carbon pools 21–24
  as catalyst for physical activities 44–46, **47**
  chemical inputs 35–36
  definition 3
  effects of proximity, scale and type 57–61, **58, 60, 66–68, 69**
  hydrological value 25
  promotion, official policies 54, 68, 70
    integrated planning needs 153–154
  reduced by paving/hard surfacing 24
  water use 28–33
  well-being benefits 49–51, 54–55
  see also infrastructure, green
green walls
  biodiversity potential 111, 116
  cooling and thermal insulation effects 13–14, 17–18
  indoor (bio-walls) 270–271
  maintenance 268, 270
  noise attenuation 20
  types and uses 260–261
  see also green façades; living walls
grey water
  for irrigation, compared with other sources 211, 247
  safety risks and treatment **32**, 32–33
  used for urban plantings 16, 29
growing media
  green roofs 276–277
  for interior plants 294–295
  peat and alternatives 130, 199, 216
  root zone mix for turf 232, **233**
  used in plug cells 196, 198–199
  water retention 29, 170, 281
growing-on 129–130, 192
growth patterns
  annual bedding plants 192–193
  grasses 225
  perennials 158–159
growth regulation
  chemical 130, 203–205, **204**, 253
  effects of light 200–202, **201**
  effects of temperature 200, **201**, 205
  by watering control 203, 205
guerrilla gardening 260

habitats
  disturbance impacts 91–92, 186
  established on green roofs 111–112, 115
  impacts of urbanization 76–77, **80**, 81–82
  networks and connectivity 77
  similarity of wild origins and planting site 165–167
  types in urban areas 76, 103
  see also key habitats for urban wildlife
hardening-off 200, 203, **209**
hay cutting 179–180, 183
hay fever 34–35, 56
health
  benefits of green spaces
    physiological 44–48, **46**, 59
    psychological 48–51, 57
    quality of evidence 54–55
  dietary factors 64–65
  impacts of heat waves 11
  risks
    air pollution 33
    injuries 55–56
    from pesticide use 36
  value of interior plant displays 287–290, **290**
heat islands, urban
  benefits for urban trees 153
  effects on wildlife 80, 88–89
  mitigation by plants 11–12
  occurrence and causes **10**, 10–11
heat waves 11, 16
hedges, for wind amelioration 18
Hemiptera (bugs), on urban trees 99–101, **100**
herbaceous plants
  canopy expansion and clump formation 158–159
  establishment 169–171
  maintenance 172–174
  palatability to slugs and snails 162–163, 169, **169**, 173
  selection criteria for planting sites 164–168, **166**
  spatial arrangements and design 159–164
  species for living walls **269**
  species for pollen/nectar for invertebrates **96–97**, 168–169
herbicides
  bioremediation, using rain gardens 282
  weed control 140, 172–173
  see also glyphosate
high-density urban development 13, 26
Hong Kong, garden tree diversity 102
hormones
  growth regulation 203, 205
  for rooting of cuttings 127–128, **129**
  and seed dormancy 124
  stress responses 210, **210**
  and waterlogging tolerance 147
horticulture
  approach contrasted with purist ecology 4–5
  therapeutic value 61–62

Index

horticulture (*continued*)
    traditional resources and values 6, 75, 95, 98
    see also environmental horticulture
houseplants 286, 289
HT (Horticultural Therapy) programmes 61
humidity 11–12
    within buildings **291, 292**, 295–296
    during transport 207, 208, **208**
hummingbirds, city food supplies 85, **86**, 103, **105**
hydro-seeding 233
hydrogels 135–136, 215
hydrology *see* water management
hydroponic systems
    indoor plant displays 295
    living walls 264, 266

incentive bonus schemes 2
indoor planting *see* interior plant displays
infiltration
    of rain in dry and moist soils 17, 25
    rates increased by turf maintenance 239
    related to soil drainage properties 280–281, 282
infrastructure, green (planned)
    city planning challenges 68, 70, 249
        analysis, information technology systems 154
    definition 3
    designed for native biodiversity 93
    importance of quality 55, 57, 66
    proximity/size and level of use 59–61, 65–66
    role in urban microclimate amelioration 10–19
    small scale, for noise control 20
inner city green space quality 55, 63
insulation, thermal 13, 14, 16, 17–18, **18**
integrated pest management 251–252
integrated pest management (IPM) 152, 219
interior plant displays
    effects on health and well-being 287–290, **290**, 298
    function and design 284–287, **285**
    living walls (bio-walls) 270–271, 299
    maintenance and management 295–296
    plant requirements 290–295, **291, 293**
    sustainability 296, 298–299
international trade, ornamental plants 38, 160
invasive alien species 36, 38, 91–92, 108
invertebrates
    impacts of pesticide/fertilizer use 36
    plants providing pollen/nectar **94–95, 96–99**, 168–169
    pollination services provided 103
    urban biodiversity 88–89, **90**
        in allotments 107–108
        aquatic 108–109
        on trees 99–101, **100**
    value of deadwood in woodlands 82

irradiance *see* solar irradiance
irrigation
    control and scheduling
        bedding plants 202–203, 205, 211, 214–215
        lawns/turf 243, 245–247, 248–249
        official regulation 250
    green roofs 112, 278–279
    hydroponic and drip systems, living walls 264, 266
    infrastructure 29, 32
    interior plant requirements 295
    needed to maintain evapotranspiration 16–17
    for newly-laid turf sods 234
    for recent transplants 134–135, **135**, 211, **214**
    for seedlings 187, 188–189
    water demands and supply 29, 244–245, 249–250
islands, ecological **79**, 111, 179
ivy, English (*Hedera helix*)13, 261–262

Japanese gardening style 67–68
Japanese knotweed, control measures 92

key habitats for urban wildlife 95, 98, **102**
    brownfield and wasteland sites 109–111
    community gardens/allotments 107–108
    parks and gardens 101–107, **106**
    transport infrastructure networks 109, **110**
    walls and roofs 111–116
    wetlands/ponds 108–109
    woodland and trees 98–101, **100**

landscapes
    human preferences 43–44, **45**, 53–54, **54**, 70
    interior, importance of plant form 284–285, **285**, **286**
    parks and gardens, English landscape style 101, 160
    restorative/therapeutic 50–51, 53, 65–68, **66**, **67**
    role of trees/woody plants 121–123, **122**, **153**
    role of turf/lawns 224, **226**
    urban, global proportions 76
    use of herbaceous plants 159–161
lawns
    age and soil organic carbon 22–23, **24**
    alternatives, for dry conditions **30**, 249–250
    maintenance practices 234–240
    management impacts 27–28, 35–36, 241
    sustainable water use 248–249
    tolerance of dicot weeds 253–254, **254**
    see also turf
layer structures
    herbaceous planting 161, 162–163, 167
    sports turf 231–232, **232**
layering, for propagation **125**
leaching, from lawns 28, 243–244, **244**

leaf area index (LAI) 16
leaf litter
    decomposition 22
    management (removal) 82, 262
    water retention 26
leca (clay pellets) 277, 294
LEDs (light emitting diodes) 202, 293–294
life cycle analyses (LCAs) 21, 22, 224, **225**, 279
light
    effects on growth and flowering 200–202, **201**
    requirements for germination 198
    requirements for indoor plants 292–294, **293**, 295
    see also solar irradiance
lighting, supplementary 202, 241, 292–294
living walls **14**, 261, 262, 264–271, **265**
load bearing calculations 274
local authorities (UK)
    communal water storage 30
    parks departments, service delivery 1–2
London Olympic Park, wildflower meadows 180, **186**, **187**
longevity
    herbaceous plants 164
    urban trees 133, 152
low-density urban development 27
low input systems 217, 252–253

maintenance
    green walls and roofs 268, 270, 276
    importance in establishment phase 171, 188–190
    lawns 234–240
    low-maintenance street trees 23–24
    monoculture block planting 161
    during plant transport/retail 206–210, **209**
    procedures for woody plants 140–144, 262
    rain gardens 281–282
    tidying for aesthetic reasons 144, 174
mammals
    dispersal value of green corridors 77
    habitat loss and woodland management 82
    urban biodiversity 84
management
    costs and rationing 6–7, 176
    in gardens, influence on biodiversity 101, 102
    intensification impacts 35–36, 82, 104–105
    interior landscapes 296
    levels of human agency 3–4, 56
    life cycle analysis 21, 22
    long-term community development 22, 173, 179–180, 185
    reduced intensity, for turf 236–237, 252–254
    skills, needs and resources 7–8, 154, 219
    'top-down' and 'bottom-up' approaches 89, 91
    see also sustainability
mat layer, grass swards 239
matrix planting arrangements 162–163

meadows
    annual flower communities 217, **218**
    biodiversity, compared with close-cut turf 104, 175–176
    carbon footprint 23
    creation from scratch 183, 185
    establishment of forbs **180**, 180–183
    mowing regimes and effects 179–180
    public perceptions 175, 180
    traditional management 175, 180, 183
media *see* growing media
Mediterranean plant types
    adaptation to dry conditions 12, **31**, 148
    BVOC emissions **35**
    preponderance of shrubs/sub-shrubs 121
    survival of waterlogging **146**, 147
mental health
    causes of problems 51
    therapeutic activities 49, 61
    variation in responses to landscapes 53, 57
microclimate
    hard-surfaced patios 194
    modified by urban vegetation 10–19, 122, **123**, 224
    protection given by tree guards 140, **140**
    urban warming effects on biodiversity 88, 109
    on walls
        challenges for plant colonization 111
        protection for fruiting plants 262
mist propagation 127, 128, **128**
models (simulations)
    effect of urban trees on crime rate 63–64
    noise attenuation in urban courtyards 20
    rainfall and runoff 26, 28, **273**
    species distribution 110
    temperature effects of green infrastructure 11, 15
    urban air flow and pollution levels 33–34
molluscs, plant palatability 219
    annuals/bedding plants 219
    grasses 176
    perennials 162–163, 169, **169**, 173
monocarpic plants 157–158
mosaics, habitat
    spatial configuration 77
    value of vacant/wasteland sites 109–111
mowing machinery
    alternative and low-energy 238
    carbon footprint of lawn maintenance 22–23, 238
    injury risks 56
    principles of action 237–238
    use of agricultural machines 179
mowing regimes
    first year, for sown meadows 189–190
    flail cutting, occasional 176–179
    gang mowing 104, 176, **237**, 238
    height of cut 234–237, **235**, **236**
    meadow cutting and baling 179–180
    spring meadows 176

mulches
- benefits and drawbacks of use 29, 140, **142**, 143
- choice of material **31**, 140–141, **141**, 171, 295
- for sown wildflower meadows 186–187
- time of application 172

mycorrhizae 137–138, **139**

native species
- ethical and scientific issues 5–6
- excluded by invasive aliens 91
- range available for planting
  - global trends and attitudes 92–95
  - in large nation states 5
  - in UK 4–5
- urban richness gradients 81, 83
- weedy grasses 185

Natural England (government body) **3**, 59

nature
- engagement with 43–44, **44**, 49, 70
- human responses to urban biodiversity 116–117
- real and artificial views, physical effects 49, 50

nectar resources
- for hummingbirds, Brazilian city study 85, **86**
- open **vs.** double flowers 195
- sugar concentration 169
- useful plant species **94–95**, **96–99**, 254
- value of non-native plants 6, **6**, 168–169

nematodes
- biocontrol agents 152, 220, 251–252, **298**
- as pests 220, **220**, 251

neonicotinoids 36

nitrogen
- physiology and fertilizers 199, 216, 242
- pollution control with rain gardens 282

noise
- attenuation by plants **19**, 19–20
- perception related to surroundings 20, **20**

noisy miners (Australian bird), urban populations 85

non-native species
- aesthetic and functional fitness 5
- biodiversity gradients in cities 81, 83
- in gardens 101
- horticultural introductions 84
- invasive aliens 36, 38, 91–92
- value as resources for local fauna 6, **6**, 93–95, **94–95**, **96–99**

nursery products
- acclimatization for indoor plants 295–296
- classification of tree types 130, **130**
- drought preconditioning 30, 134–136, **135**, 203
- growing-on before sale 129–130
- for landscaping contracts 131
- selection of product type 170
- withdrawn, due to invasiveness 92

nutrition
- biodiversity of nutrient-poor sites 110
- cation exchange capacity of media 198–199
- flail-cut rough grassland 176–177
- hemiparasitic, grassland species 182–183
- human, value of fruit and vegetables 64–65
- influence of mycorrhizae 137–138
- management
  - indoor plants 294
  - living walls 266
  - turf grasses 238, 241–244
- N:P:K fertilization, seedlings/young plants 199–200, 216

obesity 46, 48, 65

open spaces
- definition **3**
- horticultural biodiversity 76
- public, quality parameters 66

orchids, found on green roofs 112, **112**, 113

organic composts 22, 36, 171

ornamental plants
- deciduous trees, budding propagation 129
- genetically modified 195
- global homogenization 83, 102
- international trade 38, 160
- numbers available to gardeners 101
- organic production standards 206
- retail outlets 130–131, **131**

Oudolf, Piet 160, 161

overwatering, lawns 248–249

ozone 34

parasitoids, urban/rural abundance 89

parks
- biodiversity 81, 101, 104–105
- conservation and ecosystem services 103
- effect on air quality 34
- history of design styles 101, 103–104, 192
- human responses to design 53–54, **54**
- management options 105–107, **106**
- tropical 101–102

particulate matter, aerial (PM) 33, 34

patches (habitat), size and connectivity 77, **79**

pathogens
- of annuals 217, 219
- associated with grey water 32–33
- control methods 151, 152, 296
- of grass swards 241
- introduced by international plant trade 38
- of trees and shrubs 38, 133, 149, **150**

peat, as growing medium 130, 198–199

pelleted seeds 198

perennials
    life cycle and growth forms 157–159
    persistent weed species 173
    tender, treated as annuals 194
peri-urban environments 34, 76, 84, **84**
perlite, as growing medium 199
permaculture 22
pesticides 35–36, **37**, 151–152, 219, 250–251
pests
    of annuals 217, 219–220
    control methods 150–152
    indoor plant displays 296, **297–298**
    introduction and spread 38
    of trees and shrubs 50, 149–150, **151**
phenology, flowering plants 162–163, 167–168
phosphorus/phosphates 199, 242, 243
photoperiod manipulation 202
photosynthetically active radiation (PAR) 202, 241, 292
physiology
    drought acclimation 135, 138, 203
    grasses, cool- and warm-season 225–226
    human, benefits of outdoor exercise 44–48
    light requirements 294, 295
    responses to anaerobic conditions 146–147, **148**
    roles of mineral nutrients 242
    see also hormones
phytotoxicity
    chemical products used indoors 294
    excess nitrogen 242
    root products (anaerobic conditions) 147
    softwood bark organic compounds 199
    toxic components of grey/saline water 33, 247
Pictorial Meadows™ seed mixes 185–186
plant selection *see* selection of plants
planting
    carbon footprint 22
    causes of failure 93, 133
    choice of native species in UK 4–5, 93
    in competition-free gaps 181
    process protocols 134–140, 170–171
    spacing and density 161, 162, 163
    spatial arrangements 161–164
    substitute vegetation, to reduce costs 6–7
    use of non-native species 5, 123
    water requirements 29
    see also selection of plants
planting depth
    herbaceous plants/geophytes 171
    trees 138
plants, urban biodiversity 83–84, 111
plug technology
    benefits in bedding plant production 196
    nursery specialization 198
    transplantation mortality 170

    used for forb establishment in grassland 180–181
    watering and feeding 199, 202–203
poisonous/harmful plants 56, 290
pollen
    allergic reactions to (hay fever) 34–35, 56
    for invertebrates, useful plant species **94–95**, **96–99**, 169, **254**
    resources from non-native plants 6, 168–169, **195**
pollinating services 103
pollution
    air 33–35, 279
    indoor 271, **272**, 288–289
    trapped and deactivated by turf 223, 247
    water 25, **32**, 32–33, 36
        control by rain gardens 282
        excess phosphate 243–244
        urban ponds 87–88
ponds
    importance for amphibians 87–88, 102–103, **103**, 109
    management for biodiversity 108–109
potassium nutrition 242
potting on 128, 130
prairie
    communities in North America 175, 184–185
    establishment from seed mixes **188**, 190, **225**
predators
    adaptability to urban conditions 82–83, 84
    avoidance by nesting/roosting on green roofs 115
    impact of pets on wildlife **80**, 83
    ornamental fish in ponds 87, 108
priming, seeds 197–198
production, commercial
    garden plant retailing 130–131, **131**
    glasshouse/polytunnel equipment 200, 202
    interior plants 286, 295, 296
    international trade implications 38
    marketing objectives 193–196
    procurement reliability 169–170
    profit margins 129, 193
    selection for garden fitness 165
    sustainability 205–206
    see also nursery products
propagation
    bedding/annual plants 196–198
    woody plants 123–129, **125–127**
proximity and usage of green space 59, 66–67
pruning 129, 143–144, **144**
psychological health 48–51, 271, 287

railways
    microclimate effects 10
    noise 19
    wildlife bridges/tunnels **110**

rain gardens 279–282, **280**
rainfall
    infiltration in dry and moist soils 17, 25
    interception by trees **26**, 26–27, **27**
    predicted impacts of climate change 29
    simulation modelling in lawns 28
    world climatic variation 166–167
raptors, urban opportunities 84
rare species
    brownfield and wasteland sites 109
    conservation opportunities of green roofs 112, **113**, 116
    in parks and gardens 103
    in urban woodland 99
reciprocating blade mowers 238
recreational value (of green space) 66–68, **68, 69**
recycled water *see* grey water
refining, seeds 197
reinforcement systems for turf 235, 240–241
remnant vegetation, and bird populations 85
replant disease/disorder 133, 151
representative concentration pathways (RCPs) 30
reptiles, urban biodiversity 88
restorative landscapes 50–51, 53, 65–68, **66, 67**
retail
    product quality and shelf life 195–196, 206–210, **209**
    systems for woody plants 130–131
retaining living walls 261
Rhinanthus minor in grass swards 182
riparian plants 145–146, 147, 281
roads
    de-icing salts, toxicity 88, 152, 262
    microclimate effects 11
    roadkill rates 88, 109
    roadside vegetation benefits and dangers 50, 64, 109
    traffic calming schemes 64
    traffic noise 19, 20, **20**
    wildlife bridges/tunnels 109, **110**
robustness 164, **165, 178**
rogueing, weeds 189
root cuttings **125**
root zone corridors, for street trees 137
rootball and burlap transplantation 130, 134
rotary mowers 237
Royal Horticultural Society (RHS) 103, **104**, 123, 169
    Wisley Gardens 173, **287**
ruderal plants 79, 83, 172
runoff
    contamination by pollutants 25, 36
    and leaching of nutrients 243–244, **244**
    rate control 26, 27, 28, 280
    from roof types, related to rainfall **273, 274**

salinity stress 211, **215**, 247, **248**
sanitary measures in plant trade 38, 93, 133, 151

scarification
    dethatching of turf 239
    ground surface 171, 182
    seeds 124
school gardens 62, 64–65
screens
    microclimate modification 18, 20, **30**
    woody climbers, beneficial features 121, **123**, 262
seed
    cost, for grass swards 230
    limitation in urban sites 179
    production for bedding plants 196–198
    self-sowing, herbaceous plants/geophytes 159
    woody plants 124
    see also sowing
seed bed preparation 216–217, 230–232
seed mixes, components
    cornfield annuals 216, 217, **218**
    turf grasses 229, 230
    wildflower meadows 185–186, **186, 187**, 188
selection of plants
    climate change impacts on choice **31**
    community type (meadow creation) 185
    global trend to uniformity 83
    information for gardeners 92, **131**, 209
    for interior plantscapes 284–285, **285, 286**, 288
    for living walls 266–267, **267, 268, 269**–270
    native **vs.**non-native species 4–6, **6**, 75–76, 92–95
    for rain gardens 281
    shrubs for urban planting 17, 64
    species with risks to human health 34–35, 56, 290
    street trees, size and types 13, 23–24, 122, 133, 152–153
    suitability for site
        bedding plants/annuals **212–213, 214**
        forbs in grassland **179, 180**, 181–182
        herbaceous plants/geophytes 164–168, **166**
        trees and shrubs **132**, 132–133, 144–148
semi-natural landscapes
    community structure 162
    ecosystem robustness 148
    grasslands 175–176, 183–185, **184**
    herbaceous plants in public spaces 160
sense of place 121, 152, **153**, 284
sexual propagation 123–124
shading
    cooling effects of trees 11, 12–13, 23
    in gardens **30–31**
    within herbaceous plantings 162, 163, 167, 168
    by plants in green walls 14, **15**
    by tall grasses in grasslands 177, 179–180
    tolerance of indoor plants 292
    tolerance of turf grasses 241
shelter belts 17, **17, 18**, 19
shrubs
    carbon sequestration 22
    definition and ecological roles 121

pruning 143, 144, **144**
responses to water status 17, **31**
species for living walls **268**
species for pollen/nectar for invertebrates **94–95**
urban uses and species 122–123
used for xeriscaping **149**
'sick building syndrome' 289
site preparation for sowing 186–187, 216–217, 230–232
sky garden concept 260, 279
sleep patterns, effect of green spaces 59, **60**
slow release N fertilizers 242
slug/snail damage
    bedding plant susceptibility 219
    herbaceous plants 162–163, 169, **169**, 171, 173
    influence in grasslands 176, 180
    slug pellets and alternatives 219–220
smart controllers, for irrigation 246, 295
social horticulture programmes 61–62
social issues
    attitudes to green space 54–55, 68–70, 260
    benefits of green space 46, **46**, 49, 51
softwood cuttings **125**, 127, 128
soil sealing (impermeable paving) 136, **136**
soils
    carbon content 21, 22, 23, **24**
    compaction in sports pitches 239–240, **240**
    effects of grey water use on pH 33
    erosion and stabilization 25, 223–224, 261
    management practices 22, 29, **31**
        irrigation, post-planting 134–136
        'no-till' systems 107
    urban
        contamination 80, 110
        mycorrhizal colonization **139**
        structure and characteristics 28, **132**, 136
    water storage capacity 26–27
    waterlogging, plant tolerance 145–147, **148**
solar energy technologies 13, 238, 298
solar irradiance
    canopy penetration 143, 163, 167, **168**
    influence on indoor planting design 292
    reflected by plants 12, 14, 147
sowing
    direct, annual flower beds 216–217
    grass seeding 230, 232–233
    by hand, technique 187–188
    oversowing, forbs in grasslands 182
    rate calculation for seed mixes 186
    site preparation 186–187, 230–232
space travel 299
specialist nurseries 131
sports
    design of stadia 241
    greens/pitches
        chemical inputs 36, **37**, 241–243
        compaction patterns 239–240, **240**

grass species composition **229**
structural layers 231–232, **232**
performance, influence of arena landscape 45–46
playing surface properties of turf 224, 225, 234–235
spring bedding plants 194–195
spring meadows 176, **177**, 186
steppe communities
    occurrence and traditional management 183–184, **184**
    seed mixes **189**
stimpmeters, green speed 234
storage, of runoff water 29, **30**
storms
    damage to trees 18–19, 143
    stormwater management 24–25, 28, 280
stress (human)
    gender variation 57, 59
    mechanism and avoidance 51–53
    restorative value of green space 49, 50–51, 63
strimmers 238
structural form of plants 163, 167, **168**
substitution of nursery material 170
substrates
    composition, for golf greens 231–232, **233**
    cooling and thermal insulation properties 14, 15, 16
    depth on green roofs
        shallow, challenges for wildlife 116
        and water-holding capacity 16, 28, **274**, 277
    natural, importance for colonization 111
    surface temperature, and grey/hairy plants 12, **12**
    systems for living walls 264
    types and characteristics, green roofs 276–278
suburbs, species biodiversity and abundance 81, 82, **87**
surfactants 208–209, 246–247
sustainability
    approaches, evaluation and assessment 21
    interior plant landscapes 296, 298–299
    management of grass swards 224, 229–230, 254, 256
    management strategies 22, 176, 217
    in plant production 131, 205–206
    water use 211, 248–249
sustainable urban drainage systems (SUDS) 24, 26, 108, 260, 279
swales **31**, 145, 280
synanthropic species 82
syringing, turf 245

teenagers, attitudes to green space 54–55, 68
temperature
    control in bedding plant production 200, **201**, 205
    interior plant requirements 291–292
    mechanisms of cooling 11–12
    modelling, for cities 11, 15
    requirements for germination 198, 230
    soil temperature and planting depth 171

temperature (*continued*)
    substrate/ground surface, effect of different plants 12, **12**
    during transport and retailing **207**, 207–208
    urban heat island effect **10**, 10–11
    walls, with and without vegetation 13–14, **15**, 18
thatch accumulation in swards 238, 239, 246
thigmomorphogenesis 200, 205
tillage 107, 231
time of planting/sowing
    lawn/turf seed 230
    plants 170, 178, 181
    seeds 186, **196**, **197**, 216
tining (turf maintenance) 239–240
tissue culture, for propagation 123, **126**
topdressing, turf 239, 240
traffic
    calming schemes 64
    impacts on wildlife 109
    noise perceptions and stress 20
    as source of aerial pollutants 33
training
    crisis in horticultural sector 6, 8
    interior plantscape management 296
    for school gardening projects 65
    seedling identification 189
    vulnerable adults, in horticulture 62
transport
    handling of bedding plants 206–210, **207**, **209**
    pre-grown turf 233
tree guards 140, **140**
tree planting pits
    dimensions and staking 134
    use of structured soil 27, 136–137, **137**
    water infiltration role 26, 280
trees
    benefits in urban setting
        air quality effects 33, 34–35
        biodiversity value 98–101, **100**
        carbon sequestration 20–22, **21**, 23–24
        cooling effects 11, 12–13
        landscape value 121–122, **122**, 153
        rainfall interception **26**, 26–27, **27**
        species for pollen/nectar for invertebrates **94**
    nursery stock, categories 130, **130**
    pests and pathogens 38, 50, 148–152, **150**, **151**
    planting and maintenance
        planting constraints in urban sites 13, 131–133, **132**, 136
        pruning 143, 144
        transplantation and establishment 29, **30**, 130, 133–140, **137**
        weed control and mulching 140–143, **141**
    as shelter belts 17, **17**, **18**, 19
    storm damage and avoidance 18–19, 56, 143
    urban life expectancy 133, 152

trellis/cable systems, for climbers 262
turf
    artificial 250, 253
    attitudes to 224, **226**, 249–250
    reinforcement 235, 240–241
    sods, for establishing grass swards 233–234
    traditional roofing 272, **273**
    water runoff and leaching 27–28
    see also lawns

United States Golf Association (USGA) 231, 232
urban food production 65, 107, **275**
urban heat island effect *see* heat islands, urban
urban horticulture, North America 2
urban vegetation
    density and diversity 123, 179
    ecosystem benefits 9–10, 25
    effects on bird and butterfly biodiversity 81, 109
    native, importance of remnants 85
urbanization, impacts on biodiversity 77, **80**, 81–83, 88, 116–117

vacant (derelict) plots 89, 92, 109–111
vegetative propagation techniques 124–129, **125–127**
verges 223, 253
    management for wildlife 109, 181
vermiculite, as growing medium 199
verticutting, turf 239
veteran trees 99
vinegar, for weed control 172–173
volatile organic compounds (VOCs)
    biogenic (from trees) 34, **35**
    city air pollutants 33–34
    inside buildings 271, 289–290, **290**

walls
    colonizing plant species 111, 262, 264
    construction materials and noise 20
    temperature and insulation 13–14, **15**, 17, 18
    wind shelter in gardens **31**
warm-season turf grasses 226, **228–229**, 229
wastelands, biodiversity value 109–111
water hyacinth, biodiversity impacts 91
water management
    automated watering systems 211, 279, 295
    in bedding plant production 202–203, 205
    conservation strategies 230, 249–250
    during propagation of softwood cuttings 127, 128, **128**
    quality in urban ponds 87–88
    rainfall runoff and capture 24–28, 279–281

during transport and retailing 206, 207, 208–210
treatment systems for water quality 32–33
urban water use 28–29, **30–31**, 211
see also irrigation
waterlogging *see* flooding
wear tolerance, turf 235, 239, 240–241
weeds
    competition with planted material 140, **141**, 171
    in container-grown nursery stock 170–171
    control methods 140–141, 143, 172–173
        rogueing 189
        soil cultivation/hoeing 211
    in gardens 101, 253–254, **254**
    seed bank in soil 186–187
weight loading
    green roofs 274, 279
    interior plant displays 294
well-being
    affected by indoor plants 287–288
    components 44, **46**
    evidence for benefits of green space 49–51, 54–55
wetlands
    for grey water remediation 32, **32**
    temporary and permanent, amphibian diversity 88
    see also ponds
white roofs, reflective 15–16, **16**
wildflower meadow seed mixes 185–186
wildlife
    effects of pesticides 36
    impacts of watercourse management 25, 108
    risks and benefits of urban traffic networks 109
    scope of term 76
    urban diversity and abundance 77, 81–82, 93, 253
    see also key habitats for urban wildlife
wind
    canyon effects of urban buildings 11
    reduction by shelter belts/screens 17, **17**, 18, **31**
    spider ballooning and colonization 88
    storm damage to trees 18–19
    tolerance in trees and shrubs 145
winter insulation 17–18, **18**
woodland
    coppicing/pollarding management 144
    'tidying' impacts on biodiversity 82
    urban, value for wildlife 98–101, **100**
woody plants
    carbon storage in biomass 21, 21–22
    choice of species, site factors **132**, 133, 144–148
    establishment and maintenance 134–138, 140–144
    pests and pathogens 148–152, **150**, **151**
    propagation methods 123–129, **125–127**
    risk of invasiveness 91
    see also climbing plants; shrubs; trees
work environments 51, 56–57
    outdoor jobs 55

xeriscapes
    public acceptability 249–250, **250**
    suitable plants 147–148, **149**, 225